AF839479

The Biology of Blennies

The Biology of Blennies

Editors

Robert A. Patzner
Department Organismic Biology
University of Salzburg
Salzburg, Austria

Emanuel J. Gonçalves
Eco-Ethology Research Unit
Instituto Superior de Psicologia Aplicada
Lisboa, Portugal

Philip A. Hastings
Scripps Institution of Oceanography
University of California San Diego
La Jolla, California, USA

B.G. Kapoor
Formerly Professor of Zoology
The University of Jodhpur
Jodhpur, India

CRC Press
Taylor & Francis Group
Boca Raton London New York

CRC Press is an imprint of the
Taylor & Francis Group, an **informa** business
A SCIENCE PUBLISHERS BOOK

First published 2009 by Science Publishers

Published 2019 by CRC Press
Taylor & Francis Group
6000 Broken Sound Parkway NW, Suite 300
Boca Raton, FL 33487-2742

© 2009 reserved
CRC Press is an imprint of Taylor & Francis Group, an Informa business

No claim to original U.S. Government works

ISBN-13: 978-1-57808-439-5 (hbk)

This book contains information obtained from authentic and highly regarded sources. While all reasonable efforts have been made to publish reliable data and information, neither the author[s] nor the publisher can accept any legal respon-sibility or liability for any errors or omissions that may be made. The publishers wish to make clear that any views or opinions expressed in this book by individual editors, authors or contributors are personal to them and do not neces-sarily reflect the views/opinions of the publishers. The information or guidance contained in this book is intended for use by medical, scientific or health-care professionals and is provided strictly as a supplement to the medical or other professional's own judgement, their knowledge of the patient's medical history, relevant manufacturer's instructions and the appropriate best practice guidelines. Because of the rapid advances in medical science, any information or advice on dosages, procedures or diagnoses should be independently verified. The reader is strongly urged to consult the relevant national drug formulary and the drug companies' and device or material manufacturers' printed instructions, and their websites, before administering or utilizing any of the drugs, devices or materials mentioned in this book. This book does not indicate whether a particular treatment is appropriate or suitable for a particular individual. Ultimately it is the sole responsibility of the medical professional to make his or her own professional judgements, so as to advise and treat patients appropriately. The authors and publishers have also attempted to trace the copyright holders of all mate-rial reproduced in this publication and apologize to copyright holders if permission to publish in this form has not been obtained. If any copyright material has not been acknowledged please write and let us know so we may rectify in any future reprint.

Except as permitted under U.S. Copyright Law, no part of this book may be reprinted, reproduced, transmitted, or uti-lized in any form by any electronic, mechanical, or other means, now known or hereafter invented, including photocopy-ing, microfilming, and recording, or in any information storage or retrieval system, without written permission from the publishers.

For permission to photocopy or use material electronically from this work, please access www.copyright.com (http://www.copyright.com/) or contact the Copyright Clearance Center, Inc. (CCC), 222 Rosewood Drive, Danvers, MA 01923, 978-750-8400. CCC is a not-for-profit organization that provides licenses and registration for a variety of users. For organizations that have been granted a photocopy license by the CCC, a separate system of payment has been arranged.

Trademark Notice: Product or corporate names may be trademarks or registered trademarks, and are used only for identification and explanation without intent to infringe.

Visit the Taylor & Francis Web site at
http://www.taylorandfrancis.com

and the CRC Press Web site at
http://www.crcpress.com

Cover illustration: (from top to bottom)

Parablennius ruber (Blenniidae), Island of Faial, Azores, Atlantic. Photo: Robert A. Patzner

Aidablennius sphinx (Blenniidae), Island of Ibiza, Baleares, Mediterranean Sea. Photo: Robert A. Patzner

Cirriemblemaria lucasana (Chaenopsidae), Bahia Bandaras, Mexico, Pacific Ocean. Photo: Alicia Hermosillo

Helcogramma maldivensis (Tripterygiidae), Ari Atoll, Maldives, Indian Ocean. Photo: Robert A. Patzner

Library of Congress Cataloging-in-Publication Data

The biology of blennies / editors, Robert A. Patzner ... [et al.].
p. cm.
Includes bibliographical references and index.
ISBN 978-1-57808-439-5 (hardcover)
1. Blenniidae. I. Patzner, Robert A.
QL638.B6B56 2009
597'.77--dc22

2009002035

Preface

'If you ever hear of any eccentric habit among fish, you can be sure that some species of blenny is in on it'.

Will Cuppy, 1983, *How to Become Extinct*, p. 23.

Blennies are a fantastically diverse group of bony fishes. Most blennies are small and somewhat difficult to identify, so until recently these fish did not occupy the interests of many ichthyologists and even fewer ecologists. But a select number of scientists from around the world and over the decades have developed a keen eye for blennies and a determination to study them and their habits. This persistence has been richly rewarded, opening the world of blennies to a wide array of studies, many of which are reflected in this book, *The Biology of Blennies*.

With nearly 900 species, blennies are important members of most coastal marine communities. Blennies are found around the globe, from tropical shores to the Antarctic Peninsula and in an enormous variety of habitats. Most tropical and warm temperate rocky reefs include one or often several species of blennies. Blennies are also prominent members of many other coastal communities, including tropical coral reefs and mangroves, warm temperate mollusk and sea grass beds, temperate kelp communities and even brackish and freshwaters. Blennies are known to dwell in tide pools, beneath rocks and boulders, on the surface of massive corals and sea fans, within the tentacles of sea anemones, the fronds of kelps and the blades of sea grasses, and concealed within sandy and gravelly substrates and even within vacant shells and tubes of a wide variety of invertebrates. Larvae of most blennies are found in coastal plankton communities, but some are also found in the pelagic realm, while others appear to completely lack a larval stage.

The diversity of form in blennies is truly astounding. Most notable are their jaws and teeth. Many species have rows of short canine teeth, but others have broad chisel-like teeth, tiny comb-like teeth in rows with as many as 400 teeth in a single jaw, or large recurved canines used in combat and defense against predators. The diversity of food eaten by blennies is equally impressive, ranging from algae, to plankton, to small benthic invertebrates, to other fishes, to even skin and scales of other fishes for those that mimic cleaners. Predators in blennies are generally poorly documented, so their roles in most coastal food webs require further study. Some blennies are large enough, however, to be eaten by humans. In coastal Peru, remains of one species have been found within archaeological middens, while another variety, when eaten in a soup, is said to induce a mild intoxication.

Blennies include among their numbers some of the most striking cases of sexual dimorphism known in fishes. Most species have external fertilisation, but internal fertilisation and associated intromittent organs have evolved at least twice within the group. Parental care patterns range from males guarding the eggs deposited in their territory, to males carrying batches of eggs on their body, to females incubating eggs internally.

Is there anything ichthyological that blennies don't do? Well, yes, there are few. For one, they don't live in the deep sea—most species are found well above 40 meters depth. Second, blennies also do not appear to change sex. Although widespread in fishes, there are no hermaphrodite blennies.

Our understanding of the biology, ecology and evolution of blennies has grown significantly in recent years. The substance of this book has been divided into 5 sections, viz., 1. Systematics, 2. Biogeography, 3. Feeding and sensory systems, 4. Reproduction and development, and 5. Ecology and behaviour. At the very end of the book, there is a list of valid species of blennies.

Much remains to be learned of this amazing group of fishes. While the important roles of blennies in coastal communities have begun to be appreciated by more and more researchers, we hope that this book stimulates interest in blennies among a wider array of students of fishes and marine biologists in general. Blennies are especially rewarding subjects of study and once an eye for them is developed and an appreciation of their diversity is gained, they are intoxicating fishes indeed.

Robert A. Patzner
Emanuel J. Gonçalves
Philip A. Hastings
B.G. Kapoor

Contents

List of Contributors

Almada Vítor C.
Eco-Ethology Research Unit, Instituto Superior de Psicologia Aplicada, Rua Jardim do Tabaco 34, 1149-041 Lisboa, Portugal.
E-mail: valmada@ispa.pt

Barata Eduardo N.
CCMAR-CIMAR Laboratório Associado, Universidade do Algarve, Faro, Portugal and Departmento de Biologia, Universidade de Évora, Portugal.
E-mail: ebarata@ualg.pt

Cheney Karen L.
School of Integrative Biology, University of Queensland, St Lucia QLD 4072, Australia.
E-mail: k.cheney@uq.edu.au

Faria Cláudia
Eco-Ethology Research Unit, Instituto Superior de Psicologia Aplicada, Rua Jardim do Tabaco 34, 1149-041 Lisboa, Portugal.
E-mail: cbmfaria@clix.pt

Fishelson Lev
Department of Zoology, George S. Wise Faculty of Life Sciences, Tel Aviv University, Tel Aviv 69978, Israel.
E-mail: fishelv@post.tau.ac.il

Fricke Ronald
Staatliches Museum für Naturkunde in Stuttgart, Rosenstein 1, 70191 Stuttgart, Germany.
E-mail: fricke.smns@naturkundemuseum-bw.de

Gon Ofer

South African Institute for Aquatic Biodiversity, P.B. 1015, Grahamstown, 6140, South Africa.
E-mail: o.gon@ru.ac.za

Gonçalves David M.

Unidade de Investigação em Eco-Etologia, Instituto Superior de Psicologia Aplicada, Rua Jardim do Tabaco 34, 1149-041 Lisboa, Portugal.
E-mail: davidg@ispa.pt

Gonçalves Emanuel J.

Eco-Ethology Research Unit, Instituto Superior de Psicologia Aplicada, Rua Jardim do Tabaco 34, 1149-041 Lisboa, Portugal.
E-mail: emanuel@ispa.pt

Hastings Philip A.

Scripps Institution of Oceanography, University of California San Diego, La Jolla, California, USA.
E-mail: phastings@ucsd.edu

Lahnsteiner Franz

Dept. Organismic Biology, University of Salzburg, Hellbrunnerstr. 34, 5020 Salzburg, Austria.
E-mail: franz.lahnsteiner@sbg.ac.at

Lengkeek Wouter

Bureau Waardenburg bv consultants for environment & ecology, marine ecology group P.O. Box 3654100 AJ Culemborg, The Netherlands.
E-mail: w.lengkeek@buwa.nl

Levy André

Eco-Ethology Research Unit, Instituto Superior de Psicologia Aplicada, Rua Jardim do Tabaco 34, 1149-041 Lisboa, Portugal.
E-mail: andre.levy@ispa.pt

Neat Francis

Fisheries Research Services, Marine Laboratory, P.O. Box 101, 375 Victoria Road, Aberdeen AB11 9DB, Scotland.
E-mail: f.neat@marlab.ac.uk.

Oliveira Rui F.

Unidade de Investigação em Eco-Etologia, Instituto Superior de Psicologia Aplicada, Rua Jardim do Tabaco 34, 1149-041 Lisboa, Portugal.
E-mail: ruiol@ispa.pt

Patzner Robert A.

Dept. Organismic Biology, University of Salzburg, Hellbrunnerstr. 34, 5020 Salzburg, Austria.
E-mail: robert.patzner@sbg.ac.at

Robalo Joana I.

Eco-Ethology Research Unit, Instituto Superior de Psicologia Aplicada, Rua Jardim do Tabaco 34, 1149-041 Lisboa, Portugal.
E-mail: jrobalo@ispa.pt

Ros Albert

Unidade de Investigação em Eco-Etologia, Instituto Superior de Psicologia Aplicada, Rua Jardim do Tabaco 34, 1149-041 Lisboa, Portugal.
E-mail: aros@ispa.pt

Springer Victor G.

National Museum of Natural History, Smithsonian Institution Washington, D.C., USA.
E-mail: springer@si.edu

Watson William

NOAA Southwest Fisheries Science Center, 8604 La Jolla Shores Drive, La Jolla, CA 92037-1508, USA.
E-mail: william.watson@noaa.gov

Wilson Shaun K.

School of Marine Science and Technology, University of Newcastle, Newcastle upon Tyne, NE1 7RU, UK.
Present address: Marine Science Program, Dept. Environment & Conservation, 14 Dick Perry Ave, Kensington, WA 6151, Australia.
E-mail: shaun.wilson@dec.wa.gov.au

Wirtz Peter

Centro de Ciencias do Mar, Univ. Algarve, Faro, Portugal.
E-mail: peterwirtz2004@yahoo.com

SECTION 1

Systematics

CHAPTER

1.1

Systematics of the Blennioidei and the Included Families Dactyloscopidae, Chaenopsidae, Clinidae and Labrisomidae

Philip A. Hastings[1] and Victor G. Springer[2]

INTRODUCTION

The perciform suborder Blennioidei comprises approximately 883 species of mostly tropical and warm temperate marine fishes. As a group, they are worldwide in distribution except for the deep sea, occurring in most shallow water marine systems and occasionally entering freshwater areas. There is a growing appreciation of the importance of small fishes such as blennioids in the ecology of tropical coastal systems (Munday and Jones, 1998; Depczynski and Bellwood, 2003), where they are often among the

Authors' addresses: [1]Scripps Institution of Oceanography, University of California San Diego, La Jolla, California, USA. E-mail: phastings@ucsd.edu
[2]National Museum of Natural History, Smithsonian Institution, Washington, D.C., USA. E-mail: springer@si.edu

most numerous species present (e.g., Thomson and Gilligan, 1983; Smith-Vaniz *et al.*, 2006). As blennioids are typically small, benthic fishes that with care can be readily observed and studied, they have become significant models for the study of evolution and ecology.

Despite their importance in ecological systems and their increasing roles as study organisms, their systematics is not yet settled. New species continue to be recognised and described on a regular basis as a result of intensive collecting efforts, even in well-known regions such as the Caribbean (e.g., Collette *et al.*, 2003), revisionary studies of particular genera (e.g., Springer and Williams, 1994) and analyses of molecular data (e.g., Carreras-Carbonell *et al.*, 2005). Similarly the phylogenetic relationships of blennioids remain incompletely resolved. In this chapter, we will briefly review the history of classification of blennioids, the current thinking on the limits of the Blennioidei, and the systematics of four of its six included families.

A BRIEF HISTORY OF BLENNIOID CLASSIFICATION

What constitutes a true 'blenny' has been controversial because concepts of the Blennioidei have changed frequently and considerably over the past 100 years and more (Springer, 1968; Rosenblatt, 1984). Five major definitions of the Blennioidei have been proposed by Regan (1912), Jordan (1923), Hubbs (1952), Gosline (1968) and Springer (1993). All include the so-called 'tropical' blennies which are currently allocated to the families Tripterygiidae, Blenniidae, Chaenopsidae, Clinidae and Labrisomidae. However, these concepts of the Blennioidei differ greatly in the other included taxa. That of Jordan (1923) is the most inclusive, while that of Hubbs (1952) is the least so.

Regan (1912) included together with the 'tropical' blennies what are now termed the Zoarcoidei (including the zoarcids, stichaeids, pholids, cryptacanthodids, anarhichadids and ptilichthyids), the Congrogadidae and Notograptidae (now considered percoids; Godkin and Winterbottom, 1985; Gill and Mooi, 1993), and the ophidiiforms, including the ophidiids, brotulids and carapids, often excluded from the Perciformes (e.g., Rosen and Patterson, 1969; Patterson and Rosen, 1989). Jordan (1923) added several groups to Regan's list, including the Trachinoidei (Pietsch, 1989; Johnson, 1993), the Ammodytidae (questionably included in the Trachinoidei by Pietsch and Zabetian, 1990; see Johnson, 1993), the

Notothenioidei, the Bathymasteridae (included in the Zoarcoidei by Anderson, 1984), the Opistognathidae (now included in the Percoidei; Johnson, 1984, 1993), callionymoids, and the Batrachoididae (excluded from the Perciformes by Patterson and Rosen, 1989; but see Miya *et al.*, 2005).

Hubbs (1952) restricted the blennioids (as the super family Blenniicae) to those fishes now included in the families Tripterygiidae, Blenniidae, Chaenopsidae, Clinidae and Labrisomidae. Gosline (1968) retained the trachinoids, notothenioids and zoarcoids with the blennioids, but excluded the ophidiiforms, batracoidids, ammodytids and opistognathids. Over thirty years ago, Springer and Freihofer (1976) questioned the validity of Gosline's 'Blennioidea'. Subsequently, Gosline's Blennioidei has been reduced in size by the exclusion of selected taxa. Between the first (1976) and second (1984) editions of Nelson's book on fish classification, common opinion had shifted to a more restricted view of the Blennioidei. Springer (1993) formalised the common opinion that a monophyletic Blennioidei included only the 'tropical' blenny families (= Blenniicae sensu Hubbs, 1952) plus the Dactyloscopidae, long considered to be related to fishes now included with the trachinoids (e.g., Gosline, 1968).

LIMITS AND DEFINITION OF THE BLENNIOIDEI

The Blennioidei, including six currently recognised families (Table 1.1.1), has been hypothesised to comprise a monophyletic group based primarily

Table 1.1.1 Composition of the Blennioidei

Family	*Genera/Species*	*Primary occurrence*
Blenniidae	57/387	Worldwide in temperate and tropical seas with a few species in fresh waters
Chaenopsidae	14/91	Neotropical marine waters with a few temperate species, including NW Pacific species of *Neoclinus*
Clinidae	26/85	Worldwide in temperate marine waters except for the North Atlantic
Dactyloscopidae	9/48	Neotropical, marine and estuarine waters (rarely in freshwater)
Labrisomidae	14/109	Primarily Neotropical marine and New World warm temperate waters with two species in the eastern Atlantic
Tripterygiidae	31/163	Worldwide in tropical and warm temperate marine waters including the Antarctic Peninsula

on seven character complexes (Johnson, 1993; Springer, 1993). These complexes provide convincing evidence of the monophyly of a Blennioidei restricted to the Tripterygiidae, Dactyloscopidae, Blenniidae, Chaenopsidae, Clinidae and Labrisomidae (Figure 1.1.1). The sporadic occurrence of similar character states in other perciforms is indicative of either homoplasy or perhaps of the out-group relationships of this restricted Blennioidei. These complexes are briefly discussed below.

Pelvic fin and girdle: Blennioids have a unique pelvis with the convex halves united into a bean-shaped structure (see figures in Springer, 1993). The fin comprises one non-pungent, embedded and difficult to see spine and 2-4 unbranched rays. It is positioned far forward, anterior to a vertical through the anterior most pectoral-fin ray base. In other perciforms, the pelvic bones are flat to concave and the bean-shaped pelvis appears to be unique to blennioids (Springer, 1993). In most other perciforms, the pelvic fin comprises one pungent spine and usually 5 rays, one or more of which are usually branched. The pelvics are inserted anterior of the pectoral fin—as in blennioids—in several other groups of perciforms (Gosline, 1968).

Dorsal gill arches: In blennioids, the first infrapharyngobranchial is absent or represented only by cartilage (in some clinids, labrisomids and chaenopsids). The second and fourth infrapharyngobranchials are absent, the first epibranchial lacks an uncinate process, and there is no interarcual cartilage (Springer, 1993). In most other perciforms, the first infrapharyngobranchial is ossified, the second and fourth infrapharyngobranchials are present, the first epibranchial usually has an uncinate process, and the interarcual cartilage is either present or absent. The blennioid configuration of the dorsal gill arches is seen in a few other fishes, including the Gobiesocidae and Liparididae, and some specialised gobioids and cottoids (Springer, 1993). Springer and Orrell (2004) identified two potential synapomorphies of the Blennioidei in the muscles associated with the pharyngeal arches. First, the primary division of the levator externus (LE1) insertion includes all or part of the lateral one fourth of the surface of the first epibranchial (Eb1). In other acanthomorphs, LE1 inserts completely medial to the lateral one fourth of the surface of Eb1. Second, transversus ventralis four (TV4) is attached to the fifth ceratobranchial (Cb5), whereas it is free from Cb5 in most other acanthomorphs (exceptions include the labroids, pholidichthyids, grammatids, pseudochromids, leiognathids, most beloniforms and centrogenyids).

Fig. 1.1.1 Representatives of four families of the Blennioidei. **a-c**: Dactyloscopidae (a = *Dactyloscopus pectoralis*; b = *Gillellus semicinctus*; c = *Heteristius cinctus*). **d-f**: Chaenopsidae (d = *Acanthemblemaria exilispinous*); e = *Coralliozetus angelicus*; f = *Emblemaria nivipes*). **g**: Clinidae (*Pavoclinus graminis*). **h-l**: Labrisomidae (h = *Alloclinus holderi*, Cryptotremini; i = *Labrisomus striatus*, Labrisomini; j = *Paraclinus sini*, Paraclinini; k = *Starksia galapagensis*, Starksiini; l = *Dialommus fuscus*, Mnieripini). Photos by D.R. Robertson (a, e, f, l), G. Allen (b, c, d, i, j, k), J. Randall (g), and P. Hastings (h).

Pectoral fin and girdle: In blennioids, some or all of the four proximal pectoral-fin radials are longer than wide, while some or all of the fin rays are unbranched, the ventrolateral coracoid process is short, and the ventroposterior corner of the coracoid forms a process well removed from the proximal radials (Springer, 1993). All of these conditions are seen in selected perciforms, but they are not as well-developed as in blennioids (Springer, 1993).

Caudal fin: The caudal fin of blennioids is simple compared to that of other perciforms (Springer, 1993). The most complex configuration includes an autogenous ventral hypural plate, a dorsal hypural plate fused to the urostylar complex, an autogenous hypural 5, two epurals, and fewer than 16 segmented rays with 12 or fewer of them branched (Springer, 1993). Within blennioids, there is a further trend toward fusion (or loss) of independent elements including fusion of the ventral hypural plate with the urostylar complex, loss of hypural 5, loss of one of the epurals and presence of only simple, unbranched fin rays. A few other groups of perciforms exhibit similar caudal fin morphologies (Springer, 1993).

Anal fin: The anal fin of blennioids comprises 0-2 spines and a series of simple segmented rays, while that of most perciforms has three pungent spines and a series of branched rays. A condition similar to that of blennioids is, however, seen in cottoids, gobiesocids, bembropids, creediids, leptoscopids, and hoplichthyids (Springer, 1993).

First neural spine: The first vertebra of blennioids lacks a neural spine, while that of virtually all other perciforms has a well-developed spine (Johnson, 1993).

Epaxial musculature: The epaxial musculature associated with the dorsal fin pterygiophores of blennioids (exclusive of the Labrisomidae; see below) has a unique fan-shaped anterior slip of muscle that inserts on the neurocranium (Mooi and Gill, 1995). However, this feature is absent in the Labrisomidae examined in that study. Mooi and Gill (1995) interpreted this absence as a reversal, but discussed the possibility that this muscle is a synapomorphy of all blennioids exclusive of the Labrisomidae.

OUTGROUP RELATIONSHIPS OF THE BLENNIOIDEI

The Blennioidei is one of the several suborders of the Perciformes (Nelson, 2006). Concepts of the Perciformes are controversial, currently under intense scrutiny and changing rapidly (Johnson, 1984, 1993; Johnson and Patterson, 1993; Miya *et al.*, 2003; Smith and Wheeler, 2004; Dettai and Lecointre, 2005; Smith and Craig, 2007). Not surprisingly, the relationships of the Blennioidei (*sensu* Springer, 1993) within the Perciformes are not fully resolved. Historically the Blennioidei was included by Gosline (1968) in his 'Blennioidea' (see above). More recently,

Springer and Orrell (2004) included the Blennioidei *sensu stricto* within a newly recognised clade, the Benthomorpha. This clade includes four lineages of primarily marine fishes associated with the bottom: the Blennioidei, Callionymoidei, Gobiesocoidei, and tentatively the Dactylopteridae. Evidence of the monophyly of this clade includes: (1) insertion of the levator internus one (LI1) muscle only on the third pharyngobranchial (insertion on the second pharyngobranchial or on the interarcual cartilage in other acanthomorphs); and 2) the first pharyngobranchial (Pb1) cartilaginous or absent and, if present, articulating with the anterior process of the first epibranchial (Pb1 partly ossified or absent in other acanthomorphs). However, these characters exhibit considerable homoplasy in these and other acanthomorph fishes.

Within the Benthomorpha, a growing consensus, based on several lines of evidence, indicates that the sister group of the Blennioidei is the Gobiesocoidei. Rosen and Patterson (1990) noted the similarity of the gill arches in these groups. Following this lead, Springer and Orrell (2004) identified six potential synapmorphies of blennies and clingfishes. However, only one of these (1) the absence of the second pharyngobranchial (their character 38: state 2), exhibits no homoplasy among acanthomorphs. Additional morphological evidence of the sister group relationship of the Blennioidei and Gobiesocoidei presented by Springer and Orrell (2004) includes the following. (2) The transverses epibranchialis 2 muscle does not extend laterally much, if any, past the medial edge of the insertion of levator externus 2 (11:1). (3) The attachments of the transverses dorsalis muscle (TD) exclude the third epibranchial (Eb3) (14:1). (4) The pharyngoclavicularis internus (PCI) attachment on the fifth ceratobranchial (Cb5) begins at the distal end and extends medially and the muscle does not join a raphe with the obliqus posterioris (OP) ventrally (36:1). (5) The third epibranchial (Eb3) is cartilage-tipped and lacks an uncinate process (41:1).

Recently emerging molecular data also support the sister group relationship of the Gobiesocoidei and Blennioidei. Although no molecular studies have been designed to specifically address the sister group relationships of the Blennioidei, representatives of the blennies and clingfishes have been included in a number of recent studies on acanthomorph relationships. Among the studies that have included representatives of the Blennioidei and Gobiesocoidei and recovered their sister group relationship are Chen *et al.* (2003), Simmons and Miya (2004),

Miya *et al.* (2005), and Dettai and Lecointe (2005), although Smith and Wheeler (2006) did not do so.

COMPOSITION OF THE BLENNIOIDEI

The Blennioidei currently includes six families of perciform fishes (Table 1.1.1): the Blenniidae, Tripterygiidae, Dactyloscopidae, Clinidae, Chaenopsidae and Labrisomidae, although debate continues about the status of the latter three. The systematics of the Tripterygiidae and Blenniidae are covered in other chapters of this book, while that of the latter four families is discussed below. For these four families (Fig. 1.1.1), the currently recognised genera with numbers of included species are presented within a classification in Table 1.1.2, while key references discussing the composition of these lineages are summarised in Table 1.1.3.

Dactyloscopidae

Dactyloscopidae or sand stargazers are a curious group of 48 species allocated among nine genera (Table 1.1.2). This distinctive group is entirely neotropical and found most often on sandy and gravelly bottoms in marine and estuarine settings with occasional records from freshwater (e.g., Miller and Briggs, 1962). They are unusual within the Blennioidei in having eyes on the top of the head (Fig. 1.1.1), the gape oblique to subvertical, the ventral margins of the opercles overlapping below the isthmus, fimbriae on the ventral margins of the interopercles, and the lateral line strongly arched anteriad and approaching the dorsal midline (Dawson, 1982). They are also the only blennioids in which the pelvic bones extend anterior to their junction with the cleithra (Springer, 1993). Several of these features appear to be modifications for dwelling within sand and gravel, an apparent behavioral synapomorphy of the family. Although not confirmed for most included species, dactyloscopids reportedly have a unique modification of the branchial pump, where the opercular pump is replaced by a pump involving the branchiostegal rays (Todd, 1973). In addition, egg masses are defended until hatching by males, but not on the substrate, as in most other blennioids. Instead, they are held on the ventral side of males with their prominent recurved anal-fin spines in single species of *Dactyloscopus* and *Myxodagnus* (Petti, 1969). This behaviour has not been confirmed in other genera.

Dactyloscopids were long thought to be closely related to the superficially similar uranoscopid stargazers (e.g., Gosline, 1968) and, thus,

Table 1.1.2 Currently recognised genera of four families of the Blennioidei (excluding Tripterygiidae and Blenniidae). Numbers in parentheses ( ) indicate number of valid species in each genus. Numbers in brackets [ ] indicate the number of valid genera/valid species in higher taxonomic categories.

Dactyloscopidae [9/48]
- *Storrsia* Dawson, 1982 (1)
- *Heteristius* Myers and Wade, 1946 (1)
- *Platygillellus* Dawson, 1974 (6)
- Dactyloids [3/28]
  - *Dactylagnus* Gill, 1863 (3)
  - *Dactyloscopus* Gill, 1859 (20)
  - *Myxodagnus* Gill, 1861 (5)
- Gillelloids [3/12]
  - *Gillellus* Gilbert, 1890 (10)
  - *Leurochilus* Böhlke, 1968 (1)
  - *Sindoscopus* Dawson, 1977 (1)

Clinidae [26/85]
- Clinini [17/64]
  - *Blennioclinus* Gill, 1860 (2)
  - *Blenniophis* Swainson, 1839 (2)
  - *Cancelloxus* Smith, 1961 (3)
  - *Cirribarbis* Valenciennes, 1836 (1)
  - *Climacoporus* Barnard, 1935 (1)
  - *Clinoporus* Barnard, 1927 (1)
  - *Clinus* Cuvier, 1816 (17)
  - *Cologrammus* Gill, 1893 (1)
  - *Cristiceps* Valenciennes in Cuvier and Valenciennes, 1836 (3)
  - *Ericentrus* Gill, 1893 (1)
  - *Fucomimus* Smith, 1946 (1)
  - *Heteroclinus* Castelnau, 1872 (15)
  - *Muraenoclinus* Smith, 1946 (1)
  - *Pavoclinus* Smith, 1946 (9)
  - *Smithichthys* Hubbs, 1952 (1)
  - *Springeratus* Shen, 1971 (3)
  - *Xenopoclinus* Smith, 1948 (2)
- Myxodini [5/9]
  - *Clinitrachus* Swainson, 1839 (1)
  - *Gibbonsia* Cooper, 1864 (3)
  - *Heterostichus* Girard, 1854 (1)
  - *Myxodes* Cuvier, 1829 (3)
  - *Ribeiroclinus* Pinto, 1965 (1)
- Ophiclinini [4/12]
  - *Ophiclinops* Whitley, 1932 (3)

(Table 1.1.2 Contd.)

(Table 1.1.2 Contd.)

Ophiclinus Castelnau, 1872 (6)
Peronedys Steindachner, 1883 (1)
Sticharium Günther, 1867 (2)
Chaenopsidae [14/91]
Neoclinus Girard, 1858 (9)
Mccoskerichthys Rosenblatt and Stephens, 1978 (1)
Stathmonotus Bean, 1885 (7)
Chaenopsinae [11/74]
Acanthemblemaria clade [2/23]
Acanthemblemaria Metzelaar, 1919 (20)
Ekemblemaria Stephens, 1963 (3)
Chaenopsis clade [5/29]
Chaenopsis Gill (ex Poey), 1865 (10)
Emblemaria Jordan and Gilbert, 1883 (16)
Hemiemblemaria Longley and Hildebrand, 1940 (1)
Lucayablennius Böhlke, 1958 (1)
Tanyemblemaria Hastings, 1992 (1)
Coralliozetus clade [4/22]
Cirriemblemaria Hastings, 1997 (1)
Coralliozetus Evermann and Marsh, 1899 (6)
Emblemariopsis Longley, 1927 (12)
Protemblemaria Stephens 1963 (3)

Labrisomidae [14/109]
Cryptotremini [4/8]
Alloclinus Hubbs, 1927 (1)
Auchenionchus Gill, 1860 (3)
Calliclinus Gill, 1860 (2)
Cryptotrema Gilbert, 1890 (2)
Labrisomini [2/41]
Labrisomus Swainson, 1839 (20)
Malacoctenus Gill, 1860 (21)
Paraclinini [2/24]
Exerpes Jordan and Evermann in Jordan, 1896 (1)
Paraclinus Mocquard 1888 (23)
Starksiini [2/31]
Starksia Jordan and Evermann in Jordan, 1896 (30)
Xenomedea Rosenblatt and Taylor, 1971 (1)
Mnierpini [2/3]
Cottoclinus McCosker, Stephens and Rosenblatt, 2003 (1)
Dialommus Gilbert, 1891 (2)
Uncertain [2/2]
Haptoclinus Böhlke and Robins, 1974 (1)
Nemaclinus Böhlke and Springer, 1975 (1)

Table 1.1.3 Lineages within four families of blennioids and primary references discussing composition and evidence for monophyly. Taxa with weak or no morphological evidence for monophyly are marked with an asterisk (*).

Dactyloscopidae	Dawson, 1982; Springer, 1993; Doyle, 1998
Dactyloids	Dawson, 1982; Doyle, 1998
Gillelloids	Dawson, 1982; Doyle, 1998
Clinidae	Hubbs, 1952; George and Springer, 1980; Springer, 1993
Clinini*	George and Springer, 1980
Ophiclinini	George and Springer, 1980
Myxodini*	George and Springer, 1980; Stepien, 1992
Chaenopsidae	Springer, 1993; Hastings and Springer, 1994
Chaenopsinae	Stephens, 1963; Hastings and Springer, 1994
Acanthemblemaria clade	Hastings, 1990, 1992a
Coralliozetus clade	Hastings, 1997
Chaenopsis clade	Hastings, 1992b
Labrisomidae*	Hubbs, 1952; Springer, 1993
Cryptotremini*	Hubbs, 1952; Stephens and Springer, 1974
Labrisomini*	Hubbs, 1952; Springer, 1959
Mnierpini	Hubbs, 1952; McCosker *et al.*, 2003
Paraclinini	Hubbs, 1952; Rosenblatt and Parr, 1969
Starksiini	Hubbs, 1952; Rosenblatt and Taylor, 1971

were considered to be members of the Trachinioidei and/or similar higher level clades. Springer (1993) included them within a restricted Blennioidei based on several osteological features (see above) and this placement has been supported by more recent authors (e.g., Doyle, 1998; Springer and Orrell, 2004).

Springer (1993) hypothesised four synapomorphies of dactyloscopids, but Doyle (1998) argued that only one of these, the presence of bony fimbriae extending from the ventral margin of the interopercle and posterodorsal margin of the opercle, uniquely characterises all included species. Doyle (1998) further analysed the phylogenetic relationships of mostly the type species of eight of the nine genera of dactyloscopids (excepting *Storrsia*) based on osteological features and presented seven additional hypothesised synapomorphies of these fishes. However, all of these additional features exhibit in-group homoplasy and the only feature known to be unique to all species remains the fimbriae feature.

Relationships within the Dactyloscopidae based on a phylogenetic analysis of mostly osteological characters (Doyle, 1998) are similar, but not identical, to those hypothesised by Dawson (1982) based on overall similarity (Fig. 1.1.2). In short, the Doyle hypothesis places the monotypic *Heteristius* as sister to all others, and *Platygillellus* as sister to a clade made up of two lineages. One of these, the 'gillelloids' of Dawson (1982) minus *Heteristius* and *Platygillellus*, includes three genera and 12 species, whereas the 'dactyloids' includes three genera and 28 species. Both Dawson (1982) and Doyle (1998) did not hypothesise the relationships of the monotypic *Storrsia*. The monophyly of the dactyloscopid genera has not been tested since they were defined by Dawson (1974, 1975, 1976, 1977, 1982).

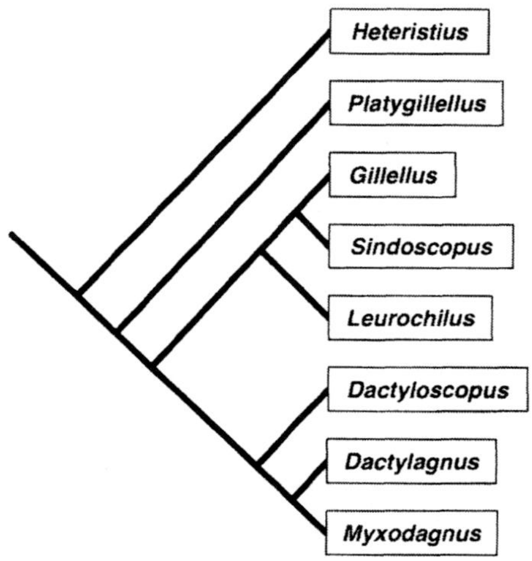

Fig. 1.1.2 Phylogeny of Dactyloscopidae based on Doyle (1998).

The alpha taxonomy of dactyloscopids was studied sporadically by a number of researchers (e.g., Böhlke, 1968) but was thoroughly reviewed by C.E. Dawson (Dawson, 1974, 1975, 1976, 1977, 1982) who, over many years, examined and collected morphological data from most specimens available to him. Since Dawson's careful and exhaustive work, only a few species have been described (e.g., Feitoza, 2002; Williams, 2002) and the basic taxonomy of the group has remained stable except that five subspecies recognised by Dawson have been elevated to full species (Hastings and Springer, in press).

Clinidae

Clinid fishes, often called kelp blennies, kelpfishes or weed fishes, include 26 genera and 85 species (Table 1.1.2). These common names reflect their propensity to occur in algae and sea grasses. Many species are variable in color, often closely matching their background (e.g., Stepien *et al.*, 1988). These fishes are found in all oceans, primarily in warm temperate waters (Springer, 1970; Stepien, 1992), although a few species occur in tropical waters (Shen, 1971; Fraser, 1972). Clinids are especially diverse in South African and southern Australian waters (Penrith, 1969, 1970; Kuiter, 1993).

The Clinidae is characterised by two synapomorphies. First, the scales of clinids (absent in one species, *Clinoporus biporosus*) are cycloid and have radii in all the fields (Hubbs, 1952; George and Springer, 1980; Springer, 1993). The ctenoid scales of tripterygiids and the cycloid scales of other blennioids (when present) have radii in only the anterior field. Second, clinids have a cordlike ligament that extends from the ceratohyal to the dentary (George and Springer, 1980). This ligament is absent in other blennioids, but a similar one is present in some pseudochromids and possibly some serranids (Springer *et al.*, 1977). Other features of clinids include a strong hook on the anterior margin of the cleithrum (George and Springer, 1980; Springer, 1993), but a similar hook also occurs in some labrisomids (e.g., *Malacoctenus* spp.).

Three tribes are recognised within the Clinidae (Table 1.1.2). The largest, the Clinini (Fig. 1.1.1), including 17 genera and 64 species, lacks any known morphological synapomorphies (George and Springer, 1980). These fishes generally have a notch between the slightly elongate first three dorsal-fin spines and the remaining, shorter spines, and usually possess orbital and nasal cirri. Clinines also have internal fertilisation and males have an intromittent organ (also in the Ophiclinini). Relationships within the Clinini have been discussed by several authors, including Hubbs (1952) and Penrith (1969) but have not been hypothesised based on a thorough analysis of all included taxa and appropriate outgroups.

A second tribe, the Ophiclinini, with four genera and 12 species, is restricted to southern Australia. It was included within the Clinidae by George and Springer (1980) who reported that ophiclinines are unique in having the dorsal and anal fins joined to the caudal fin by membranes (also in the chaenopsid *Stathmonotus*) and the membranes between the anal-fin elements usually not incised. They also exhibit internal fertilisation and males have an intromittent organ (George and Springer, 1980). Sometimes called snake-blennies, ophiclinines are elongate fishes with small or vestigial pectoral fins.

A third tribe, the Myxodini, with five genera and nine species, lacks any known morphological synapomorphies (George and Springer, 1980). Stepien (1992), however, reported nine allozymic synapomorphies of the myxodines. Unlike other clinids, myxodines exhibit external fertilisation and males lack an intromittent organ. Myxodines are found in temperate waters of the northeastern Pacific (*Gibbonsia* and *Heterostichus*), the southeastern Pacific (*Myxodes*), the southwestern Atlantic (*Ribeiroclinus*) and the Mediterranean (*Clinitrachus*; Springer, 1970; Stepien, 1992). Relationships within the Myxodini were hypothesised by Stepien (1992) based on molecular data.

George and Springer (1980) hypothesized the relationships among the tribes of the Clinidae based on morphological characters (Fig. 1.1.3). This was corroborated by Stepien (1992) based on allozyme data. The Clinini and Ophiclinini are sister groups (often referred to as the Clininae), sharing the reproductive pattern of internal fertilisation with males possessing an intromittent organ (George and Springer, 1980).

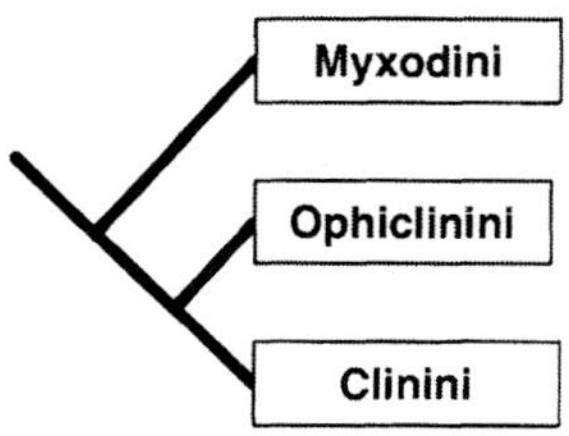

Fig. 1.1.3 Phylogeny of Clinidae tribes based on George and Springer (1980).

Chaenopsidae

Chaenopsids are commonly known as tube blennies because of their habit of occupying the vacant tests of various species of invertebrates. The Chaenopsidae was recently expanded to include the Chaenopsidae of Stephens (1963, 1970), plus *Stathmonotus*, *Mccoskerichthys*, and *Neoclinus* (Hastings and Springer, 1994). This redefined Chaenopsidae, with 14 genera and 91 species, is characterised by eight synapomorphies (Hastings and Springer, 1994), all but one of which are homoplastic (i.e., reverse within chaenopsids or seen in other groups of blennioids). (1) Chaenopsids are unique in having the palatine long relative to the length of the vomer. (2) The ventral arm of the post-temporal is free from the neurocranium rather than close to it. This condition is also seen in dactyloscopids and in some clinids. (3) Chaenopsids lack embedded, tubed scales in the posterior portion of the lateral line (also absent in many blenniids). (4) The upper jaw of both males and females is long, extending well past the posterior margin of the orbit (also in some labrisomids and some blenniids). (5) The insertion of the hyomandibula on the neurocranium is shifted posteriorly away from the orbit. This condition is also seen in dactyloscopids, some labrisomids and some blenniids. (6) A small lateral spine is present on the sphenotic. A similar spine is present in some blenniids and some labrisomids. (7) The dorsal arm of the scapula is reduced and free from the cleithrum in all chaenopsids except *Mccoskerichthys* and at least one species of *Neoclinus*. This arm is larger and abuts the cleithrum in other blennioids.

Hypothesised relationships within the Chaenopsidae were most recently summarised by Hastings and Springer (1994). *Neoclinus*, *Mccoskerichthys* and *Stathmonotus* are sequentially outgroups of the remaining species, the Chaenopsinae (Fig. 1.1.4). The Chaenopsinae, including 11 genera and 74 species (= Chaenopsidae of Stephens, 1963, 1970), is defined by three unique and four homoplastic characters (Hastings and Springer, 1994). (1) The mesopterygoid does not parallel the ectopterygoid, but is restricted posteriorly. (2) The testicular tissue is in one rather than two lobes. (3) A single-lobed testicular accessory gland is present. (4) Two rather than four or more infraorbitals are present (also in the blenniid *Medusablennius*). (5) The dorsal and anal-fin spines are ossified proximally, but the ossification ends abruptly and the distal portion is unossified (also in dactyloscopids and some blenniids). (6) The posterior most branchiostegal is broad (also in dactyloscopids). (7) The proximal pterygiophores supporting the dorsal-fin spines have 1 rather than 2 lateral ridges (also in blenniids, tripterygiids and dactyloscopids).

Within the Chaenopsinae, three lineages have been hypothesised (Fig. 1.1.1), the *Acanthemblemaria* clade (Hastings, 1990, 1992a) with two genera and 23 species, the *Chaenopsis* clade (Hastings, 1992b) with five genera and 29 species and the *Coralliozetus* clade (Hastings, 1997) with four genera and 22 species (Fig. 1.1.4). These hypothesised relationships are currently under evaluation based on molecular data for the

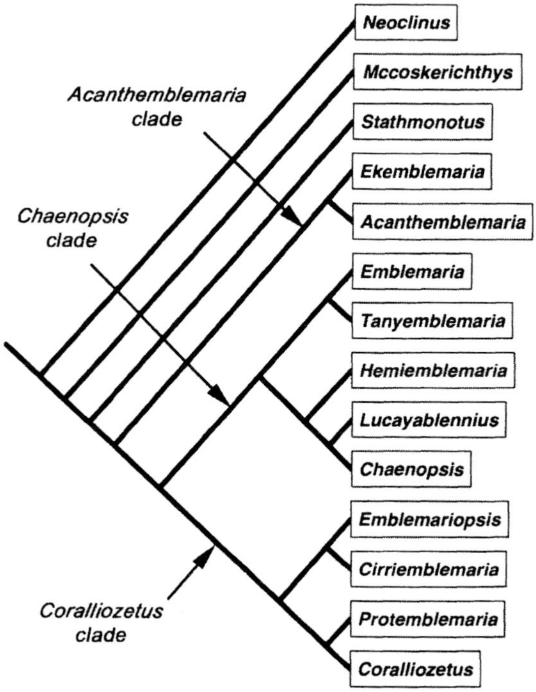

Fig. 1.1.4 Phylogeny of Chaenopsidae based on Hastings and Springer (1994).

Acanthemblemaria clade (Eytan, Helberg, Lin and Hastings, in preparation) and for the entire family (Lin and Hastings, in preparation).

Labrisomidae

The Labrisomidae as now understood includes 14 genera and 109 species (Table 1.1.2). For the most part, labrisomids are generalised blennioids that do not fall within the limits of the other, relatively well-defined families. The monophyly of the Labrisomidae has been questioned by a number of recent workers based on morphological (e.g., Springer and Orrell, 2004) and molecular data (e.g., Stepien *et al.*, 1993, 1997). Mooi and Gill (1995), however, reported that the seven species representing four labrisomid genera that they examined lacked the fan-shaped anterior extension of the epaxial musculature that characterises other members of the Blennioidei. If the labrisomids are nested within the Blennioidei (see below), the loss of this muscle may be interpreted as a synapomorphy of the Labrismidae (Mooi and Gill, 1995). Alternatively, if labrisomids are sister to the remaining blennioid families, then the lack of this feature is a symplesiomorphy in labrisomids and its presence is potentially a synapomorphy of the remaining lineages of blennioids, rather than a synapomorphy of the Blennioidei as interpreted by Mooi and Gill (1995; see above).

Hubbs (1952) first recognised and defined the Labrisomidae (as the Labrisominae) and it was elevated to familial status by George and Springer (1980) in their analysis of the Clinidae. The limits of Labrisomidae have been expanded by adding newly described genera (*Haptoclinus* Böhlke and Robins, 1974, *Nemaclinus* Böhlke and Springer, 1975 and *Cottoclinus* McCosker, Stephens and Rosenblatt, 2003) and including the genera *Stathmonotus* and *Neoclinus* (Springer, 1993). The latter two genera were subsequently reallocated from the Labrisomidae to the Chaenopsidae (Hastings and Springer, 1994). As a consequence, the Labrisomidae now includes 14 genera allocated among five tribes (Table 1.1.2; Fig. 1.1.1). With one exception, the limits of these tribes have changed little since they were named and discussed by Hubbs (1952) but their monophyly has generally not been tested using the techniques of phylogenetic systematics. However, at least three of them appear to be monophyletic, based on morphological synapomorphies.

The Paraclinini, with two genera and 24 species, is unique among the labrisomids in having a spine on the posterior margin of the opercle and 0-2 segmented dorsal-fin rays (Hubbs, 1952; Rosenblatt and Parr, 1969; Brooks, 1992). It is restricted to tropical and warm temperate waters of the western Atlantic and eastern Pacific.

The Starksiini, with two genera and 31 species, is unique among the labrisomids in having the anterior anal-fin of males modified as an intromittent organ (Hubbs, 1952; Rosenblatt and Taylor, 1971). This lineage is found in tropical waters of the eastern Pacific and Caribbean, where several new species have been recently named (Williams and Mounts, 2003).

The distinctive Mnierpini was defined by Hubbs (1952) by the presence of: (1) a fleshy bar across the pupil of the eye; (2) an elongate body; (3) anterior rays of the spinous dorsal fin soft and flexible, with no incision between the spinous and segmented portions of the fin; (4) enlarged fleshy tips on the anal-fin rays (also in males of some blenniids); and (5) a blunt head with the upper jaw projecting. McCosker *et al.* (2003) added; (6) posterior dentary teeth abruptly smaller and set lower than the anterior teeth; (7) lower lip interrupted by a pair of grooves outlining a frenum on the chin; (8) lips thick and inflated; (9) scales expanded posteriorly, forming pockets; (10) membrane of anal-fin rays incised almost to the base of the rays; and (11) belly and pectoral base naked. McCosker *et al.* (2003) synonymised *Mnierpes* with *Dialommus* and added a new genus *Cottoclinus*. Their new genus lacked synapomorphies (1) and (6), thus, the Mnierpini, now with two genera and three species, is characterised by the remaining nine synapomorphies listed above. Mnierpines are found only in the tropical eastern Pacific.

The Labrisomini, with two genera and 41 species (Table 1.1.2; Hubbs, 1952; Springer, 1959), is not defined by synapomorphies. Species of *Labrisomus* and *Malacoctenus* are conspicuous members of tropical reef communities in the western Atlantic and eastern Pacific although a few species do occur outside of these regions in the eastern Atlantic (Oliveira *et al.*, 1992; Floeter *et al.*, 2007) and the southeastern Pacific (Stephens and Springer, 1974).

Hubbs' (1952) Cryptotremini, comprising *Alloclinus* and *Cryptotrema*, found in the temperate waters of the northeastern Pacific, was expanded by Stephens and Springer (1974), who included *Auchenionchus* and *Calliclinus* from the temperate waters of the southeastern Pacific. This tribe, with four genera and eight species (Table 1.1.2), is not definable based on synapomorphies. However, all of the included species are characterized by the apparently plesiomorphic condition of branched caudal-fin rays (unbranched in other labrisomids, many dactyloscopids, many blenniids, all chaenopsids, and all clinids).

Finally, two rare and poorly known monotypic genera, *Haptoclinus* Böhlke and Robins (1974) and *Nemaclinus* Böhlke and Springer (1975) have not been assigned to a tribe.

Relationships within the Labrisomidae are unknown and essentially unhypothesised, based on morphological characters. The Cryptotremini are assumed to be relatively plesiomorphic, based on the presence of branched caudal-fin rays (Stephens and Springer, 1974). However, polarisation of this character is questionable without the supportive evidence of the relationships of these fishes within the Blennioidei. The single cryptotremin included in a study of allozyme variation (Stepien *et al.*, 1993), was hypothesised to be the sister group of the Clinidae independent of the remaining Labrisomidae. Other relationships within the Labrisomidae are essentially unhypothesised, although the Stepien *et al.* (1997) analysis of molecular data provided limited support for the labrisomid tribes Labrisomini, Paraclinini and Starksiini, but recovered a monophyletic Labrisomidae only with the inclusion of the Chaenopsidae (see discussion in the following paragraphs).

RELATIONSHIPS WITHIN THE BLENNIOIDEI

Phylogenetic relationships of the six families included in the Blennioidei (the four discussed above and the Blenniidae and Tripterygiidae) are poorly resolved. Recently, Springer and Orrell (2004) analysed the relationships of selected representatives of various blennioid lineages in the context of a broad survey of acanthomorphs based mostly on details of the pharyngeal skeleton and associated musculature. A strict consensus tree for the Blennioidei (Fig. 1.1.5) supports the sister-group relationship of the Tripterygiidae with other blennioids (ignoring the non-monophyly of the Tripterygiidae based solely on this data set). This was based on the remaining blennioids sharing the condition of the transverses epibranchialis 2 muscle (TEb2) extending laterally well past the medial end of the insertion of levator externus 2 (LE2; character 11, state 0). This

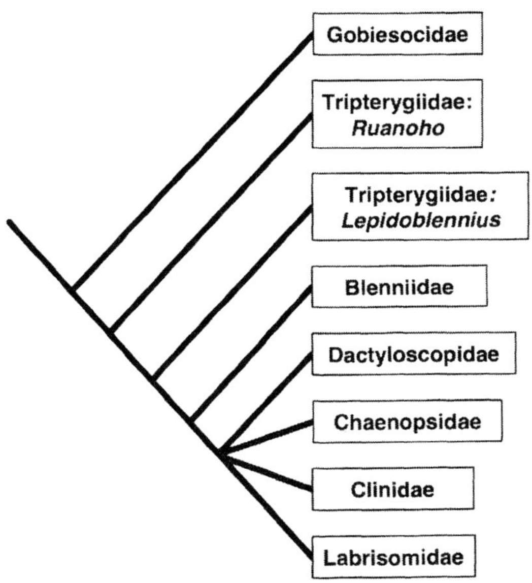

Fig. 1.1.5 Phylogenetic relationships of the Gobiesocidae and Blennioidei based on consensus tree of Springer and Orrell (2004, Fig. 5).

phylogenetic hypothesis (Fig. 1.1.5) also implies that the Blenniidae is sister to the remaining four families based on two characters, the loss of the adductor muscles of the first arch (Ad1; 30, 0) and the loss of the adductor muscles on the second and third arches (Ad2 and Ad3; 31, 0). Relationships of the remaining families are unresolved in their analysis (Fig. 1.1.5).

Several other morphological characters that may have a bearing on the relationships of blennioid families have been mentioned sporadically in the existing literature but these offer conflicting evidence. A few of these characters are highlighted below to illustrate the degree of conflicting character data regarding the relationships of these fishes. The Tripterygiidae is the only family with branched rays in the dorsal and pectoral fins. If this is a plesiomorphic condition, then unbranched rays would unite the remaining 5 families (Springer, 1993). All blennioids except for the Labrisomidae have a fan-shaped anterior extension of the epaxial musculature (Mooi and Gill, 1995). This could represent either a synapomorphy of the remaining families or a loss of the muscle in the Labrisomidae. All blennioids except for the Tripterygiidae and the Dactyloscopidae lack pectoral-fin rays articulating with the scapula (Springer, pers. obs.) supporting the monophyly of a clade including the remaining families. The Chaenopsidae, Clinidae and Labrisomidae share a unique shape of the dorsal-fin pterygiophores and the presence of mandibular sensory pore 1B (Hastings and Springer, 1994). Many members of the Clinidae and Labrisomidae have the ventral tip of the ventral post-cleithrum expanded (Hastings and Springer, 1994). A rod-shaped swelling is present in the upper pharyngeal of the Dactyloscopidae, Clinidae and Labrisomidae (Williams, 1990). Hubbs (1953) united chaenopsins and blenniids based on their eye position, stout infraorbitals, and reduced or absent scales. Many blenniids and most chaenopsids have muscles of the adductor mandibularis A2 complex covering the dorsal portion of the neurocranium (Kotrschal, 1988; Hastings and Springer, 1994). Blenniids, chaenopsids and starksiins (Labrisomidae) have a spinous process on the sphenotic (Hastings and Springer, 1994). Blenniids, chaenopsids, dactyloscopids and starksiins have the insertion of the hyomandibula relatively far posterior (Hastings and Springer, 1994). Some blenniids, some chaenopsids and dactyloscopids have the distal portion of their median-fin spines unossified (Hastings and Springer, 1994).

Clearly, these and other morphological features offer conflicting evidence of relationships within the Blennioidei. Detailed osteological descriptions are now available for several blennioids (e.g., Thiele, 1963; Springer, 1968; Smith-Vaniz, 1976; Bath, 1977, 1996, 2001; Bock and Zander, 1986; Hastings and Springer, 1994; Doyle, 1998; Ruck, 1998; Cancion *et al.*, 1999). These studies need to be expanded to include other lineages of blennioids; characters need to be described and compared across all blennioids and then included in parsimony analyses based on a broad sampling of all blennioid lineages and appropriate outgroups.

Two studies on the phylogenetic relationships of an array of blennioids based on molecular data have been published by Stepien and colleagues. The first (Stepien *et al.*, 1993) was based primarily on allozyme data from 40 presumptive gene loci as well as sequence data from the ITS1 of rDNA for 23 species of blennioids. This analysis supported the monophyly of a clade comprising the Chaenopsidae, Labrisomidae and Clinidae, when the trees were rooted with the Tripterygiidae. This rooting was based on Springer's (1993) hypothesis of relationships based on the absence of branched rays in dorsal and pectoral fins in all families except the Tripterygiidae. Within this clade, Stepien *et al.* (1993) found that the two included species of the Chaenopsidae were sister to a paraphyletic Labrisomidae and monophyletic Clinidae.

The second analysis (Stepien *et al.*, 1997) included a significantly broader taxonomic sampling of blennioid species and an array of potential outgroup taxa (59 species total) and was based on a 400 base pair segment of 12s rDNA. That study (that did not include the Gobiesociformes, now thought to be the sister group of the Blennioidei), reported the following. First, a monophyletic Clinidae was recovered, albeit with low bootstrap support (55%). Within the Clinidae, the Clinini and Ophiclinini were sister groups and the Myxodini was paraphyletic. Second, the four included species of the Chaenopsinae (= Chaenopsidae of Stephens, 1963) were recovered as a monophyletic group that was nested within the Labrisomidae, while two genera (*Neoclinus* and *Stathmonotus*) that had been included in the Chaenopsidae based on morphological characters (Hastings and Springer, 1994) did not cluster with other chaenopsids. Third, the Labrisomidae was monophyletic only with the inclusion of all chaenopsids present in the analysis. Fourth, the Clinidae and Labrisomidae clades (the latter including the Chaenopsidae) were recovered as sister groups relative to three 'outgroup' blennioid taxa (a triplefin, a dactyloscopid and a blenniid). This study lays the groundwork for future analyses that should include a broader sampling of taxa, including representatives of the Gobiesociformes, and a greater number of sequence regions, including both mitochondrial and nuclear markers.

FUTURE OF BLENNIOID SYSTEMATICS

Much remains to be learned regarding the systematics of blennioid fishes. In addition to continued work on their alpha taxonomy, it is necessary to expand upon the morphological studies designed to resolve relationships within the group. This includes the traditionally studied systems such as osteology (Bock and Zander, 1986) as well as other morphological systems such as muscles associated with the branchial apparatus (Springer and Johnson, 2004; Springer and Orrell, 2004) and the feeding apparatus (Kotrschal, 1988). Similarly, other soft tissue systems may also prove informative. For example, the Blenniidae are well known for the strongly developed fleshy structures surrounding the anterior anal-fin spines of breeding males (Zander, 1975; Kotrschal *et al.*, 1984). These so-called 'fin glands' were hypothesised by Williams (1990) to represent a synapomorphy of the Blenniidae. Superficially similar but probably not homologous fleshy thickenings on these and other fin elements occur in other blennioids (Northcott and James, 1996). The reproductive anatomy of male blennioids is extraordinarily diverse (Eggert, 1931; Seiwald and Patzner, 1987; Patzner, 1989, 1991; DeJonge *et al.*, 1989) and may well have many clues to their phylogenetic relationships. Finally, despite a number of recent studies (Cano *et al.*, 1982; Nieder and Busse, 1992; Stepien *et al.*, 1993, 1997; Almada *et al.*, 2005), the promise of molecular systematics in recovering the evolutionary history of blennioids is yet to be fully realised. Studies in all of these areas are now underway, providing optimism that continued progress will be achieved in the study of the systematics of blennioid fishes.

References

Almada, F., V.C. Almada, T. Guillemaud and P. Wirtz. 2005. Phylogenetic relationships of the north-eastern Atlantic and Mediterranean blenniids. *Biological Journal Linnean Society* 86: 283-295.

Anderson, M.E. 1984. *On the Anatomy and Phylogeny of Zoarcidae (Teleostei: Perciformes)*. Unpubl. Ph.D. dissertation. College of William and Mary, Williamsburg, Virginia.

Bath, H. 1977. Revision der Blenniini. *Senckenbergiana Biologica* 57: 167-234.

Bath, H. 1996. Beitrag zur Osteologie der Arteen der Tribus Parablenniini Die Beziehungen der Knochen des Schädeldaches zum Seitenorgan-System und zu den Weichteilbildungen der Kopfoberseite sowie die systematische Bedeutung der Befund nebst Bemerkungen zu *Lupinoblennius dispar* Herre 1942 (Pisces: Blenniidae). *Senckenbergiana Biologica* 76: 65-92.

Bath, H. 2001. Osteology and morphology of fishes of the subfamily Salariinae and its junior synonym Parablenniinae (Pisces: Blenniidae). *Stuttgarter Beiträge zur Naturkunde* A628: 1-42.

Bock, M. and C.D. Zander. 1986. Osteological characters as tool for blenniid taxonomy—A generic revision of European Blenniidae (Percomorphi; Pisces). *Zeitschrift für Zoologische Systematik und Evolutionsforschung* 24: 138-143.

Böhlke, J.E. 1968. The descriptions of three new stargazers (Dactyloscopidae) from the tropical west Atlantic. *Notulae Naturae (Philadelphia)* No. 414: 1-16.

Böhlke, J.E. and C.H. Robins. 1974. Description of a new genus and species of clinid fish from the western Caribbean, with comments on the families of the Blennioidea. *Proceedings of the Academy of Natural Sciences Philadelphia* 126: 1-8.

Böhlke, J.E. and V.G. Springer. 1975. A new genus and species of fish (*Nemaclinus atelestos*) from the western Atlantic (Perciformes: Clinidae). *Proceedings of the Academy of Natural Sciences Philadelphia* 127: 57-61.

Brooks, M.J. 1992. *The Ontogeny and Evolution of Sexual Dimorphism in Paraclinin Blennies (Teleostei, Labrisomidae)*. Unpubl. Ph.D dissertation. University of Arizona, Tucson.

Cancion, C., E. de la Hoz and K. Farias. 1999. Morfología cefálica de *Hypsoblennius sordidus* (Bennet, 1828)(Perciformes, Blenniidae): Un blénido omnívoro. *Revista de Biología Marina y Oceanografía* 34: 167-180.

Cano, J., M.C. Alverez, G. Thode and E. Munoz. 1982. Phylogenetic interpretation of chromosomal and nuclear-DNA-content data in the genus *Blennius* (Blennidae, Perciformes). *Genetica* 58: 11-16.

Carreras-Carbonell, E. Macpherson and M. Pascual. 2005. Rapid radiation and cryptic speciation in Mediterranean triplefin blennies (Pisces: Tripterygiidae) combining multiple genes. *Molecular Phylogenetics and Evolution* 37: 751-761.

Chen, W., C. Bonillo and G. Lecointre. 2003. Repeatability of clades as a criterion of reliability: A case study for molecular phylogeny of Acanthomorpha (Teleostei) with larger number of taxa. *Molecular Phylogenetics and Evolution* 26: 262-288.

Collette, B.B., J.T. Williams, C.E. Thacker and M.L. Smith. 2003. Shore fishes of Navassa Island, West Indies: A case study on the need for rotenone sampling in reef fish biodiversity studies. *Aqua* 6: 89-131.

Dawson, C.E. 1974. Studies on eastern Pacific sand stargazers (Pisces: Dactyloscopidae) 1. *Platygillelus* new genus, with descriptions of new species. *Copeia* 1974: 39-55.

Dawson, C.E. 1975. Studies on eastern Pacific sand stargazers (Pisces: Dactyloscopidae). 2. Genus *Dactyloscopus*, with descriptions of new species and subspecies. *Natural History Museum of Los Angeles County Science Bulletin* 22: 1-61.

Dawson, C.E. 1976. Studies on eastern Pacific sand stargazers. 3. *Dactylagnus* and *Myxodagnus*, with description of a new species and subspecies. *Copeia* 1976: 13-43.

Dawson, C.E. 1977. Studies on eastern Pacific sand stargazers (Pisces: Dactyloscopidae). 4. *Gillellus*, *Sindoscopus* new genus, and *Heteristius* with description of new species. *Proceedings of the California Academy of Sciences*, Series 4, 41: 125-160.

Dawson, C.E. 1982. Atlantic sand stargazers (Pisces: Dactyloscopidae), with description of one new genus and seven new species. *Bulletin of Marine Science* 32: 14-85.

De Jonge, J., A.J.H. De Ruiter and R. Van Den Hurk. 1989. Testis-testicular gland complex of two *Tripterygion* species (Blennioidei, Teleostei): Differences between territorial and non-territorial males. *Journal of Fish Biology* 35: 497-508.

Depcznski, M. and D.R. Bellwood. 2003. The role of crytobenthic reef fishes in coral reef trophodynamics. *Marine Ecology Progress Series* 256: 183-191.

Dettai, A. and G. Lecointre. 2005. Further support for the clades obtained by multiple molecular phylogenies in the acanthomorph bush. *C. R. Biologies* 328: 674-689.

Doyle, K.D. 1998. Phylogeny of the sand stargazers (Dactyloscopidae: Blennioidei). *Copeia* 1998: 76-96.

Eggert, B. 1931. Die Geschlechtsorgane der Gobiiformes und der Blenniiformes. *Zeitschrift für wissenschaftliche Zoologie* 139: 249-558.

Feitozoa, B.M. 2002. *Platygillellus brasiliensis* n. sp. (Perciformes: Dactyloscopidae), the third species of the genus from the Atlantic. *Aqua* 6: 21-28.

Floeter, S.R., L.A. Rocha, D.R. Robertson, J.C. Joyeux, W.F. Smith-Vaniz, P. Wirtz, A.J. Edwards, J.P. Barreiros, C.E.L. Ferreira, J.L. Gasparini, A. Brito, J.M. Falcón, B.W. Bowen and G. Bernardi. 2007. Atlantic reef fish biogeography and evolution. *Journal of Biogeography* 2007: 1-26.

Fraser, T.H. 1972. A new species of the klipfish genus *Springeratus* (Clinidae) from the Indian Ocean. *J. L. B. Smith Institute of Ichthyology*, Special Publication 9: 1-14.

George, A. and V.G. Springer. 1980. Revision of the clinid fish tribe Ophiclinini, including five new species, and definition of the family Clinidae. *Smithsonian Contributions to Zoology* 307: 1-31.

Gill, A.C. and R.D. Mooi. 1993. Monophyly of the Grammatidae and of the Notograptidae, with evidence for the phylogenetic position among perciforms. *Bulletin of Marine Science* 52: 327-350.

Godkin, C.M. and R. Winterbottom. 1985. Phylogeny of the family Congrogadidae (Pisces: Perciformes) and its placement as a subfamily of the Pseudochromidae. *Bulletin of Marine Science* 36: 633-671.

Gosline, W.A. 1968. The suborders of perciform fishes. *Proceedings of the United States National Museum* 124: 1-78.

Hastings, P.A. 1990. Phylogenetic relationships of the species of *Acanthemblemaria*. *Bulletin of Marine Science* 47: 725-738.

Hastings, P.A. 1992a. *Ekemblemaria lira*, a new blennioid fish from Ecuador, with comments on sexual dimorphism and relationships in *Ekemblemaria* (Teleostei: Chaenopsidae). *Copeia* 1992: 769-776.

Hastings, P.A. 1992b. Phylogenetic relationships of *Tanyemblemaria alleni*, a new genus and species of chaenopsid (Pisces: Blennioidei) from the Gulf of Panama. *Bulletin of Marine Science* 51: 147-160.

Hastings, P.A. 1997. Phylogenetic relationships of the *Coralliozetus* clade of chaenopsid blennies, with description of a new genus (Teleostei, Blennioidei). *Bulletin of Marine Science* 61: 743-761.

Hastings, P.A. and V.G. Springer. 1994. A review of *Stathmonotus*, with redefinition and phylogenetic analysis of the Chaenopsidae (Pisces: Blennioidei). *Smithsonian Contributions to Zoology* 558: 1-48.

Hubbs, C.L. 1952. A contribution to the classification of blennioid fishes of the family Clinidae, with a partial revision of the eastern Pacific forms. *Stanford Ichthyological Bulletin* 4: 41-165.

Hubbs, C.L. 1953. Revision and systematic position of blenniid fishes of the genus *Neoclinus*. *Copeia* 1953: 11-23.

Johnson, G.D. 1984. Percoidei: development and relationships. In: Ontogeny and Systematics of Fishes. *American Society of Ichthyologists and Herpetologists*, Special Publication *1*, pp. 464-496.

Johnson, G.D. 1993. Percomorph phylogeny: progress and problems. *Bulletin of Marine Science* 52: 3-28.

Johnson, G.D. and C.W. Patterson. 1993. Percomorph phylogeny: A survey of acanthomorphs and a new proposal. *Bulletin of Marine Science* 52: 554-626.

Jordan, D.S. 1923. A classification of fishes including families and genera as far as known. *Stanford University Publications, Biological Sciences* 3: 77-243.

Kotrschal, K. 1988. A catalogue of skulls and jaws of eastern tropical Pacific blennioid fishes (Blennioidei: Teleostei): A proposed evolutionary sequence of morphological change. *Zeitschrift für Zoologische Systematik und Evolutionsforschung* 26: 442-466.

Kotrschal, K., H. Weisl and A. Goldschmid. 1984. Mehrzellige Drusen in der Peidermis der unpaaren Flossen bei den Blenniidae (Blenniidae, Teleostei). *Zeitschrift mikroskopisch-anatomisch Forschung, Leipzig* 98: 184-192.

Kuiter, R.H. 1993. *Coastal Fishes of South-eastern Australia.* University of Hawaii Press, Honolulu.

McCosker, J.E., J.S. Stephens and R.H. Rosenblatt. 2003. *Cottoclinus canops*, a new genus and species of blenny (Perciformes: Labrisomidae) from the Galápagos Islands. *Proceedings of the California Academy of Sciences* 54: 155-160.

Miller, R.R. and J.C. Briggs. 1962. *Dactyloscopus amnis*, a new sand stargazer from rivers of the Pacific slope of southern Mexico. *Occasional Papers of the Museum of Zoology, University of Michigan* 627: 1-11.

Miya, M., A. Kawaguchi and M. Nishida. 2003. Mitogenomic exploration of higher teleostean phylogenies: A case study for moderate-scale evolutionary genomics with 38 newly determined complete mitochondrial DNA sequences. *Molecular Biology and Evolution* 18: 1993-2009.

Miya, M., T.P. Satoa and M. Nishida. 2005. The phylogenetic position of toadfishes (order Batrachoidiformes) in the higher ray-finned fish as inferred from partitioned Bayesian analysis of 102 whole mitochondrial genome sequences. *Biological Journal of the Linnean Society* 85: 289-306.

Mooi, R.D. and A.C. Gill. 1995. Association of epaxial musculature with dorsal-fin pterygiophores in acanthomorph fishes, and its phylogenetic significance. *Bulletin of the Natural History Museum London (Zoology)* 62: 121-137.

Munday, P.L. and G.P. Jones. 1998. The ecological implications of small body size among coral-reef fishes. *Annual Review of Oceanography and Marine Biology* 36: 373-411.

Nelson, J.S. 1976. *Fishes of the World.* John Wiley & Sons, New York. 1st Edition.

Nelson, J.S. 1984. *Fishes of the World.* John Wiley & Sons, New York. 2nd Edition.

Nelson, J.S. 2006. *Fishes of the World.* John Wiley & Sons, New York. 4th Edition.

Nieder, J. and K. Busse. 1992. Remarks on the systematics of the tribe Parablenniini based on blood serum electrophoresis (Pisces: Blenniidae). *Zeitschrift für Zoologische Systematik und Evolutionsforschung* 30: 123-128.

Northcott, S.J. and M. A. James. 1996. Ultrastructure of the glandular epidermis on the fins of male estuarine triplefins *Forsterygion nigripenne*. *Journal of Fish Biology* 49: 95-107.

Oliveira, R.F., V.C. Almada, A.J. Almeida, R.S. Santos and E.J. Gonçalves. 1992. A checklist of the blennioid fishes (Teleostei, Blennioidei) occurring in Portuguese waters. *Arquipélago. Life and Earth Sciences* 10: 23-37.

Patterson, C. and D.E. Rosen. 1989. The Paracanthopterygii revisited: Order and disorder. *Natural History Museum of Los Angeles County, Science Series* 32: 5-39.

Patzner, R.A. 1989. Morphology of the male reproductive systems of two IndoPacific blenniid fishes. *Zeitschrift für Zoologische Systematik und Evolutionsforschung* 27: 135-141.

Patzner, R.A. 1991. Morphology of the male reproductive system of *Coralliozetus angelica* (Pisces, Blennioidei, Chaenopsidae). *Journal of Fish Biology* 39: 867-872.

Penrith, M. 1969. The systematics of the fishes of the family Clinidae in South Africa. *Annals of the South African Museum* 55: 1-121.

Penrith, M. 1970. The distribution of the fishes of the family Clinidae in southern Africa. *Annals of the South African Museum* 55: 135-150.

Petti, J.C. 1969. *Behavioral and Morphological Adaptations to Burrowing of two Species of Dactyloscopid Fishes from the Northern Gulf of California.* Unpublished Master's Thesis. University of Arizona, Tucson.

Pietsch, T.W. 1989. Phylogenetic relationships of the trachinoid fishes of the family Uranoscopidae. *Copeia* 1989: 253-303.

Pietsch, T.W. and C.P. Zabetian. 1990. Osteology and interrelationships of the sand lances (Teleostei: Ammodytidae). *Copeia* 1990: 78-100.

Regan, C.T. 1912. The classification of the blennioid fishes. *The Annals and Magazine of Natural History*, Series 8, 36: 265-280.

Rosen, D.E. and C. Patterson. 1969. The structure and relationships of the paracanthopterygian fishes. *Bulletin of the American Museum of Natural History* 141: 357-474.

Rosen, D.E. and C. Patterson. 1990. On Müller's and Cuvier's concepts of pharyngognath and labyrinth fishes and the classification of percomorph fishes, with an atlas of percomorph dorsal gill arches. *American Museum Novitates* 2983: 1-57.

Rosenblatt, R.H. 1984. Blennioidei: Introduction. In: Ontogeny and Systematics of Fishes. *American Society of Ichthyologists and Herpetologists* Special Publication *1*, pp. 551-552.

Rosenblatt, R.H. and T.D. Parr. 1969. The Pacific species of the clinid fish genus *Paraclinus*. *Copeia* 1969: 1-20.

Rosenblatt, R.H. and L.R. Taylor, Jr. 1971. The Pacific species of the clinid fish tribe Starksiini. *Pacific Science* 25: 436-463.

Ruck, J.G. 1998. Larval, prejuvenile and adult osteology of *Forsterygion varium* Schneider (Pisces: Tripterygiidae) from New Zealand. *Wellington Polytechnic Occasional Paper* 98/2: 1-72.

Seiwald, M. and R.A. Patzner. 1987. Ultrastructure of the testicular gland of *Blennius pavo* (Pisces, Teleostei). *Zoomorphology* 107: 26-32.

Shen, S.C. 1971. A new genus of clinid fishes from the Indo-West-Pacific, with a redescription of *Clinus nematopterus*. *Copeia* 1971: 697-707.

Simmons, M.P. and M. Miya. 2004. Efficiently resolving the basal clades of a phylogenetic tree using Bayesian and parsimony approaches: A case study using mitogenomic data from 100 higher teleost fishes. *Molecular Phylogenetics and Evolution* 31: 351-362.

Smith, W.L. and W.C. Wheeler. 2004. Polyphyly of the mail-cheeked fishes (Teleostei: Scorpaeniformes): Evidence from mitochondrial and nuclear sequence data. *Molecular Phylogenetics and Evolution* 32: 627-646.

Smith, W.L. and W.C. Wheeler. 2006. Venom evolution is widespread in fishes: A phylogenetic road map for the bioprospecting of piscine venoms. *Journal of Heredity* 97: 206-217.

Smith, W.L. and M.T. Craig. 2007. Casting the percomorph net widely: The importance of broad taxonomic sampling in the search for the placement of serranid and percid fishes. *Copeia* 2007: 35-55.

Smith-Vaniz, W.F. 1976. The saber-toothed blennies, tribe Nemophini (Pisces: Blenniidae). *Monographs of the Academy of Natural Sciences of Philadelphia* 19: 1-196.

Smith-Vaniz, W.F., H.L. Jelks and L.A. Rocha. 2006. Relevance of cryptic fishes in biodiversity assessments: A case study at Buck Island Reef National Monument, St. Croix. *Bulletin of Marine Science* 79: 17-48.

Springer, V.G. 1959. Systematics and zoogeography of the clinid fishes of a subtribe Labrisomini Hubbs. *Publication of the Institute of Marine Science, University of Texas* 5: 417-492.

Springer, V.G. 1968. Osteology and classification of fishes of the family Blenniidae. *United States National Museum Bulletin* 284: 1-85.

Springer, V.G. 1970. The western south Atlantic clinid fish *Ribeiroclinus eigenmanni*, with discussion of the intrarelationshps and zoogeography of the Clinidae. *Copeia* 1970: 430-436.

Springer, V.G. 1993. Definition of the suborder Blennioidei and its included families (Pisces: Perciformes). *Bulletin of Marine Science* 52: 472-495.

Springer, V.G. and W.C. Freihofer. 1976. Study of the monotypic fish family Pholidichthyidae (Perciformes). *Smithsonian Contributions to Zoology* 216: 1-43.

Springer, V.G. and J.T. Williams. 1994. The Indo-Pacific blenniid fish genus *Istiblennius* reappraised: A revision of *Istiblennius*, *Blenniella*, and *Paralticus*, new genus. *Smithsonian Contributions to Zoology* 565: 1-193.

Springer, V.G. and G.D. Johnson. 2004. Study of the dorsal gill-arch musculature of Teleostome fishes, with special reference to the Actinopterygii. *Bulletin of the Biological Society of Washington* 11: 1-235.

Springer, V.G. and T.M. Orrell. 2004. A phylogenetic analysis of 147 families of acanthomorph fishes based primarily on dorsal gill-arch muscles and skeleton. *Bulletin of the Biological Society of Washington* 11: 237-260.

Springer, V.G., C.L. Smith and T.H. Fraser. 1977. *Anisochromis straussi*, a new species of protogynous hermaphroditic fish, and synonymy of Anisochromidae, Pseudoplesiopidae, and Pseudochromidae. *Smithsonian Contributions to Zoology* 252: 1-15.

Stephens, J.S. 1963. A revised classification of the blennioid fishes of the American family Chaenopsidae. *University of California Publications in Zoology* 68: 1-165.

Stephens, J.S. 1970. Seven new chaenopsid blennies from the western Atlantic. *Copeia* 1970: 280-309.

Stephens, J.S. and V.G. Springer. 1974. Clinid fishes of Chile and Peru, with description of a new species, *Myxodes ornatus*, from Chile. *Smithsonian Contributions to Zoology* 159: 1-24.

Stepien, C.A. 1992. Evolution and biogeography of the Clinidae (Teleostei: Blennioidei). *Copeia* 1992: 375-392.

Stepien, C.A., M.T. Dixon and D.M. Hillis. 1993. Evolutionary relationships of the blennioid fish families Clinidae, Labrisomidae, and Chaenopsidae: Congruence among DNA sequence and allozyme data. *Bulletin of Marine Science* 52: 496-515.

Stepien, C.A., A.K. Dillon, M.J. Brooks, K.L. Chase and A.N. Hubers. 1997. The evolution of blennioid fishes based on an analysis of mitochondrial 12S rDNA. In: *Molecular Systematics of Fishes*, T.T. Kocher and C.A. Stepien (eds:). Academic Press, San Diego, pp. 245-270.

Stepien, C.A., M. Glattke and K.M. Fink. 1988. Regulation and significance of color patterns of the spotted kelpfish, *Gibbonsia elegans* Cooper, 1864 (Blennioidei: Clinidae). *Copeia* 1988: 7-15.

Thiele, H. 1963. Vergleichend-morphologische Untersuchungen uber die Funktion der Nahrungserwerbsapparate von *Anarrhichas lupus* L. und einigen Blenniidae (Teleostei, Blennioidei). *Zoologische Beiträge, Neue Folge* 9: 275-440.

Thomson, D.A. and M. Gilligan. 1983. The rocky-shore fishes. In: *Island Biogeography in the Sea of Cortez*, T.J. Case and M.L. Cody (eds.). University of California Press, Berkeley, pp. 98-129.

Todd, E.S. 1973. A preliminary report on the respiratory pump in the Dactyloscopidae. *Copeia* 1973: 115-119.

Williams, J.T. 1990. Phylogenetic relationships and revision of the blenniid fish genus *Scartichthys*. *Smithsonian Contributions to Zoology*. 492: 1-30.

Williams, J.T. 2002. Three new species of blennioid shore fishes discovered at Navassa Island, Caribbean Sea. *Aqua* 6: 11-16.

Williams, J.T. and J.H. Mounts. 2003. Descriptions of six new Caribbean fish species in the genus *Starksia* (Labrisomidae). *Aqua* 6: 145-164.

Zander, C.D. 1975. Secondary sex characteristics of blennioid fishes (Perciformes). *Pubblicazione della Stazione Zoologica di Napoli* 39: 717-727.

CHAPTER

1.2

Systematics of the Tripterygiidae (Triplefins)

Ronald Fricke

INTRODUCTION

Tripterygiid fishes (triplefins or threefin blennies) are a family of bottom-living blennioid fishes, inhabiting cold, temperate, subtropical and tropical sea shores and offshore islands. They are usually associated with hard substrates, mainly rock. The surface of the substrate is ideally rough, so that triplefins can adhere to it using their pelvic fin rays. Most species live subtidally on rocky or coral reefs or in intertidal rock pools; a few occur deeper, on the continental shelf and other slopes down to at least 550 m depth. Very few species live in fresh water. The use of modern collecting techniques like rotenone collections has revealed that the Tripterygiidae is a large and abundant group. A total of 32 genera and 164 species is currently known worldwide (Table 1.2.1). All species are small; the largest does not exceed 150 mm SL (*Blennodon dorsale*, New Zealand), while the smallest attains 20 mm SL as its maximum size.

Author's address: Staatliches Museum für Naturkunde in Stuttgart, Rosenstein 1, 70191 Stuttgart, Germany. E-mail: fricke.smns@naturkundemuseum-bw.de

Table 1.2.1 Currently recognised genera and species of the Tripterygiidae

Taxon	*References*	*Distribution*	*Remarks*
1 Subfamily Notoclininae			
1.1 Tribe Notoclinini	present paper		
1.1.1 Genus *Brachynectes* Scott, 1957	Fricke (1994b: 145)		
1.1.1.1 *Brachynectes fasciatus* Scott, 1957	Fricke (1994b: 146-151)	Australia (southern-western Australia to Tasmania)	
1.1.2 Genus *Notoclinus* Gill, 1893	Fricke (1994b: 500)		
1.1.2.1 *Notoclinus compressus* (Hutton, 1872)	Fricke (1994b: 503-507)	New Zealand main islands, Stewart Island and Chatham Islands	
1.1.2.2 *Notoclinus fenestratus* (Bloch and Schneider [ex Forster], 1801)	Fricke (1994b: 507-512)	New Zealand main islands, Stewart, Snares and Chatham islands	
2. Subfamily Tripterygiinae			
2.1 Tribe Trianectini			
2.1.1 Genus *Acanthanectes* Holleman and Buxton, 1993	Holleman and Buxton (1993: 329-331)		
2.1.1.1 *Acanthanectes hystrix* Holleman and Buxton, 1993	Holleman and Buxton (1993: 332-336)	South Africa	
2.1.1.2 *Acanthanectes rufus* Holleman and Buxton, 1993	Holleman and Buxton (1993: 336-338)	South Africa	
2.1.2 Genus *Apopterygion* Kuiter, 1986	Kuiter (1986: 90); Fricke (1994b: 111-112)		
2.1.2.1 *Apopterygion alta* Kuiter, 1986	Kuiter (1986: 90-91); Fricke (1994b: 115-117)	Australia (Victoria, Tasmania)	

Table 1.2.1 Contd.

Table 1.2.1 Contd.

2.1.2.2 *Apopterygion oculus* Fricke and Roberts in Fricke, 1994	Fricke (1994b: 118-122)	Southern New Zealand region	
2.1.3 Genus *Gilloblennius* Whitley and Phillipps, 1939	Hardy (1986: 145-147); Fricke (1994b: 350-352)		
2.1.3.1 *Gilloblennius abditus* Hardy, 1986	Hardy (1986: 154-157); Fricke (1994b: 353-357)	New Zealand region	
2.1.3.2 *Gilloblennius tripennis* (Bloch and Schneider [ex Forster], 1801)	Hardy (1986: 147-154); Fricke (1994b: 358-362)	New Zealand region	*Tripterygium forsteri* Valenciennes in Cuvier and Valenciennes (1836: 415) is a junior synonym (see Fricke, 1994b: 358)
2.1.4 Genus *Ruanoho* Hardy, 1986	Hardy (1986: 157-158)		
2.1.4.1 *Ruanoho decemdigitatus* (Clarke, 1879)	Fricke (1994b: 521-531)	New Zealand region	
2.1.4.2 *Ruanoho whero* Hardy, 1986	Hardy (1986: 161-164); Fricke (1994b: 521-531, *whero*-morph); Jawad and Clements (2004: 87)	New Zealand region	Considered as a colour morph of *Ruanoho decemdigitatus* by Fricke (1994b: 521-531), but treated as separate species by Jawad and Clements (2004: 87)
2.1.5 Genus *Trianectes* McCulloch and Waite, 1918	Fricke (1994b: 532)		*Verconectes* Whitley (1931: 324) is a junior synonym (see Fricke, 1994b: 532)
2.1.5.1 *Trianectes bucephalus* McCulloch and Waite, 1918	Fricke (1994b: 533-534)	Southwestern Australia to Tasmania	
2.2 Lineage Norfolkiini			
2.2.1 Genus *Cremnochorites* Holleman, 1982	Holleman (1982: 125-131)		
2.2.1.1 *Cremnochorites capensis* (Gilchrist and Thompson, 1908)	Holleman (1982: 131-132)	South Africa	

Table 1.2.1 Contd.

Table 1.2.1 Contd.

2.2.2 Genus *Lepidonectes* Rosenblatt in Bussing, 1991	Bussing (1991: 81)		*Taboguilla* Allen and Robertson (1991: 81) is a junior synonym (see Fricke, 1997: 582)
2.2.2.1 *Lepidonectes bimaculatus* Allen and Robertson, 1992	Allen and Robertson (1992: 55-56)	Isla Malpelo/Colombia	Eschmeyer (2007) spells the species name *bimaculata* but as the genus name *Lepidonectes* is masculine, the species name must be *bimaculatus*
2.2.2.2 *Lepidonectes clarkhubbsi* Bussing, 1991	Bussing (1991: 81-84)	Pacific coast of Costa Rica to Panama	*Taboguilla signata* Allen and Robertson (1991: 81-82) is a junior synonym (see Fricke, 1997: 582)
2.2.2.3 *Lepidonectes corallicola* (Kendall and Radcliffe, 1912)	Kendall and Radcliffe (1912: 153); Fricke (1997: 582)	Galapagos Islands/Ecuador	
2.2.3 Genus *Norfolkia* Fowler, 1953	Holleman (1991: 227-243); Fricke (1994b: 461-464)		
2.2.3.1 *Norfolkia brachylepis* (Schultz in Schultz *et al.*, 1960)	Fricke (1994b: 465-470; 1997: 504-512; 2002: 194-195)	Red Sea and East Africa to northern Australia, Japan, Bonin and Marshall Islands, and Fiji	*Norfolkia springeri* Clark (1980: 95-96) is a junior synonym (see Holleman, 1991: 233-236)
2.2.3.2 *Norfolkia leeuwin* Fricke, 1994	Fricke (1994b: 471-472)	Southwestern Australia	
2.2.3.3 *Norfolkia squamiceps* (McCulloch and Waite, 1916)	Fricke (1994b: 475-475; 1997: 512-515; 2002: 195)	Queensland/Australia to Lord Howe and Norfolk islands, New Caledonia and Loyalty Islands	*Norfolkia lairdi* Fowler (1953: 264-266) is a junior synonym (see Fricke, 1997: 473)
2.2.3.4 *Norfolkia thomasi* Whitley, 1964	Holleman (1991: 238-239); Fricke (1994b: 468-478; 1997: 516-519; 2002: 195-196)	Ryukyu Islands; tropical Australia to Gambier and Austral Islands	

Table 1.2.1 Contd.

Table 1.2.1 Contd.

2.2.4 Genus *Trinorfolkia* Fricke, 1994	Fricke (1994b: 537-540)		
2.2.4.1 *Trinorfolkia clarkei* (Morton, 1888)	Fricke (1994b: 541-547)	Southern half of Australia	*Tripterygium striaticeps* Ramsay and Ogilby (1889: 419-420), *T. macleayanum* Lucas (1891: 12), and *T. whitleyi* Scott [ex Johnson] (1977: 148-150) are junior synonyms (see Fricke, 1994b: 544)
2.2.4.2 *Trinorfolkia cristata* (Kuiter, 1986)	Fricke (1994b: 547-549)	South Australia	
2.2.4.3 *Trinorfolkia incisa* (Kuiter, 1986)	Fricke (1994b: 549-553)	Western Australia to Victoria	
2.3 Tribe Tripterygiini			
2.3.1 Genus *Axoclinus* Fowler, 1944	Smith and Williams (2002: 8)		
2.3.1.1 *Axoclinus cocoensis* Bussing, 1991	Bussing (1991: 78-81); Fricke (1997: 565, *Enneanectes c.*); Smith and Williams (2002: 8)	Isla de Coco, Costa Rica	
2.3.1.2 *Axoclinus lucillae* Fowler, 1944	Fricke (1997: 565, as *Enneanectes lucillae*); Smith and Williams (2002: 8)	Panama	
2.3.1.3 *Axoclinus multicinctus* Allen and Robertson, 1992	Allen and Robertson (1992: 54-55); Fricke (1997: 565, *Enneanectes m.*); Smith and Williams (2002: 8)	Revillagigedo Islands	
2.3.1.4 *Axoclinus nigricaudus* Allen and Robertson, 1991	Allen and Robertson (1991: 79); Fricke 1997: 565, *Enneanectes n.*); Smith and Williams (2002: 8)	Baja California/Mexico	

Table 1.2.1 Contd.

Table 1.2.1 Contd.

2.3.1.5 *Axoclinus rubinoffi* Allen and Robertson, 1992	Allen and Robertson (1992: 53-54); Fricke (1997: 565, *Enneanectes r.*); Smith and Williams (2002: 8)	Colombia	
2.3.1.6 *Axoclinus storeyae* (Brock, 1940)	Smith and Williams (2002: 8)	Baja California/Mexico	
2.3.2 Genus *Ceratobregma* Holleman, 1987	Holleman (1987: 174-175); Fricke (1994b: 152-156; 1997: 127-128)		
2.3.2.1 *Ceratobregma acanthops* (Whitley, 1964)	Holleman (1987: 177-179); Fricke (1994b: 157-159; 1997: 129-133)	Australia (Queensland) and West Coral Sea	
2.3.2.2 *Ceratobregma helenae* Holleman, 1987	Holleman (1987: 175-177); Fricke (1997: 133-149)	Philippines to Pohnpei and Samoa, north to Ryukyu Islands, south to northern Australia	*Ceratobregma striata* Fricke (1991: 337-339) is a junior synonym (see Fricke, 1997: 137)
2.3.3 Genus *Crocodilichthys* Allen and Robertson, 1991	Allen and Robertson (1991: 79-80)		
2.3.3.1 *Crocodilichthys gracilis* Allen and Robertson, 1991	Allen and Robertson (1991: 79-80)	Gulf of California/Mexico	
2.3.4 Genus *Enneanectes* Jordan and Evermann in Jordan, 1895	Fricke (1997: 564)		*Axoclinus* Fowler (1944: 288-289) is a junior synonym (see Fricke, 1997: 564). Fricke (1997: 574) considered *Gillias* Evermann and Marsh, 1900 as a separate genus
2.3.4.1 *Enneanectes altivelis* Rosenblatt, 1960	Rosenblatt (1960: 19-23); Fricke (1997: 574 as *Gillias a.*); Smith and Williams (2002: 9)	Florida and Bahamas to Central America	

Table 1.2.1 Contd.

Table 1.2.1 Contd.

2.3.4.2 *Enneanectes atrorus* Rosenblatt, 1960	Rosenblatt (1960: 9-12); Fricke (1997: 574 as *Gillias a.*); Smith and Williams (2002: 9)	Bahamas to Belize	
2.3.4.3 *Enneanectes boehlkei* Rosenblatt, 1960	Rosenblatt (1960: 5-9); Fricke (1997: 575 as *Gillias b.*); Smith and Williams (2002: 9)	Florida, Bahamas and Yucatan to Venezuela	
2.3.4.4 *Enneanectes carminalis* (Jordan and Gilbert, 1882)	Fricke (1997: 564-565)	Baja California to Panama	*Gillias sexmaculatus* Fowler (1944: 286-288) is a junior synonym (Fricke, 1997: 564)
2.3.4.5 *Enneanectes jordani* (Evermann and Marsh, 1899)	Rosenblatt (1960: 16-19); Fricke (1997: 575, as *Gillias j.*); Smith and Williams (2002: 9)	Florida to Puerto Rico	
2.3.4.6 *Enneanectes pectoralis* (Fowler, 1941)	Rosenblatt (1960: 12-16); Fricke (1997: 575 as *Gillias p.*); Smith and Williams (2002: 9)	Florida, Bahamas and Yucatan to Venezuela	
2.3.4.7 *Enneanectes reticulatus* Allen and Robertson, 1991	Allen and Robertson (1991: 80-81)	Baja California/Mexico	
2.3.4.8 *Enneanectes smithi* Lubbock and Edwards, 1981	Lubbock and Edwards (1981: 148-149); Fricke (1997: 575 as *Gillias s.*); Smith and Williams (2002: 9)	Saint Paul's Rock/Central Atlantic	

Table 1.2.1 Contd.

Table 1.2.1 Contd.

2.3.5 Genus *Enneapterygius* Rüppell, 1835	Fricke (1994b: 173-186; 1997: 142-146)		*Vauclusella* Whitley (1931: 324) and *Rosenblatella* Shen in Shen and Wu (1994: 25) are junior synonyms (see Fricke 1997: 142)
2.3.5.1 *Enneapterygius abeli* (Klausewitz, 1960)	Clark (1980: 97-99)	Red Sea south to South Africa, Comores and Saint Brandon's Shoals	
2.3.5.2 *Enneapterygius atriceps* (Jenkins, 1903)	Fricke (2007: 161-165)	Hawaiian Islands, north to Midway Atoll	
2.3.5.3 *Enneapterygius atrogulare* (Günther, 1873)	Fricke (1994b: 187-194; 1997: 165-170)	Eastern Australia and West Coral Sea	*Tripterygium annulatum* Ramsay and Ogilby (1888: 1021-1022) is a junior synonym; see Fricke (1994b: 187, 189)
2.3.5.4 *Enneapterygius bahasa* Fricke, 1997	Fricke (1997: 170-176)	Southern Japan to Queensland/ Australia	
2.3.5.5 *Enneapterygius cheni* Wang, Shao and Shen, 1996	Wang, Shao and Shen (1996: 80-82)	Japan to Taiwan, northern Philippines and Pohnpei	This was previously supposed to be a synonym of *Enneapterygius niger*; considered as valid by Shibukawa, Suzuki and Hosokawa (2004: 2-6)
2.3.5.6 *Enneapterygius clarkae* Holleman, 1982	Holleman (2005: 6)	Western Indian Ocean	This was supposed to be a junior synonym of *E. destai*, but considered as valid by Holleman (2005: 6)
2.3.5.7 *Enneapterygius clea* Fricke, 1997	Fricke (1997: 176-179)	Queensland/Australia	
2.3.5.8 *Enneapterygius destai* Clark, 1980	Clark (1980: 102-103)	Red Sea	
2.3.5.9 *Enneapterygius elegans* (Peters, 1876)	Fricke (1994b: 203-208; 1997: 179-185)	East Africa to Samoa	

Table 1.2.1 Contd.

Table 1.2.1 Contd.

2.3.5.10 *Enneapterygius elaine* Holleman, 2005	Holleman (2005: 7)	Rodrigues	
2.3.5.11 *Enneapterygius etheostomus* (Jordan and Snyder, 1902)	Fricke (1997: 185-192)	Japan to northern Vietnam	
2.3.5.12 *Enneapterygius fasciatus* (Weber, 1909)	Fricke (1997: 193-199)	East Africa to Taiwan and Solomon Islands	
2.3.5.13 *Enneapterygius flavoccipitis* Shen in Shen and Wu, 1994	Fricke (1997: 199-209; 2002: 184)	Ryukyu Islands to northern Australia and Loyalty Islands	*Enneapterygius bichrous* Fricke (1994b: 195-202) is a junior synonym (see Fricke, 1997: 203)
2.3.5.14 *Enneapterygius fuscoventer* Fricke, 1997	Fricke (1997: 210-213)	Taiwan and Philippines to Society Islands	
2.3.5.15 *Enneapterygius genamaculatus* Holleman, 2005	Holleman (2005: 11)	Saint Brandon's Shoals	
2.3.5.16 *Enneapterygius gracilis* Fricke, 1994	Fricke (1994b: 209-214)	Northern Australia	
2.3.5.17 *Enneapterygius gruschkai* Holleman, 2005	Holleman (2005: 11)	Islands of western and central Indian Ocean	
2.3.5.18 *Enneapterygius hemimelas* (Kner and Steindachner, 1867)	Fricke (1997: 214-224)	Taiwan and Bali/Indonesia east to Samoa	
2.3.5.19 *Enneapterygius hollemani* Randall, 1995	Randall (1995: 27-29); Holleman (2005: 13)	Oman	
2.3.5.20 *Enneapterygius howensis* Fricke, 1997	Fricke (1997: 224-227; 2002: 185)	Lord Howe Island to New Caledonia and Loyalty Islands	
2.3.5.21 *Enneapterygius kermadecensis* Fricke, 1994	Fricke (1994b: 230-234; 1997: 227-230)	Kermadec Islands	

Table 1.2.1 Contd.

Table 1.2.1 Contd.

2.3.5.22 *Enneapterygius kosiensis* Holleman, 2005	Holleman (2005: 13)	South Africa	
2.3.5.23 *Enneapterygius larsonae* Fricke, 1994	Fricke (1994b: 235-241; 1997: 230-234)	Northern Australia and Papua New Guinea	
2.3.5.24 *Enneapterygius melanospilus* Randall, 1995	Randall (1995: 29-31); Holleman (2005: 14)	Oman	
2.3.5.25 *Enneapterygius mirabilis* Fricke, 1994	Fricke (1994b: 251-255; 1997: 234-237)	Northern Queensland/Australia to Papua New Guinea and Vanuatu	
2.3.5.26 *Enneapterygius miyakensis* Fricke, 1987	Fricke (1987: 333-337; 1997: 237-240)	Izu Islands/Japan	
2.3.5.27 *Enneapterygius namarrgon* Fricke, 1997	Fricke (1997: 240-242)	Northern Territory/Australia	
2.3.5.28 *Enneapterygius nanus* (Schultz in Schultz *et al.*, 1960)	Fricke (1994b: 255-259; 1997: 242-248; 2002: 185-186)	Taiwan to west Coral Sea, Vanuatu and Marshall Islands	
2.3.5.29 *Enneapterygius niger* Fricke, 1994	Fricke (1994b: 259-263; 1997: 248-254; 2002: 186-187)	New Caledonia and Vanuatu	*Enneapterygius cheni* from Taiwan to Pohnpei is a separate species (see above)
2.3.5.30 *Enneapterygius nigricauda* Fricke, 1997	Fricke (1997: 255-261)	Taiwan and Vanuatu east to Society Islands	
2.3.5.31 *Enneapterygius obscurus* Clark, 1980	Clark (1980: 105-106); Holleman (2005: 15)	Red Sea	
2.3.5.32 *Enneapterygius ornatus* Fricke, 1997	Fricke (1997: 262-264)	Henderson/Pitcairn Group	
2.3.5.33 *Enneapterygius pallidoserialis* Fricke, 1997	Fricke (1997: 264-271)	Japan and Micronesia to Vanuatu	

Table 1.2.1 Contd.

Table 1.2.1 Contd.

2.3.5.34 *Enneapterygius paucifasciatus* Fricke, 1994	Fricke (1994b: 264-266; 1997: 271-274; 2002: 187)	Grande Terre/New Caledonia	
2.3.5.35 *Enneapterygius philippinus* (Peters, 1868)	Fricke (1994b:242-250; 1997: 274-288; 2002: 187-188)	Western Indian Ocean to Ryukyu and Marshall Islands, and Samoa	*Tripterygium minutum* Günther (1877: 211), *Enneapterygius tusitalae* Jordan and Seale (1906: 416-417), *E. pardochir* Jordan and Seale (1906: 417-418), *E. cerasinus* Jordan and Seale (1906: 419), *Tripterygium callionymi* Weber (1909: 147-148), *Enneapterygius punctulatus* Herre (1935: 432-433), *E. waigiensis* Herre (1935: 434) and *E. fuligicauda* Fowler (1946: 188-189) are junior synonyms (see Fricke, 1997: 274-281)
2.3.5.36 *Enneapterygius pusillus* Rüppell, 1835	Clark (1980: 108-110); Randall (1995: 308)	Red Sea and South Arabian coast	*Enneapterygius pallidus* Clark (1980: 107) is a junior synonym (see Fricke, 1997: 570)
2.3.5.37 *Enneapterygius pyramis* Fricke, 1994	Fricke (1994b: 266-269; 1997: 289-295)	Micronesia and Queensland/ Australia to Pitcairn Group	
2.3.5.38 *Enneapterygius randalli* Fricke, 1997	Fricke (1997: 295-298)	Austral Islands	
2.3.5.39 *Enneapterygius rhabdotus* Fricke, 1994	Fricke (1994b: 270-272; 1997: 298-303; 2002: 188)	Gulf of Thailand and Taiwan to New Caledonia and Marquesas Islands	
2.3.5.40 *Enneapterygius rhothion* Fricke, 1997	Fricke (1997: 304-310; 2002: 188-189)	New Caledonia and Vanuatu	

Table 1.2.1 Contd.

Table 1.2.1 Contd.

2.3.5.41 *Enneapterygius rubicauda* Shen in Shen and Wu, 1994	Fricke (1997: 310-314; 2002: 189)	Ogasawara and Ryukyu islands to Taiwan and Philippines; New Caledonia and Vanuatu	*Enneapterygius erythrosoma* Shen in Shen and Wu (1994: 7-8) is a junior synonym; *E. macrobranchium* Fowler (1946: 189) may be a senior synonym (see Fricke, 1997: 313), but the holotype is missing and the original description imprecise
2.3.5.42 *Enneapterygius rufopileus* (Waite, 1904)	Fricke (1997: 314-321; 2002: 189-190)	New Caledonia and Lord Howe Island to Fiji and Tonga	
2.3.5.43 *Enneapterygius senoui* Motomura, Harazaki and Hardy, 2005	Motomura, Harazaki and Hardy (2005: 6-14)	Izu and Ogasawara Islands/Japan	
2.3.5.44 *Enneapterygius signicauda* Fricke, 1997	Fricke (1997: 322-326)	Vanuatu to Tonga and Samoa	
2.3.5.45 *Enneapterygius similis* Fricke, 1997	Fricke (1997: 326-330; 2002: 190)	Philippines to eastern Australia and Vanuatu	
2.3.5.46 *Enneapterygius triserialis* Fricke, 1994	Fricke (1994b: 280-285; 197: 331-336; 2002: 190-191)	Eastern Australia to Society and Austral Islands	
2.3.5.47 *Enneapterygius trisignatus* Fricke, 2001	Fricke (2001: 6-8; 2002: 191-192)	Northern Grande Terre/ New Caledonia	
2.3.5.48 *Enneapterygius tutuilae* Jordan and Seale, 1906	Fricke (1994b: 285-297; 1997: 337-351; 2002: 191)	Red Sea and western Indian Ocean to Taiwan, northern Australia, Marshall and Society Islands	*Enneapterygius altipinnis* Clark (1980: 99-101) is a junior synonym (see Fricke, 1997: 1994b: 285-297)
2.3.5.49 *Enneapterygius unimaculatus* Fricke, 1994	Fricke (1994a: 5-8; 1997: 352-355)	Taiwan and eastern Indonesia to Palau and Papua New Guinea	

Table 1.2.1 Contd.

Table 1.2.1 Contd.

2.3.5.50 *Enneapterygius ventermaculus* Holleman, 1982	Holleman (1982: 123-125; 2005: 21)	Western Indian Ocean	*Enneapterygius nasimae* Hoda (1981: 116-122) is a junior synonym (see Holleman, 2005: 21)
2.3.5.51 *Enneapterygius vexillarius* Fowler, 1946	Fricke (1997: 355-361)	Ryukyu Islands and Taiwan to southern China	*Enneapterygius hsiojenae* Shen in Shen and Wu (1994: 11-12) and *E. leucopunctatus* Shen in Shen and Wu (1994: 12-14) are junior synonyms (see Fricke, 1997: 358)
2.3.5.52 *Enneapterygius williamsi* Fricke, 1997	Fricke (1997: 361-366; 2002: 192)	New Caledonia to Tonga	
2.3.5.53 *Enneapterygius ziegleri* Fricke, 1994	Fricke (1994a: 8-12; 1997: 366-370)	Philippines and eastern Indonesia	
2.3.6 Genus *Springerichthys* Shen in Shen and Wu, 1994	Fricke (1997: 519-520)		*Graciloptergion* Fricke (1994b: 363-365) is a junior synonym (see Fricke, 1997: 520)
2.3.6.1 *Springerichthys bapturus* (Jordan and Snyder, 1902)	Fricke (1997: 521-525)	Japan to Taiwan	
2.3.6.2 *Springerichthys kulbickii* (Fricke and Randall in Fricke, 1994)	Fricke (1997: 525-530)	Queensland/Australia to Samoa	
2.3.7 Genus *Tripterygion* Risso, 1827	Wirtz (1980: 83-84)		One additional fossil species known, *Tripterygion pronasus* Arambourg (1927: 207-208, pl. 42, Figs 3 – 5, pl. 43, Fig. 1) from Algeria
2.3.7.1 *Tripterygion delaisi* Cadenat and Blache, 1970	Wirtz (1980: 81-89, part)	English Channel south to Senegal, including Canary Islands, Madeira and Azores	*Tripterygion atlanticus* Wheeler and Dunne (1975: 641-650) is a junior synonym

Table 1.2.1 Contd.

Table 1.2.1 Contd.

2.3.7.2 *Tripterygion melanurum* Guichenot, 1850	Bath (1973: 48-53); Zander in Whitehead *et al.* (1986: 1120, part, as *T. melanurus melanurus*); Fricke, Bilecenoglu and Sari (2007: 96)	Southern Mediterranean Sea	Gender of genus neuter; therefore species name must be *melanurum*
2.3.7.3 *Tripterygion minor* Kolombatović, 1892	Zander in Whitehead *et al.* (1986: 1120, part, as *T. melanurus minor*)	Northern Mediterranean Sea	Raised to species level by Fricke, Bilecenoglu and Sari (2007: 96-97). Treated as valid subspecies by authors.
2.3.7.4 *Tripterygion tartessicum* Carreras-Carbonell, Pascual and Macpherson, 2007	Zander in Whitehead *et al.* (1986: 1120-1121, part, as *T. tripteronotus*); Carreras-Carbonell, Pascual and Macpherson (2007: 80-82)	South western Mediterranean Sea	
2.3.7.5 *Tripterygion tripteronotum* (Risso, 1827)	Zander in Whitehead *et al.* (1986: 1120-121, part); Fricke, Bilecenoglu and Sari (2007: 97)	Mediterranean Sea	*Tripterygion nasus* Risso 1827: 241-242), *Tripterygium melaenocephalus* Cocco (1829: 141), and *T. nikolskii* Maksimov (1909: 59-60) are junior synonyms (see Fricke, 1997: 587). Gender of genus neuter; therefore species name must be *tripteronotum*
2.3.7.6 *Tripterygion xanthosoma* Zander and Heymer, 1970	Wirtz (1980: 81-89, part, as *T. delaisi*)	Mediterranean Sea	Raised to species level by Fricke, Bilecenoglu and Sari (2007: 97). Treated as valid subspecies by authors.
2.4. Tribe Forsterygiini	present chapter		
2.4.1 Genus *Forsterygion* Whitley and Phillipps, 1939	Hardy (1989a: 493); Fricke (1994b: 302-310)		

Table 1.2.1 Contd.

Table 1.2.1 Contd.

2.4.1.1 *Forsterygion flavonigrum* Fricke and Roberts in Fricke, 1994	Fricke (1994b: 314-319)	New Zealand region	
2.4.1.2 *Forsterygion lapillum* Hardy, 1989	Hardy (1989a: 504-509); Fricke (1994b: 320-330)	New Zealand and Stewart Island	
2.4.1.3 *Forsterygion malcolmi* Hardy, 1987	Hardy (1987d: 48-52; 1989a: 49); Fricke (1994b: 331-336)	New Zealand region	
2.4.1.4 *Forsterygion varium* (Bloch and Schneider [ex Forster], 1801)	Hardy (1989a: 494-499); Fricke (1994b: 341-349)	New Zealand region; Tasmania/Australia	*Trypterigium robustum* Clarke (1879: 292-293) and *Forsterygium multiradiatum* Scott (1977: 153-158) are junior synonyms (see Fricke, 1994b: 347), the former by selection of a neotype (Fricke, 1994b: 347); Clements, Jawad and Stewart (2000: 380) consider *T. robustum* as a *nomen dubium*
2.4.2 Genus *Grahamina* Fricke and Roberts, 1993	Fricke and Roberts (1993: 4-6); Fricke (1994b: 373-377)		
2.4.2.1 *Grahamina capito* (Jenyns, 1842)	Fricke and Roberts (1993: 6-11); Fricke (1994b: 378-386)	New Zealand region	*Tripterygium jenningsi* Hutton (1879: 339-340) and *Enneapterygius mortenseni* Rendahl (1926: 11-12, part) are junior synonyms (see Fricke, 1994b: 378)
2.4.2.2. *Grahamina gymnota* (Scott, 1977)	Fricke and Roberts (1993: 12-14); Fricke (1994b: 387-389); Clements, Jawad and Stewart (2000: 373-379)	New Zealand region; Tasmania/Australia	*Grahamina signata* Fricke and Roberts (1993: 158-161) is a junior synonym (see Clements, Jawad and Stewart, 2000: 374-379)

Table 1.2.1 Contd.

Table 1.2.1 Contd.

2.4.2.3 *Grahamina nigripennis* (Valenciennes in Cuvier and Valenciennes, 1836)	Fricke and Roberts (1993: 14-16); Fricke (1994b: 391-394)	New Zealand region	The gender of *Grahamina* is feminine; therefore, the species name must be *nigripennis* rather than *nigripenne*, as previously used by authors
2.4.3 Genus *Matanui* Jawad and Clements, 2004	Jawad and Clements (2004: 96)		
2.4.3.1 *Matanui bathytaton* (Hardy, 1989)	Hardy (1989a: 500-504); Fricke (1994b: 311-314 as *Forsterygion* b.); Jawad and Clements (2004: 97)	Southern New Zealand region	
2.4.3.2. *Matanui profundum* (Fricke and Roberts in Fricke, 1994)	Fricke (1994b: 336-341, *Forsterygion p.*); Jawad and Clements (2004: 100)	New Zealand region	
2.4.4 Genus *Notoclinops* Whitley, 1930	Hardy (1987b: 166); Fricke (1994b: 479-481)		
2.4.4.1 *Notoclinops caerulepunctus* Hardy, 1989	Hardy (1989b: 328-331); Fricke (1994b: 482-487)	New Zealand main islands and Stewart Island	
2.4.4.2 *Notoclinops segmentatus* (McCulloch and Phillipps, 1923)	Hardy (1987b: 166-170); Fricke (1994b: 487-493)	New Zealand main islands and Stewart Island	*Tripterygion bucknilli* Griffin (1926: 544-546) and *Enneapterygius mortenseni* Rendahl (1926: 11-12, part) are junior synonyms (see Fricke (1994b: 491)
2.4.4.3 *Notoclinops yaldwyni* Hardy, 1987	Hardy (1987b: 170-174); Fricke (1994b: 494-499)	New Zealand main islands, Stewart Island and Chatham Islands	
2.4.5 Genus *Obliquichthys* Hardy, 1987	Hardy (1987d: 52); Fricke (1994b: 513)		

Table 1.2.1 Contd.

Table 1.2.1 Contd.

2.4.5.1 *Obliquichthys maryannae* Hardy, 1987	Hardy 1987d: 53-58); Fricke (1994b: 514-519)	New Zealand, including Three Kings, Stewart and Snares islands	
2.5 Tribe Karalepini			
2.5.1 Genus *Helcogrammoides* Rosenblatt in Gon in Gon and Heemstra, 1991	Gon (1991: 400)		
2.5.1.1 *Helcogrammoides antarcticus* (Tomo, 1982)	Gon (1991: 400-401)	Antarctic Peninsula	
2.5.1.2 *Helcogrammoides chilensis* (Cancino in Buen, 1960)	Castillo and Pequeño (1998: 115-118)	Peru and Chile	
2.5.1.3 *Helcogrammoides cunninghami* (Smitt, 1898)	Smitt (1898: 26-29); Fricke (1997: 581); Castillo and Pequeño (1998: 113-115)	Southernmost South America, including Falkland Islands	
2.5.2 Genus *Karalepis* Hardy, 1984	Hardy (1984: 176); Fricke (1994b: 443)		
2.5.2.1 *Karalepis stewarti* Hardy, 1984	Hardy (1984: 177-180); Fricke (1994b: 444-449)	New Zealand region	
2.6 Tribe Helcogrammini			
2.6.1 Genus *Helcogramma* McCulloch and Waite, 1918	Hansen (1986: 313-329); Fricke (1994b: 398-403; 1997: 370-382)		
2.6.1.1 *Helcogramma albimacula* Williams and Howe, 2003	Williams and Howe (2003: 154)	Philippines	
2.6.1.2. *Helcogramma alkamr* Holleman, 2007	Holleman (2007:56)	East Africa to Seychelles and Mauritius	

Table 1.2.1 Contd.

Table 1.2.1 Contd.

2.6.1.3 *Helcogramma aquila* Williams and McCormick, 1990	Williams and McCormick (1990: 1021-1026) Fricke (1997: 409-413, as *H. aquilum*)	Philippines to Guam	Genus name gramma is based on Greek γραμμ for line (thus, feminine); the species name must be *aquila*, and not *aquilum* as used by authors
2.6.1.4 *Helcogramma ascensionis* Lubbock, 1981	Lubbock (1980: 293-296); Holleman (2007: 51)	Ascension Island and Saint Helena, southeastern Atlantic Ocean	This species was considered as a junior synonym of *H. obtusirostre* by Fricke (1997: 462), but treated as a valid species by Holleman (2007: 51)
2.6.1.5 *Helcogramma billi* Hansen, 1986	Hansen (1986: 329-330)	Sri Lanka to southern India	
2.6.1.6 *Helcogramma capidata* Rosenblatt in Schultz *et al.*, 1960	Hansen (1986: 330-331); Fricke (1994b: 410-413; 1997: 413-421)	Sabah/Malaysia to Marshall Islands and Samoa	Genus name 'gramma' is based on Greek γραμμ for line (thus feminine); the species name must be *capidata*, and not *capidatum* as used by authors
2.6.1.7 *Helcogramma cerasina* Williams and Howe, 2003	Williams and Howe (2003: 159)	Fiji and Tonga	
2.6.1.8 *Helcogramma chica* Rosenblatt in Schultz *et al.*, 1960	Hansen (1986: 331-333); Fricke (1994b: 414-417; 1997: 422-429)	Andaman Sea to Marshall and Society Islands	
2.6.1.9 *Helcogramma decurrens* McCulloch and Waite, 1918	Hansen (1986: 333-335); Fricke (1994b: 417-420)	South-western Australia	*Vauclusella calva* Whitley (1944: 272) is a junior synonym (see Fricke (1994b: 420)
2.6.1.10 *Helcogramma desa* Williams and Howe, 2003	Williams and Howe (2003: 161)	Philippines to Vietnam	
2.6.1.11 *Helcogramma ellioti* (Herre, 1944)	Hansen (1986: 335-337); Fricke (1997: 578)	Southern India	*Helcogramma indicus* Talwar and Sen (1971: 248-250) is a junior synonym (see Fricke, 1997: 578)

Table 1.2.1 Contd.

Table 1.2.1 Contd.

2.6.1.12 *Helcogramma ememes* Holleman, 2007	Holleman (2007: 61)	East coast of Africa, Seychelles	
2.6.1.13 *Helcogramma fuscipectoris* (Fowler, 1946)	Fricke (1997: 429-437)	Southern Japan and Thailand to Vanuatu	*Enneapterygius personatus* Fowler (1946: 185-186) and *E. quadrimaculatus* Fowler (1946: 189-190) are junior synonyms (see Fricke, 1997: 433)
2.6.1.14 *Helcogramma fuscopinna* Holleman, 1982	Hansen (1986: 337-339, part); Fricke (1997: 115-120); Williams and Howe (2003: 163)	Western Indian Ocean	
2.6.1.15 *Helcogramma gymnauchen* (Weber, 1909)	Fricke (1994b: 421-425; 1997: 437-442)	Eastern Indonesia to northern Australia and New Guinea	
2.6.1.16 *Helcogramma hudsoni* (Jordan and Seale, 1906)	Fricke (1994b: 426-428; 1997: 442-445; 2002: 193)	New Caledonia to Gilbert Islands and Samoa	
2.6.1.17 *Helcogramma inclinata* (Fowler, 1946)	Fricke (1997: 446-452)	Ryukyu Islands to Taiwan and northern Philippines	*Lepidoblennius marmoratus ishigakiensis* Aoyagi (1954: 241-242) and *Helcogramma habena* Williams and McCormick (1990: 1026-1029) are junior synonyms (see Fricke, 1997: 449). The genus name 'gramma' is based on Greek γραμμ for line (thus feminine); the species name must be *inclinata*, and not *inclinatum* as used by authors
2.6.1.18 *Helcogramma kranos* Fricke, 1997	Fricke (1997: 452-455)	Nusa Tenggara/Indonesia	
2.6.1.19 *Helcogramma lacuna* Williams and Howe, 2003	Williams and Howe (2003: 165)	Andaman Sea	

Table 1.2.1 Contd.

Table 1.2.1 Contd.

2.6.1.20 *Helcogramma larvata* Fricke and Randall, 1992	Fricke and Randall (1992: 579)	Maldives	
2.6.1.21 *Helcogramma maldivensis* Fricke and Randall, 1992	Fricke and Randall (1992: 9-11)	Maldives	
2.6.1.22 *Helcogramma microstigma* Holleman, 2006	Holleman (2006: 92)	Comores, Madagascar and Mozambique	
2.6.1.23 *Helcogramma nesion* Williams and Howe, 2003	Williams and Howe (2003: 167)	Southern Japan including Ogasawara Islands	
2.6.1.24 *Helcogramma nigra* Williams and Howe, 2003	Williams and Howe (2003: 169)	Papua New Guinea to Rotuma	
2.6.1.25 *Helcogramma novaecaledoniae* Fricke, 1994	Fricke (1994b: 429-430; 1997: 456-460; 2002: 193)	Solomon Islands to New Caledonia and Loyalty Islands	
2.6.1.26 *Helcogramma obtusirostris* (Klunzinger, 1871)	Hansen (1986: 341-344, part); Fricke (1997: 461-467)	Red Sea, Oman; Southeastern Atlantic	Genus name '...gramma' is based on Greek γραμμ for line (thus, feminine)
2.6.1.27 *Helcogramma rharhabe* Holleman, 2007	Holleman (2007: 69)	East coast of South Africa and southern Mozambique	
2.6.1.28 *Helcogramma randalli* Williams and Howe, 2003	Williams and Howe (2003: 171)	Komodo and Bali/Indonesia	
2.6.1.29 *Helcogramma rhinoceros* Hansen, 1986	Hansen (1986: 344-347); Fricke (1994b: 431-433; 1997: 467-471)	Vietnam and Philippines to Vanuatu	
2.6.1.30 *Helcogramma rosea* Holleman, 2006	Holleman (2006: 95)	Sri Lanka and Andaman Sea	
2.6.1.31 *Helcogramma serendip* Holleman, 2007	Holleman (2007: 73)	Sri Lanka	

Table 1.2.1 Contd.

Table 1.2.1 Contd.

2.6.1.32 *Helcogramma shinglensis* Lal Mohan, 1971	Lal Mohan (1971: 219-233); Holleman (2007: 74)	Sri Lanka, southeastern India and Laccadive Islands	This species was considered as a junior synonym of *H. obtusirostre* by Fricke (1997: 462), but treated as a valid species by Holleman (2007: 74)
2.6.1.33 *Helcogramma solorensis* Fricke, 1997	Fricke (1997: 472-474)	Solor/Indonesia	
2.6.1.34 *Helcogramma springeri* Hansen, 1986	Hansen (1986: 345-347); Fricke (1994b: 434-436; 1997: 475-480)	Malaysia and Philippines to northern Australia and Vanuatu	
2.6.1.35 *Helcogramma steinitzi* Clark, 1980	Clark (1980: 88-93); Hansen (1986: 347-349)	Red Sea and southern Arabia	
2.6.1.36 *Helcogramma striata* Hansen, 1986	Hansen (1986: 349-351, part); Fricke (1994b: 437-442; 1997: 480-489)	Japan and Indonesia to Australia and Line Islands	Genus name '...gramma' is based on Greek γραμμ for line (thus feminine); the species name must be *striata*, and not *striatum* as used by authors
2.6.1.37 *Helcogramma trigloides* (Bleeker, 1858)	Fricke (1997: 489-496; 2002: 193-194)	Andaman Sea to New Caledonia and Vanuatu	
2.6.1.38 *Helcogramma vulcana* Randall and Clark, 1993	Randall and Clark (1993: 27-31); Fricke (1997: 496-500)	Banda Sea/Indonesia	
2.6.1.39 *Helcogramma* sp.	Fricke (1997: 403-408, *Helcogramma* sp. 7)	Andaman Sea to Ryukyu Islands and Tonga	Undescribed species of the *Helcogramma-ellioti*-group; to be described by J. T. Williams
2.6.2 Genus *Lepidoblennius* Steindachner, 1867			
2.6.2.1 *Lepidoblennius haplodactylus* Steindachner, 1867	Hardy (1987b: 254-257); Fricke (1994b: 453-456)	Queensland to Victoria/Australia	*Lepidoblennius geminatus* Macleay (1881: 13-14) is a junior synonym (see Fricke, 1984b: 456)

Table 1.2.1 Contd.

Table 1.2.1 Contd.

2.6.2.2 *Lepidoblennius marmoratus* (Macleay, 1878)	Hardy (1987b: 257-259); Fricke (1994b: 457-460)	Southwestern Australia	
2.6.3 Genus *Ucla* Holleman, 1993	Holleman (1993: 2-4); Fricke (1994b: 553-554; 1997: 530-531)		
2.6.3.1 *Ucla xenogrammus* Holleman, 1993	Holleman (1993: 5-8); Fricke (1994b: 554-562; 1997: 531-543; 2002: 196-197)	Saint Brandon's Shoals; Christmas Island and Australia east to Caroline and Austral islands	
2.7 Tribe Blennodontini	present paper		
2.7.1 Genus *Bellapiscis* Hardy, 1987	Hardy (1987c: 259-260)		
2.7.1.1 *Bellapiscis lesleyae* Hardy, 1987	Hardy (1987c: 263-266); Fricke (1994b: 126-131)	New Zealand and Chatham Islands	
2.7.1.2 *Bellapiscis medius* (Günther, 1861)	Hardy (1987c: 260-263); Fricke (1994b: 132-137)	New Zealand and Chatham Islands	
2.7.2 Genus *Blennodon* Hardy, 1987	Hardy (1987a: 158-159); Fricke (1994b: 138-139)		
2.7.2.1 *Blennodon dorsale* (Clarke, 1879)	Hardy (1987a: 159-163); Fricke (1994b: 140-144)	New Zealand and Chatham, Islands	
2.7.3 Genus *Cryptichthys* Hardy, 1987	Hardy (1987c: 26); Fricke (1994b: 167)		
2.7.3.1 *Cryptichthys jojettae* Hardy, 1987	Hardy (1987c: 26-271); Fricke (1994b: 168-172)		

DEFINITION OF THE TRIPTERYGIIDAE

Tripterygiidae are typically elongate fishes, characterised by scales on the sides of the body and three dorsal fins. Several genera have blunt, rounded heads (e.g., *Helcogramma, Blennodon*), while others have more slender, pointed heads (e.g., *Enneapterygius, Ceratobregma*). One species (*Helcogramma rhinoceros*) has a long median epidermal extension on the tip of the upper lip; another species (*Ucla xenogrammus*) has enlarged jaws, with the lower jaw protruding. The eyes are generally high on the head, often bearing supraorbital cirri (e.g., some *Enneapterygius, Apopterygion, Brachynectes, Ceratobregma, Norfolkia, Trinorfolkia*). Unlike the Blenniidae, nape cirri are generally missing. The anterior nostril may bear a tubular extension. The body is scaled with ctenoid (rarely cycloid) scales, but the scalation may be incomplete with naked strips along the dorsal-fin base and the belly (e.g., *Helcogramma*). The head is often scaleless, but may bear patches of scales in some genera. The lateral line is typically discontinuous, consisting of an anterior series of pored scales, and a posterior, lower series of notched scales. In some genera, the lateral line is reduced into a single, pored series (e.g., *Bellapiscis, Blennodon, Cryptichthys, Helcogramma, Karalepis, Lepidoblennius, Ucla*). Typically, the anterior and middle dorsal fins consist of flexible spines, while the posterior dorsal fin has unbranched, segmented rays, the last divided at its base. The anal fin consists of 0-2 (rarely to 4) spines and 14-32 segmented rays.

The Tripterygiidae is characterised by the following morphological character complexes (mostly following Springer, 1993):

1. No dorsal-fin spine articulating with the pterygiophore serially associated with first segmented ray ('posterior most' dorsal fin spine is absent; as many as four of the posterior most spines may be absent) (Springer, 1993; Fricke, 1997).
2. Dorsal fin divided, spinous and soft dorsal fins separate (Rosenblatt, 1959; Springer, 1993; Fricke, 1994b, 1997); the dorsal fin divided into three sections, two spinous, one segmented (Rosenblatt, 1959; Fricke, 1994b).
3. Dorsal fin spines are more numerous than the segmented rays (Springer, 1993) (except *Brachynectes* and *Notoclinus*; see Fricke, 1997).
4. Segmented dorsal-fin rays branched or simple (Springer, 1993; Fricke, 1997).

5. Posterior most first dorsal-fin pterygiophore supporting 1 or 2 fin ray elements (Springer, 1993; Fricke, 1997).
6. Autogenous bony stay present or absent following the posterior most dorsal-fin pterygiophore (Springer, 1993; Fricke, 1997).
7. Anal-fin spines 0, 1 or 2.
8. Anal-fin spines of mature males without fleshy bulbous swellings (Springer, 1993; Fricke, 1997).
9. Posterior most anal fin pterygiophore supporting 1 or 2 fin ray elements (Springer, 1993; Fricke, 1997).
10. Autogenous bony stay present or absent following the posterior most anal-fin pterygiophore (Springer, 1993; Fricke, 1997).
11. Branched pectoral-fin rays present or absent (Hubbs, 1952; Springer, 1993; Fricke, 1994b, 1997).
12. Dorsal most pectoral-fin ray articulating entirely, or in part, with scapula (Springer, 1993; Fricke, 1997).
13. Coracoid autogenous (Springer, 1993; Fricke, 1997).
14. Some caudal-fin rays branched (Hubbs, 1952; Springer, 1993; Fricke, 1994b, 1997).
15. Ventral hypural plate autogenous (Springer, 1993; Fricke, 1997).
16. Hypural 5 present (most genera) or absent (Springer, 1993; Fricke, 1997).
17. Scales present, some or most ctenoid, bearing only one row of cteni (except scales all cycloid in the two species of *Notoclinus*), with the radii only in anterior field.
18. Lateral line contained on scales with free posterior margins (scales not embedded, not covered by a bone) (Hubbs, 1952; Springer, 1993; Fricke, 1994b, 1997).
19. Lateral line extending half or more of the length of body (except restricted anteriorly in *Ucla*).
20. Rostral cartilage absent.
21. Septal bone present or absent.
22. Ecto- and mesopterygoids autogenous.
23. Posterior end of interopercle extending posteriorly, past the posterior end of epihyal.
24. Premaxillae protractile.
25. No noticeably enlarged canine teeth posteriorly in jaws.

26. Free margins of lips entire (as opposed to fimbriate, crenulate, or with lappets).
27. First gill arch united to the operculum by a membrane (Hubbs, 1952; Fricke, 1994b).
28. No cordlike ligament extending from the dorsoposterior portion of each ceratohyal to anteriormedial end of its respective dentary.
29. Urohyal lacking vertical pair of processes on each side.
30. Gill membranes broadly attached to isthmus.
31. Free bony margins of opercular bones not fimbriate.
32. Infraorbital bones 4 or 5.
33. Palatine teeth present or absent.
34. Nape cirri absent.
35. Anterior ends of pelvises not extending anteriorly past their juncture with the cleithra.
36. Sensory canals of infraorbital bones and preopercle unroofed by bone (except *Notoclinus*).

Character (1) is the single nonhomoplastic synapomorphy that defines the Tripterygiidae as a monophyletic group (Springer, 1993; Fricke, 1997). All other character states exhibit variations with the Tripterygiidae or occur in other lineages of blennioid fishes.

LINEAGES WITHIN THE TRIPTERYGIIDAE

Limited research on tripterygiid relationships and classification has been published. The most comprehensive approach is by Fricke (1994b), who recognised 8 lineages within the family, mainly based on morphological characters. These are, here, raised to a tribe level similar to the approach in the Blenniidae (Figs. 1.2.1 and 1.2.2): (1) Notoclinini (= the 'Notoclinus/Brachynectes Branch'), (2) Trianectini (= the 'Trianectes/Gilloblennius Branch'), (3) Norfolkiini (= the 'Norfolkia/Cremnochorites Branch'), (4) Tripterygiini (= the 'Enneapterygius/Ceratobregma Branch'), (5) Forsterygiini (= the 'Grahamina/Forsterygion Branch'), (6) Karalepini (= the 'Karalepis/Helcogrammoides Branch'), (7) Helcogrammini (= the 'Lepidoblennius/Helcogramma/Ucla Branch') and (8) Bellapiscini (= the 'Bellapiscis/Cryptichthys/Blennodon Branch'). The tribe Notoclinini is sufficiently distinct and supported by synapomorphies to be proposed as a subfamily Notoclininae; the remaining tribes are, consequently, placed in another subfamily, the Tripterygiinae.

Fig. 1.2.1 Lineages within the Tripterygiidae (modified version based on data of Fricke, 1994b), part A. (**a**) Notoclinini: *Notoclinus fenestratus*, Hauraki Gulf, New Zealand, © Kendall Clements, 30 Mar. 2001. **(b)** Trianectini: *Apopterygion oculus*, Stewart Island, New Zealand, © Kendall Clements, Apr. 2004. (**c**) Norfolkiini: *Norfolkia squamiceps*, Lord Howe Island, © John E. Randall, 3 Feb. 1973. **(d)** Tripterygiini: *Tripterygion melanurus*, Cyprus, Mediterranean, © Robert A. Patzner, Nov. 2004.

Fig. 1.2.2 Lineages within the Tripterygiidae (modified version based on data of Fricke, 1994b), part B. **(a)** Forsterygiini: *Forsterygion lapillum*, Great Barrier Island, New Zealand, © Kendall Clements, 18 Apr. 2000. **(b)** Karalepini: *Karalepis stewarti*, Mokohinau Islands, New Zealand, © Kendall Clements, 7 Aug. 2001. **(c)** Helcogrammini: *Helcogramma striata*, Milne Bay, Papua New Guinea, © Robert A. Patzner, Sep. 2000. **(d)** Blennodontini: *Bellapiscis lesleyae*, Great Barrier Island, New Zealand, © Kendall Clements, 14 July 1999.

Notoclinini

The Notoclinini is restricted to two genera from temperate waters of the southern hemisphere, the monotypic *Brachynectes*, and *Notoclinus* with 2 species. This clade is characterised by a laterally compressed body, only 10-12 pectoral fin rays, and the second dorsal fin base shorter than the third. This lineage is found only in southern Australia (*Brachynectes*) and New Zealand (*Notoclinus*). These fishes live among sea grass and kelp.

Trianectini

The Trianectini comprises five genera from temperate waters of the southern hemisphere: the genera *Acanthanectes*, *Apopterygion*, *Gilloblennius* and *Ruanoho* with two species each, and the monotypic *Trianectes*. The clade is characterised by the combination of 10-11 abdominal vertebrae with a depressed body, more than 14 pectoral fin rays, the third dorsal-fin based shorter than the base of the second dorsal. Species of this lineage are found in New Zealand (*Apopterygion*, *Gilloblennius*, *Ruanoho*), southern Australia (*Trianectes*) and South Africa (*Acanthanectes*). These fishes live on shallow or moderately deep gravel or rock bottoms; some are associated with sponges (e.g., *Apopterygion*).

Norfolkiini

The Norfolkiini includes four genera, some from the temperate waters of the southern hemisphere, others from the tropics: the monotypic genus *Cremnochorites*, the genera *Lepidonectes* and *Trinorfolkia* with three species each, and *Norfolkia* with four species. Members of this clade share a heavily scaled head in combination with more than 12 abdominal vertebrae, a discontinuous and complete lateral line, and ctenoid body scales. This species is distributed in South Africa (*Cremnochorites*), southern Australia (*Trinorfolkia*), the tropical eastern Pacific (*Lepidonectes*), and the tropical Indo-West Pacific with three variety confined to Australia and the Southwest Pacific, and another more widespread species (*Norfolkia*). These fishes are associated with rocky reefs, some (*Norfolkia*) also with coralline rock in coral reefs, in shallow waters from the intertidal zone to depths of 25 m.

Tripterygiini

The Tripterygiini is the most speciose lineage with 78 species, and includes the genus *Enneanectes* with 8 species, the genera *Axoclinus* and *Tripterygion* comprising 6 species each, *Ceratobregma* and *Springerichthys* with 2 species each, the monotypic *Crocodilichthys*, and *Enneapterygius* with 53 species. The clade is characterised by 3 spines in the first dorsal fin, in combination with reduced or absent head scalation, more than 12 abdominal vertebrae, a discontinuous and complete lateral line, and ctenoid body scales. Species are found in the tropical East Pacific (*Axcoclinus*, *Crocodilichthys*, *Enneanectes*), the tropical West and Central Atlantic (*Enneanectes*), the temperate northeastern Atlantic and Mediterranean (*Tripterygion*), and the tropical and warm temperate Indo-West Pacific (*Ceratobregma*, *Enneapterygius*, *Springerichthys*). They live in shallow waters from the intertidal zone to 40 m depth, associated with rocky reefs or coralline rock in coral reefs.

Forsterygiini

The Forsterygiini comprises the genus *Forsterygion* with 4 species, *Grahamina* and *Notoclinops* with 3 species each, *Matanui* with 2 species, and the monotypic *Obliquichthys*. The clade is characterised by possessing four or more spines in the first dorsal fin and lacking a supraorbital tentacle, in combination with the head scalation absent, and more than 12 abdominal vertebrae. *Obliquichthys* is a unique neotenous genus that retains the post-larval characteristics as an adult, including the free-swimming behaviour. Species of the lineage are found in the New Zealand region, with two exceptions in southeastern Tasmania (at least one of them, *Grahamina gymnota*, introduced there by human activity according to Clements, Jawad and Stewart, 2000). Its members are found in shallow to deep waters, from the intertidal zone to 50 m, but *Matanui* adults are found from 80-550 m depth. They typically occur on hard rock substrates or on gravel bottom, with some (e.g., *Notoclinops*) associated with encrusted vertical or overhanging rock surfaces.

Karalepini

The Karalepini includes two genera: *Helcogrammoides* with three species, and the monotypic *Karalepis*. The genera share a heavily scaled head and belly and the presence of a supraorbital tentacle, in combination with a continuous lateral line. They are distributed around southern South America and the Antarctic Peninsula (*Helcogrammoides*), and in the New Zealand region (*Karalepis*). The species are found on rocks, under overhangs or large boulders, in shallow waters to 33 m depth.

Helcogrammini

The Helcogrammini consists of three genera, *Helcogramma* with 39 species, *Lepidoblennius* with two species, and the monotypic *Ucla*. The clade shares three spines in the first dorsal fin in combination with a continuous and often reduced lateral line, the presence of cycloid body scales, the head and belly scaleless, and no supraorbital tentacle. Species are found in southern Australia (*Lepidoblennius*), or widespread in the tropical and warm temperate Indo-West Pacific (*Helcogramma, Ucla*). They live on rock, mainly coralline, while some also on live corals (e.g., *Helcogramma striata, Ucla xenogrammus*).

Blennodontini

The Blennodontini are a small clade of three genera and four species: *Bellapiscis* with 2 species, and the monotypic genera *Cryptichthys* and *Blennodon*. They are characterised by four dorsal fin spines in combination with a continuous and reduced lateral line, cycloid scales on the body, the head and belly without scales, and no supraorbital tentacle. The lineage is endemic to the New Zealand region. The species are found on shallow rocky substrates, from the intertidal zone to 12 m depth. *Blennodon* is common only on extremely exposed rocky reefs and shorelines in the zone of high surf.

RELATIONSHIPS WITHIN THE TRIPTERYGIIDAE

The relationships within the Tripterygiidae were studied by Fricke (1994b) on the basis of morphological characters. The Notocliniini are a monophyletic lineage sharing a single apomorphy, an elongate third dorsal fin longer than the second dorsal fin, and are here raised to subfamily level as the Notoclininae. The remaining tripterygiids share two synapomorphies, a body which is not laterally compressed, and pectoral fin with more than 14 rays. They are raised to the subfamily Tripterygiinae.

Within the Tripterygiinae, the three lineages Karalepini, Helcogrammini and Blennodontini are a monophyletic group based on two synapomorphies. First, the lateral line is continuous, in a single, pored row, which is often reduced, while the second lateral line with notched scales is absent. Second, the body scales are mainly cycloid. Within this group, the Helcogrammini and Blennodontini share three synapomorphies, namely a reduced or absent head scalation, a reduced or absent belly scalation, and a reduced or absent supraorbital tentacle.

The clades Norfolkini, Tripterygiini and Forsterygiini share a single apomorphy, more than 12 abdominal vertebrae. Within this group, the Tripterygiini and Forsterygiini share a reduced head scalation. The whole complex of the Trianectini, Norfolkini, Tripterygiini and Forsterygiini, however, is so far only defined by symplesiomorphies and the relationships of and within this group need additional study as it cannot be ruled out that this group is not monophyletic.

The genus *Sauvagea* Jordan and Seale, 1906, which was based on *Lepidoblennius* Sauvage, 1874 (preoccupied by *Lepidoblennius* Steindachner, 1867), was assumed to be a tripterygiid or clinid by authors (e.g., Hoese, 2006). However, an examination of MNHN 8158, the holotype of *Lepidoblennius caledonicus* Sauvage, 1884 from New Caledonia (the type species of the genus) revealed that it is not a blennioid, but a microdesmid.

CLASSIFICATION OF THE TRIPTERYGIIDAE

The Tripterygiidae is here classified as having two subfamilies, the Notoclininae new subfamily (definition as above; type genus *Notoclinus* Gill, 1893; see above), and the Tripterygiinae. The latter subfamily comprises seven tribes. As research on tripterygiid classification has been limited (Fricke, 1994b, 1997), future studies of the relationships within the Tripterygiinae are necessary.

Several morphological character systems are well documented for most tripterygiids. It will be important in the future to use molecular analyses to test tripterygiid relationships. An initial study by Hickey and Clements (2005) comparing haploid genome sizes of 18 New Zealand tripterygiid species resulted in a tree similar to Figures 1.2.1 and 1.2.2, except that *Matanui* was arranged in a clade with *Karalepis*, *Blennodon* and *Bellapiscis*, not with *Forsterygion* and *Grahamina* as in the present study.

References

Allen, G.R. and D.R. Robertson. 1991. Description of two new genera and four new species of triplefins (Pisces: Tripterygiidae) from the tropical eastern Pacific. *Revue Française d'Aquariologie* 18: 79-82.

Allen, G.R. and D.R. Robertson. 1992. Three new species of triplefins (Pisces: Tripterygiidae) from Malpelo and Socorro islands, in the tropical eastern Pacific. *Revue Française d'Aquariologie* 19: 53-56.

Aoyagi, H. 1954. Additional notes on the new and the rare fishes of the family Blenniidae from the Riu-Kiu Islands. *Dobutsugaku Zasshi (Zoological Magazine) (Tokyo)* 63: 239-242.

Arambourg, C. 1927. Les poissons fossiles d'Oran. Matériaux pour la Carte Géologique de l'Algérie, (1) *Paléontologie* 6: 1-298.

Bath, H. 1973. Wiederbeschreibung und neuer Nachweis von *Tripterygion melanurus* Guichenot (Pisces: Blennioidea: Tripterygiidae). *Senckenbergiana Biologica* 54: 47-56.

Brock, V.E. 1940. Three new blennioid fishes from the west coast of Mexico. *Stanford Ichthyological Bulletin, Stanford University* 2: 29-35.

Bussing, W.A. 1991. A new genus and two new species of tripterygiid fishes from Costa Rica. *Revista de Biología Tropical* 39: 77-85.

Carreras-Carbonell, J., M. Pascua and E. Macpherson. 2007. Review of the *Tripterygion tripteronotus* (Risso, 1810) complex, with a description of a new species from the Mediterranean Sea (Teleostei: Tripterygiidae). *Scientia Marina* 71: 75-86.

Castillo, M.R. and R.G. Pequeño. 1998. Sinopsis de Tripterygiidae de Chile (Osteichthys: Perciformes). *Gayana Zoología* 62: 109-133.

Clark, E. 1980. Red Sea fishes of the family Tripterygiidae with descriptions of eight new species. *Israel Journal of Zoology* 28: 63-113.

Clarke, F.E. 1879. On some new fishes. *Transactions and Proceedings of the New Zealand Institute* 11 (for 1878): 291-295.

Clements, K.D., L.A. Jawad and A.L. Stewart. 2000. The New Zealand triplefin *Grahamina signata* (Teleostei: Tripterygiidae): A junior synonym of G. *gymnota* from Tasmania. *Journal of the Royal Society of New Zealand* 30: 373-384.

Cocco, A. 1829. Su di alcuni nuovi pesci de' mari di Messina. *Giornali Scienzia Lettera Arti Sicilia Anno* 7(26): 138-147.

Cuvier, G.L.C.F.D. and A. Valenciennes. 1836. *Histoire naturelle des poissons*. Tome onzième (11). Paris (F.G. Levrault), xx + 506 + 2 pp., pls. 307-343.

Eschmeyer, W.N. 2007. *The Catalog of Fishes on-line*. Updated October 9, 2007. California Academy of Sciences. http://www.calacademy.org/research/ichthyology/catalog/fishcatsearch.html

Fowler, H.W. 1944. The fishes. Results of the 5th George Vanderbilt Expedition (1941). *The Academy of Natural Sciences of Philadelphia, Monographs* 6: 57-529.

Fowler, H.W. 1946. A collection of fishes obtained in the Riu Kiu Islands by Captain Ernest R. Tinkham A.U.S. *Proceedings of the Academy of Natural Sciences of Philadelphia* 98: 123-218.

Fowler, H.W. 1953. On a collection of fishes made by Dr Marshall Laird at Norfolk Island. *Transactions of the Royal Society of New Zealand* 81: 257-267.

Fricke, R. 1987. A new species of *Enneapterygius* from Miyake-jima, Izu Islands, Japan (Teleostei. Tripterygiidae). *Senckenbergiana Maritima* 19: 331-338.

Fricke, R. 1991. *Ceratobregma striata*, a new triplefin (Tripterygiidae) from northern Australia, and a record of *Norfolkia brachylepis* from Western Australia. *Japanese Journal of Ichthyology* 37: 337-343.

Fricke, R. 1994a. Tripterygiid fishes of the genus *Enneapterygius* from Bali, Indonesia, with descriptions of two new species (Teleostei: Blennioidei). *Stuttgarter Beiträge zur Naturkunde* A512: 1-13.

Fricke, R. 1994b. *Tripterygiid Fishes of Australia, New Zealand and the Southwest Pacific Ocean (Teleostei)*. Koenigstein (Koeltz Scientific Books).

Fricke, R. 1997. *Tripterygiid fishes of the Western and Central Pacific, with Descriptions of 15 new Species, including an annotated Checklist of World Tripterygiidae (Teleostei)*. Koenigstein (Koeltz Scientific Books).

Fricke, R. 2001. *Enneapterygius trisignatus*, a new species from northern Grande Terre, with a key to New Caledonian tripterygiid fishes (Teleostei). *Stuttgarter Beiträge zur Naturkunde* A630: 1-9.

Fricke, R. 2002. Tripterygiid fishes of New Caledonia, with zoogeographical remarks. *Environmental Biology of Fishes* 65: 175-198.

Fricke, R. and J.E. Randall. 1992. Tripterygiid fishes of the Maldives Islands, with descriptions of two new species (Teleostei: Blennioidei). *Stuttgarter Beiträge zur Naturkunde*, A484: 1-13. pl. 1.

Fricke, R. and C.D. Roberts. 1993. *Grahamina*, a new genus for robust-bodied triplefins (Teleostei: Tripterygiidae) from New Zealand and Australia, with description of a new species. *Stuttgarter Beiträge zur Naturkunde* A504: 1-21.

Fricke, R., M. Bilecenoglu and H.M. Sari. 2007. Annotated checklist of fish and lamprey species (Gnathostomata and Petromyzontomorphi) of Turkey, including a Red List of threatened and declining species. *Stuttgarter Beiträge zur Naturkunde* A706: 1-169.

Gon, O. 1991. Tripterygiidae. In: *Fishes of the Southern Ocean*. O. Gon and P.C. Heemstra (eds.) J.L.B. Smith Institute of Ichthyology, Grahamstown, pp. 400-401.

Griffin, L.T. 1926. Descriptions of New Zealand fishes. *Transactions and Proceedings of the New Zealand Institute* 56: 538-546.

Günther, A. 1877. Andrew Garrett's Fische der Südsee. VI. *Journal des Museums Godeffroy* 13: 169-216.

Hansen, P.E.H. 1986. Revision of the tripterygiid fish genus *Helcogramma*, including descriptions of four new species. *Bulletin of Marine Sciences* 38: 313-354.

Hardy, G.S. 1984. A new genus and species of triplefin (Pisces: family Tripterygiidae) from New Zealand. *Records of the National Museum of New Zealand* 2: 175-180.

Hardy, G.S. 1986. Redescription of *Gilloblennius* Whitley and Phillipps, 1939 (Pisces: Tripterygiidae), and description of a new genus and two new species from New Zealand. *Journal of the Royal Society of New Zealand* 16: 145-168.

Hardy, G.S. 1987a. A new genus for *Tripterygium dorsalis* Clarke, 1879, an unusual triplefin (Pisces: Tripterygiidae) from New Zealand. *Journal of the Royal Society of New Zealand* 17: 157-164.

Hardy, G.S. 1987b. Revision of *Notoclinops* Whitley, 1930 (Pisces: Tripterygiidae), and description of a new species from New Zealand. *Journal of the Royal Society of New Zealand* 17: 165-176.

Hardy, G.S. 1987c. Revision of some triplefins (Pisces: Tripterygiidae) from New Zealand and Australia, with descriptions of two new genera and two new species. *Journal of the Royal Society of New Zealand* 17: 153-174.

Hardy, G.S. 1987d. Descriptions of a new genus and two new species of tripterygiid fishes from New Zealand. *Records, National Museum of New Zealand* 3: 47-58.

Hardy, G.S. 1989a. The genus *Forsterygion* Whitley and Phillipps, 1939 (Pisces: Tripterygiidae) in New Zealand and Australia, with descriptions of two new species. *Journal of Natural History* 23: 491-512.

Hardy, G.S. 1989b. The 'blue-dot' triplefin; a new species of *Notoclinops* Whitley, 1930 (Pisces: Tripterygiidae) from New Zealand. *Journal of the Royal Society of New Zealand* 19: 327-332.

Herre, A.W.C.T. 1935. New fishes obtained by the Crane Pacific Expedition. *Publications of the Field Museum of Natural History, 355, Zoological Series* 18: 383-438.

Hickey, A.J.R. and K.D. Clements. 2005. Genome size evolution in New Zealand triplefin fishes. *Journal of Heredity* 96: 356-362.

Hoda, S.M.S. 1981. *Enneapterygius nasimae*, a new species of tripterygiid fish from Karachi coast, northern Arabian Sea. *Indian Journal of Fisheries* 30: 116-123.

Hoese, D.F. 2006. *Clinidae*. In: Zoological Catalogue of Australia. Volume 35. Fishes. The Australian Museum, Sydney.

Holleman, W. 1982. Three new species and a new genus of tripterygiid fishes (Blennioidei) from the Indo-West Pacific Ocean. *Annals of the Cape Provincial Museums (Natural History)* 14: 109-137.

Holleman, W. 1987. Description of a new genus and species of tripterygiid fish (Perciformes: Blennioidei) from the Indo-Pacific, and the reallocation of *Vauclusella acanthops* Whitley, 1965. *Cybium* 11: 173-181.

Holleman, W. 1991. A revision of the tripterygiid fish genus *Norfolkia* Fowler, 1953 (Perciformes: Blennioidei). *Annals of the Cape Provincial Museums (Natural History)* 18: 227-243.

Holleman, W. 1993. *Ucla xenogrammus*, a new genus and species of Indo-Pacific fishes (Perciformes, Tripterygiidae). *J.L.B. Smith Institute of Ichthyology*, Special Publication 55: 1-10.

Holleman, W. 2005. A review of the triplefin fish genus *Enneapterygius* (Blennioidei: Tripterygiidae) in the western Indian Ocean, with descriptions of four new species. *Smithiana Bulletin* 5: 1-25, pls 1-2.

Holleman, W. 2006. Fishes of the *Helcogramma steinitzi* species group (Blennioidei: Tripterygiidae) from the Indian Ocean, with descriptions of two new species. *Aqua* 11: 89-104.

Holleman, W. 2007. Fishes of the genus *Helcogramma* (Blennioidei: Tripterygiidae) in the western Indian Ocean, including Sri Lanka, with descriptions of four new species. *Smithiana Bulletin* 7: 51-81.

Holleman, W. and C.D. Buxton. 1993. *Acanthanectes*, a new genus of triplefin with two new species from the southern coast of South Africa (Blennioidei: Tripterygiidae). *Cybium* 17: 327-342.

Hubbs, C.L. 1952. Antitropical distribution of fishes and other organisms. *Proceedings of the 7th Pacific Science Congress* 3: 324-329.

Hutton, F.W. 1879. Notes on a collection from the Auckland Islands and Campbell Island. *Transactions and Proceedings of the New Zealand Institute* 11 (for 1878): 337-343.

Jawad, L.A. and K.D. Clements. 2004. *Matanui*, a new genus of deepwater triplefin fishes (Pisces: Tripterygiidae) from New Zealand. *Journal of the Royal Society of New Zealand* 34: 81-103.

Jordan, D.S. and A. Seale. 1906. The fishes of Samoa. Descriptions of the species found in the archipelago, with a provisional check-list of the fishes of Oceania. *Bulletin of the United States Bureau of Fisheries* 25 (for 1905): 175-455.

Kendall, W.C. and Radcliffe. 1912. Report on the scientific results of the expedition to the eastern Pacific from October, 1904, to March, 1905. XXV. The shore fishes. *Memoirs of the Museum of Comparative Zoology, Harvard College* 35: 77-170.

Kuiter, R.H. 1986. A new genus and three new species of tripterygiid fishes of Australia's south coast. *Revue Française d'Aquariologie* 12: 89-96.

Lal Mohan, R.S. 1971. *Helcogramma shinglensis*, a new species of tripterygiid fish from Gulf of Mannar with a key to the fishes of the family Tripterygiidae of eastern and central Indian Ocean. *Senckenbergiana Biologica* 52: 219-223.

Lubbock, R. 1980. The shore fishes of Ascension Island. *Journal of Fish Biology* 17: 283-303.

Lubbock, R. and A.J. Edwards. 1981. The fishes of Saint Paul's Rocks. *Journal of Fish Biology* 18: 135-157.

Lucas, A.H.S. 1891. On the occurrence of certain fish in Victorian seas, with descriptions of some new species. *Proceedings of the Royal Society of Victoria*, New Series, 3: 8-14.

Macleay, W. 1881. Descriptive catalogue of the fishes of Australia. Part 3. *Proceedings of the Linnean Society of New South Wales* 6: 1-138.

Maksimov, N.E. 1909. Two species of the genus *Tripterygium* from the Black Sea. *Trudy Obshesthva Ispytately Prirody Kharkov University* 42 (1907-1908): 59-63. (In Russian).

Motomura, H., S. Harazaki and G.S Hardy. 2005. A new species of triplefin (Perciformes: Tripterygiidae), *Enneapterygius senoui*, from Japan with a discussion of its *in situ* colour pattern. *Aqua* 10: 5-14.

Ramsay, E.P. and J.D. Ogilby. 1888. Descriptions of two new fishes from Port Jackson. *Proceedings of the Linnean Society of New South Wales* (2) 2 (1887): 1021-1023.

Ramsay, E.P. and J.D. Ogilby. 1889. Description of a new *Tripterygium* from Port Jackson. *Proceedings of the Linnean Society of New South Wales* (3) 2 (1888): 419-420.

Randall, J.E. 1995. A review of the triplefin fishes (Perciformes: Blennioidei: Tripterygiidae) from Oman, with descriptions of two new species of *Enneapterygius*. *Revue Française d'Aquariologie* 22: 27-34.

Randall, J.E. and E. Clark. 1993. *Helcogramma vulcana*, a new tripterygiid fish (Blennioidei: Tripterygiidae) from the Banda Sea, Indonesia. *Revue Française d'Ichthyologie* 20: 27-32.

Rendahl, H. 1926. Papers from Dr Th. Mortensen's Pacific Expedition 1914-16. XXX. Fishes from New Zealand and the Auckland-Campbell Islands. *Videnskabelige Meddelelser fra Dansk Naturhistorisk Forening* 81: 1-14.

Risso, A. 1827. *Histoire naturelle des principales productions de l'Europe Méridionale et particulièrement de celles des environs de Nice et des Alpes Maritimes. Tome 3. Poissons.* pp. 97-480, Levrault, Paris.

Rosenblatt, R.H. 1959. *A Revisionary Study of the Blennioid Fish Family Tripterygiidae*. Doctoral Dissertation, University of California, Los Angeles.

Rosenblatt, R.H. 1960. The Atlantic species of the blennioid fish genus *Enneanectes*. *Proceedings of the Academy of Natural Sciences of Philadelphia* 112: 1-23.

Scott, E.O.G. 1977. Observations on some Tasmanian fishes: Part XXIII. *Papers and Proceedings of the Royal Society of Tasmania* 111: 111-180.

Shen, S.-C. and K.-Y. Wu. 1994. A revision of the tripterygiid fishes from coastal waters of Taiwan with descriptions of two new genera and five new species. *Acta Zoologica Taiwanica* 5: 1-32.

Shibukawa, K., T. Suzuki and M. Hosokawa. 2004. First record of a triplefin, *Enneapterygius cheni*, from Japan (Perciformes: Tripterygidae). *Izu Oceanic Park Diving News* 15: 2-6.

Smith, D.G. and J.T. Williams. 2002. History and status of the genera *Enneanectes* and *Axoclinus* (Teleostei: Blennioidei: Tripterygiidae). *Zootaxa* 105: 1-10.

Smitt, F.A. 1898. Poissons de l'expédition scientifique à la Terre de Feu sous direction du Docteur O. Nordenskiöld, recueillis par le Docteur A. Ohlin et Mr. H. Akerman. Bihang till Kungliga Svenska Vetenskaps-Akademia Handlingar, 24 (Section 4): 1-80.

Springer, V.G. 1993. Definition of the suborder Blennioidei and its included families (Pisces: Perciformes). *Bulletin of Marine Science* 52: 472-495.

Talwar, P.K. and T.K. Sen. 1971. On some fishes from the Madras coast with description of a new species of the family Clinidae. *Records of the Zoological Survey of India* 65: 243-251.

Wang, S.-C., K.-T. Shao and S.-C. Shen. 1996. *Enneapterygius cheni*, a new triplefin fish (Pisces: Tripterygiidae) from Taiwan. *Acta Zoologica Taiwanica* 7: 79-83.

Weber, M. 1909. Diagnosen neuer Fische der Siboga-Expedition. *Notes of the Leyden Museum* 31: 143-169.

Wheeler, A. and J. Dunne. 1975. *Tripterygion atlanticus* sp. nov. (Teleostei-Tripterygiidae) the first record of a tripterygiid fish in North-Western Europe. *Journal of Fish Biology* 7: 639-649.

Whitehead, P.J.P., M.-L. Bauchot, J.-C. Hureau, J. Nielsen and E. Tortonese. 1986. *Fishes of the North-eastern Atlantic and the Mediterranean.* Vol. 3. Paris (UNESCO), pp. 1008-1473.

Whitley, G.P. 1931. New names for Australian fishes. *Australian Zoologist* 6: 310-334.

Whitley, G.P. 1944. New sharks and fishes from Western Australia. *Australian Zoologist* 10: 252-273.

Williams, J.T. and J.C. Howe. 2003. Seven new species of the triplefin genus *Helcogramma* (Tripterygiidae) from the Indo-Pacific. *Aqua* 7: 151-176.

Williams, J.T. and C.J. McCormick. 1990. Two new species of the triplefin genus *Helcogramma* (Tripterygiidae) from the western Pacific Ocean. *Copeia* 1990: 1020-1030.

Wirtz, P. 1980. A revision of the eastern Atlantic Tripterygiidae (Pisces, Blennioidei) and notes on some West African blennioid fish. *Cybium* 11: 83-101.

CHAPTER

1.3

Systematics of the Blenniidae (Combtooth Blennies)

Philip A. Hastings[1] and Victor G. Springer[2]

INTRODUCTION

The Blenniidae comprises 57 genera and 387 species (Table 1.3.1) of small benthic fishes. The family is distributed worldwide, with most species occurring in shallow coastal, tropical and warm temperate marine waters, along with a few species known from fresh and/or brackish water. Despite being relatively well-known taxonomically, the phylogenetic relationships of the blenniids with the other five families of blennioids are poorly known (Hastings and Springer, Chapter 1.1). The Blenniidae has been the most studied among any of the blennioid families, in part because it is by far the most speciose, but also because it is distributed widely, specimens can be easily collected, and the species occur in a variety of habitats: rocky intertidal, coral reefs, mangroves, oyster beds and in the lower reaches of most rivers. Six putative lineages have been identified within the

Authors' addresses: [1]Scripps Institution of Oceanography, University of California San Diego, La Jolla, California, USA. E-mail: phastings@ucsd.edu
[2]National Museum of Natural History, Smithsonian Institution, Washington, D.C., USA. E-mail: springer@si.edu

Table 1.3.1 Currently recognised tribes and genera of the Blenniidae in a classification. Numbers in parentheses ( ) indicate number of valid species in each genus. Numbers in brackets [ ] indicate number of valid genera/valid species in higher taxonomic categories. * = allied with the Salariini by Bath (2001).

Blenniinae [15/91]

Blenniini [2/4]

Blennius Linnaeus, 1757 (2)
Spaniblennius Bath and Wirtz, 1989 (2)

Nemophini [5/52]

Aspidontus Cuvier in Quoy and Gaimard, 1834 (3)
Meicanthus Norman, 1943 (25)
Petroscrites Ruppell, 1830 (11)
Plagiotremus Gill, 1865 (11)
Xiphasia Swainson, 1839 (2)

Omobranchini [7/34]

Enchelyurus Peters, 1868 (5)
Haptogenys Springer, 1972 (1)
Laiphognathus Smith, 1955 (2)
Oman Springer, 1985 (1)
Omobranchus Ehrenberg in Cuvier and Valenciennes, 1836 (21)
Omox Springer, 1972 (2)
Parenchelyurus Springer, 1972 (2)

Phenablenniini [1/1]

Phenablennius Springer and Smith-Vaniz, 1972 (1)

Salariinae [42/296]

Parablenniini [14/82]

Aidablennius Whitley, 1947 (1)
Bathyblennius Bath, 1977 (1)
Chalaroderma Norman, 1943 (2)
Chasmodes Valenciennes in Cuvier and Valencinnes, 1836 (3)
Coryphoblennius Norman, 1943 (1)*
Hypleurochilus Gill, 1861 (10)
Hypsoblennius Schultz, 1941 (14)
Lipophrys Gill, 1896 (2)
Lupinoblennius Herre, 1942 (3)
Microlipophrys Almada, Almada, Guillemaud and Wirtz, 2005 (7)
Parablennius Miranda-Ribeiro, 1915 (27)
Parahypsos Bath, 1982 (1)
Salaria Forsskål, 1775 (4)
Scartella Jordan, 1886 (6)

Table 1.3.1 Contd.

Table 1.3.1 Contd.

Salariini [28/214]
Salarias group [20/187]
Alticus Lacepède, 1800 (10)
Andamia Blyth, 1858 (7)
Atrosalarias Whitley, 1933 (3)
Blenniella Reid, 1943 (9)
Cirripectes Swainson, 1839 (22)
Crossosalarias Smith-Vaniz and Springer, 1971 (1)
Dodekablennos Springer and Spreitzer, 1978 (1)
Ecsenius McCulloch, 1923 (53)
Entomacrodus Gill, 1859 (25)
Exallias Jordan and Evermann, 1905 (1)
Glyptoparus Smith, 1959 (1)
Istiblennius Whitley, 1943 (14)
Nannosalarias Smith-Vaniz and Springer, 1971 (1)
Ophioblennius Gill, 1860 (5)
Paralticus Springer and Williams, 1994 (1)
Pereulixia Smith, 1959 (1)
Praealticus Schultz and Chapman in Schultz *et al.*, 1960 (13)
Salarias Cuvier, 1817 (13)
Scartichthys Jordan and Evermann, 1898 (4)
Stanulus Smith, 1959 (2)
Rhabdoblennius group [8/27]
Alloblennius Smith-Vaniz and Springer, 1971 (4)
Antennablennius Fowler, 1931 (9)
Cirrisalarias Springer, 1976 (1)
Hirculops Smith, 1959 (1)
Litobranchus Smith-Vaniz and Springer, 1971 (1)
Medusablennius Springer, 1966 (1)
Mimoblennius Smith-Vaniz and Springer, 1971 (5)
Rhabdoblennius Whitley, 1930 (5)

Blenniidae (Fig. 1.3.1). These have usually been considered tribes, but Bath (2001) elevated five of them to subfamilies, a change that we do not fully support (see below), and included the sixth (Parablenniini) within his Salariinae. These lineages differ greatly in diversity, with only one species in the Phenablenniini and well over 200 in the Salariini (*sensu stricto*).

Fig. 1.3.1 Representatives of the six tribes of the Blenniidae. (a) *Blennius ocellaris* (Blenniini); (b) *Petroscirtes ancyclodon* (Nemophini); (c) *Omobranchus fasciolatus* (Omobranchini); (d) *Phenablennius heylegeri* (Phenablenniini); (e) *Parablennius opercularis* (Parablenniini); and (f) *Istiblennius chrysospilus* (Salariini). Photos by G. Neto (a), J. Randall (b, c, e), N. McNear (d), and J. Williams (f).

DEFINITION OF THE BLENNIIDAE

Blenniids are typically elongate fishes with long dorsal and anal fins. Many species have blunt, rounded heads, although some varieties, such as *Omox lupus* (Springer, 1981) and *Phenablennius heyligeri* (Smith-Vaniz, 1975; Fig. 1.3.1) have somewhat prolonged snouts. The eyes of blenniids are generally high on the head and often bear supraorbital cirri. Cirri are also present near the nape of many species, usually on the anterior nostril, variously on the posterior nostril, and near one or more of the preoperculomandibular sensory canal pores (particularly in the nemophines and the salarine *Crossosalarias*). Blenniids lack scales and many have the lateral line reduced on the body, or even absent in the Phenablenniini and some Omobranchini. Teeth on the dentaries and premaxillae are in comb-like rows, but many genera have a separate, usually enlarged caniniform tooth posteriorly on each side of one or both of these bones. Teeth are either present or absent on the vomer, but always absent on the pterygoids and palatines. Absence of teeth on the palatines is a heretofore unreported hypothesised synapomorphy of the family (see below).

The Blenniidae is hypothesised to be a monophyletic group based on eight morphological character complexes.

1. The coracoid is reduced and fused to the cleithrum (Springer, 1968, 1993), whereas in other perciforms, it is free from the cleithrum.
2. The interopercle is reduced in size and does not extend past the posterior end of the posterior ceratohyal, except in the Omobranchini, which have a secondarily developed spur on the posterior end of the interopercle (Springer, 1968, 1993).
3. The teeth are incisoriform and in a single comb-like row on the dentaries and premaxillaries (VGS, pers. obs.) except in those taxa that have a posterior canine in each jaw. Jaw teeth in other blennioids are conical, acute and caniniform and may be in one or more rows in the jaws.
4. Palatine teeth are always absent (VGS, pers. obs.), whereas palatine teeth are variously present or absent in other blennioids (except palatines are edentate in all dactyloscopids).
5. The urohyal has two lateral projections on each side with strong ligamentous attachments to the respective hypohyals (Williams, 1990; Springer, 1993).
6. Adult male blenniids generally have fleshy rugosities on the anal-fin spines (Williams, 1990). These have been demonstrated to secrete antimicrobial mucus that is applied to eggs guarded by males of *Ophioblennius atlanticus* (Giacomello *et al.*, 2006). Similar but apparently non-homologous features are present on the fin-ray elements of other blennioid lineages (e.g., Tripterygiidae; Northcott and James, 1996).
7. All blenniids lack scales (VGS, pers. obs.). All other blennioids, except most chaenopsids are scaled. Of those chaenopsids with scales (all but one species of *Neoclinus* and three species of *Stathmonotus*), *Neoclinus* has been hypothesised as the sister group of all other chaenopsids and the condition in the *Stathmonotus* species is hypothesised to be a reversal (Hastings and Springer, 1994).
8. The premaxilla of all blenniids is not protractile (VGS, pers. obs.), while that of most other blennioids is protractile. Exceptions

include the monotypic chaenopsid genus *Mccoskerichthys* (Hastings and Springer, 1994).

LINEAGES WITHIN THE BLENNIIDAE

Most recent classifications recognise six tribes of the Blenniidae (Fig. 1.3.1). The genera currently allocated to each tribe are listed in Table 1.3.1. Bath (2001) provided an alternative classification (see below), in which he recognised only five major lineages because he included the Parablenniini within the Salariini. He also elevated the remaining tribes to subfamilial status. We present a slightly modified classification of the Blenniidae after discussing these lineages.

Blenniini

The Blenniini has been restricted to two species of *Blennius* and two species of *Spaniblennius* (Smith-Vaniz, 1976; Bock and Zander, 1986; Bath and Wirtz, 1989). In our opinion, the monophyly of this clade has not been convincingly demonstrated as it includes species with dentaries that form a tight interdigitating joint, but lack the specialisations diagnostic of related lineages. Bath and Wirtz (1989) characterised it as having the left and right dentaries well sutured to one another and the ascending premaxillary processes relatively broad-based, but the distribution of this condition among other blenniids has not been well documented. Members of this lineage also have the gill opening restricted to the side of the head and a lateral line consisting of short, unbranched tubules (Bath and Wirtz, 1989), features also seen in omobranchines. The phylogenetic status of the Blenniini, found only in the eastern Atlantic, including the Mediterranean Sea, needs further study.

Nemophini

The Nemophini, or so-called sabertooth blennies, includes five genera and 52 species (Table 1.3.1; Smith-Vaniz, 1987). Their common name derives from the presence of prominent canine teeth in the lower jaw that function in defense against predators and in interspecific aggressive interactions. Nemophines are mainly found on coral reefs of the Indo-Pacific, but one species (*Plagiotremus azaleus*) occurs in the eastern Pacific and at least one species (*Meiacanthus anema* (Bleeker), enters estuaries and freshwaters (Smith-Vaniz *et al.*, 2001). Unlike most blennioids, nemophins often hover above the bottom or swim in the water column. Associated with this behavior, most adult nemophins possess a swimbladder, which is absent in adults of other blenniids. Exceptions are the two species of the genus *Xiphasia* that lack a swimbladder in adults. These extraordinary blenniids are elongate with a long dorsal fin that begins well over the neurocranium. They are nocturnal and semipelagic, although one species, *X. setifer*, has been observed to occupy tubes in sandy substrates (Smith-Vaniz, 1976; Kuiter, 1992, pg. 241, Fig. D).

The active lifestyle of nemophins is associated with the prominent mimetic associations exhibited by a variety of the included species (Springer and Smith-Vaniz, 1972a; Smith-Vaniz *et al.*, 2001). Accordingly, nemophins vary greatly in coloration, both across and sometimes within species. They most often mimic wrasses and a few other free-swimming fishes, using this mimicry to gain access to fishes from which they take bites of skin, scales and fins. On the other hand, the species of *Meiacanthus* are unique among all fishes in having venom glands associated with their canine teeth that clearly afford them protection from predators (Springer and Smith-Vaniz, 1972a; Fishelson, 1974, 1975; Smith-Vaniz *et al.*, 2001; Smith and Wheeler, 2006). Several species of *Meiacanthus*, as a consequence, serve as models in mimetic associations with other fishes including other blennioids, cardinalfishes and threadfin breams (Springer and Smith-Vaniz, 1972a; Smith-Vaniz *et al.*, 2001). Some species of *Plagiotremus*, which prey on the skin and scales of other fishes, are protected, either when alive or dead, from some predators because they have an objectionable taste (Springer and Smith-Vaniz, 1972a; VGS, pers. obs.). However, evidence that *Petroscirtes* and *Meiacanthus* feed on other fishes is lacking.

The systematics of nemophines has been studied extensively by William F. Smith-Vaniz (1976, 1987; Smith-Vaniz *et al.*, 2001). Five synapomorphies have been reported for the Nemophini although two have exceptions within the tribe (Smith-Vaniz, 1976).

1. Nemophines are unique among blennies in having cirri associated with the symphyseal mandibular sensory canal.
2. The supracleithrum of nemophines articulates with the epiotic and post-temporal while it articulates only with the post-temporal in other blenniids.
3. Nemophines lack a basisphenoid.

4. All nemophines except *Dasson* (a subgenus of *Petroscirtes*), lack an intercalar bone.
5. In all nemophines except *Xiphasia*, the cranial bones of adults are ornamented with depressions or deep pits.

Smith-Vaniz (1976) provided a hypothesis of the relationships of genera within the Nemophini (Fig. 1.3.2). He recognised two lineages, one with *Meiacanthus* and *Petroscirtes* and the other with *Plagiotremus*, *Aspidontus* and *Xiphasia*, with the latter two as sister groups.

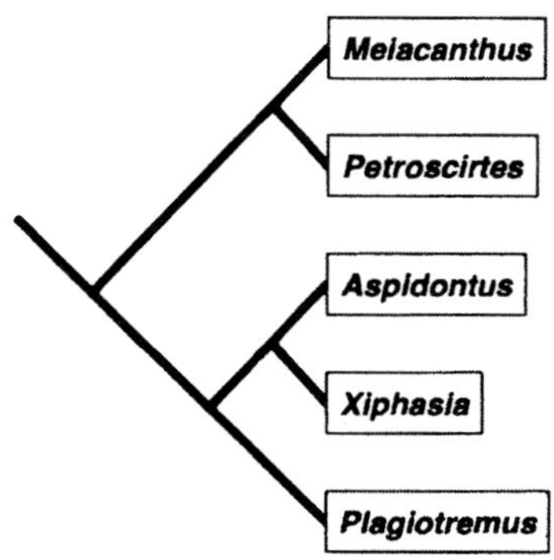

Fig. 1.3.2 Phylogenetic relationships of the genera of Nemophini according to Smith-Vaniz (1976).

Omobranchini

The Omobranchini includes seven genera and 34 species (Table 1.3.1). Omobranchines are relatively small (< 100 mm SL) and are typically found on rocks, corals and shell rubble substrates throughout the Indo-West Pacific (Springer, 1972a; Springer and Gomon, 1975). Members of this lineage are generally found in relatively shallow waters including the intertidal. Several species are reported to occur in brackish and/or freshwaters. By far the largest genus is *Omobranchus* with 23 species, many of which exhibit considerable sexual dimorphism (Springer and Gomon, 1975). Several species of *Omobranchus* have been introduced, probably in fouling on boats or in ballast water, in geographic areas in the Indo-West and Central Pacific, where they were previously unknown (Springer and Gomon, 1975; Mundy, 2005) and one species of *Omobranchus* has been introduced into the southern Caribbean and coastal Brazil (Springer and Gomon, 1975; Gerhardinger *et al.*, 2006).

Omobranchines share two apparent apomorphies. First, all have a ventroposterior spur on the interopercle. Second, omobranchines have reduced pelvic fins with only two segmented rays (Springer, 1968, 1972a). Relationships within the Omobranchini have not been hypothesised (Springer, 1985).

Phenablenniini

The Phenablenniini includes a single genus and species, *Phenablennius heyligeri* (Bleeker). Springer and Smith-Vaniz (1972b) erected the genus and tribe based on this species described by Bleeker, who had placed it in the genus *Petroscirtes* (Nemophini). They also described its osteology, noting its similarity to nemophines, as well as its uniqueness among blenniids apropos to four features.

1. *Phenablennius* has a single, long and thin postcleithrum that is well separate from the cleithrum. This position implies that the upper postcleithrum has been lost because other blenniids typically have two postcleithra with the upper one lying near the cleithrum. A few species of Salariini and Omobranchini have reduced postcleithra, but most often as fragments, rather than one single postcleithrum. An exception is *Enchelyurus* in which the ventral postcleithrum is present and complete, while the dorsal postcleithrum is either absent or fragmentary (Springer, 1985).
2. The basisphenoid of *Phenablennius* is reduced and loosely attached to the neurocranium. This condition is intermediate between the absence of a basisphenoid in the Nemophini and a well-developed basisphenoid of other blenniids (Springer and Smith-Vaniz, 1972b).
3. A swimbladder is present in adults of *Phenablennius* (Smith-Vaniz, 1975), a condition seen in some Nemophini.
4. In *Phenablennius*, the margin of the lower jaws bears well-developed labial flaps similar to those seen in the Omobranchine genus *Omox* (Smith-Vaniz, 1975).

This poorly known species has been recorded from fresh and brackish waters of Cambodia, Sumatra and Borneo and exhibits conspicuous sexual dimorphism (Smith-Vaniz, 1975). In males, the distal tips of the dorsal and anal-fin rays are expanded and have flaplike structures, with a fleshy flap near the base of the anterior margin of the first dorsal-fin spine, features that are absent in females. Males lack the conspicuous banding pattern seen in females.

Salariini

The Salariini (sensu Williams, 1990) is the largest of the blenniid tribes, including 28 genera and 214 species (Table 1.3.1). The monophyly of the Salariini is supported by three synapomorphies (Smith-Vaniz and Springer, 1971; Springer and Spreitzer, 1978; Williams, 1990).

1. The premaxilla of salariines is an open capsule with a relatively thick band of connective tissue supporting the teeth along its anteroventral margin.
2. The infrapharyngobranchial plate of salariines is uniquely shaped with a broad flattened bony flange projecting medially from the tooth-bearing portion of the bone and articulating with the second epibranchial. In other blenniids and tripterygiids, this flange is narrow and bears an anterior protuberance for articulation with the second epibranchial (Williams, 1990).
3. In salariines, the gap between the coracoid and scapula is narrow and partially indented; this indentation is absent in other blennies (Bath, 2000a).

Within the Salariini, Williams (1990) recognised two lineages, the *Salarias* group and the *Rhabdoblennius* group. The *Salarias* group has a highly modified dentary (Smith-Vaniz and Springer, 1971) in that its anterodorsal wall is not ossified. Members of the *Salarias* group also have numerous premaxillary teeth, ranging from 55 to nearly 400 with most species having over 100. The *Salarias* group includes 20 genera and 187 species (Table 1.3.1). Williams (1990; Fig. 1.3.3) hypothesised the phylogenetic relationships of a portion of the *Salarias* group. Bath (2000a, 2002) divided members of this lineage into two groups based on their dentition (see below). The *Rhadoblennius* group of Williams (1990) lacks known synapomorphies, but includes eight genera and 27 species (Table 1.3.1). It corresponds, in part, to Group 2 of Bath (2000a, 2001, 2002).

Bath (2001) provided a different interpretation of his expanded Salariini, recognising four groups. His Group 1, characterised by relatively low numbers of relatively immovable teeth closely attached to the bony surface of the jaw bones, includes the genera of the Parablenniini of Bock and Zander (1986) except for the genus *Coryphoblennius*. Bath's Group 2 includes genera with dentition intermediate between that of Group 1 and the more numerous and closely spaced teeth of Groups 3 and 4. He recognised three subgroups: (1) the genera *Coryphoblennius* and

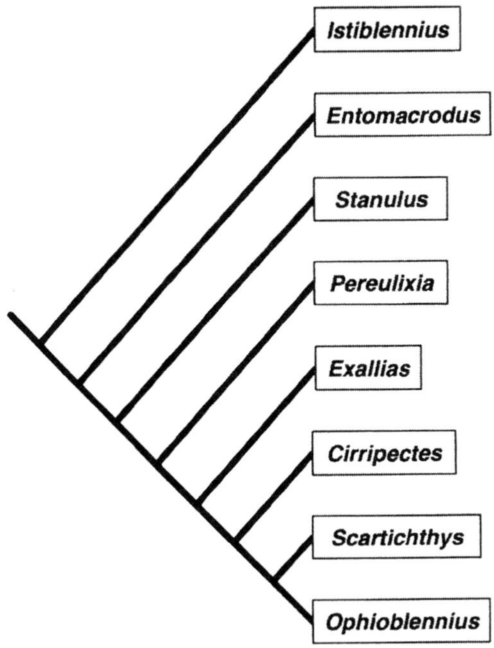

Fig. 1.3.3 Phylogenetic relationships of selected genera of the Salariini hypothesised by Williams (1990).

Cirrisalarias; (2) the genus *Hirculops*; and (3) the genera *Antennablennius*, *Mimoblennius* and *Rhabdoblennius*. Bath's Group 3, with incomplete premaxillary and dentary capsules, includes *Nannosalaias*, *Dodekablennos* and *Ecsenius*. Finally, Bath's Group 4, with open premaxillary and dentary capsules and inferior excavation of the premaxillary and superior excavation of the dentary, includes *Alticus*, *Blenniella*, *Entomacrodus*, *Exallias*, *Istiblennius*, *Ophioblennius*, *Praealticus* and *Salarias* (Bath, 2001, 2002).

Revisionary and evolutionary studies of the Salariini include: Bath, 1992 (*Praealticus*), Springer, 1962, Muss *et al.*, 2001 (*Ophioblennius*), Springer, 1967, 1972b, Springer and Fricke, 2000 (*Entomacrodus*), Springer and Spreitzer, 1978 (several genera including a key to genera), Smith-Vaniz and Springer, 1971 (several genera), Williams, 1990 (*Scartichthys*), Springer and Williams, 1994 (*Istiblennius*, *Blenniella* and *Paralticus*), Springer, 1971, 1972b, McKinney and Springer, 1976, Springer, 1988 (*Ecsenius*), Williams, 1988 (*Cirripectes*), Bath, 1983 (*Antennablennius*) and

Bath, 2004 (*Rhabdoblennius*). Several recently described species of Salariines are included in the list of valid species of blennioids (Patzner *et al.*, Chapter 6.1).

Parablenniini

Historically, the Blenniini (sensu Springer, 1968) included a variety of generalised blenny species that did not fit well with other lineages. Many of these had been included in the large, polyphyletic genus *Blennius* by Norman (1943) and others. Smith-Vaniz (1976) recognised that this represented an unnatural group (i.e., not monophyletic) because it included a variety of species that shared a number of plesiomorphic characters but no apomorphies. He also recognised that two of its members, *Blennius ocellaris* and *B. normani*, were distinctive in having the left and right dentaries especially strongly sutured to one another. This feature was similar to that of the blenniid tribes Omobranchini, Nemophini and Phenablenniini (but not the Salariini) in the sense that these lineages have sutured dentaries, although less strongly so. He, thus, restricted the genus *Blennius* to these two species that together with two subsequently described species of *Spaniblennius*, are now the sole members of the tribe Blenniini. Smith-Vaniz (1976) placed the remaining members of the Blenniini (i.e., those without sutured dentaries) in the 'Blenniini group' with uncertain relationships (Fig. 1.3.4). Based on skull and cephalic sensory canal characters, Bock and Zander (1986) erected the tribe Parablenniini for the members of the 'Blenniini group' (sensu Smith-Vaniz, 1976). They also presented a hypothesis of the phylogenetic relationships of several genera in their Parablenniini and hypothesised its

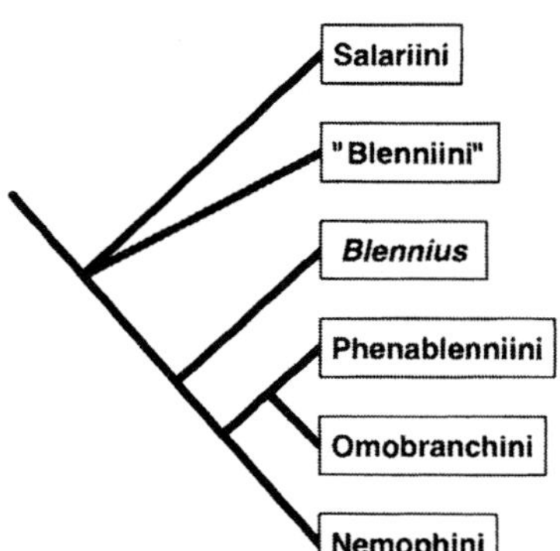

Fig. 1.3.4 Phylogenetic relationships of the Blenniidae according to Smith-Vaniz (1976).

sister-group relationship with the Salariini within the context of the overall relationships of other blenniid tribes. However, they failed to report any synapomorphies supporting either the monophyly of the Parablenniini or its sister-group relationship with the Salariini.

Shortly thereafter, Williams (1990) hypothesised the monophyly of the Parablenniini based on two synapomorphies. First, the first basibranchial is U-shaped, while in other blennioids, it is straight in lateral aspect or, in a few, slightly upturned on its anterior end. Second, the fourth epibranchial is relatively smooth with a flattened longitudinal bony flange projecting dorsally from the lateral half of the shaft. In other blennioids, the fourth epibranchial has one or two dorsally directed bony projections dorsal and anterior of the articulation with the infrapharyngobranchial. Bath (1996) discussed the neurocranial osteology of the Parablenniini and provided a list of the included genera and species.

After studying a variety of blenniid genera, Bath (2001) reported that both synapomorphies hypothesised by Williams (1990) exhibit a high degree of homoplasy both inside and outside of the Parablenniini, with presumed ancestral and derived character states occurring in species allocated to both the Parablenniini and the Salariini. Pointing to the lack of unique features of the Parablenniini, as well as the continuous nature of variation in tooth patterns in these blennies, Bath (2001) rejected the monophyly of the Parablenniini and included this tribe within an expanded Salariini (as the Salariinae).

As with many taxonomic categories of fishes, the tribe Parablenniini has served as a convenient way to isolate a number of lineages that do not fit within other relatively well-defined clades. Although the monophyly of this group is in doubt, we feel that this taxon will continue to serve a useful purpose until the relationships of its members with other blenniids have been resolved. Thus, we prefer to retain the Parablenniini, currently with 14 genera and 82 species (Table 1.3.1), recognising that the relationships of its members remain one of the primary unresolved issues in the systematics of blenniid fishes.

Bock and Zander (1986) recognised three subgroups in their Parablenniini based on details of the neurocranium, cephalic sensory canals and other features. The genera *Hypsoblennius*, *Chasmodes* and *Hypleurochilus* share restricted gill openings; the genera *Aidablennius*, *Parablennius* and *Scartella* have multiple branches of the supratemporal sensory canal present on the parietals; and the genera *Lipophrys* and *Coryphoblennius* possess similar neurocrania.

Bath (2001) recognised four groups within his expanded Salariini, two of which include members of the Parablenniini. His Group 1 is characterised by relatively few teeth that are firmly attached to the jaws and includes the genera of the Parablenniini (sensu Bock and Zander, 1986) except for the genus *Coryphoblennius*. Bath placed the latter genus in his Group 2, based on its dentition which, in some ways, is intermediate between that of Group 1 and Groups 3 and 4 (Bath, 1994a). Within Group 2, Bath hypothesised a close relationship between *Coryphoblennius* and *Cirrisalarias* (Bath, 2001). Bath's Group 1 lineage most closely resembles the Parablenniini of Bock and Zander (1986). However, this is based on a single character complex (dentition) and needs to be tested in a phylogenetic analysis of all morphological characters relevant to the relationships of blenniid fishes. Interestingly, a recent phylogenetic analysis of blenniids based on molecular sequence data placed *Coryphoblennius* inside the Parablenniini as sister to *Lipophrys*, not within the Salariini as would be predicted by Bath's (2001) conclusions.

Revisionary studies of the Parablenniini include Bath, 1977 (entire tribe), Bath, 1989, 1990 (*Parablennius*), Krejsa, 1960, Smith-Vaniz, 1980, Bath, 2000b (*Hypsoblennius*), Bath, 1994b (*Hypleurochilus*), Springer, 1959a and Williams, 1983 (*Chasmodes*).

RELATIONSHIPS WITHIN THE BLENNIIDAE

Inter-relationships of the blenniid tribes have been studied by several workers (Smith-Vaniz, 1976; Bock and Zander, 1986; summarised and expanded by Williams, 1990). The Salariini and Parablenniini (sensu Bock and Zander, 1986) form a monophyletic group (Fig. 1.3.5) based on two hypothesised synapomorphies (Williams, 1990). First, the ascending process of the premaxilla is only weakly attached to the main body of the bone as opposed to being broadly attached in the Blenniini and other blenniids. Second, the second from the dorsal most pectoral radial inserts only on the scapula rather than partially or completely on the coracoid. Bath (2001, 2002) examined the osteology of these lineages, concluding that they were indeed closely related, placing them together in his Salariinae because he found no clear distinction between them.

The Blenniini, Nemophini, Omobranchini and Phenablenniini form a monophyletic group (Fig. 1.3.5) based on a single synapomorphy; the left and right dentaries are sutured to one another (Smith-Vaniz, 1976). This suture is especially well developed in the Blenniini. The Nemophini,

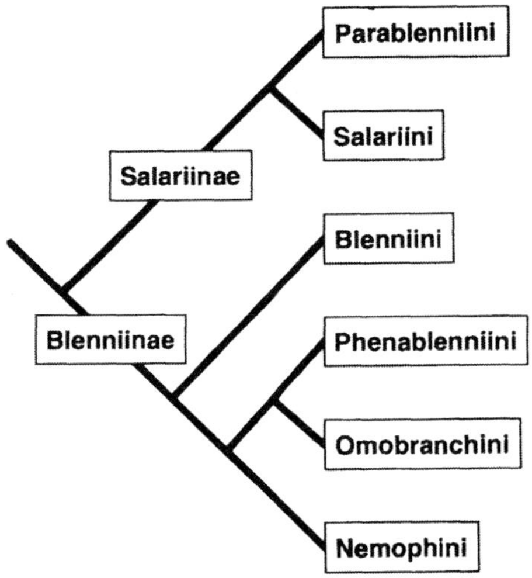

Fig. 1.3.5 Phylogenetic relationships of the tribes of the Blenniidae with the two main lineages designated as subfamilies.

together with the Omobranchini and Phenablenniini, form a monophyletic group (Fig. 1.3.5) based on two synapomorphies (Smith-Vaniz, 1976). First, the central caudal-fin rays are unbranched rather than branched as is the case in most other blenniids (exceptions include *Ecsenius*, *Atrosalarias*, *Andamia*, *Alticus*, *Cirrisalarias*, *Litobranchus* and *Medusablennius* that have the central caudal rays unbranched and *Dodekablennos* that has only two to six branched rays, but these appear to represent multiple reversals). Second, the number of epurals is reduced from two to one. The Phenablenniini appears to be the sister group of the Omobranchini (Fig. 1.3.5) based a reduction of the pterosphenoid and its exclusion from the external surface of the neurocranium (Springer and Smith-Vaniz, 1972b).

CLASSIFICATION OF THE BLENNIIDAE

Blenniid classification has a long and complicated history that has been reviewed by several recent authors (e.g., Williams, 1990; Bath, 2001, 2002, 2004). Briefly, Norman (1943) provided one of the first intrafamilial classifications of the Blenniidae as currently understood, recognising three subfamilies, the Ophioblenniinae, Blenniinae and Salariinae (Norman's Ophioblenniinae included only larval stages, among which were species of three salariinine genera and one genus of the Labrisomidae; see Springer, 1959b, 1962). Springer (1968) defined the Blenniidae with respect to other blennioids, recognising two subfamilies: the Nemophinae (first used as a family group name, Nemophidae, by Kaup, 1858) and the Blenniinae, the latter including three tribes, the Blenniini, Omobranchini (first appearance), and Salariini. Springer and Smith-Vaniz (1972b) erected the tribe Phenablenniini and provided evidence that it was related to the Nemophini which they designated as a tribe, not a subfamily. Smith-Vaniz (1976) recognised six lineages of blenniids (Fig. 1.3.4): the Salariini, the 'Blenniini group', the species of *Blennius*, the Phenablenniini, the Omobranchini and the Nemophini. Bock and Zander (1986) erected the tribe Parablenniini for the 'Blenniini group' of Smith-Vaniz (1976). Since then, six tribes have been recognised by most authors (Nelson, 1994).

Bath (2001) provided an alternative classification scheme of the Blenniidae based in part on a survey of their tooth structures. Noting the continuity of character states in the dentition of the Salariini and Parablenniini, together with his reevaluation and rejection of the hypothesised synapomorphies of the Parablenniini (see above), he synonymised the Parablenniini with the Salariini and elevated this entire lineage to subfamily level (i.e., the Salariinae). Bath also elevated each one of the remaining tribes to subfamily status, thus recognising the Blenniinae, Omobranchinae, Phenablenniinae and Nemophinae in addition to the greatly expanded Salariinae. Nelson (2006) continued to recognise these lineages as tribes rather than subfamilies.

While we do not question Bath's conclusion that the species included in the Parablenniini lack any known synapomorphies, we are of the opinion that until relationships of this lineage are resolved, the Parablenniini continues to serve as a convenient way to allocate genera that are allied with the Salariini but not a part of it. Further, Bath's substitution of one taxonomic level (subfamilies) for another (tribes) does little to clarify the phylogenetic relationships among blenniids. Instead, we retain the widely recognised tribal classification (reviewed above) and group these tribes into two subfamilies that reflect the current consensus of their evolutionary relationships (Fig. 1.3.5). Consequently, we recognise two subfamilies, the Blenniinae that includes four tribes (Blenniini, Phenablenniini, Omobranchini and Nemophini), 15 genera and 91 species, and the Salariinae that includes two tribes (Salariini and Parablenniini), 42 genera and 296 species (Table 1.3.1). As discussed above, the relationships of the members of the non-monophyletic Parablenniini need to be further evaluated in the light of Bath's (2001) important observations. However, recognition of two major lineages within the Blenniidae as subfamilies highlights the widely accepted hypothesis for the phylogenetic relationships of these fishes (Fig. 1.3.5).

RELATIONSHIPS TO OTHER BLENNIOID FAMILIES

Phylogenetic relationships of the families included in the Blennioidei are poorly resolved (Hastings and Springer, Chapter 1.1). A few characters that may have bearing on the issue have been mentioned sporadically in the literature. Hubbs (1953) united chaenopsins and blenniids based on their eye position, stout infraorbitals, and reduced or absent scales. Many blenniids and most chaenopsids have muscles of the adductor mandibularis A2 complex covering the dorsal portion of the neurocranium (Kotrschal, 1988; Hastings and Springer, 1994). Blenniids, chaenopsids and starksiins (Labrisomidae) have a spinous process on the sphenotic (Hastings and Springer, 1994). Blenniids, chaenopsids, dactyloscopids and starksiins have the insertion of the hyomandibula relatively far posterior, well separated from the posterior margin of the orbit (Hastings and Springer, 1994). Some blenniids, chaenopsins and dactyloscopids have the distal portion of their median-fin spines unossified (Hastings and Springer, 1994).

These and other characters offer conflicting evidence of the relationships of the Blenniidae within the remainder of the Blennioidei. Clearly, they should not be studied in isolation or without knowledge of states in appropriate out groups. Recently, some progress has been made in determining the relationships of the Blennioidei within the Perciformes (see Hastings and Springer, Chapter 1.1). Springer and Orrell (2004) included the Blennioidei *sensu stricto* within a newly recognised clade, the Benthomorpha that also includes the Callionymoidei, Gobiesocoidei and, tentatively, the Dactylopteridae and provided evidence of its monophy based on details of pharyngeal anatomy. There is a growing consensus, based on both morphological (Rosen and Patterson, 1990; Springer and Orrell, 2004) and molecular data (Chen *et al.*, 2003; Simmons and Miya, 2004; Dettai and Lecointe, 2005; Miya *et al.*, 2005), that the sister group of the Blennioidei is the Gobiesocoidei (clingfishes). This, together with emerging studies on the relationships within and between the remaining blennioid families, provides the context for a thorough analysis of relationships within the combtooth blennies.

FUTURE STUDIES

The most pressing need within the Blenniidae is study of the relationships of the genera of the Parablenniini. Given Bath's (2001) conclusion that this is not a monophyletic group—together with the broad consensus that its members are related to the Salariini—the stage is set for a rigorous cladistic analysis of these blennies using both molecular and morphological data.

Several morphological character systems are increasingly well documented for blenniids. First, several more or less complete descriptions of the osteology of various blenniids are available. These include several members of the Salariini and Parablenniini (Thiele, 1963; Bath, 1977, 1996, 2001; Bock and Zander, 1986) including *Entomacrodus* (Springer, 1968), *Ecsenius* (Springer, 1988) and *Hypsoblennius* (Cancion *et al.*, 1999) and selected Nemophini (Smith-Vaniz, 1976). These, together with several studies on selected osteological characters in selected lineages (cited above), form a firm foundation for a morphologically based phylogenetic analysis of the Blenniidae. Second, Springer and Orrell (2004) included selected blenniids in their survey of the musculature associated with the gill arches of acanthomorphs. This character-rich anatomical region deserves more in-depth study within the Blenniidae as well as among other blennioid families. Third, knowledge of larval development of blenniids has increased significantly (Watson, Chapter 4.4). Several features unique to larvae of particular groups of blenniids may hold clues to their phylogenetic relationships. Fourth, the Blenniidae are well known for their strongly developed fleshy structures surrounding the anterior anal-fin spines of breeding males (Zander, 1975; Kotrschal *et al.*, 1984). These so-called 'fin glands' were hypothesised by Williams (1990) to represent a synapomorphy of the Blenniidae. However, these structures are absent in the Omobranchini, Phenablenniini and some members of the Salariini (Smith-Vaniz, 1976; Williams, 1990). Superficially similar fleshy thickenings on these and other fin elements are present in some other blennioids (e.g., Northcott and James, 1996). Fifth, most male blenniids have accessory structures associated with their testes (Eggert, 1931; Seiwald and Patzner, 1987; Patzner and Seiwald, 1988; Patzner, 1989; Richtarski and Patzner, 2000). Within the Blenniidae, all species studied have a bilobate testis, but several types of accessory structures are known, including 'seminal vesicles', 'testicular glands', and 'testicular blind pouches'. Patzner and Lahnsteiner (Chapter 4.1) are exploring these structures in greater detail.

Few researchers have applied the tools of molecular biology to resolve the relationships of lineages within the Blenniidae. Most recently, molecular data of selected species from the eastern Atlantic region (Almada *et al.*, 2005) did not support some aspects of earlier morphologically based phylogenetic hypotheses. First, the genus *Blennius* was nested inside the Parablenniini, albeit with poor bootstrap support, not as a sister to the remaining lineages of the Blenniinae (Fig. 1.3.5). Second, *Coryphoblennius* was nested inside the Parablenniini as sister to *Lipophrys*, not within the Salariini, as hypothesised by Bath (2004). Analysis of molecular sequence data from both mitochondrial and nuclear genes from a broad array of blennioid taxa is clearly needed. These studies are currently underway in a number of laboratories and will hopefully provide new insights into the phylogenetic relationships of the Blenniidae.

References

Almada, F., V.C. Almada, T. Guillemaud and P. Wirtz. 2005. Phylogenetic relationships of the north-eastern Atlantic and Mediterranean blenniids. *Biological Journal Linnean Society* 86: 283-295.

Bath, H. 1977. Revision der Blenniini. *Senckenbergiana biologica* 57: 167-234.

Bath, H. 1983. Revision der Gattung *Antennablennius* Fowler 1931 mit Beschreibung einer neuen Art und Untersuchung der taxonomischen Stellung von *Antennablennius anuchalis* Springer and Spreitzer 1978. *Senckenbergiana biologica* 64: 47-80.

Bath, H. 1989. Die Arten der Gattung *Parablennius* Ribeiro 1915 im Roten Meer, Indischen und NW des Pazifischen Ozeans. *Senckenbergiana biologica* 69: 301-343.

Bath, H. 1990. Taxonomie und Verbreitung von *Parablennius* Ribeiro 1915 an der W-Küste Afrikas und de Kapverdischen Inseln mit Revalidation von *P. verryckeni* (Poll 1959) und Beschreibung drei neuer Arten (Pisces: Blenniidae). *Senckenbergiana biologica* 70: 15-69.

Bath, H. 1992. Revision der Gattung *Praealticus* Schultz and Chapman 1960 (Pisces: Blenniidae). *Senckenbergiana biologica* 72: 237-316.

Bath, H. 1994a. Die Bezahnung von *Coryphoblennius galerita* (Linneasu 1785) im Vergleich zu anderen Arten der Tribus Parablenniini und Salariini (Pisces: Blenniidae). *Senckenbergiana biologica* 74: 47-58.

Bath, H. 1994b. Untersuchung der Arten *Hypleurochilus geminatus* (Wood 1825), *H. fissicornis* (Quoy and Gaimard 1824) und *H. aequipinnis* (Günther 1861), mit Revalidation von *Hypleurochilus multifilis* (Girard 1858) und Beschreibung von zwei neuen Arten. *Senckenbergiana biologica* 74: 59-85.

Bath, H. 1996. Beitrag zur Osteologie der Arten der Tribus Parablenniini Die Beziehungen der Knochen des Schädeldaches zum Seitenorgan-System und zu den Weichteilbildungen der Kopfoberseite sowie die systematische Bedeutung der Befund nebst Bemerkungen zu *Lupinoblennius dispar* Herre 1942 (Pisces: Blenniidae). *Senckenbergiana biologica* 76: 65-92.

Bath, H. 2000a. Untersuchung des Schultergürtels der Arten der Tribus Parablenniini unter Berücksichtigung von Arten weiterer Tribus der Familie Blenniidae (Pisces, Blenniidae). *Senckenbergiana biologica* 80: 189-203.

Bath, H. 2000b. Verwandtschaftliche Beziehungen der Arten der Gattung *Hypsoblennius* Gill 1861 (Pisces, Blenniidae). *Senckenbergiana biologica* 80: 175-187.

Bath, H. 2001. Osteology and morphology of fishes of the subfamily Salariinae and its junior synonym Parablenniinae (Pisces: Blenniidae). *Stuttgarter Beiträge zur Naturkunde* A 628: 1-42.

Bath, H. 2002. Dentition of the blenniid fishes in the subfamily Salariinae with premaxillaries and dentaries forming a complete shell (Pisces: Blenniidae). *Stuttgarter Beiträge zur Naturkunde* A 635: 1-17.

Bath, H. 2004. Revision of the genus *Rhabdoblennius* Whitley (Pisces: Blenniidae: Salariinae), with descriptions of two new species. *Stuttgarter Beiträge zur Naturkunde* A 669: 1-26.

Bath, H. and P. Wirtz. 1989. *Spaniblennius clandestinus* n. g., n. sp. der tribus Blenniini von W-Africa und ihre Beziehungen zur Gattung *Blennius* Linnaeus 1758. *Senckenbergiana biologica* 69: 277-291.

Bock, M. and C.D. Zander. 1986. Osteological characters as tool for blenniid taxonomy — A generic revision of European Blenniidae (Percomorphi; Pisces). *Zeitschrift für Zoologische Systematik und Evolutionsforschung* 24: 138-143.

Cancion, C., E. de la Hoz and K. Farias. 1999. Morfología cefálica de *Hypsoblennius sordidus* (Bennet, 1828) (Perciformes, Blenniidae): un blénido omnívoro. *Revista de Biología Marina y Oceanografía* 34: 167-180.

Chen, W., C. Bonillo and G. Lecointre. 2003. Repeatability of clades as a criterion of reliability: A case study for molecular phylogeny of Acanthomorphoa (Teleostei) with larger number of taxa. *Molecular Phylogenetics and Evolution* 26: 262-288.

Dettai, A. and G. Lecointre. 2005. Further support for the clades obtained by multiple molecular phylogenies in the acanthomorph bush. *C. R. Biologies* 328: 674-689.

Eggert, B. 1931. Die Geschlechtsorgane der Gobiiformes und der Blenniiformes. *Zeitschrift für wissenschaftliche Zoologie* 139: 249-558.

Fishelson, L. 1974. Histology and ultrastructure of the recently found buccal toxic gland in the fish *Meiacanthus nigrolineatus* (Blenniidae). *Copeia* 1974: 386-392.

Fishelson, L. 1975. Observations on behaviour of the fish *Meiacanthus nigrolineatus* Smith-Vaniz (Blenniidae) in nature (Red Sea) and in captivity. *Australian Journal of Marine and Freshwater Research* 26: 329-341.

Gerhardinger, L.C., M.O. Freitas, A.B. Andrade and C.A. Rangel. 2006. *Omobranchus punctatus* (Teleostei: Blenniidae), an exotic blenny in the Southwestern Atlantic. *Biological Invasions* 8: 941-946.

Giacamello, E., D. Marchini and M.B. Rasotto. 2006. A male sexually dimorphic trait provides antimicrobials to eggs in a blenny fish. *Biology Letters* 2: 330-333.

Hastings, P.A. and V.G. Springer. 1994. A review of *Stathmonotus*, with redefinition and phylogenetic analysis of the Chaenopsidae (Pisces: Blennioidei). *Smithsonian Contributions to Zoology* 558: 1-48.

Hubbs, C.L. 1953. Revision and systematic position of blenniid fishes of the genus *Neoclinus*. *Copeia* 1953: 11-23.

Kaup, J. 1858. On *Nemophis*, a new genus of riband-shaped fishes. *Proceedings of the Zoological Society of London* 26: 168-169.

Kotrschal, K. 1988. A catalogue of skulls and jaws of eastern tropical Pacific blennioid fishes (Blennioidei: Teleostei): A proposed evolutionary sequence of morphological change. *Zeitschrift für Zoologische Systematik und Evolutionsforschung* 26: 442-466.

Kotrschal, K., H. Weisl and A. Goldschmid. 1984. Mehrzellige Drüsen in der Epidermis der unpaaren Flossen bei den Blenniidae (Blenniidae, Teleostei). *Zeitschrift für mikroskopisch-anatomisch Forschung, Leipzig* 98: 184-192.

Krejsa, R.J. 1960. The eastern tropical Pacific fishes of the genus *Blenniolus*, including a new island endemic. *Copeia* 1960: 322-336.

Kuiter, R.H. 1992. *Tropical Reef-Fishes of the Western Pacific Indonesia and Adjacent Waters*. Pustaka Utama, Jakarta, Indonesia.

McKinney, J.F. and V.G. Springer. 1976. Four new species of the fish genus *Ecsenius* with notes on other species of the genus (Blenniidae: Salariini). *Smithsonian Contributions to Zoology* 236: 1-27.

Miya, M., T.P. Satoa and M. Nishida. 2005. The phylogenetic position of toadfishes (order Batrachoidiformes) in the higher ray-finned fish as inferred from partitioned Bayesian analysis of 102 whole mitochondrial genome sequences. *Biological Journal of the Linnean Society* 85: 289-306.

Mundy, B.C. 2005. Checklist of the fishes of the Hawaiian Archipelago. *Bishop Museum Bulletin in Zoology* 6: 1-703.

Muss, A., D.R. Robertson, C.A. Stepien, P. Wirtz and B.W. Bowen. 2001. Phylogeography of *Ophioblennius*: The role of ocean currents and geography in reef fish evolution. *Evolution* 55: 561-572.

Nelson, J.S. 1994. *Fishes of the World*. John Wiley & Sons, New York. 3rd Edition.

Nelson, J.S. 2006. *Fishes of the World*. John Wiley & Sons, New York. 4th Edition.

Norman, J.R. 1943. Notes on the blennioid fishes. I. A provisional synopsis of the genera of the family Blenniidae. *Annals and Magazine of Natural History* (Series 11) 10: 793-812.

Northcott, S.J and M.A. James. 1996. Ultrastructure of the glandular epidermis on the fins of male estuarine triplefins *Forsterygion nigripenne*. *Journal of Fish Biology* 49: 95-107.

Patzner, R.A. 1989. Morphology of the male reproductive systems of two Indopacific blenniid fishes. *Zeitschrift für Zoologische Systematik und Evolutionsforschung* 27: 135-141.

Patzner, R.A. and M. Seiwald. 1988. Different types of testicular accessory organs in blennioid fishes. *VI Congress of European Ichthyology, Stockholm* (Abstract).

Richtarski, U. and R.A. Patzner. 2000. Comparative morphology of male reproductive systems in Mediterranean blennies (Blenniidae). *Journal of Fish Biology* 56: 22-36.

Rosen, D.E. and C. Patterson. 1990. On Müller's and Cuvier's concepts of pharyngognath and labyrinth fishes and the classification of percomorph fishes, with an atlas of percomorph dorsal gill arches. *American Museum Novitates* 2983: 1-57.

Seiwald, M. and R.A. Patzner. 1987. Ultrastructure of the testicular gland of *Blennius pavo* (Pisces, Teleostei). *Zoomorphology* 107: 26-32.

Simmons, M.P. and M. Miya. 2004. Efficiently resolving the basal clades of a phylogenetic tree using Bayesian and parsimony approaches: A case study using mitogenomic data from 100 higher teleost fishes. *Molecular Phylogenetics and Evolution* 31: 351-362.

Smith, W.L. and W.C. Wheeler. 2006. Venom evolution is widespread in fishes: A phylogenetic road map for the bioprospecting of piscine venoms. *Journal of Heredity* 97: 206-217.

Smith-Vaniz, W.F. 1975. Supplemental description of rare blenniid fish *Phenablennius heyligeri* (Bleeker). *Proceedings of the Academy Natural Sciences Philadelphia* 127: 53-55.

Smith-Vaniz, W.F. 1976. The saber-toothed blennies, tribe Nemophini (Pisces: Blenniidae). *Monographs of the Academy of Natural Sciences of Philadelphia* 19: 1-196.

Smith-Vaniz, W.F. 1980. Revision of western Atlantic species of the blenniid fish genus *Hypsoblennius*. *Proceedings of the Academy Natural Sciences Philadelphia* 132: 285-305.

Smith-Vaniz, W.F. 1987. The saber-toothed blennies, tribe Nemophini (Pisces: Blenniidae): An update. *Proceedings of the Academy Natural Sciences Philadelphia* 139: 1-52.

Smith-Vaniz, W.F. and V.G. Springer. 1971. Synopsis of the tribe Salariini, with description of five new genera and three new species (Pisces: Blenniidae). *Smithsonian Contributions to Zoology* 73: 1-72.

Smith-Vaniz, W., U. Satapoomin and G.R. Allen. 2001. *Meiacanthus urostigma*, a new fangblenny from the northeastern Indian Ocean, with discussion and examples of mimicry in species of *Meiacanthus* (Teleostei: Blenniidae: Nemophini). *Aqua* 5: 25-43.

Springer, V.G. 1959a. Blenniid fishes of the genus *Chasmodes*. *Texas Journal of Science* 11: 321-334.

Springer, V.G. 1959b. A new species of *Labrisomus* from the Caribbean Sea, with notes on other fishes of the subtribe Labrisomini. *Copeia* 1959: 289-292.

Springer, V.G. 1962. A review of the blenniid fishes of the genus *Ophioblennius* Gill. *Copeia* 1962: 426-433.

Springer, V.G. 1967. Revision of the circumtropical shorefish genus *Entomacrodus* (Blenniidae: Salariinae). *Proceedings of the United States National Museum* 122: 1-150.

Springer, V.G. 1968. Osteology and classification of fishes of the family Blenniidae. *United States National Museum Bulletin* 284: 1-85.

Springer, V.G. 1971. Revision of the fish genus *Ecsenius* (Blenniidae, Blenniinae, Salariini). *Smithsonian Contributions to Zoology* 72: 1-74.

Springer, V.G. 1972a. Synopsis of the tribe Omobranchini with descriptions of three new genera and two new species (Pisces: Blenniidae). *Smithsonian Contributions to Zoology* 130: 1-31.

Springer, V.G. 1972b. Additions to revisions of the blenniid fish genera *Ecsenius* and *Entomacrodus*, with descriptions of three new species of *Ecsenius*. *Smithsonian Contributions to Zoology* 134: 1-13.

Springer, V.G. 1981. Notes on blenniid fishes of the tribe Omobranchini, with descriptions of two new species. *Proceedings of the Biological Society of Washington* 94: 699-707.

Springer, V.G. 1985. *Oman ypsilon*, a new genus and species of blenniid fish from the Indian Ocean. *Proceedings of the Biological Society of Washington* 98: 90-97.

Springer, V.G. 1988. The Indo-Pacific blenniid fish genus *Ecsenius*. *Smithsonian Contributions to Zoology* 465: 1-134.

Springer, V.G. 1993. Definition of the suborder Blennioidei and its included families (Pisces: Perciformes). *Bulletin of Marine Science* 52: 472-495.

Springer, V.G. and R.A. Fricke. 2000. Description of two new blenniid fish species: *Entomacrodus lemuria* from the western Indian Ocean and *E. williamsi* from the western Pacific Ocean. *Proceedings of the Biological Society of Washington* 113: 386-396.

Springer, V.G. and M.F. Gomon. 1975. Revision of the blenniid fish genus *Omobranchus* with descriptions of three new species and notes on other species of the tribe Omobranchini. *Smithsonian Contributions to Zoology* 177: 1-135.

Springer, V.G. and T.M. Orrell. 2004. A phylogenetic analysis of 147 families of acanthomorph fishes based primarily on dorsal gill-arch muscles and skeleton. *Bulletin of the Biological Society of Washington* 11: 237-260.

Springer, V.G. and W.F. Smith-Vaniz. 1972a. Mimetic relationships involving fishes of the family Blenniidae. *Smithsonian Contributions to Zoology* 112: 1-36.

Springer, V.G. and W.F. Smith-Vaniz. 1972b. A new tribe (Phenablenniini) and genus (*Phenablennius*) of blenniid fishes based on *Petroscirtes heyligeri* Bleeker. *Copeia* 1972: 64-71.

Springer, V.G. and A.E. Spreitzer. 1978. Five new species and a new genus of Indian Ocean blenniid fishes, tribe Salariini, with a key to genera of the tribe. *Smithsonian Contributions to Zoology* 268: 1-20.

Springer, V.G. and J.T. Williams. 1994. The Indo-Pacific blenniid fish genus *Istiblennius* reappraised: a revision of *Istiblennius*, *Blenniella*, and *Paralticus*, new genus. *Smithsonian Contributions to Zoology* 565: 1-193.

Thiele, H. 1963. Vergleichend-morphologische Untersuchungen über die Funktion der Nahrungserwerbsapparate von *Anarrhichas lupus* L. und einigen Blenniidae (Teleostei, Blennioidei). *Zoologische Beiträge*, Neue Folge 9: 275-440.

Williams, J.T. 1983. Taxonomy and ecology of the genus *Chasmodes* (Pisces: Blenniidae) with a discussion of its zoogeography. *Bulletin of the Florida State Museum, Biological Sciences* 29: 65-100.

Williams, J.T. 1988. Revision and phylogenetic relationships of the blenniid fish genus *Cirripectes*. *Indo-Pacfic Fishes* 17: 1-78.

Williams, J.T. 1990. Phylogenetic relationships and revision of the blenniid fish genus *Scartichthys*. *Smithsonian Contributions to Zoology* 492: 1-30.

Zander, C.D. 1975. Secondary sex characteristics of blennioid fishes (Perciformes). *Pubblicazione della Stazione Zoologica di Napoli* 39: 717-727.

SECTION 2

Biogeography

CHAPTER

2.1

Biogeography of New World Blennies

Philip A. Hastings

INTRODUCTION

The coastal marine waters surrounding North and South America have long been recognised as biogeographically distinct. While most biogeographers have traditionally considered tropical and temperate marine waters separately (e.g., Ekman, 1953; Briggs, 1974), this analysis explores the diversity and distributions of both tropical and temperate blennioid fishes within this vast region. Blennioids are ideal subjects for studying the biogeography of coastal marine waters because they exhibit significant diversity, occur in most coastal habitats and, importantly, exhibit a range of dispersal potentials. Blennioids include species with long larval development, spending over 50 days in the plankton (Riginos and Victor, 2001; Robertson *et al.*, 2004; Watson, Chapter 4.4), as well as those that give birth to well-developed juveniles that may lack a significant dispersal stage (Rosenblatt and Taylor, 1971; Moser, 2007). In addition,

Author's address: Scripps Institution of Oceanography, University of California San Diego, La Jolla, California, USA. E-mail: phastings@ucsd.edu

variation in dispersal potential is predicted based on the behavior and location in which larval development occurs. In some species, larvae remain in inshore waters, often very close to reef substrates throughout their entire development, while in others, larval development occurs in the open ocean, far from reefs (Brogan, 1994). As a consequence, various blennies are predicted to react differently to potential barriers to dispersal, revealing a variety of patterns and associated factors that may have led to their isolation and diversification.

This chapter presents a brief overview of the diversity and summary of the distributional patterns for blennioids of the New World. As will be seen, this region has an extraordinarily high degree of endemism and is a major global center of diversification of blennioid fishes.

METHODS

A list of all blennioid species occurring in New World waters was assembled along with known distributions of each. These were compiled from the primary literature, summaries of regional faunas (e.g., Lubbock and Edwards, 1981; Robins and Ray, 1986; Böhlke and Chaplin, 1994; Bearez, 1996; Grove and Lavenberg, 1997; Smith-Vaniz *et al.*, 1999; Garrison, 2000; Robertson and Allen, 2002; Feitzoa *et al.*, 2003; Menezes *et al.*, 2003; Rocha, 2003; Love *et al.*, 2005; Floeter *et al.*, 2008) and from museum records. Two undescribed species of chaenopsids (*Chaenopsis* and *Acanthemblemaria*) and three undescribed species of triplefins (*Enneanectes*) from the tropical eastern Pacific, under study by the author and colleagues but not appearing in the list of valid species (Patzner *et al.*, Chapter 6), as well as Atlantic insular species listed by Floeter *et al.* (2008), are included. Undescribed continental species listed by Floeter *et al.* (2008) are not included because of uncertainty about their distributions. Introduced species and recent dispersal events via the Panama Canal are also not considered. Diversity (number of species recorded) and percent endemism (proportion recorded only in a particular area) were tabulated for the eastern Pacific and western Atlantic Oceans and several included subregions (Fig. 2.1.1).

Subregions were designated for the eastern Pacific following Brusca and Wallerstein (1979) and Hastings (2000) as follows. The Pacific North Temperate subregion includes outer Baja California from Bahia Magdalena northward along the west coast of North America. The mainland portion of the Tropical Eastern Pacific extends from the

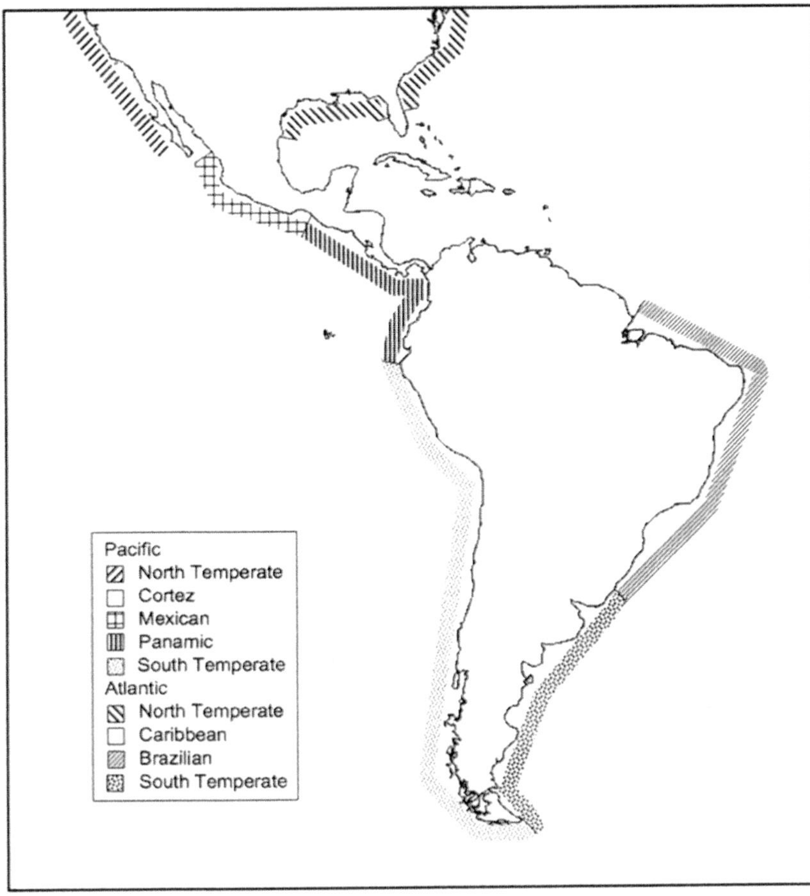

Fig. 2.1.1 Coastal subregions of the New World.

northern Gulf of California to northern Peru and includes three subregions. The Cortez subregion includes the Gulf of California southward to Topolobampo, as well as outer Baja California northward to Bahia Magdalena. The Mexican subregion includes southern Mexico from Mazatlán southward to the Gulf of Tehuantepec. The Panamic subregion extends from Guatemala southward to northern Peru. The Pacific South Temperate subregion extends from Peru southward along the coast of South America. For this general overview, the Oceanic Islands subregion includes the tropical islands (Islas Revillagigedos, Isla del Coco, Isla Malpelo, Clipperton Atoll, and Islas Galápagos) as well as the temperate islands of Isla Guadalupe off Baja California and Islas Juan Fernandez off Chile. Subregions for the western Atlantic were designated as follows. The Atlantic North Temperate subregion includes the northern Gulf of Mexico and Atlantic coastal areas of North America from central Florida northward. The Caribbean subregion includes the entire basin from the Bahamas, southern Florida and coastal Mexico southward to the northern coast of South America and northern Brazil. The Brazilian subregion includes the coast of that country south of the Amazon River. The Atlantic South Temperate subregion extends from the northern border of Argentina southward. Oceanic islands of the western Atlantic include Bermuda, Trindade, St. Paul's Rocks, Rocas Atoll and Fernando de Noronha.

RESULTS AND DISCUSSION

Regional Diversity and Endemism

Species

Three hundred twenty seven species of blennioid fishes are known from coastal waters of the New World (Table 2.1.1). This represents approximately 36% of the species of blennioid fishes. Only three species known from New World waters also naturally occur beyond this region. These are exclusively trans-Atlantic and include two blenniids and one labrisomid. The blenniid *Scartella cristata* is known from throughout the New World (North Temperate subregion to Brazil) as well as the west coast of Africa, the Mediterranean and the islands of Sao Tome and Canary (Robins and Ray, 1986; Rangel *et al.*, 2004; Floeter *et al.*, 2008). *Parablennius pilicornis* has been reported from the South Temperate

Table 2.1.1 Species diversity of New World (NW) blennioid fish families compared to global (G) species diversity. * = number includes undescribed species not listed in Chapter 6.1.

	NW	G	*Percent NW*
Blenniidae	46*	389*	12
Chaenopsidae	89*	95*	94
Clinidae	8	85	9
Dactyloscopidae	48	48	100
Labrisomidae	113*	114*	99
Tripterygiidae	23*	166*	14
Total	327*	897*	37

subregion of the New World, and in the eastern Atlantic from temperate South Africa northward to the Bay of Biscay and in the western Mediterranean (Springer, 1986; Bath, 1990). The labrisomid *Labrisomus nuchipinnis* occurs throughout the Caribbean and coastal Brazil, at several oceanic islands in the Atlantic (e.g., Sao Tome, Cape Verde, Canary, Madera) and along the west coast of Africa (Springer, 1959a; Floeter *et al.*, 2008). There is evidence that, similar to Atlantic representatives of the blenniid genus *Ophioblennius* (see Muss *et al.*, 2001), *L. nuchipinnis*, may represent a species complex that includes one or more species restricted to the New World (Springer, pers. comm.). The same may be true for the remaining trans-Atlantic blenny species and studies of their genetics are needed to resolve their status. None of the indigenous eastern Pacific blennioid species are known from outside the New World. Thus, New World blennies show an extraordinarily high degree of endemism, with 99% (324 of 327) of the species restricted to this region.

Families

The diversity of New World blennioids is not evenly spread across the six families (Table 2.1.1). The Labrisomidae contributes the greatest number of species (113), followed by the Chaenopsidae (89), Dactyloscopidae (48), Blenniidae (46), Tripterygiidae (23) and Clinidae (8).

The three most diverse New World families are endemic or nearly endemic to this region and together comprise 76% (250 of 327) of the species of New World blennioids. All dactyloscopid species are confined to the New World (Dawson, 1982). Only two species of labrisomids occur outside this region. In addition to the *Labrisomus nuchipinnis* complex discussed above, *Malacoctenus africanus* is found at Senegal in the eastern Atlantic (Springer, 1959a; Wirtz, 1990). The only chaenopsids occurring outside of the New World are six of the nine species of *Neoclinus* that are found in the northwestern Pacific (Stephens and Springer, 1971; Fukao, 1987, 1990).

The center of diversification of these largely endemic lineages is within the New World and their evolution is undoubtedly intimately tied to the breakup of the Tethys Sea (Rosenblatt, 1963). However, the timing of their origination and diversification remains unclear owing to the lack of a well-resolved phylogeny of blennioids as well as their sparse fossil record. A fossil blenniid from the early middle Eocene (50 mya; Bellwood, 1996) significantly predates the terminal Tethyan event (12 to 18 mya) that completely separated the Atlantic and Indian Oceans (Rögl and

Steininger, 1983). There is evidence that tropical fish faunas showed significant diversification between the eastern and western Tethys Sea prior to that time (Bellwood and Wainwright, 2002). It is tempting to interpret the presence of the two labrisomids in the eastern Atlantic as relatively recent dispersal events subsequent to the evolution of this group in the New World. However, the fossil species *Labrisomus pronuchipinnis* Arambourg, 1927 dates from the upper Miocene (7.2-5.3 mya) of Algeria and resembles *Labrisomus nuchipinnis*, implying that the similarities in the blennioid faunas of the eastern and western Atlantic may be considerably older (Carnevale and Bannikov, 2006). A detailed understanding of the phylogenetic relationships of the Blennioidei, currently under study (Hastings and Springer, Chapter 1.1), should provide significant insights into the events that may have led to the present-day distributions of these fishes.

The relative diversity of the remaining three families of blennioids in the New World is roughly equal when compared to their global species diversity. Although differing greatly in overall diversity, the proportion of the New World representatives of these families ranges from 9–14% of the world fauna (Table 2.1.1). Clearly, these lineages are dominant in other regions of the world, with blenniids and tripterygiids being especially diverse in the Indo-Pacific region, clinids and tripterygiids being especially diverse in temperate regions of South Africa, Australia and New Zealand, and blenniids dominating the blennioid fauna of the eastern Atlantic including the Mediterranean region (Hastings and Springer, Chapter 1.1).

Genera

The extraordinarily high degree of endemism seen in blennioid species and families, not surprisingly, is also reflected in the distributions of blennioid genera. Of the 57 total New World genera, 47 (82%) are restricted to this region (Table 2.1.2). Of the 10 genera that include species occurring outside this region, five have their center of diversity in the New World, with more species found there than in other regions of the world (Table 2.1.3). Only five genera of blennioids with representatives in the New World include more species in other areas.

The global distributions of species included in the non-endemic genera follow one of five patterns (Table 2.1.3).

(1) Two non-endemic blenniid genera are broadly distributed in all the oceans and have their greatest diversity in the tropical Indo-Pacific.

Table 2.1.2 Species diversity of blennioid genera in the eastern Pacific (P), western Atlantic (A) and combined waters of the New World (NW). *Calliclinus* (Labrisomidae) and *Helcogrammoides* (Tripterygiidae) include species found in both oceans. # = includes undescribed species (see text). * = genera not endemic to the New World (see Table 2.1.3).

	P	A	NW
Blenniidae	19	27	46
Chasmodes	0	3	3
*Entomacrodus**	1	3#	4
*Hypleurochilus**	0	7	7
Hypsoblennius	10	4	14
Lupinoblennius	0	3	3
*Ophioblennius**	2	2	4
*Parablennius**	0	2	2
Parahypsops	1	0	1
*Plagiotremus**	1	0	1
Scartichthys	4	0	4
*Scartella**	0	3#	3
Chaenopsidae	38	51	89
Acanthemblemaria	10#	11	21
Chaenopsis	5#	6	11
Cirriemblemaria	1	0	1
Coralliozetus	5	1	6
Ekemblemaria	2	1	3
Emblemaria	5	11	16
Emblemariopsis	0	14	14
Hemiemblemaria	0	1	1
Lucayablennius	0	1	1
Mccoskerichthys	1	0	1
*Neoclinus**	3	0	3
Protemblemaria	2	1	3
Stathmonotus	3	4	7
Tanyemblemaria	1	0	1
Clinidae	7	1	8
Gibbonsia	3	0	3
Heterostichus	1	0	1
Myxodes	3	0	3
Ribeiroclinus	0	1	1
Dactyloscopidae	29	19	48
Dactylagnus	2	1	3
Dactyloscopus	13	7	20
Gillellus	5	5	10

Table 2.1.2 Contd.

Table 2.1.2 Contd.

Heteristius	1	0	1
Leurochilus	0	1	1
Myxodagnus	4	1	5
Platygillellus	3	3	6
Sindoscopus	1	0	1
Storrsia	0	1	1
Labrisomidae	56	58	113
Alloclinus	1	0	1
Auchnionchus	3	0	3
Calliclinus	2	1	2
Cottoclinus	1	0	1
Cryptotrema	2	0	2
Dialommus	2	0	2
Exerpes	1	0	1
Haptoclinus	0	1	1
*Labrisomus**	10	11#	21
*Malacoctenus**	12	11#	23
Nemaclinus	0	1	1
Paraclinus	12	11	23
Starksia	9	22#	31
Xenomedea	1	0	1
Tripterygiidae	17	7	23
Axoclinus	6	0	6
Crocodilichthys	1	0	1
Enneanectes	5#	6	11
*Helcogrammoides**	2	1	2
Lepidonectes	3	0	3
All Families	166	163	327

The genus *Entomacrodus* includes at least 26 species, four of which are known from the New World. These are members of the circumtropical '*nigricans* species group' (Springer, 1967; Springer and Fricke, 2000). The circumtropical genus *Parablennius* includes 27 species, two of which are found in the New World (Springer, 1986; Bath, 1989, 1990, 1996). As the species-level relationships of these two genera have not been hypothesised, the history of their occupation of New World waters remains unclear.

(2) The blenniid genus *Plagiotremus*, is found throughout the tropical Indo-Pacific and includes 11 species, one of which (*Plagiotremus azaleus*) is restricted to the eastern Pacific. This species is well

Table 2.1.3 Diversity and distribution of species in non-endemic New World blennioid genera. NW = number of New World species; G = number of global species.

	NW	G	*Distributions outside the New World*
Blenniidae			
Entomacrodus	4	26	Eastern Atlantic & Indo-Pacific
Hypleurochilus	7	10	Eastern Atlantic
Ophioblennius	4	5	Eastern Atlantic
Parablennius	2	27	Eastern Atlantic & Indo-Pacific
Plagiotremus	1	11	Indo-Pacific
Scartella	3	7	Eastern Atlantic & Indian Ocean
Chaenopsidae			
Neoclinus	3	9	Northwest Pacific
Labrisomidae			
Labrisomus	21	21	Eastern Atlantic
Malacoctenus	23	24	Eastern Atlantic
Tripterygiidae			
Helcogrammoides	2	3	Antarctic Peninsula

nested within the hypothesised phylogeny of the genus along with a widespread Indo-Pacific species and another that is endemic to the Hawaiian Islands (Smith-Vaniz, 1976). These relationships suggest that, unique among blennioids, the ancestor of *P. azaleus* dispersed to the eastern Pacific across the vast East Pacific barrier (Leis, 1984; Robertson *et al.*, 2004).

(3) Three species of the chaenopsid genus *Neoclinus* are found in the eastern Pacific North Temperate subregion, while six species occur in the northwest Pacific (Hubbs, 1953; Fukao, 1987). Assuming the monophyly of *Neoclinus* (that has not been demonstrated based on synapomorphies), this distribution implies one or more dispersal events across the temperate waters of the north Pacific, probably during interglacial periods.

(4) The tripterygiid genus *Helcogrammoides* uniquely includes two species in the temperate regions of South America and a third in the Antarctic Peninsula (Williams and Springer, 2001). This implies dispersal of this lineage across the Antarctic Convergence.

(5) The remaining non-endemic New World blennioid genera are primarily trans-Atlantic. This includes two genera of labrisomids (*Labrisomus* and *Malacoctenus*), discussed above, and three blenniid genera. The genus *Hypleurochilus*, with seven New World and three

eastern Atlantic species (Bath and Wirtz, 1981; Bath, 1994), has its center of diversity in the New World, as does the genus *Ophioblennius* that includes four New World species and one or more in the central and eastern Atlantic (Muss *et al.*, 2001; Floeter *et al.*, 2008). The genus *Scartella* includes one widely distributed species that occurs in the western Atlantic, two western Atlantic island endemics, one species restricted to the Indian Ocean and four species restricted to the eastern Atlantic (Springer, 1986; Bath, 1996).

Western Atlantic versus Eastern Pacific

Among New World blennioids, only two species naturally occur in both the Atlantic and Pacific Oceans. Both have been reported from along the eastern and western coasts of southern South America. These are the temperate triplefin *Helcogrammoides cunninghami*, known from Peru to Argentina (Williams and Springer, 2001), and the labrisomid *Calliclinus geniguttatus*, known from Chile to Argentina (Stephens and Springer, 1974; Menni *et al.*, 1984). Ignoring recent dispersal events through the Panama Canal, no other species of blennioid occurs in both the eastern Pacific and western Atlantic.

The past connection of these areas is, however, evident in a number of hypothesised geminate species pairs of fishes on either side of the Isthmus of Panama (Jordan, 1908; Rosenblatt, 1963; Collins, 1996), especially among the Chaenopsidae (Stephens, 1963; Hastings, 2000). The prevalence of geminate species pairs among other blennioids remains unclear, in part as a consequence of the tendency among many students of New World blenny systematics to treat Pacific and Atlantic lineages in separate studies (e.g., Dawson, 1974 – 1977, 1982). However, the role of the rise of the isthmus in the evolution of New World blennioids is clearly evident in the number of genera shared between the regions (Rosenblatt, 1967; Table 2.1.2) and in the hypothesised phylogenetic relationships of several New World lineages (e.g., Rosenblatt, 1963; Hastings, 2000; Muss *et al.*, 2001).

The overall diversity of blennioid fishes in the two oceans of the New World is similar: 166 occur in the eastern Pacific and 163 occur in the western Atlantic (Table 2.1.2). Although some authors have asserted that fish species diversity is higher in the western Atlantic (e.g., Ekman, 1953), this broader analysis of blennioids substantiates the contention of Springer

(1959a), based on a subset of the labrisomid blennies, that the diversity of many lineages of fishes is comparable in these two oceans. The similarity in diversity among certain genera is striking. The chaenopsid genus *Acanthemblemaria* includes 10 Pacific and 11 Atlantic species, the labrisomid genus *Labrisomus* includes 10 and 11, the labrisomid genus *Paraclinus* includes 12 and 11, and the tripterygiid genus *Enneanectes* includes 5 and 6 (Table 2.1.2).

On the other hand, certain lineages (genera) of blennies exhibit dramatic differences in diversity between the two oceans. In some cases, this appears to be attributable to differences in the relative abundance of types of reef habitat in tropical areas of the two oceans. Reefs of the Caribbean are predominantly coralline while those of the eastern Pacific are predominantly rocky (Glynn and Ault, 2000), a difference that is reflected in the diversity of selected blennioid lineages. For example, species in the chaenopsid genus *Emblemariopsis* are coral reef specialists (Butter *et al.*, 1980; Tyler and Tyler, 1999) and this genus includes 14 Atlantic and no Pacific species (Table 2.1.2). Similarly, certain lineages that prefer rocky substrates are considerably more diverse in the eastern Pacific. For example, three of the New World genera of triplefins (*Axoclinus*, *Crocodilichthys* and *Lepidonectes*) together include 10 Pacific and no Atlantic species (Table 2.1.2). Other rocky reef specialists include the blenniid genus *Hypsoblennius*, with 10 Pacific and four Atlantic species (Bath, 2000), and the chaenopsid genus *Coralliozetus*, with five Pacific and one Atlantic species (Hastings, 1997).

Temperate Subregions

Blennioid fishes are well represented in the temperate waters of the New World. The Atlantic North Temperate subregion (northern Gulf of Mexico and Atlantic coast of North America) has a blenny fauna of 22 species, with about one third of them endemic (Table 2.1.4). Most of the endemics are restricted to the northern Gulf of Mexico (e.g., Williams, 1983; Bath, 1994). The Atlantic South Temperate subregion (coastal South America south of Brazil) includes only five species of blennies, none of which are endemic.

The blenny faunas of the temperate areas of the Pacific are comparable in overall diversity to that of the Atlantic North Temperate subregion. The Pacific North Temperate subregion includes 22 species, eight of which are endemic (36%). The endemics are three chaenopsids of

Table 2.1.4 Diversity of New World blennioid fishes by biogeographic subregions. Number of endemic species are indicated in parentheses. B = Blenniidae; Ch = Chaenopsidae; Cl = Clinidae; D = Dactyloscopidae; L = Labrisomidae; T = Tripterygiidae; % End. = percent endemic in subregions.

	B	Ch	Cl	D	L	T	Total	% End.
Atlantic								
North Temperate	14(7)	3(1)	0	3(0)	2(0)	0	22(8)	36%
Caribbean	12(4)	47(45)	0	17(11)	46(39)	5(4)	127(103)	81%
Brazil	8(1)	2(2)	1(0)	5(1)	8(5)	1(0)	25(9)	36%
South Temperate	2(0)	0	1(0)	0	1(0)	1(0)	5(0)	0
Oceanic Islands	11(4)	2(1)	1(0)	3(0)	10(6)	2(1)	29(12)	41%
Pacific								
North Temperate	5(0)	4(3)	4(2)	0	9(3)	0	22(8)	36%
Cortez	7(0)	14(7)	2(0)	11(3)	25(6)	6(2)	65(18)	28%
Mexican	4(0)	10(3)	0	15(2)	20(4)	5(1)	54(10)	19%
Panamic	12(4)	15(8)	0	16(4)	14(7)	3(1)	60(24)	40%
South Temperate	5(1)	1(1)	3(3)	1(1)	6(6)	2(1)	18(13)	72%
Oceanic Islands	7(3)	8(5)	2(0)	8(5)	12(9)	6(6)	43(28)	65%

the genus *Neoclinus* (Hubbs, 1953), two clinids (Stepien, 1992) and three labrisomids (Hubbs, 1952). The Pacific South Temperate subregion includes 18 species, 13 (72%) of which are endemic. Representatives of all blennioid families occur there, but the subregion's endemic blennies are dominated by clinids and labrisomids of the tribe Cryptotremini (Stephens and Springer, 1974).

Tropical Subregions

Western Atlantic

Not surprisingly, the large Caribbean subregion houses the greatest diversity of western Atlantic blennioids. This subregion includes 127 species, 103 (81%) of which are endemic. Many of these species are found throughout the entire Caribbean, consistent with the long-held notion that the Caribbean operates as a biogeographic unit, with relatively little subdivision for many groups (Ekman, 1953; Springer, 1959a; Rosenblatt, 1967; Briggs, 1974). Other authors have discussed evidence for subdivisions within the greater Caribbean such as differences between continental and islands fish faunas (Robins, 1971; Gilbert, 1972). This pattern is complicated by the presence of large, continent-like islands such as Cuba, together with the presence of island-like habitats, such as the Mesoamerican barrier reef system, adjacent to continental margins. Recently, Floeter *et al.* (2008) divided the region into the 'eastern Caribbean' (islands from the Bahamas southward through the Greater and Lesser Antilles) and the 'western Caribbean' (continental areas of northern South America, Central America and the entire Gulf of Mexico). These designations were not followed in this study primarily because they confound distributions of warm temperate species in the northern Gulf of Mexico with those of tropical Caribbean species. Although limited reef fish faunas exist offshore in the northern Gulf of Mexico (Smith *et al.*, 1975; Rooker *et al.*, 1997), these include relative few blennioids. In contrast, the inshore waters of the northern Gulf include several species of blennioids restricted there or found also along the northeastern coast of North America, but generally not in the Caribbean. This is consistent with the recent connection of these warm temperate areas during periods of elevated sea level (e.g., Springer, 1959b; Williams, 1983).

Further study of the distributional limits of blennioid species within the Caribbean, beyond the scope of this general overview, may reveal distinctive biogeographic areas within the greater Caribbean. For example, several narrow-range blennioid species are found in the western Caribbean, especially along the Mesoamerican barrier reef (Greenfield, 1975; Greenfield and Johnson, 1981; Tyler and Tyler, 1997; Tyler and Hastings, 2004), along the Atlantic coast of Costa Rica and Panama (Hastings and Robertson, 1999a), and along the northwestern coast of South America (Böhlke and Cervigon, 1967; Stephens, 1970; Acero, 1984a, b). Also, a faunal break occurs for some blennioids at Mona Pass, west of Puerto Rico (Hastings and Springer, 1994; Bellwood and Wainwright, 2002) but its prevalence among other fishes is unclear.

The Brazilian subregion includes 25 species of blennies, nine of which (36%) are unique to this region. This points to the significant isolation of the area from the Caribbean, an apparent consequence of the outflow of the Amazon River (Rocha, 2003). Floeter *et al.* (2008) subdivided this region into northeastern Brazil (from the mouth of the Amazon to southern Bahia) and southeastern Brazil, from Bahia southward, but this division does not appear warranted for blennioids. No blenny species are restricted to northeastern Brazil, and 20 species are common to both regions.

Eastern Pacific

The mainland coast of the Tropical Eastern Pacific, extending from the Gulf of California southward to Peru (Hastings, 2000), includes 115 species of blennioid fishes. Together with those found on the tropical oceanic islands, 140 species of blennies are known from the entire Tropical Eastern Pacific. All but 17 (or 88%) are endemic and do not occur outside of this tropical region. As with other areas, additional species are expected to come to light with further collection and study of the Tropical Eastern Pacific fish fauna (Zapata and Robertson, 2006).

The mainland Tropical Eastern Pacific has been divided into three subregions, often called provinces, based in part on the distributions of blennioids (Fig. 2.1.2). The Cortez, Mexican and Panamic Provinces are separated from one another by sections of coastline that are largely devoid of rocky reefs (Springer, 1959a; Walker, 1960; Brusca and Wallerstein, 1979; Hastings, 2000; Mora and Robertson, 2005). These are the so-called 'Central American gap' between the Panamic and Mexican Provinces and the 'Sinaloan gap' between the Mexican and Cortez Provinces (Hastings, 2000). This analysis of the distributions of all eastern

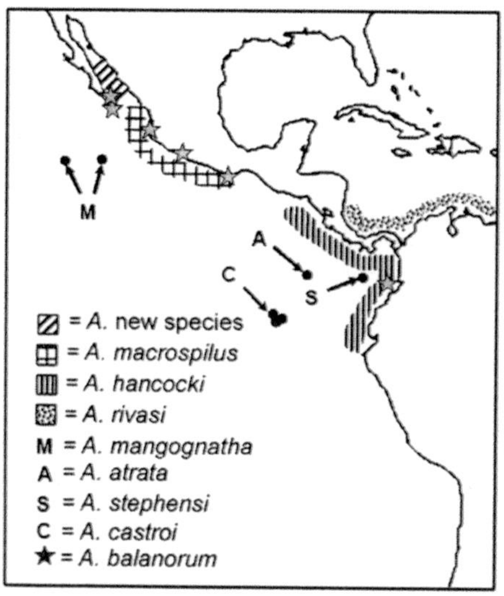

Fig. 2.1.2 Distributions of nine species of barnacle blennies of the chaenopsid genus *Acanthemblemaria*. Eight species have relatively narrow ranges, while *A. balanorum* has been recorded from the Gulf of California to Colombia (selected collecting sites indicated by stars).

Pacific blennioid fishes supports these prior designations. The three mainland subregions have similar overall levels of diversity, with 63 species known from the Cortez Province, 54 from the Mexican Province and 60 from the Panamic Province. However, these subregions differ in levels of endemism. Endemism is highest (40%) in the Panamic Province, intermediate (28%) in the Cortez Province and lowest (19%) in the Mexican Province. This is consistent with the conclusions based only on chaenopsid distributions that the intervening Mexican subregion represents a mixture of species from the Cortez and Panamic Provinces together with relatively few endemics (Hastings, 2000).

Oceanic Islands

Most of the oceanic islands of the eastern Pacific have diverse fish faunas (e.g., Grove and Lavenberg, 1997; Garrison, 2000) that together include 43 blenny species. One exception is distant Clipperton Atoll that includes only two species of blennies, the endemic blenniid *Ophioblennius clippertonensis* and *Entomacrodus chiostichus* (Robertson and Allen, 1996). The blennioid faunas of the eastern Pacific islands are often quite distinct from those of adjacent mainland faunas, but this varies with island and with the group of blennies. Overall 65% (28 of 43) of the blenny species occurring on these islands are endemic to one or more islands (Table 2.1.4). Oceanic islands in the western Atlantic generally have fewer blenny species than most islands in the eastern Pacific. The islands off Brazil include a number of endemic species (Gasparini and Floeter, 2001; Menezes *et al.*, 2001; Floeter *et al.*, 2008) while none of the 11 blennies known from Bermuda are endemic (Smith-Vaniz *et al.*, 1999).

Narrow versus Broad Range Species

A significant number of apparently narrow range blenny species occur in the New World. Some of these are known from few specimens, especially those collected from relatively deep waters. Examples include the labrisomids *Haptoclinus apectolophus*, known from two specimens from between 174 and 366 meters depth at Arrowsmith Bank off the Yucatan Peninsula (Böhlke and Robins, 1974), and *Cryptotrema seftoni*, known from a single collection of two specimens from the northern Gulf of California from between 27 and 39 meters depth (Hubbs, 1954). In addition, recent intensive collections of reef fishes in relatively shallow waters at Navassa Island in the West Indies using the ichthyocide rotenone revealed, for the first time, specimens of several new species of blennies (Williams, 2002; Collette *et al.*, 2003). However, the extent of their distributions may remain unclear until similar intensive rotenone-based fish surveys are conducted at other sites (Robertson and Smith-Vaniz, 2008).

Oceanic islands, not surprisingly, harbor a large number of blenny species that are restricted to one or a few islands (see above: Robertson, 2001; Mora and Robertson, 2005; Floeter *et al.*, 2008). In the eastern Pacific, other islands or island groups that are located closer to the mainland harbor some apparently endemic blennioids. These include the Islas Tres Marias off mainland Mexico, the only known locality for the labrisomid *Starksia lepidogaster* (Rosenblatt and Taylor, 1971) and Islas Perlas in the Bay of Panama, the only known site for two species of chaenopsids, *Tanyemblemaria alleni* and *Protemblemaria perlas* (Hastings, 1992a, 2001). Narrow range species along the tropical mainland include the chaenopsid *Mccoskerichthys sandae,* known only from Costa Rica and western Panama (Rosenblatt and Stephens, 1978), and the labrisomids *Labrisomus pomaspilus* and *Paraclinus fehlmani*, and the chaenopsid *Ekemblemaria lira,* known only from northern Ecuador (Springer and Rosenblatt, 1965; Springer and Trist, 1969; Hastings, 1992b). In addition, a few narrow range blennies (e.g., the labrisomids *Paraclinus magdalenae* and *P. walkeri*) are found only along the outer coast of Baja California (Rosenblatt and Parr, 1969). In the tropical western Atlantic, the Mesoamerican barrier reef system is notable for having a significant number of narrow range endemic blennies. For example, the chaenopsids *Acanthemblemaria paula*, *Emblemaria hyltoni*, *Emblemariopsis pricei*, and *E. dianae* (Johnson and Greenfield, 1976; Johnson and Brothers, 1989; Tyler and Hastings, 2004) are known only from this region.

A few lineages of New World blennioids appear especially prone to restricted ranges. Most notable among these are the labrisomids of the tribe Starksiini. The monotypic genus *Xenomedea* is endemic to the Gulf of California and many species of *Starksia* in both oceans have extremely small distributions (Rosenblatt and Taylor, 1971; Greenfield, 1979; Williams and Mount, 2003). This may be related to the fact that these species exhibit internal fertilisation and a reduced or absent larval stage that may lead to reduced dispersal, accelerated population differentiation, and ultimately allopatric speciation (Riginos and Victor, 2001).

On the other extreme, several New World blennioids, especially among the Blenniidae, have surprisingly broad distributions. Those transcending the Atlantic are discussed above. In some species, such as *Ophioblennius steindachneri*, found throughout the Tropical Eastern Pacific, their broad distributions may be attributable to their potential for dispersal as long-lived larvae that develop in offshore waters (Riginos and Victor 2001; Watson, Chapter 4.4). Other instances of broadly distributed species are more difficult to explain given our current knowledge of their biology and ecology. This is true for ecologically similar species such as the barnacle blennies of the genus *Acanthemblemaria* (Hastings and Robertson, 1999b). This lineage includes eight allopatric species, each with relatively small distributions, as well as *A. balanorum* that is found from the southern Gulf of California southward to at least Isla Gorgona, Colombia (Fig. 2.1.2). Whether this species and others like it with broad distributions compared to close relatives possess extraordinary dispersal abilities, slow rates of divergence, or include one or more cryptic species awaits further study.

CONCLUSIONS

Although biogeographers have frequently considered the tropical and temperate coastal waters of the New World separately, this analysis of blennioid fishes provides a slightly different perspective by analysing blennies in both temperate and tropical areas. It is clear that the barriers separating the New World from the Western Pacific and the Eastern Atlantic are highly effective for both tropical and temperate blennies. The tropical east Pacific barrier is especially formidable for blennioids (Robertson *et al.*, 2004), with evidence of only a single crossing of this barrier that led to the evolution of a New World endemic species (Smith-Vaniz, 1976). Although less effective, the open waters of the Atlantic Ocean also present a significant barrier to the dispersal of most lineages of blennies. Within the New World, the temperature barriers to the north and south of the tropical regions are also quite effective, restricting the majority of blennioid species to tropical waters. However, a few of these tropical lineages have successfully invaded temperate areas of the New World, where they occur with the predominantly temperate lineages Clinidae (Springer, 1970; Stepien, 1992) and Cryptotremini (Labrisomidae; Stephens and Springer, 1974), and selected genera from other families. The Tropical Eastern Pacific exhibits considerably more regional subdivision than does the Caribbean, but a finer scale analysis may reveal more biogeographic structure within Caribbean blennioids.

References

Acero, P.A. 1984a. A new species of *Emblemaria* (Pisces: Clinidae: Chaenopsinae) from the southwestern Caribbean with comments on two other species of the genus. *Bulletin of Marine Science* 35: 187-194.

Acero, P.A. 1984b. The chaenopsine blennies of the southwestern Caribbean (Pisces: Clinidae: Chaenopsinae). 1. Systematic analysis and zoogeography. *Anales de Instituto de Investigaciones Marinas de Punta de Betin* 14: 29-46.

Bath, H. 1989. Die Arten der Gattung *Parablennius* Ribeiro 1915 im Roten Meer, Indischen und NW des Pazifischen Ozeans (Pisces: Blenniidae). *Senckenbergiana biologica* 69: 301-343.

Bath, H. 1990. Taxonomie und Verbreitung von *Parablennius* Ribeiro 1915 an der W-Küste Afrikas und den Kapverdischen Inseln mit Revalidation von *P. verryckeni* (Poll 1959) und Beschreibung drei neuer Arten (Pisces: Blenniidae). *Senckenbergiana biologica* 70: 15-69.

Bath, H. 1994. Untersuchung der Arten *Hypleurochilus geminatus* (Wood 1825), *H. fissicornis* (Quoy und Gaimard 1824) und *H. aequipinnis* (Günther 1861), mit Revalidation von *Hypleurochilus multifilis* (Girard 1858) und Beschreibung von zwei neuen Arten. *Senckenbergiana biologica* 74: 59-85.

Bath, H. 1996. Beitrag zur Osteologie der Arten des Tribus Parablenniini. Die Beziehungen der Knochen des Schädeldaches zum Seitenorgan-System und zu den Weichteilbildungen der Kopfoberseite sowie die systematische Bedeutung der Befund nebst Bemerkungen zu *Lupinoblennius dispar* Herre 1942 (Pisces: Blenniidae). *Senckenbergiana biologica* 76: 65-92.

Bath, H. 2000. Verwandtschaftliche Beziehungen der Arten der Gattung *Hypsoblennius* Gill 1861 (Pisces, Blenniidae). *Senckenbergiana biologica* 80: 175-187.

Bath, H. and P. Wirtz. 1981. Rediscovery of *Hypleurochilus aequipinnis* (Günther 1861) in West Africa (Pisces: Blenniidae). *Senckenbergiana biologica* 61: 349-355.

Bearez, P. 1996. Lista de los peces marinos del Ecuador continental. *Revista Biologia Tropical* 44: 731-741.

Bellwood, D.R. 1996. The Eocene fishes of Monte Bolca: The earliest coral reef fish assemblage. *Coral Reefs* 15: 11-19.

Bellwood, D.R. and P.C. Wainwright. 2002. The history and biogeography of fishes on coral reefs. In: *Coral Reef Fishes*, P.F. Sale (ed.), Academic Press, San Diego, pp. 5-32.

Böhlke, J.E. and F. Cervigon. 1967. Redescription, illustration, and type selection of the Venezuelan chaenopsine blenny, *Protemblemaria punctata* Cervigon. *Notulae Naturae* 401: 1-8.

Böhlke, J. and C. Chaplin. 1994. *Fishes of the Bahamas and Adjacent Tropical Waters*. Published for the Academy of Natural Sciences of Philadelphia by Livingston.

Böhlke, J.E. and C.H. Robins. 1974. Description of a new genus and species of clinid fish from the western Caribbean, with comments on the families of the Blennioidea. *Proceedings of the Academy of Natural Sciences Philadelphia* 126: 1-8.

Briggs, J.C. 1974. *Marine Zoogeography*. McGraw-Hill, New York.

Brogan, M.W. 1994. Distribution and retention of larval fishes near reefs in the Gulf of California. *Marine Ecology Progress Series* 115: 1-13.

Brusca, R.C. and B.R. Wallerstein. 1979. Zoogeographic pattern of idoteid isopods in the northeast Pacific, with a review of shallow water zoogeography of the area. *Bulletin of the Biological Society, Washington* 3: 67-105.

Butter, M.E., M. Wapstra and E. van Dijk. 1980. *Meandrina meandrites* and *Emblemariopsis diaphana*, first record of an association between a stony coral and a fish, similar to anemone/fish relationships. *Bijdragen tot de Dierkunde* 50: 87-95.

Carnevale, G. and A.F. Bannikov. 2006. Description of a new stromateoid fish from the Miocene of St. Eugène, Algeria. *Acta Palaeontologica Polonica* 51: 489-497.

Collette, B.B., J.T. Williams, C.E. Thacker and M.L. Smith. 2003. Shore fishes of Navassa Island, West Indies: A case study on the need for rotenone sampling in reef fish biodiversity studies. *Aqua* 6: 89-131.

Collins, T. 1996. Molecular comparisons of transisthmian species pairs: rates and patterns of evolution. In: *Evolution and Environment in Tropical America*, J.B.C. Jackson, A.F. Budd and A.G. Coates (eds.). University of Chicago Press, Chicago, pp. 303-334.

Dawson, C.E. 1974. Studies on eastern Pacific sand stargazers (Pisces: Dactyloscopidae) 1. *Platygillelus* new genus, with descriptions of new species. *Copeia* 1974: 39-55.

Dawson, C.E. 1975. Studies on eastern Pacific sand stargazers (Pisces: Dactyloscopidae). 2. Genus *Dactyloscopus*, with descriptions of new species and subspecies. *Natural History Museum of Los Angeles County Science Bulletin* 22: 1-61.

Dawson, C.E. 1976. Studies on eastern Pacific sand stargazers. 3. *Dactylagnus* and *Myxodagnus*, with description of a new species and subspecies. *Copeia* 1976: 13-43.

Dawson, C.E. 1977. Studies on eastern Pacific sand stargazers (Pisces: Dactyloscopidae). 4. *Gillellus*, *Sindoscopus* new genus, and *Heteristius* with description of new species. *Proceedings of the California Academy of Sciences*, Series 4, 41: 125-160.

Dawson, C.E. 1982. Atlantic sand stargazers (Pisces: Dactyloscopidae), with description of one new genus and seven new species. *Bulletin of Marine Science* 32: 14-85.

Ekman, S. 1953. *Zoogeography of the Sea*. Sidgwick and Jackson Ltd., London.

Feitzoa, B.M., L.A. Rocha, O.J. Luiz-Júnior, S.R. Floeter and J.L. Gasparini. 2003. Reef fishes of St. Paul's Rocks: New records and notes on biology and zoogeography. *Aqua* 7: 61-82.

Floeter, S.R., L.A. Rocha, D.R. Robertson, J.C. Joyeux, W.F. Smith-Vaniz, P. Wirtz, A.J. Edwards, J.P. Barreiros, C.E.L. Ferreira, J.L. Gasparini, A. Brito, J.M. Falcón, B.W. Bowen and G. Bernardi. 2008. Atlantic reef fish biogeography and evolution. *Journal of Biogeography* 35: 22-47.

Fukao, R. 1987. Fishes of *Neoclinus bryope* species complex from Shirahama, Japan, with description of two new species. *Japanese Journal of Ichthyology* 34: 291-308.

Fukao, R. 1990. Fishes of *Neoclinus* from Okinawa with notes on the traits of their habitats. *Japanese Journal of Ichthyology* 37: 116-126.

Garrison, G. 2000. *Pesces de la Isla del Coco*. Santo Domingo de Heredia: Instituto Nacional de Biodiversidad, Costa Rica.

Gasparini, J.L. and S.R. Floeter. 2001. The shore fishes of Trindade Island, southwestern Atlantic. *Journal of Natural History* 35: 1639-1656.

Glynn, P.W. and J.S. Ault. 2000. A biogeographic analysis and review of the far eastern Pacific coral reef region. *Coral Reefs* 19: 1-23.

Gilbert, C.R. 1972. Characteristics of the Western Atlantic reef-fish fauna. *Quarterly Journal of the Florida Academy of Sciences* 35: 130-144.

Greenfield, D.W. 1975. *Emblemariopsis pricei*, a new species of chaenopsid blenny from Belize. *Copeia* 1975: 713-715.

Greenfield, D.W. 1979. A review of the western Atlantic *Starksia ocellata*-complex (Pisces: Clinidae) with the description of two new species and proposal of superspecies status. *Fieldiana: Zoology* 73: 9-48.

Greenfield, D.W. and R.K. Johnson. 1981. The blennioid fishes of Belize and Honduras, Central America with comments on their systematics, ecology and distribution (Blenniidae, Chaenopsidae, Labrisomidae, Tripterygiidae). *Fieldiana, Zoology*, New Series 8: 1-106.

Grove, J.S. and R.J. Lavenberg. 1997. *Fishes of the Galápagos Islands.* Stanford University Press, Stanford.

Hastings, P.A. 1992a. Phylogenetic relationships of *Tanyemblemaria alleni*, a new genus and species of chaenopsid (Pisces: Blennioidei) from the Gulf of Panama. *Bulletin of Marine Science* 51: 147-160.

Hastings, P.A. 1992b. *Ekemblemaria lira*, a new blennioid fish from Ecuador, with comments on sexual dimorphism and relationships in *Ekemblemaria* (Teleostei: Chaenopsidae). *Copeia* 1992: 769-776.

Hastings, P.A. 1997. Phylogenetic relationships of the *Coralliozetus* clade of chaenopsid blennies, with description of a new genus (Teleostei, Blennioidei). *Bulletin of Marine Science* 61: 743-761.

Hastings, P.A. 2000. Biogeography of the tropical eastern Pacific: Distribution and phylogeny of chaenopsid blennies. *Zoological Journal of the Linnean Society* 128: 319-335.

Hastings, P.A. 2001. *Protemblemaria perla*, a new species of tube blenny (Perciformes: Chaenopsidae) from the tropical eastern Pacific. *Revista Biologia Tropical* 49 (Supplement 1): 125-132.

Hastings, P.A. and D.R. Robertson. 1999a. Notes on a collection of chaenopsid blennies from Bahia Azul, Bocas del Toro, Caribbean, Panama. *Revue francaise Aquariologie* 26: 33-38.

Hastings, P.A. and D.R. Robertson. 1999b. *Acanthemblemaria atrata* and *Acanthemblemaria mangognatha*, new species of eastern Pacific barnacle blennies (Chaenopsidae) from Isla del Coco, Costa Rica, and Islas Revillagigedo, Mexico, and their relationships with other barnacle blennies. *Revue francaise Aquariologie* 25: 107-118.

Hastings, P.A. and V.G. Springer. 1994. A review of *Stathmonotus*, with redefinition and phylogenetic analysis of the Chaenopsidae (Pisces: Blennioidei). *Smithsonian Contributions to Zoology* 558: 1-48.

Hubbs, C. 1952. A contribution to the classification of blennioid fishes of the family Clinidae, with a partial revision of the eastern Pacific forms. *Stanford Ichthyological Bulletin* 4: 41-165.

Hubbs, C.L. 1953. Revision and systematic position of blenniid fishes of the genus *Neoclinus*. *Copeia* 1953: 11-23.

Hubbs, C.L. 1954. Additional records of clinid fishes, with the description of a new species *Cryptotrema* from the Gulf of California. *Copeia* 1954: 17-19.

Johnson, G.D. and E.B. Brothers. 1989. *Acanthemblemaria paula*, a new diminutive chaenopsid (Pisces: Blennioidei) from Belize, with comments on life history. *Proceedings of the Biological Society of Washington* 102: 1018-1030.

Johnson, R.K. and D.W. Greenfield. 1976. A new chaenopsid fish, *Emblemaria hyltoni*, from Isla Roatán, Honduras. *Fieldiana, Zoology* 70: 13-28.

Jordan, D.S. 1908. The law of geminate species. *American Naturalist* 42: 73-80.

Leis, J. 1984. Larval dispersal and the east Pacific barrier. *Oceanography Tropical* 19: 181-192.

Love, M.S., C.W. Mecklenburg, T.A. Mecklenburg and L.K. Thorsteinson. 2005. *Resource inventory of marine and estuarine fishes of the West Coast and Alaska: A checklist of North Pacific and Arctic Ocean species from Baja California to the Alaska–Yukon Border.* U.S. Department of the Interior, U.S. Geological Survey, Biological Resources Division, Seattle, Washington.

Lubbock, H.R. and A.J. Edwards. 1981. The fishes of Saint Paul's Rocks. *Journal of Fish Biology* 18: 135-157.

Menezes, G.M., P.A. Buckup, J.L.Figueiredo and R.L. Moura. 2003. *Catálogo das Espécies de Peixes Marinhos do Brasil.* Museum de Zoological da Univerisidada de São Paulo, São Paulo.

Menni, R.C., R.A. Ringuelet and R.A. Arámburu. 1984. *Peces marinos de la Argentina y Uruguay.* Editorial Hemisferio Sur, Buenos Aires.

Mora, C. and D.R. Robertson. 2005. Factors shaping the range-size frequency distribution of the endemic fish fauna of the Tropical Eastern Pacific. *Journal of Biogeography* 32: 277-286.

Moser, H.G. 2007. Reproduction in the viviparous South African clinid fish *Fucomimus mus*. *African Journal of Marine Science* 29: 423-436.

Muss, A., D.R. Robertson, C.A. Stepien, P. Wirtz and B.W. Bowen. 2001. Phylogeography of *Ophioblennius*: The role of ocean currents and geography in reef fish evolution. *Evolution* 55: 561-572.

Rangel, C.A., J.L. Gasparini and R.Z.P. Guimarães. 2004. A new species of combtooth blenny *Scartella* Jordan, 1886 (Teleostei: Blenniidae) from Trindade Island, Brazil. *Aqua* 8: 89-96.

Riginos, C. and B.C. Victor. 2001. Larval spatial distributions and other early life-history characteristics predict genetic differentiation in eastern Pacific blennioid fishes. *Proceeding of the Royal Society London* B268: 1-6.

Robertson, D.R. 2001. Population maintenance among tropical reef fishes: Inferences from small-island endemics. *Proceedings of the National Academy of Sciences of Philadelphia* 98: 5667-5670.

Robertson, D.R. and G.R. Allen. 1996. Zoogeography of the shorefish fauna of Clipperton Atoll. *Coral Reefs* 15: 121-131.

Robertson, D.R. and G.R. Allen. 2002. *Shorefishes of the Tropical Eastern Pacific: An Information System.* CD-ROM, Smithsonian Tropical Research Institute, Balboa.

Robertson, D.R., J.S. Grove and J.E. McCosker. 2004. Tropical transpacific shore fishes. *Pacific Science* 58: 507-565.

Robertson, D.R. and W.F. Smith-Vaniz. 2008. Rotenone: An essential but demonized tool for assessing marine diversity. *Bioscience* 58: 165-170.

Robins, C.R. 1971. Distributional patterns of fishes from coastal and shelf waters of the tropical western Atlantic. *FAO, Fisheries Report* 71-2: 249-255.

Robins, C.R. and G.C. Ray. 1986. *A Field Guide to Atlantic Coast Fishes of North America*. Houghton Mifflin Company, Boston.

Rocha, L.A. 2003. Patterns of distribution and processes of speciation in Brazilian reef fishes. *Journal of Biogeography* 30: 1161-1171.

Rögl, F. and F.F. Steininger. 1983. Neogene Paratethys, Mediterranean and Indo-Pacific seaways: Implications for the paleobiogeography of marine and terrestrial biotas. *Geological Journal*, Special Issue 11: 171-200.

Rooker, J.R., Q.R. Dokken, C.V. Pattengill and G.J. Holt. 1997. Fish assemblages on artificial and natural reefs in the Flower Garden Banks National Marine Sanctuary, USA. *Coral Reefs* 16: 83-92.

Rosenblatt, R.H. 1963. Some aspects of speciation in marine shore fishes. *Systematics Association Publications* 5: 171-180.

Rosenblatt, R.H. 1967. The zoogeographic relationships of the marine shore fishes of tropical America. *Studies in Tropical Oceanography* 5: 579-592.

Rosenblatt, R.H. and T.D. Parr. 1969. The Pacific species of the clinid fish genus *Paraclinus*. *Copeia* 1969: 1-20.

Rosenblatt, R.H. and J.S. Stephens, Jr. 1978. *Mccoskerichthys sandae*, a new and unusual chaenopsid blenny from the Pacific coast of Panama and Costa Rica. *Contributions to Science, Los Angeles County Museum* 293: 1-22.

Rosenblatt, R.H. and L.R. Taylor, Jr. 1971. The Pacific species of the clinid fish tribe Starksiini. *Pacific Science* 25: 436-463.

Smith, G.B., H.M. Austin, S.A. Bortone, R.W. Hastings and L.H. Ogren. 1975. Fishes of the Florida Middle Ground with comments on ecology and zoogeography. *Florida Marine Research Publications* 9: 1-14.

Smith-Vaniz, W.F. 1976. The saber-toothed blennies, tribe Nemophini (Pisces: Blenniidae). *Monographs of the Academy of Natural Sciences of Philadelphia* 19: 1-196.

Smith-Vaniz, W.F., B.B. Collette and B.E. Luckhurst. 1999. Fishes of Bermuda: history, zoogeography, annotated checklist, and identification keys. *American Society of Ichthyologists and Herpetologists*, Special Publication 4: 1-424.

Springer, V.G. 1959a. Systematics and zoogeography of the clinid fishes of the subtribe Labrisomini Hubbs. *Institute of Marine Science Publications* 5: 417-492.

Springer, V.G. 1959b. Blenniid fishes of the genus *Chasmodes*. *Texas Journal of Science* 11: 321-334.

Springer, V.G. 1967. Revision of the circumtropical shorefish genus *Entomacrodus* (Blenniidae: Salariinae). *Proceedings of the United States National Museum* 122: 1-150.

Springer, V.G. 1970. The western south Atlantic clinid fish *Ribeiroclinus eigenmanni*, with discussion of the intrarelationshps and zoogeography of the Clinidae. *Copeia* 1970: 430-436.

Springer, V.G., 1986. Blenniidae. In: *Smiths' Sea Fishes*. M.M. Smith and P.C. Heemstra (eds.). Springer-Verlag, Berlin, pp. 742-755.

Springer, V.G. and R.A. Fricke. 2000. Description of two new blenniid fish species: *Entomacrodus lemuria* from the western Indian Ocean and *E. williamsi* from the western Pacific Ocean. *Proceedings of the Biological Society of Washington* 113: 386-396.

Springer, V.G. and R.H. Rosenblatt. 1965. A new blennioid fish of the genus *Labrisomus* from Ecuador, with notes on the Caribbean species *L. filamentosus*. *Copeia* 1965: 25-27.

Springer, V.G. and R.E. Trist. 1969. A new clinid fish species, *Paraclinus fehlmanni*, from Ecuador. *Proceedings of the Biological Society of Washington* 82: 323-328.

Stephens, J.S., Jr. 1963. A revised classification of the blennioid fishes of the American family Chaenopsidae. *University of California Publications in Zoology* 68: 1-165.

Stephens, J.S. 1970. Seven new chaenopsid blennies from the western Atlantic. *Copeia* 1970: 280-309.

Stephens, J.S., Jr. and V.G. Springer. 1971. *Neoclinus nudus*, new scaleless clinid fish from Taiwan with a key to *Neoclinus*. *Proceedings of the Biological Society of Washington* 84: 65-72

Stephens, J.S. and V.G. Springer. 1974. Clinid fishes of Chile and Peru, with description of a new species, *Myxodes ornatus*, from Chile. *Smithsonian Contributions to Zoology* 159: 1-24.

Stepien, C.A. 1992. Evolution and biogeography of the Clinidae (Teleostei: Blennioidei). *Copeia* 1992: 375-392.

Tyler, D.M. and J.C. Tyler. 1997. A new species of chaenopsid fish, *Emblemariopsis ruetzleri*, from the western Caribbean off Belize (Blennioidei), with notes on its life history. *Proceedings of the Biological Society of Washington* 110: 24-38.

Tyler, J.C. and P.A. Hastings. 2004. *Emblemariopsis dianae*, a new species of chaenopsid fish from the western Caribbean off Belize (Blennioidei). *Aqua* 8: 49-60.

Tyler, J.C. and D.M. Tyler. 1999. Natural history of the sea fan blenny, *Emblemariopsis pricei* (Teleostei: Chaenopsidae), in the western Caribbean. *Smithsonian Contributions to Zoology* 601: 1-24.

Walker, B.W. 1960. The distribution and affinities of the marine fish fauna of the Gulf of California. *Systematic Zoology* 9: 123-133.

Williams, J.T. 1983. Taxonomy and ecology of the genus *Chasmodes* (Pisces: Blenniidae) with a discussion of its zoogeography. *Bulletin of the Florida State Museum, Biological Sciences* 29: 65-100.

Williams, J.T. 2002. Three new species of blennioid shore fishes discovered at Navassa Island, Caribbean Sea. *Aqua* 6: 11-16.

Williams, J.T. and J.H. Mounts. 2003. Descriptions of six new Caribbean fish species in the genus *Starksia* (Labrisomidae). *Aqua* 6: 145-164.

Williams, J.T. and V.G. Springer. 2001. Review of the South American-Antarctic triplefin fish genus *Helcogrammoides* (Perciformes:Tripterygiidae). *Revista Biologia Tropical* 49 (Supplement 1): 117-123.

Wirtz, P. 1990. Clinidae. In: *Check-list of the Fishes of the Eastern Tropical Atlantic (CLOFETA)*. J.C. Quero, J.C. Hureau, C. Karrer, A. Post and L. Saldanha (eds.) JNICT, Lisbon; SEI, Paris; and UNESCO, Paris, Vol. 2, pp. 918-919.

Zapata, F.A. and D.R. Robertson. 2006. How many species of shore fishes are there in the Tropical Eastern Pacific? *Journal of Biogeography* 34: 38-51.

CHAPTER

2.2

Blennies in Temperate Seas

Vítor C. Almada[1], Joana I. Robalo[2], Emanuel J. Gonçalves[3], André Levy[4] and Robert A. Patzner[5]

INTRODUCTION

Temperate seas are those that lie between tropical and polar waters. Seasonality is typically marked both in terms of temperature and photoperiod variations, which entrain a wide spectrum of physical and biological changes throughout the annual cycle. On the contrary, tropical seas are much less affected by these kind of seasonal variations, although seasonality is often expressed by other factors such as changes in current directions and intensity, frequency of storms, changes in salinity in coastal waters and many other parameters, so that the old idea of the tropics as devoid of seasonality is no longer accepted (Russell *et al.*, 1977; Lowe-McConnell, 1979; Munro *et al.*, 1990; Conover, 1992). In polar waters, seasonality reaches extremes of severity, with freezing temperatures and ice cover being prominent features in the winter. Temperate waters have

Authors' addresses: [1–4]Eco-Ethology Research Unit, Instituto Superior de Psicologia Aplicada, Rua Jardim do Tabaco 34, 1149-041 Lisboa, Portugal. E-mails: [1]valmada@ispa.pt; [2]jrobalo@ispa.pt; [3]emanuel@ispa.pt (*Corresponding author*); [4]andre_levy@ispa.pt

[5]Dept. Organismic Biology, University of Salzburg, Austria. E-mail: robert.patzner@ sbg.ac.at

an intermediate situation both in latitude and physical, chemical and biological conditions between the warm tropics and the extreme polar environments. These conditions allow the co-occurrence in temperate regions of tropical and subtropical fishes that manage to survive and often breed there, along with fishes that are more abundant in boreal environments.

Most blennioids are widespread in tropical and subtropical waters. Therefore, when considering the blennioids that live in temperate seas, the question arises: Are they recent migrants that invaded these areas from the tropics due to their tolerance to colder conditions? Or did they evolve for some time in temperate waters, acquiring specific adaptations to these conditions which distinguish them from their tropical counterparts?

In this chapter, we will address some questions that bear directly on this issue: what is the evolutionary history of the blennies that live and breed in temperate seas? What is the proportion of eurythermal tropical species (species of tropical origin that are able to survive and reproduce in colder conditions) when compared to that of true temperate blennies (species which evolved in temperate conditions)? Do the temperate blennioid faunas of different regions differ in their places of origin so that the different families and genera are represented in discordant proportions in different areas?

These questions are a starting point to enable researchers to address a number of new questions in the future: Are there recognisable traits of morphology, physiology and life-history that are especially useful for blennies living in temperate conditions?

We begin the chapter with a brief presentation on the faunal history of the present-day temperate seas. In the following section, we attempt to answer the questions we had raised in the preceding paragraphs, using a simple biogeographical approach and the available phylogenetic information. In the final section, we will discuss a number of questions that may be addressed in the future concerning the biological traits that are specially favoured in temperate blennies, when compared with their tropical counterparts. These questions require an extensive comparative database, adequately supported both by phylogenetic and biogeographic information, which only a global effort of data compilation will make possible.

THE PRESENT-DAY TEMPERATE SEAS

The short presentation that follows is largely based on Briggs (1995). In the ocean, like in terrestrial and freshwater habitats, diversity reaches its maximum in the tropics and decreases towards the poles (Briggs, 1974, 1995). Although many explanations have been proposed to account for this pattern (Begon *et al.*, 1990), a factor that certainly needs to be considered is the history of the different regions of the planet.

As land masses moved and climate changed, so did the locations of different ecological settings. During warm periods, polar conditions were likely absent or very reduced in area and the temperate seas occupied much smaller areas than today. For instance, before the cooling period initiated in the Neogene, the Arctic was a temperate sea, while tropical or subtropical conditions prevailed in most of the North Atlantic.

Although the tropical seas suffered very severe disturbances, well documented by the mass extinctions which are detected in the fossil record, some world regions maintained tropical conditions for very long periods and tropical seas were always present on a larger or smaller scale throughout geological time. This means that during most geological periods, organisms specialised for life in the tropical seas must have been able to find suitable habitats, at least in some places. On the contrary, cold water faunas and floras must have been much more disturbed when the latitudinal ocean temperature gradient was much less pronounced than today.

The Perciformes, the order to which the blennies belong, experienced a very important radiation in the first part of the Tertiary, when modern reefs evolved (Cantatore *et al.*, 1994; Bellwood and Wainwright, 2002). Like many other groups of marine organisms, blennies also show a peak of diversity in the tropics with a very large number of species being found on coral and other tropical reef environments (see Gonçalves and Faria, this book, Chapter 5.3). They are almost absent from polar waters, with only one species recorded in these environments (Table 2.2.1). This pattern strongly suggests that blennies are a group of tropical origin that invaded cooler waters with varied success.

During the first part of the Tertiary (65 to about 20 mya), the tropical seas formed a wide belt along the equator and were not divided in separate oceans as is the case today. The so-called Tethys Sea was a huge tropical ocean that occupied what are now the Atlantic, the Indian and the Pacific oceans, forming a broad band along the equator but extending

Table 2.2.1 Number of species in each biogeographical area per family. Legend: polar[p], tropical[tr] and temperate[t]

	Arctic Ocean[p]	*Atlantic, Northwest*[t]	*Atlantic, Northeast*[t]	*Atlantic, Western central*[tr]	*Atlantic, Eastern central*[tr]	*Mediterranean and Black Sea*[t]	*Atlantic, Southwest*[t]	*Atlantic, Southeast*[t]	*Atlantic, Antarctic*[p]	*Indian Ocean, Western*[tr]	*Indian Ocean, Eastern*[tr]	*Indian Ocean, Antarctic*[p]	*Pacific, Northwest*[t]	*Pacific, Northeast*[t]	*Pacific, Western central*[tr]	*Pacific, Eastern central*[tr]	*Pacific, Southwest*[t]	*Pacific, Southeast*[t]	*Pacific, Antarctic*[p]
Blenniidae	0	4	19	25	33	20	11	25	0	115	88	0	86	0	189	103	25	15	0
Chaenopsidae	0	0	0	46	0	0	4	0	0	0	0	0	6	1	0	36	0	12	0
Tripterygiidae	0	0	2	5	3	4	2	4	1	31	36	0	36	0	68	35	38	6	0
Labrisomidae	0	2	0	46	4	0	15	0	0	0	0	0	0	0	0	38	0	22	0
Clinidae	0	0	1	0	1	1	1	40	0	10	27	0	2	2	6	6	22	3	0
Dactyloscopidae	0	1	0	18	0	0	7	0	0	0	0	0	0	0	0	20	0	10	0

considerably to the north and south. Polar seas were totally absent during long periods and temperate conditions were restricted to high latitudes.

In the Miocene, about 20 mya, the collision of the Arabic Peninsula with Africa closed the connection between the Indian Ocean and the Mediterranean Sea, which was already an extension of the Atlantic, effectively separating the tropical regions of the two oceans. The Mediterranean suffered a dramatic salinity crisis in the Messinian (5.5 mya) with the closure of the entrance to the Atlantic when Africa and the Iberian Peninsula became connected. This event turned the Mediterranean into a closed sea, without enough freshwater input to compensate for evaporation. It became reduced to a number of fragmentary lakes and lagoons, many of them likely hypersaline so that most of its marine fauna may have gone extinct. When the straight of Gibraltar opened in the end of the Messinian (5.2 mya), the Mediterranean was mostly recolonised with fishes coming from the Atlantic, promoting an increase in its distinctiveness from the neighbour Red Sea which is connected to the Indian Ocean. The Atlantic still communicated with the Pacific through the gap that existed between North and South America, but the Pacific Ocean was already so wide that its fauna and flora from the western and eastern margins were becoming increasingly distinct. While the western Pacific was widely connected to the Indian Ocean (the two faunas still share many features), the American margin of the Pacific Ocean was becoming more and more distinct, sharing many features with the western Atlantic.

During the Pliocene, from 5.2 to 1.64 mya, a general trend of cooling of the world oceans was established. In the extreme north and south, polar conditions gradually developed with the formation of increasingly important ice caps, and temperate conditions became gradually established at mid latitudes. In the same period, the Isthmus of Panama emerged, originating a continuous land mass from North to South America, thus separating the Atlantic and Pacific tropical regions. Roughly at the same time, the Arctic and the Pacific, which were isolated, became connected through the Bering Strait, causing an influx of temperate species from the Pacific into the Atlantic, which was open to the Arctic.

In the Pleistocene, the succession of glacial cycles caused repeated regressions of the tropics and shifts of temperate conditions to areas closer to the Equator. These processes were repeated several times, causing local extinctions of temperate faunas. During this period, the importance of polar habitats as a more permanent feature of the world oceans was accentuated and an increased differentiation of the temperate faunas of the different ocean margins occurred.

To summarise, in the last 20 million years the Tethys Sea, which linked all tropical marine faunas, was divided into the oceans we know today. The middle latitudes became temperate and polar conditions became established at the highest latitudes. Although some connections prevailed between the different temperate faunas, for instance between the eastern and western Atlantic, between the two margins of the Pacific or even around the tip of South Africa, the present-day temperate seas evolved largely independently in each ocean margin.

PATTERNS OF DISTRIBUTION OF TEMPERATE BLENNIES

A Brief Methodological Note

The biogeographical analysis that follows is based on a compilation of information available on FishBase (Froese and Pauly, 2007) with some taxonomic amendments resulting from the works of Almada *et al.* (2001, 2005) and a survey of more recent taxonomic literature. The distribution of each species for all families recognised as belonging to the Blennioidei (sensu Nelson, 2006), namely the Blenniidae, Chaenopsidae, Clinidae, Dactyloscopidae, Labrisomidae and Tripterygiidae, was recorded.

The distributions of genera were derived by combining the distributions of their constituent species. The geographic areas considered were those referred to in FishBase (Froese and Pauly, 2007): Arctic Ocean, northwest Atlantic, northeast Atlantic, western central Atlantic, eastern central Atlantic, southwest Atlantic, southeast Atlantic, Antarctic Atlantic, Mediterranean and Black Sea, western Indian Ocean, eastern Indian Ocean, Antarctic Indian Ocean, northwest Pacific, northeast Pacific, western central Pacific, eastern central Pacific, southwest Pacific, southeast Pacific and Antarctic Pacific.

The patterns derived from this analysis should be viewed as preliminary for two main reasons: (1) the zoogeographical classifications adopted in FishBase do not always agree with the areas defined by biogeographic systems (e.g., Briggs, 1974), with regard to its classification in tropical, temperate or polar regions. For instance, several Blenniidae that are listed as being present in the central eastern Atlantic are not found to the south of Morocco or the north of Mauritania. Therefore, they cannot be considered as being present in the tropics. Indeed, the northern limit of the tropical eastern Atlantic is considered by Briggs (1995) to be located at Cape Verde, well to the south; (2) in addition, several recently described or redefined taxa are not yet included in FishBase. As we could not get precise distribution information for many of the recently described taxa, we decided to leave them out of the analysis, while being aware that this will imply a re-analysis of the obtained patterns, when reliable information is available for this species. It is also important to note that recent immigrants into the Mediterranean from the Suez Canal were not included in the analysis. The raw database can be requested from the authors.

We believe, however, that in spite of the limitations noted above, the analysis of the distributional data presented in the following lines reveals a number of interesting patterns that are worth considering and may serve as a starting point to a more refined study in the future.

Distributional data were compared analysing the numbers of species in tropical and temperate areas using a Man-Whitney U Test with the software Statistica 5.5 (StatSoft). Apart from the distributional information, the phylogenetic data of Almada *et al.* (2005) for the northeastern Atlantic and Mediterranean blenniids were re-analysed to answer the question: are there some clades in the phylogeny of blenniids that can be considered truly temperate? For this purpose, a group of species was considered temperate if all its members occur in temperate seas. Mapping of the ecological characters on the phylogenetic tree was performed with the software Mesquite v. 4.5.2 (Maddison and Maddison, 2005). Further details of the phylogenetic analysis may be found in Almada *et al.* (2005).

Patterns of Distribution in Bienniids

In Tables 2.2.1 to 2.2.4 we present, respectively, the number of species, genera, endemic species and endemic genera of blennioids along the geographical areas considered in this study. Taken together, the presented data allows the identification of a number of patterns described in the following paragraphs.

Blennies have a peak of diversity in tropical seas, present a considerable number of species and genera in temperate waters and—with the exception of a single species (the tripterygiid *Helcogrammoides antarcticus*; Tomo, 1981; Gon, 1990; Williams and Springer, 2001)—are absent from polar waters. Indeed, when we compare the numbers of species found in tropical regions with those found in temperate regions (Table 2.2.1), there is a significant difference between the two groups (Man-Whitney U Test, $N1 = 6$, $N2 = 9$, $Z = 2.71$, $p < 0.01$), with most tropical regions containing more blennioids than temperate ones. This is just the pattern one would expect from a group of tropical origin that has radiated considerably to temperate seas (Tables 2.2.1 and 2.2.2).

Temperate seas are, however, not inhabited exclusively by eurythermal tropical blennies. Indeed, both the numbers of endemic species and genera (Tables 2.2.3 and 2.2.4, respectively), which by definition are not shared with other geographic areas, are substantial in temperate waters of all oceans. For a total of 539 species endemic to one of the considered areas, 139 (26%) are temperate (Table 2.2.3), while the corresponding value for genera is 20% (10 genera in 49, Table 2.2.4).

The different families are not equally represented in the different ocean margins. For instance, the blenniids have a peak of diversity in the Indo-West Pacific (western Indian Ocean, eastern Indian Ocean, and western central Pacific), occur in large numbers in the temperate areas adjacent to this centre of diversity and, although in much smaller numbers, are the dominant blennioid family in the temperate waters of the north-eastern Atlantic and Mediterranean. The tripterygiids have a surprisingly high number of species in the temperate waters of the western Pacific. The chaenopsids and dactyloscopids seem to have evolved in the American tropics prior to the formation of the Isthmus of Panama, having colonised adjacent temperate waters in both temperate South American shores. Only six chaenopsid species (genus *Neoclinus*) occur in the temperate northwest Pacific, and the dactyloscopids are endemic to the Americas (Table 2.2.1). The clinids present a very interesting pattern, which suggests a comparatively ancient diversification with complex vicariant processes shaping their present-day distribution (Stepien, 1992). They are more numerous in temperate seas than in the tropics and they reach their highest values of diversity in temperate areas widely separated from each other, namely, the shores of South Africa, New Zealand and Australia. These contrasting distributional patterns strongly support the view that the blennioids of different ocean margins were undergoing distinctive evolutionary trajectories even before the closure of the Tethys Sea so that the different regions of the Tethys Sea must have evolved their own blennioid faunas. After the closure of the Tethys Sea and when the

Table 2.2.2 Number of genera in each biogeographical area per family. Legend: polar[p], tropical[tr] and temperate[t]

	Arctic Ocean[p]	*Atlantic, Northwest*[t]	*Atlantic, Northeast*[t]	*Atlantic, Western central*[tr]	*Atlantic, Eastern central*[tr]	*Mediterranean and Black Sea*[t]	*Atlantic, Southwest*[t]	*Atlantic, Southeast*[t]	*Atlantic, Antarctic*[p]	*Indian Ocean, Western*[tr]	*Indian Ocean, Eastern*[tr]	*Indian Ocean, Antarctic*[p]	*Pacific, Northwest*[t]	*Pacific, Northeast*[t]	*Pacific, Western central*[tr]	*Pacific, Eastern central*[tr]	*Pacific, Southwest*[t]	*Pacific, Southeast*[t]	*Pacific, Antarctic*[p]
Blenniidae	0	4	8	9	13	9	6	14	0	32	29	0	29	0	34	29	12	7	0
Chaenopsidae	0	0	0	10	0	0	2	0	0	0	0	0	1	1	0	11	0	6	0
Tripterygiidae	0	0	1	1	1	1	2	3	1	4	12	0	6	0	7	10	18	3	0
Labrisomidae	0	2	0	6	3	0	4	0	0	0	0	0	0	0	0	9	0	7	0
Clinidae	0	0	1	0	1	1	1	12	0	6	7	0	2	1	4	3	7	1	0
Dactyloscopidae	0	1	0	6	0	0	4	0	0	0	0	0	0	0	0	6	0	7	0

Table 2.2.3 Number of endemic species in each biogeographical area per family. Legend: polar[p], tropical[tr] and temperate[t]

	Arctic Ocean[p]	*Atlantic, Northwest*[t]	*Atlantic, Northeast*[t]	*Atlantic, Western central*[tr]	*Atlantic, Eastern central*[tr]	*Mediterranean and Black Sea*[t]	*Atlantic, Southwest*[t]	*Atlantic, Southeast*[t]	*Atlantic, Antarctic*[p]	*Indian Ocean, Western*[tr]	*Indian Ocean, Eastern*[tr]	*Indian Ocean, Antarctic*[p]	*Pacific, Northwest*[t]	*Pacific, Northeast*[t]	*Pacific, Western central*[tr]	*Pacific, Eastern central*[tr]	*Pacific, Southwest*[t]	*Pacific, Southeast*[t]	*Pacific, Antarctic*[p]
Blenniidae	0	0	0	13	12	3	3	2	0	42	13	0	8	0	69	21	4	6	0
Chaenopsidae	0	0	0	42	0	0	1	0	0	0	0	0	5	0	0	24	0	2	0
Tripterygiidae	0	0	0	4	1	2	0	2	1	15	9	0	5	0	18	9	24	4	0
Labrisomidae	0	0	0	36	1	0	6	0	0	0	0	0	0	0	0	30	0	14	0
Clinidae	0	0	0	0	0	0	1	32	0	1	10	0	0	0	1	3	5	3	0
Dactyloscopidae	0	0	0	11	0	0	2	0	0	0	0	0	0	0	0	14	0	5	0

Table 2.2.4 Number of endemic genera in each biogeographical area per family. Legend: polar[p], tropical[tr] and temperate[t]

	Arctic Ocean[p]	*Atlantic, Northwest*[t]	*Atlantic, Northeast*[t]	*Atlantic, Western central*[tr]	*Atlantic, Eastern central*[tr]	*Mediterranean and Black Sea*[t]	*Atlantic, Southwest*[t]	*Atlantic, Southeast*[t]	*Atlantic, Antarctic*[p]	*Indian Ocean, Western*[tr]	*Indian Ocean, Eastern*[tr]	*Indian Ocean, Antarctic*[p]	*Pacific, Northwest*[t]	*Pacific, Northeast*[t]	*Pacific, Western central*[tr]	*Pacific, Eastern central*[tr]	*Pacific, Southwest*[t]	*Pacific, Southeast*[t]	*Pacific, Antarctic*[p]
Blenniidae	0	0	0	1	2	0	0	0	0	4	0	0	0	0	2	0	0	0	0
Chaenopsidae	0	0	0	2	0	0	0	0	0	0	0	0	0	0	0	3	0	0	0
Tripterygiidae	0	0	0	0	0	0	0	1	0	0	2	0	0	0	0	1	9	0	0
Labrisomidae	0	0	0	1	0	0	0	0	0	0	0	0	0	0	0	4	0	2	0
Clinidae	0	0	0	0	0	0	1	7	0	0	0	0	0	0	0	1	2	1	0
Dactyloscopidae	0	0	0	1	0	0	1	0	0	0	0	0	0	0	0	0	0	1	0

marine faunas of the modern temperate regions formed, the signature of the ancient Tethyan faunas—from which they had originated—was retained. For a general overview of the biogeography of the Blennioidei see Hastings (Hastings, this book, Chapter 2.1).

The overall level of diversity of different temperate ocean margins differs markedly. For instance, when combining the Mediterranean Sea and the northeastern Atlantic, the number of blennioid species and genera inhabiting this region are surprisingly higher than those found in the north-western Atlantic (28 versus 7 species and 11 versus 7 genera, respectively). This difference may well reflect differential intensities and geographical extent of the glacial peaks in both sides of the north Atlantic (Briggs, 1970), together with the fact that the Mediterranean must have served as a refugium for many warm water species of blennioids, since during these glacial periods some warm water pockets were preserved, as suggested by Almada *et al.* (2001).

Another interesting pattern supported by clear evidence derived from the existing data indicates that the south temperate shores are richer in blennies than their northern counterparts. In the future, it will be important to assess if this difference reflects a more moderate effect of the glaciations in the southern hemisphere.

These observations reinforce the idea already expressed when discussing distributions that, in our view, it is likely that different margins of the same ocean were, at least in part, already distinct before the closure of the Tethys Sea. The possible exceptions are the western Atlantic and the eastern Pacific, which only became separated much later, and the western Pacific and eastern and western Indian oceans which are still connected in the present and, together, form the largest tropical marine region of the planet. Obviously, the transoceanic barriers were not impossible to cross and there are several examples of temperate species which occur in both sides of the Atlantic and even in different hemispheres. As mentioned above, there is also evidence that the temperate South African waters are not a complete impediment to the movement of fish between the Indian and Atlantic oceans (Floeter *et al.*, 2008). These transoceanic and interoceanic movements seem not to have been sufficiently strong to erase differences between margins and possible ancient Tethyan regionalisms.

In Figure 2.2.1, we present a phylogenetic tree of the north-eastern Atlantic and Mediterranean blenniids, showing each species'

Fig. 2.2.1 Phylogenetic Bayesian tree of the north-eastern Atlantic and Mediterranean blenniids with each species distributional information (temperate, tropical or mixed, i.e., having tropical and temperate species and species that occur both in tropical and warm temperate seas) noted on the branches and tips of the tree. *Tripterygion delaisi* was used as an outgroup and the labrisomid *Labrisomus nuchipinnis* was included to help to assess the monophyly of the ingroup.

distributional information (temperate, tropical or mixed). This dataset has the limitation of being restricted geographically and deals with only one of the six blennioid families. Nevertheless, it is interesting to note that, except for the clades that are mixed (i.e., have tropical and temperate species and species that occur both in tropical and warm temperate seas), one can identify purely temperate clades. While, for instance, *Scartella cristata* is typically a warm water species that reaches the north of the Mediterranean, the genus *Salaria*, *Aidablennius sphynx*, most species of *Microlipophrys*, *Lipophrys*, *Coryphoblennius galerita* and the clade *Parablennius ruber/Parablennius gattorugine*, are only found in temperate waters. Interestingly, genera like *Parablennius* and *Scartella* are also widely represented in the eastern tropical Atlantic, have a few members present in South Africa, the western Atlantic and in the Indo-Pacific, but did not reach the eastern Pacific. This observation supports the view that the radiation of blenniids took place mainly in the Indo-West Pacific, spreading to the adjacent areas, with only a small fraction of the family reaching the American shores.

CONCLUSIONS AND PERSPECTIVES FOR FUTURE RESEARCH

Despite the limitations of the data analysed above, there is strong evidence supporting the view that the evolution of some lineages of blenniids in temperate waters occurred for considerable amounts of time, so that these lineages cannot be viewed as mere tropical invaders of cooler seas. Another result suggested by the present work is the idea that the evolution of blennioids may bear the signature of old regional faunas of the Tethys Sea and that different ocean margins possess specific lineages that apparently independently evolved there over a considerable amount of time.

A promising area of research for the future will be to test these historical signatures and patterns with more encompassing phylogenies and more refined biogeographical frameworks. The demonstration that there are 'true' temperate blennies provides in itself a very interesting testing ground to investigate pre-adaptations and adaptations of blennies to temperate conditions. For instance, are there consistent differences in life-histories between tropical and temperate blennies? Do blenniids that evolved at middle and high latitudes for different amounts of time differ in their basic biological traits? To give only an example, *Lipophrys pholis*, which is adapted to the cooler waters of the northeastern Atlantic, being virtually absent from the Mediterranean, breeds in the late spring and summer in northern Europe, while in Portugal it breeds from late autumn to late spring (Almada *et al.*, 1990). At these more southern locations, fish grow faster and reach sexual maturity six months or earlier than in Britain (Faria *et al.*, 1996). These differences are probably caused by the fact that at higher latitudes, longer periods of reduced activity and feeding caused by the cold winters, slow both growth and maturation rates. Moreover, at the same region (Portugal), fish that evolved either in the Mediterranean like *Salaria pavo*, or in the subtropical Atlantic like *Parablennius pilicornis*, and do not reach high latitudes, breed only in the spring and summer (Almada *et al.*, 1994; Gonçalves, 1997), contrasting with what was described for *Lipophrys pholis*. Moreover, some Mediterranean blenniids and some subtropical *Parablennius* have shorter lives when compared with some blenniids that reach higher latitudes. The extent to which this pattern is widespread remains unknown.

The examples mentioned above are mere illustrations of the many interesting questions that can be addressed when the detailed phylogenetic and ecological life histories of the different blennioids are better known. The fact that multiple lineages of true temperate blennioids evolved independently provides an ideal ground to test hypotheses concerning biological traits that may have pre-adapted these fish to life in temperate seas or evolved as adaptations after they colonised temperate water. Thus, a combination of phylogenetic and ecological approaches will in the future help us to identify the traits that are true adaptations to life in temperate seas.

References

Almada, F., V.C. Almada, T. Guillemaud and P. Wirtz. 2005. Phylogenetic relationships of the north-eastern Atlantic and Mediterranean blenniids. *Biological Journal of the Linnean Society* 86: 283-295.

Almada, V.C., R.N. Barata, E.J. Gonçalves and R.F. Oliveira. 1990. On the breeding season of *Lipophrys pholis* (Pisces: Blenniidae) at Arrábida, Portugal. *Journal of the Marine Biological Association of the United Kingdom* 70: 913-916.

Almada, V.C., E.J. Gonçalves, A.J. Santos and C. Baptista. 1994. Breeding ecology and nest aggregations in a population of *Salaria pavo* (Pisces: Blenniidae) in an area where nest sites are very scarce. *Journal of Fish Biology* 45: 819-830.

Almada, V.C., R.F. Oliveira, E.J. Gonçalves, A.J. Almeida, R.S. Santos and P. Wirtz. 2001. Patterns of diversity of the north-eastern Atlantic blenniid fish fauna (Pisces: Blenniidae). *Global Ecology and Biogeography* 10: 411-422.

Begon, M., J.L. Harper and C.R. Townsend. 1990. *Ecology: Individuals, Populations and Communities*. Blackwell Scientific Publications, Cambridge, Massachusetts, 2nd Edition.

Bellwood, D.R. and P.C. Wainwright. 2002. The history and biogeography of fishes on coral reefs. In: *Coral Reef Fishes,* P.F. Sale (ed.), Academic Press, San Diego, pp. 5-32, 2nd Edition.

Briggs, J.C. 1970. A faunal history of the North Atlantic Ocean. *Systematic Zoology* 19: 19-34.

Briggs, J.C. 1974. *Marine Zoogeography*. McGraw-Hill Book Co., London.

Briggs, J.C. 1995. *Global Biogeography. Developments in Palaeontology and Stratigraphy*. Elsevier, Amsterdam.

Cantatore, P., M. Roberti, G. Pesole, A. Ludovico, F. Milella, M.N. Gadaletal and C. Saccone. 1994. Evolutionary analysis of cytochrome b sequences in some Perciformes: Evidence for a slower rate of evolution than in mammals. *Journal of Molecular Evolution* 39: 589-597.

Conover, D.O. 1992. Seasonality and the scheduling of life history at different latitudes. *Journal of Fish Biology* 41 (Supplement B): 161-178.

Faria, C., V.C. Almada and E.J. Gonçalves. 1996. Juvenile recruitment, growth and maturation of *Lipophrys pholis* (Pisces: Blenniidae), from the west coast of Portugal. *Journal of Fish Biology* 49: 727-730.

Floeter, S.R., L.A. Rocha, D.R. Robertson, J.C. Joyeux, W.F. Smith-Vaniz, P. Wirtz, A.J. Edwards, J.P. Barreiros, C.E.L. Ferreira, J.L. Gasparini, A. Brito, J.M. Falcón, B.W. Bowen and G. Bernardi. 2008. Atlantic reef fish biogeography and evolution. *Journal of Biogeography* 35: 22-47.

Froese, R. and D. Pauly. (eds.) 2007. *FishBase*. World Wide Web electronic publication. www.fishbase.org, version (10/2007).

Gon, O. 1990. Tripterygiidae. In: *Fishes of the Southern Ocean*, O. Gon and P.C. Heemstra (eds.). J.L.B. Smith Institute of Ichthyology, Grahamstown, South Africa, pp. 400-401.

Gonçalves, E.J. 1997. *Etologia, Ecologia e Biologia da Reprodução de Blenióides (Teleostei, Blennioidei)*. Ph.D. Thesis. University of Lisbon, Lisbon, Portugal.

Lowe-McConnell, R.H. 1979. Ecological aspects of seasonality in fishes of tropical waters. In: *Fish Phenology: Anabolic Adaptiveness in Teleost Fishes*, P.J. Miller (ed.). Academic Press, London, pp. 219-241.

Maddison, W.P. and D.R. Maddison. 2005. Mesquite: A modular system for evolutionary analysis. Version 1.06 <http://mesquiteproject.org>.

Munro, A.D., A.P. Scott and T.J. Lam (eds.). 1990. *Reproductive Seasonality in Teleosts: Environmental Influences*. CRC Press, Boca Raton.

Nelson, J.S. 2006. *Fishes of the World*. John Wiley & Sons, New York, 4th Edition.

Russell, B.C., G.R.V. Anderson and F.H. Talbot. 1977. Seasonality and recruitment of coral reef fishes. *Australian Journal of Marine and Freshwater Research* 28: 521-528.

Stepien, C.A. 1992. Evolution and biogeography of the Clinidae (Teleostei: Blennioidei). *Copeia* 1992: 375-392.

Tomo, A.P. 1981 Contribución al conocimiento de la fauna ictiological del sector Antárctico Argentino. *Publicacion Dirección Nacional Instituto Antárctico Argentina* 14: 1-242.

Williams, J.T. and V.G. Springer. 2001. Review of the South American-Antarctic triplefin fish genus *Helcogrammoides* (Perciformes: Tripterygiidae). *Revista Biologia Tropical* 49 (Supplement 1): 117-123.

SECTION 3

Feeding and Sensory Systems

CHAPTER

3.1

Diversity in the Diet and Feeding Habits of Blennies

Shaun K. Wilson

INTRODUCTION

Blennioid fish display a range of feeding modes, trophic niches and ecological roles that belie their small body size. Some species feed on algae and detritus, providing a vital link between primary consumers and higher trophic levels, whilst at the other end of the spectrum, some species are piscivores, feeding on smaller fish, or the scales, mucus and fins of larger fish. Blennioids are often abundant relative to other larger bodied fish and represent a significant portion of fish communities on coral reefs (Townsend and Tibbetts, 2000; Wilson, 2001a), rocky shores (Muñoz and Ojeda, 1997; Prochazka, 1998) and sandy sediments (Prochazka and Griffiths, 1992). Small body size (most species have total lengths < 100 mm) incurs high metabolic costs for many species, which are

Author's address: School of Marine Science and Technology, University of Newcastle, Newcastle upon Tyne, NE1 7RU, UK.
Present address: Marine Science Program, Dept. Environments & Conservation, 14 Dick Perry Ave, Kensington, WA 6151, Australia.
E-mail: shaun.wilson@dec.wa.gov.au

satisfied by a combination of selective and intensive feeding. These feeding strategies, combined with high abundances, can have a profound impact on the environment in which the blennies live. For example, on rocky shores, selective feeding by the herbivore, *Scartichthys viridis*, alters the composition of the algal community (Ojeda and Muñoz, 1999). Conversely, carnivorous blennies with generalist diets can reduce the abundance of herbivorous invertebrates, inducing a trophic cascade where biomass and diversity of algae increases (Bruno and O'Connor, 2005). On coral reefs, top down control of algal biomass by intensive feeding of herbivores is considered particularly important, as this maintains a benthic environment suitable for settlement and growth of corals. Blennioid feeding on algal assemblages can be intense on some reefs (Hatcher, 1981). However, many of the blennies on coral reefs ingest relatively small amounts of algae and comparatively large amounts of detritus (Wilson, 2000), implying that their role as algal-removing herbivores is negligible and emphasising the importance of correctly assessing diet before making implications about their ecological role.

Diet and feeding of the blennioid has been used to explore and substantiate several ecological ideas and theories. Inter-specific differences in the diet of sympatric species have been used to demonstrate niche partitioning of dietary resources (e.g., Nieder, 1997), and selective feeding (Ojeda and Muñoz, 1999), indicative of diet specialists. Correlating feeding intensity of the herbivorous blenny, *Parablennius sanguinolentus*, to the quality of dietary resources led to the diel feeding hypothesis (Taborsky and Limberger, 1980), which provided an explanation for the reason why feeding rates of herbivorous fish vary throughout the day. Small body size also makes blennies ideal for laboratory based experiments, such as measuring oxygen consumption and metabolism. Such experiments have been used to examine relationships between feeding rates, metabolism and temperature (Wallace, 1973) as well as the specific dynamic action that food has on metabolism (Vahl and Davenport, 1979). Studies on the jaw and tooth structure of blennioids also suggest morphology has evolved to exploit particular dietary resources (e.g., Goldschmid and Kotrschal, 1985; Lindquist and Dillaman, 1986). Combined, these studies and others have contributed to the scientific understanding of the interaction between blennioid feeding and their environment.

This chapter examines the existing diversity in the diet and feeding of blennioid, investigating explanations for spatial and temporal variation.

Generalisations regarding variation in diet between blenny taxa will be initially made using data collated from the literature, highlighting the peculiarities in the feeding traits of some species. Variation of diet within species will also be investigated, focusing on ontogenetic shifts in diet, and changes to the availability of dietary resources. At the community level, differences in the diet of blennies in two habitat types, coral and rocky reefs, will be used to assess how the environment can influence the diet of blennioid fish. Finally, selective feeding of blennies will be examined with respect to resource partitioning, diets of low quality and the diel feeding hypothesis. Thus, this chapter will provide an overview of how and why the feeding patterns and diets of blennioid fish vary.

INTER-SPECIFIC DIVERSITY IN DIET

To assess differences in the diet of different Blennioid taxa, gut content data from 23 published reports were collated. Quantitative information from reports was standardised by expressing quantities of ingested items as a percentage of the total gut contents and analysed using principal component analysis. Overall, information on 106 species, representing 5 families: Blenniidae, Chaenopsidae, Clinidae, Labrisomidae and Tripterygiidae were analysed. Due to their high taxonomic diversity, species within the family Blenniidae were further divided into one of the tribes Blenniini, Nemophini or Salariini, as described by Springer (1968). Unfortunately, no quantitative data of species from the tribe Omobranchini or the family Dactyloscopidae was found in the literature, precluding these taxa from analysis.

Diets of blennioid fish were primarily separated based on the amount of algae, detritus or sediment they ingest versus animal matter. These variables correlated strongly with the first principal axis, which explained 15% of the overall variation and effectively separated carnivorous species from herbivores and detritivores (Fig. 3.1.1). Clustering of species within some families is apparent and infers that generalisations can be made regarding the diet at this broad taxonomic level. Members of the families Chaenopsidae, Clinidae, Labrisomidae and Tripterygiidae are all carnivorous, with most species feeding on small crustaceans. In general, the clinids feed predominantly on amphipods and isopods, the labrasomids decapods and the chaenopsids copepods. Many of the tripterygids also feed on amphipods, although there were several species that feed primarily on copepods. Within the family Blenniidae, members of the tribe Blenniini

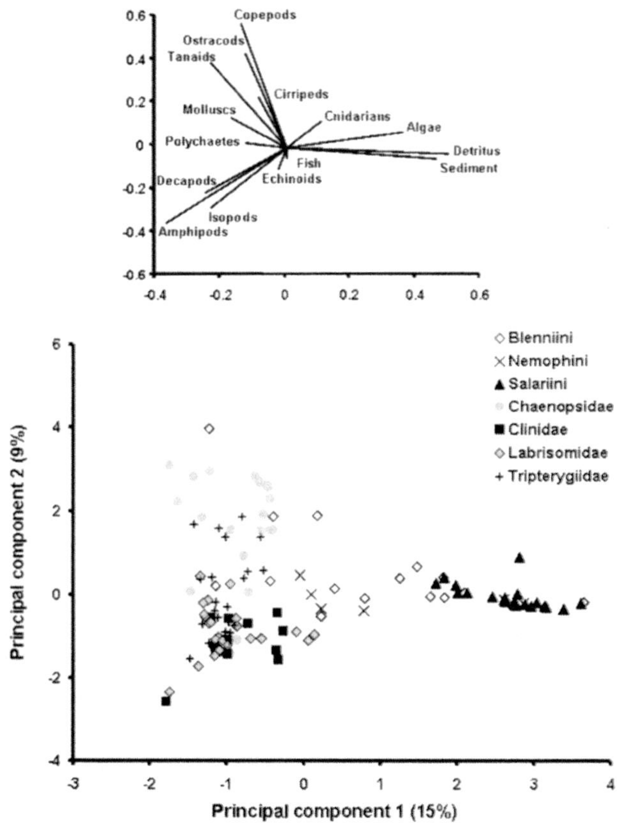

Fig. 3.1.1 Principal component analysis of Blennioid diets. Data for analysis taken from Randall (1967); Stoner and Livingstone (1980); Zander (1982); Russell (1983); Sano *et al.* (1984); Kotrschal and Thomson (1986); Lindquist and Dillaman (1986); Lindquist and Kotrschal (1987); Stepien (1990); Clarke (1992, 1999); Prochazka and Griffiths (1992); Zander and Heymer (1992); Muñoz and Ojeda (1997, 1998); Nieder (1997); Prochazka (1998); Mobley and Fleeger (1999); Wilson (2000); Silberschneider and Booth (2001); Depczensky and Bellwood (2003); Longenecker and Langston (2005).

display a diverse range of diets, some species feeding almost exclusively on algae and others targeting benthic invertebrates. Similarly, the fang blennies, tribe Nemophini, are represented by species that can be broadly categorised as carnivores, omnivores or herbivores/detritivores. In contrast, with few exceptions, the Salariini are a taxonomic group that feed on detritus and, to lesser extent algae, simultaneously ingesting large amounts of inorganic sediment.

Generalisations about the food ingested by different blennioid taxa help to identify the trophic role of these fish, however the above analysis only explained 24% of the variation in data and does not include all blennioid species. Furthermore, the means by which species forage and procure dietary items has not been considered, nor has the finer taxonomic details of gut content analyses. Thus, within each of the aforementioned taxa, there is likely to be considerable variation in diet that is not apparent when data are summarised at higher taxonomic levels.

Blenniini

Members of the tribe Blenniini are trophically diverse and can be placed anywhere within the range of herbivores to carnivores. Similarly, the manner by which these fish procure their food is variable. For example, among the herbivores, species such as *Parablennius sanguinolentus* are grazers and have numerous elongate flexible teeth (Gibson, 1968; Goldschmid and Kotrschal, 1985), whilst others such as *Hypsoblennius ionthas* have short chisel like teeth, which are used to scrape algae and detritus from surfaces (Lindquist and Dillaman, 1986). Among the omnivorous species, the nutritional utility of ingested algae is also questionable, as it appears to be undigested in the lower guts of *Parablennius incognitus* and *Lipophrys pholis,* yet partially digested in the hindgut of *Coryphoblennius galerita* (Goldschmid and Kotrschal, 1981; Milton, 1983). For some species of Blenniini, it is possible that algae are ingested incidentally whilst feeding on benthic invertebrates and do not contribute to their diet. However, *L. pholis* assimilate nitrogen and carbon from algae (Horn and Gibson, 1990), indicating ingested algae does have nutritional value in some omnivorous species.

A wide variety of benthic invertebrates are eaten by omnivorous and carnivorous Blenniini. Many species feed primarily on crustaceans, in particular decapods such as crabs and small shrimp, as well as smaller prey such as amphipods. However, polychaetes, hydrozoans and molluscs, in particular barnacles, are a major part of the diet for several marine species (Gibson, 1968, 1972; Milton, 1983). In fresh and brackish waters, *Salaria fluviatilis* and *S. pavo,* ingest large amounts of aquatic insects and pupae (Gibson, 1968; Vinoyles *et al.*, 1999), demonstrating the trophic versatility of fish within the Blenniini tribe.

Nemophini

The sabre-toothed blennies (tribe Nemophini) are highly specialised blennies, with enlarged canines for defence and well-developed swimbladders that enable semi-pelagic foraging. Species from the genus *Plagiotremus* are renowned for feeding on the scales and mucus of other fish (Randall *et al.*, 1997; Myers, 1999). They typically approach prey from behind, targeting species that are not piscivores, thereby reducing their probability of being preyed upon (Smith-Vaniz, 1976). Many *Plagiotremus* spp. also mimic non-threatening fish species (Springer and Smith-Vaniz, 1972; Moland and Jones, 2004), allowing them to casually approach their prey, before rapidly attacking them. This aggressive form of mimicry is also used by *Aspidontus taeniatus,* which is almost identical in appearance to the cleaner wrasse, *Labroides dimidiatus.* This allows *A. taeniatus* to move towards naive victims; however, unlike *Plagiotremus* spp. they tend to take bites from prey fins, rather than removing the scales (Smith-Vaniz, 1976). For both *Plagiotremus* spp. and *A. taeniatus*, gut content analyses reveal that their unusual diet is supplemented with benthic invertebrates, zooplankton and eggs (Hiatt and Strasburg, 1960; Sano *et al.*, 1984; Kotrschal and Thomson, 1986), suggesting that this mode of feeding is not always effective or is unable to satisfy all of the fish's nutritional requirements.

Other genera within the tribe feed on benthic invertebrates or plankton. Many are semi-pelagic and have large canines, which, in the case of *Meiacanthus* spp., deliver a toxic bite, deterring potential predators whilst they forage in open waters. Members of the genus *Petroscirtes* also ingest significant amounts of detritus, algae, sediment and diatoms (Sano *et al.*, 1984) and have comparatively longer guts than other Nemophini (Smith-Vaniz, 1976), implying that they consume relatively more 'low quality' food items than other species.

Salariini

The salariin blennies have a single row of close-set, elongate incisiform teeth, loosely held in connective tissue or weakly attached to the jaw bone (Smith-Vaniz and Springer, 1971). Consequently, their teeth are better suited to grazing or brushing rather than biting hard surfaces. On coral reefs, gut contents from salariin blennies are characterised by large amounts of detritus and sediment which have been brushed from epilithic algal communities (Hiatt and Strasburg, 1960; Randall, 1967; Sano *et al.*,

1984; Wilson, 2000; Depczensky and Bellwood, 2003). Nutritional evaluation of the detritus collected from blenny territories has revealed C : N (Wilson, 2000) and protein : energy (Wilson, 2002) values are comparable to those of filamentous algae and, in some areas, the protein content of detritus may exceed that of the algae (Crossman *et al.*, 2001). Lipid biomarkers extracted from detritus collected from blenny territories indicate that detritus is primarily derived from algae, although bacteria and microflora contribute to the organic content of detritus, augmenting detrital aggregates with protein and essential fatty acids (Wilson *et al.*, 2001a). Comparisons of lipid biomarkers in the body tissues of *Salarias patzneri* with detritus and algae also signify that detritus is assimilated (Wilson *et al.*, 2001b) and as it is the major organic item ingested by most salariin species on coral reefs, they should be trophically classified as detritivores rather than herbivores. In contrast to coral reef-dwelling species, it appears that sub tropical rocky reef salariin blennies primarily ingest algae and can be considered to be herbivores (Kotrschal and Thomson, 1986; Stepien, 1990; Ojeda and Muñoz, 1999). Such trophic differences among locations imply there may be variations in the quality and availability of detritus relative to algae on coral versus rocky reefs, although this remains to be examined in the field.

Although most salariin blennies can be classified as either detritivores or herbivores, there are some exceptions. For example, *Exallias brevis* resides among living coral and gut content analyses indicate that coral polyps and mucous are the main item to be ingested by this species (Hobson, 1974; Sano *et al.*, 1984). Randall *et al.* (1997) report that at least one species of *Ecsenius* also feeds on coral and *Ecsenius stictus* have been observed occasionally feeding on soft corals (pers. obs.), although it is unclear whether they are removing coral tissue and mucous or detritus that has collected on the coral surface. Another salariin blenny with an atypical diet is *Ecsenius midas*. This species has enlarged anterior canines, fewer incisors than other *Ecsenius* spp. and frequently swims 2-3 m above the benthos (Springer, 1988), where it reportedly feeds on plankton (Randall *et al.*, 1997).

Chaenopsidae

Many of the species within the family Chaenopsidae have maximum total lengths less than 5 cm and are among the smallest of all blennioid fish. Thus, they are highly susceptible to predation and tend to forage in close proximity to their refuges so they can retreat to safety, if threatened

(Kotrschal and Thomson, 1986). They generally feed on small crustaceans, in particular copepods; darting out from refuge holes to collect passing prey. However, chaenopsids also attack other invertebrates and have extremely large jaws for their body size (Kotrschal and Lindquist, 1986; Clarke, 1992, 1999), which facilitates the opportunistic ingestion of larger prey such as amphipods, shrimp, crabs, and even small fish that enter their feeding zone (Lindquist and Kotrschal, 1987). Prey can be taken from the benthos or plankton (Lindquist and Kotrschal, 1987; Clarke, 1999) and for *Acanthemblemaria* spp., sessile invertebrates are often removed with chisel-like incisors (Kotrschal and Lindquist, 1986).

Clinidae, Dactyloscopidae, Labrisomidae and Tripterygiidae

Fish from these families are carnivores; most species feed on crustaceans such as amphipods, isopods and crabs. The smaller-bodied species have a tendency to feed on smaller prey items such as copepods or amphipods whilst larger fish tend to have greater amounts of decapods in their diet. Several species will also feed on other fish, including the labrisomid *Auchenionchus microcirrhis* (Muñoz and Ojeda, 1997) and clinid, *Clinus nematopterus* (Prochazka, 1998). Most species forage sub tidally on rocky shores. However, the labrisomids *Dialommus fuscus* and *D. macrocephalus* make short yet frequent forays onto the wet rocks above the water line to forage for crabs and other prey. Unlike many of the other amphibious blennioids, these fish have split corneas, a flattened section allowing acute vision to detect their prey in the air (Graham, 1970; Nieder, 2001).

Prochazka and Griffiths (1992) described a community of sand-dwelling clinids that partially bury themselves within the sediments. Many of these species ingested sediment, implying prey are also taken from sandy habitats although the feeding mechanism of these fish is not described. The sand stargazers, family Dactyloscopidae, also bury themselves in the sand, where they wait for unsuspecting prey such as invertebrates and small fish. Morphologically, they are adapted to ambushing prey from within sediments as they have an oblique mouth and dorsally located protrusive eyes, which enables visual detection of prey when buried.

VARIATION OF DIET WITHIN SPECIES

Spatial and temporal variation in the availability of dietary resources and competition between species can result in changes to the diet of some blennioid species (Gibson, 1972; Stoner and Livingston, 1980; Milton, 1983; Zander and Heymer, 1992). Variation is expected to be greatest in feeding generalists, as these species readily exploit a diverse array of dietary resources. However, species with a narrow diet breadth can also exhibit variation in diet. In a detailed analysis of *Paraclinus fasciatus* ecology, Stoner and Livingston (1980) demonstrated that although this carnivorous species only preys on a few species of amphipod and shrimp, its diet varied according to seasonal and spatial changes in prey species abundance. Seasonal increases in the abundance of prey may also result in species within the same community, temporally targeting the same dietary resource; even when diets are normally partitioned between species (Nieder, 1997). Plasticity in diet, therefore, permits species to exploit resources as they become available and is an important facet in the subsistence of both feeding specialists and generalists.

Ontogenetic Shifts in Diet

For many species, the most dramatic change in diet is associated with the shift from the planktonic larval environment to the benthic post-settlement life history phase. This shift represents a change in dietary resources and is often associated with appropriate changes to the morphology of feeding apparatus. For example, larvae of the redlip blenny, *Ophioblennius atlanticus* have terminal mouths and fang-like teeth, suitable for planktonic feeding. However, once settled, comb teeth erupt and the mouth becomes more ventrally located, so that it is better suited to benthic feeding (Labelle and Nursall, 1985). Interestingly, during this metamorphosis, there is no feeding, and fish rely on lipids stored in the liver till they start grazing on algae and detritus (Labelle and Nursall, 1985).

Ontogenetic shifts in diet of blennies are also common during the transition from juveniles to adults. In 11 studies, where both juvenile and adult diets were recorded, ontogenetic variation in diet was recognised in 93% (13/14) of species, with only one species showing no definitive change in diet (Table 3.1.1). Explanations for these ontogenetic dietary shifts vary among species. In carnivorous blennies, increased gape of adults allows access to larger prey, which becomes a prominent component of their diet (Zander and Heymer, 1992; Muñoz and Ojeda, 1998). Similarly, development of jaw strength may allow the adults to prise sessile organisms from the benthos that juveniles are not capable of removing (Milton,

Table 3.1.1 Ontogenetic shifts in juvenile to adult diet of blennioid

Species	*Diet shift*	
	Increase	*Decrease*
Auchenionchus microcirrhis[1]	Decapods	Amphipods
Auchenionchus variolosus[1]	Decapods	Amphipods
Calliclinus geniguttatus[1]	Decapods	Amphipods
Coryphoblennius galerita[2]	Algae	Copepods, ostracods
Enneapterygius rufopileus[3]	Gastropods	
Lipophrys pholis[4]	Crabs, polychaetes, barnacles	Copepods
Lipophrys pholis[2]	Algae, limpets, barnacles	Copepods, small molluscs
Lipophrys pholis[5]	Algae, barnacles, mussels	Isopods, copepods
Myxodes viridis[1]	Amphipods	Copepods
Parablennius sanguinolentus[6]	Green algae, *Ulva*	Turf algae
Paraclinus fasciatus[7]	Crabs, shrimp, isopods	Amphipods
Scartella cristata[8]	Algae	Invertebrates
Scartichthys viridis[9]	Red algae, *Gelidium*	Green algae *Ulva*, Enteromorpha
Tripterygion chilensis[1]	Amphipods	Copepods
Tripterygion delaisi[10]	Amphipods	Copepods
Salarias patzneri[11]	None	None

[1]Muñoz and Ojeda (1998); [2]Milton (1983); [3]Silberschneider and Booth (2001); [4]Gibson (1972); [5]Monteiro *et al.* (2005); [6]Horn *et al.* (1990); [7]Stoner and Livingston (1980); [8]Mobley and Fleeger (1999); [9]Muñoz and Ojeda (2000); [10]Zander and Heymer (1992); [11]Wilson (2004).

1983; Muñoz and Ojeda, 2000), increasing the diversity of adult diets. Shifts in diet can also be associated with ontogenetic shifts in habitat, or foraging range, which alters the type of dietary resources available (Monteiro *et al.*, 2005).

At least two species, *Lipophrys pholis* and *Coryphoblennius galerita*, exhibit an ontogenetic trend of increased algal consumption by adults (Milton, 1983; Monteiro *et al.*, 2005). Shifts to herbivory by fish are often associated with morphological changes that favour procurement and digestion of algae (Bellwood, 1988; Chen, 2002). Although not strictly herbivores, Monteiro *et al.* (2005) suggested that at a critical body size scraping substratum for sessile organisms, including algae, by *L. pholis* becomes more energetically profitable than searching for motile invertebrates. Increased gut volume of adults (Muñoz and Ojeda, 2000) may also allow greater volumes of algae to be processed, making it a viable dietary resource once the fish attain a certain body size.

Although, ontogenetic diet shifts appear to be the norm in blennioids, at least one species, *Salarias patzneri*, maintains the same diet throughout its post-recruitment life (Wilson, 2004). It appears that *S. patzneri* are morphologically equipped to feed on detritus from an early age and that detritus is of a high enough nutritional quality that it sustains growth and development of fish from 1-2 cm standard length upwards.

Opportunistic Feeding

Short-term changes to diets of mature fish can be instigated when blennioids are attracted to food resources of high nutritional value that are not normally available within their habitat. Kotrschal *et al.* (1991) demonstrated opportunistic feeding of crushed urchins by herbivorous, omnivorous and carnivorous blennioids, indicating that blennies will feed on rich dietary resources, irrespective of their trophic level. However, not all blennies were attracted to the urchin tissue, competition with larger-bodied fish, or predator avoidance, reducing the likelihood that smaller species partake in opportunistic feeding behaviour. In general, females were more likely to partake in opportunistic feeding, as temporal and spatial foraging by mature males is often constrained by their need to guard eggs.

Reduced foraging of males and increased energy requirements of egg-laying females may explain the subtle differences in the diet between sexes (Goldschmid and Kotrschal, 1981; Mobley and Fleeger, 1999). A decrease in body condition during nesting is also a likely explanation for filial cannibalism by males. Males from several species have been recorded with a comparatively greater amount of eggs in their guts than females (Goldschmid and Kotrschal, 1981; Gonçalves and Almada, 1997; Vinoyles *et al.*, 1999), and it is hypothesised that eggs are taken from their own nests to supplement diet, whilst motility is restricted due to guarding of nests. Alternatively, ingested eggs may come from attacks on neighbouring nests, or represent removal of non-viable eggs from their own nests.

DIETARY TRENDS WITHIN HABITATS: ROCKY SHORES VERSUS CORAL REEFS

Scientific research on blennioid fish has been primarily conducted on rocky shores and coral reefs where blennies are a prominent component of fish assemblages. These two habitats offer the same basic dietary resources; however, the trophic composition of blennioid communities in each habitat differs. On rocky shores, most species are micro-carnivores or omnivores (Gibson, 1968; Stepien, 1990; Muñoz and Ojeda, 1997), whilst on coral reefs detritivory and herbivory are more common (Wilson, 2000; Harmelin-Vivien, 2002). The trend is most obvious when looking at the trophic distribution of species from the family Blenniidae (Fig. 3.1.2), as many of the other blennioids are carnivores, irrespective of habitat. However, even within this family, it is apparent that results are confounded by differences in community phylogeny within each habitat. Rocky shore communities are dominated by the Blenniini, whilst on coral reefs detritivorous Salariin blennies are more common (Fig. 3.1.2). Thus, the fish may have evolved to best exploit resources in their respective habitats.

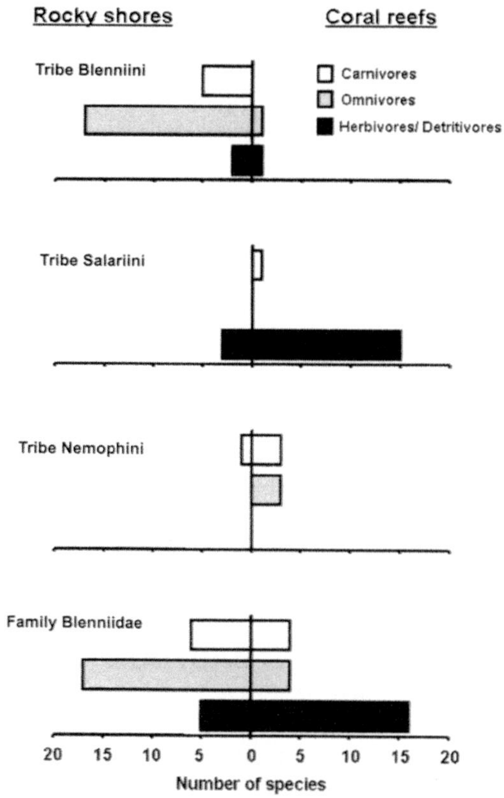

Fig. 3.1.2 Trophic composition of blenny communities on rocky shores and coral reefs. Fish assigned trophic groups based on gut content data from references listed in Figure 3.1.1.

An increase in the use of 'low quality' food (e.g., algae and detritus) by fish on tropical coral reefs compared to temperate rocky reefs is evident among other taxa (Harmelin-Vivien, 2002) and the density of herbivorous fish is greater on coral reefs on the Great Barrier Reef and Caribbean, than it is on rocky reefs in New Zealand (Meekan and Choat, 1997). The number of species and abundance of fish that feed on low quality food also increases at lower latitudes and higher sea surface temperatures, suggesting that water temperature may influence the diversity and abundance of herbivorous and detritivorous fish (Floeter *et al.*, 2004, 2005).

Harmelin-Vivien (2002) proposed that increased temperatures in tropical waters combined with environmental stability leads to increased diversity and competition within coral reef fish communities, resulting in an evolutionary movement towards the use of low quality food. The two major limitations to feeding on a low quality diet are obtaining sufficient protein and energy (Bowen *et al.*, 1995). Higher water temperatures on coral reefs encourage productivity of turf-forming algae, the dietary resource that most herbivorous fish feed upon in this habitat (Choat, 1991). Turf algae is also the primary source of detritus on coral reefs (Wilson *et al.*, 2001a) and increased water temperature will increase the productivity of bacteria and micro-flora, that enrich detritus with protein (Wilson *et al.*, 2003). Consequently, detrital feeding fish on coral reefs ingest levels of amino acids which are comparable to levels ingested by carnivorous species (Crossman *et al.*, 2005).

Detritus is plentiful within the territories of coral reef blennies and forms the basis of most species diet (Wilson, 2000, 2002). The high protein content of detritus in blenny territories (Wilson, 2000, 2002) enables rapid growth of coral reef species (Russ and St John, 1988; Wilson, 2004) underlining the dietary value of detritus. The salariin blennies have morphological traits that favour the procurement and digestion of detritus. Elongate flexible teeth facilitate brushing of filamentous algae and selective removal of nutritionally rich detritus. Salariin blennies also have extremely long guts, a comparison among 65 blennioid species, indicating the Salariini have relative gut lengths 2 to 6 times greater than other blennioid groups (Fig. 3.1.3). Relative gut length is indicative of diet (Goldschmid *et al.*, 1984; Fig. 3.1.3), longer guts allowing large volumes of detritus and algae to be held and digested. In addition, feeding intensity and consumption rates of blennies typically increases with temperature

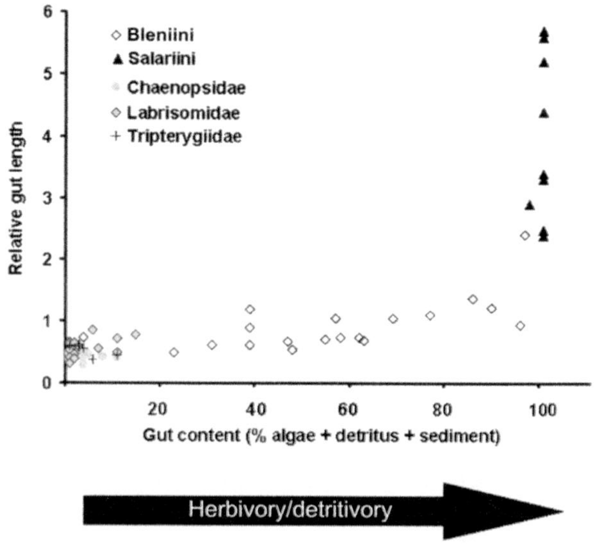

Fig. 3.1.3 Relative gut length (gut length divided by total length of fish) of blennioid taxa. Species arranged with respect to the amount of detritus algae and sediment in their diet. Data collated from: Goldschmid and Kotrschal (1981); Goldschmid *et al.* (1984); Kotrschal and Lindquist (1986); Kotrschal and Thomson (1986); Nieder (1997); Muñoz and Ojeda (2000); Depczensky and Bellwood (2003); Wilson, unpublished data.

(Horn and Gibson, 1990; Wilson, 2001b) and increased consumption is associated with faster gastric evacuation rates (Grove and Crawford, 1980; Horn and Gibson, 1990). However, assimilation efficiency is not necessarily affected, nitrogen assimilation in the herbivorous blenny, *Parablennius sanguinolentus*, remaining stable, even when temperature increases and gut transit time is reduced (Horn and Gibson, 1980). Consequently, large gut volumes, high consumption rates and fast gut transit times enable fish to process enough material to satisfy high energetic requirements associated with both living in warm waters and feeding on low quality resources. Warm water also increases the nutritional quality of detritus on coral reefs and has undoubtedly contributed to a greater prevalence of detritivorous blennies on coral reefs.

FEEDING SELECTIVITY

Many fish feed within a trophic level, targeting prey of certain dimensions or selectively feeding upon more nutritious prey. This optimises the energetic return associated with foraging for specific prey. However, when dietary resources are limited, selective feeding by fish can be the basis of resource partitioning. In cryptic fish communities, where blennies are specious and prevalent, the partitioning of dietary resources has been used to explain the co-existence of sympatric species (Milton, 1983; Nieder, 1997; Prochazka, 1998). Indeed, in describing the trophic composition of blennioids in the Gulf of California, Kotrschal and Thomson (1986) suggested that diet specialisation was an evolutionary response to competition, whereby primitive species were micro-carnivore generalists and more recently evolved species were specialised for feeding on sessile items. Furthermore, partitioning of dietary resources within the blennioid communities may occur among ontogenetic feeding guilds, Muñoz and Ojeda (1998) demonstrated that ontogenetic diet shifts result in reduced overlap of diet among carnivorous blenny species.

In assessing diet niches and resource partitioning of blennioids, studies often examine diet breadth, describing species with a low diversity of items in guts as specialists. A true specialist will, however, target items so they represent a disproportionately larger amount of their diet than is present in the environment. Most blennies forage over small areas (Gonçalves and Almada, 1998) and species tend to be found in distinct zones, depths or microhabitats (Illich and Kotrschal, 1990; Patzner, 1999; Wilson, 2001). Availability and composition of dietary resources are likely to differ among these habitats, which can influence blenny diet. For example, Clarke (1992, 1999) noted that the tube blenny *Acanthemblemaria spinosa* fed mainly on planktonic calanoid copepods, whilst another species, *A. aspera*, fed on benthic harpacticoid copepods. The diet of these blennies relates partially to the depth distribution of the two species, *A. aspera* using shelter holes closer to the benthos, where they are more likely to encounter harpacticoid copepods, whilst *A. spinosa* has better access to planktonic copepods. Partitioning of dietary resources may, therefore, relate to subtle differences in competition and habitat use, which may be important for maintaining diversity in some communities.

Selective feeding may also be utilised by herbivores and detritivores, as nitrogen is often presumed to be the diet-limiting resource of these fish (Horn, 1989; Choat, 1991). Fish are, therefore, expected to target certain algae or detritus forms with high protein content. For example, a comparison of particles within the feeding territory and guts of the detritivorous blenny, *Salarias patzneri*, found that these fish selectively feed on particles $< 125\ \mu m$ (Wilson, 2000). These small particles typically have greater nitrogen content (Wilson and Bellwood, 1997) and, as they are amorphous, are easier to digest than larger morphic detritus, which contain refractory material (Bowen, 1984; D'Avanzo *et al.*, 1991). Detrital aggregates <125 μm also have a higher organic content than larger particles (Wilson and Bellwood, 1997; Wilson, 2000) and by selectively feeding on organic rich particles, small-bodied blennies maximise energy intake. Similarly, herbivorous *Scartichthys viridis* selectively feed on the green algae *Ulva* (Ojeda and Muñoz, 1999), although it is hypothesised that selectivity is related to ease of removal from rocky reefs, rather than nutritional quality (Muñoz and Ojeda, 2000). Conversely, some fish avoid certain types of algae, because chemical and morphological characteristics of the algae act as deterrents to the herbivory (Hay, 1991).

Selectivity can also relate to the area in which the fish choose to forage. Detritivorous coral reef blennies feed in areas where the nutritional quality of detritus is high. On a broad spatial scale, abundance and biomass of salariin blennies is often greatest on exposed reef crests (Odum and Odum, 1955; Townsend and Tibbetts, 2000; Wilson, 2001), the reef zone where organic content of sediments is the greatest (Purcell and Bellwood, 2001). However, distribution of blennioids at this spatial scale is likely to be influenced by ecological parameters other than food quality. To assess if detritivorous blennies selectively feed in certain areas to access food of greater nutritional value, dietary resources within their territories were manipulated and feeding observations carried out to determine how variations in the spatial availibity of detritus influences feeding.

Dietary resources in the territories of *Salarias patzneri* were experimentally manipulated in two ways. In one experiment sediment and associated detritus was brushed from approximately 500 cm^2 of fish territories, leaving only filamentous algae. In the second experiment, 500 cm^2 of territory was shaded with an opaque plastic container, excluding sunlight and feeding by blennies for 24 hours. Holes in containers permitted water exchange with the environment; however, reduced light under opaque containers will theoretically decrease the productivity of microalgae and bacteria in detritus, reducing the nutritional quality of detritus. After both manipulations were completed and containers removed, fish were left to settle for 10 minutes, before three 5-minute feeding observations were carried out. Feeding observations were separated by at least 10 minutes and recorded the number of bites taken from the treated area (sediment removed or sunlight excluded) and a control area (sediment dabbed but not removed or a transparent container in place for 24 hours) of the same dimensions. Each experiment was carried out on four fish and repeated on each fish on 3 separate days.

In each experiment, the fish fed more frequently in control areas (Fig. 3.1.4; Table 3.1.2), demonstrating a preference for feeding on turf assemblages where detritus is present and exposed to sunlight. The preference for feeding in areas where detritus is a component of the epilithic algal matrix indicates that blennies discriminate between algal assemblages of low and high detrital content, whilst a preference for sun-illuminated sediments implies a disposition towards detritus enriched with autotrophic microbes. This clearly demonstrates that feeding is not random within the territories, with blennies targeting areas where food is more abundant and presumably more nutritious.

Table 3.1.2 Summary of statistical analysis of feeding experiments conducted within the territories of *Salarias patzneri*. Differences in the number of feeding bites on treatment and control sites were compared using repeated measures ANOVA.

		Detritus removal		*Shading*	
	df	F	*p*	F	*p*
Treatment (intercept)	1	8.80	0.018	19.20	0.002
Fish	3	3.95	0.053	1.10	0.403
Day	2	3.77	0.046	0.14	0.867
Fish * Day	6	3.89	0.014	0.69	0.663

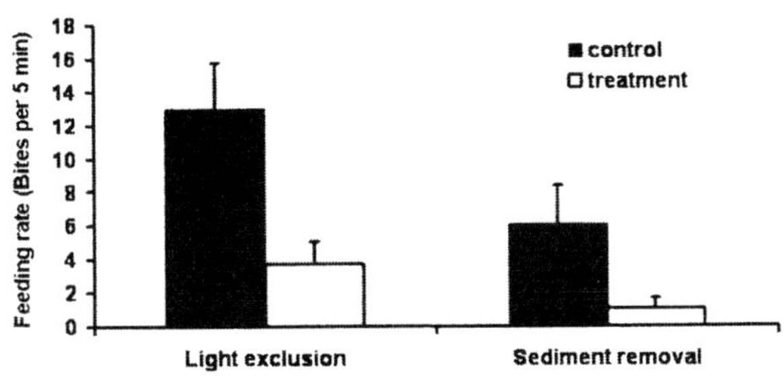

Fig. 3.1.4 Experimental manipulations of food resources within the territory of *Salarias patzneri*. Standard errors calculated from feeding observations of four fish on three days. Experiments were carried out in the Lizard Island lagoon, September 2000.

Diel Feeding Hypothesis

By selectively feeding on resources when energy levels are greatest, fish optimise their energy intake. This is particularly important for small-bodied blennies as they have a higher metabolic rate than the larger-bodied fish (Clarke and Johnston, 1999) and need to ingest relatively more energy per unit of body mass to meet basic metabolic requirements.

Temporal variation in the feeding intensity of blennies—whereby the time allocated to feeding increases steadily throughout the day, peaking in the mid afternoon—was first observed for *Ophioblennius atlanticus* (Nursall, 1981). Subsequent studies on other herbivorous and detritivorous blenny species concur with this pattern (Table 3.1.3) and feeding rates of some of these species have been directly related to diel changes in diet quality. For example, biochemical analyses of algae in the diet of the herbivorous blenny *Parablennius sanguinolentus* indicate that photosynthetic products accumulate throughout the day, peaking mid afternoon when feeding rates of blennies are the greatest (Horn *et al.*, 1990; Zoufal and Taborsky, 1991). Algae collected from coral reefs show similar patterns, starch and fluridoside increasing, until photoinhibition occurs mid afternoon (Zemke-White *et al.*, 2002). However, most blennies on coral reefs, including those listed in Table 3.1.3, are detritivores and do not ingest large amounts of turf or macroalgae. For these species, it is possible that increased productivity by detrital microbes in the afternoon gradually enhances the nutritional quality of detritus (Wilson *et al.*, 2003), although this theory is yet to be examined.

Table 3.1.3 Maximum feeding intensity of blennies on coral and rocky reefs

Species	*Bites* min^{-1} *(max)*	*Time of max feeding*	*Habitat*
Ophioblennius atlanticus[1]		1630	Coral reef
Salarias fasciatus[2]	8–12	1400–1600	Coral reef
Salarias fasciatus[3]	6–8	1300, 1400–1600	Coral reef
Salarias patzneri[4]	8–12	1200–1500	Coral reef
Parablennius sanguinolentus[5]		1500	Rocky reef
Parablennius sanguinolentus[6]	14–15	1400–2000	Rocky reef
Parablennius sanguinolentus[7]	10–20	1200–1530	Rocky reef

[1]Nursall (1981); [2]Klumpp and Polunin (1990); [3]Townsend and Tibbetts (2004); [4]Wilson *et al.* (2003); [5]Taborsky and Limberger (1980); [6]Horn *et al.* (1990); [7]Zoufal and Taborsky (1991).

Acknowledgements

I am grateful to R. Fisher, who helped with the field experiments carried out at Lizard Island and, along with N.A.J. Graham, provided general comments that greatly improved the chapter. Field work carried out at Lizard Island was supported by a Doctoral merit grant from James Cook University, Australia.

References

Bellwood, D.R. 1988. Ontogenetic changes in the diet of early post-settlement *Scarus* species (Pisces: Scaridae). *Journal of Fish Biology* 33: 213-219.

Bowen, S.H. 1984. Evidence of a detritus food chain based on consumption of organic precipitates. *Bulletin of Marine Science* 35: 440-448.

Bowen, S.H., E.V. Lutz and M.O. Ahlgren. 1995. Dietary protein as determinants of food quality: Trophic strategies compared. *Ecology* 76: 899-907.

Bruno, J.F. and M.I. O'Connor. 2005. Cascading effects of predator diversity and omnivory in a marine food web. *Ecology Letters* 8: 1048-1056.

Chen, L.S. 2002. Post-settlement diet shift of *Chlorurus sordidus* and *Scarus schlegeli* (Pisces: Scaridae). *Zoological Studies* 41: 47-58.

Choat, J.H. 1991. The biology of herbivorous fishes on coral reefs. In: *The Ecology of Fishes on Coral Reefs*, P.F. Sale (ed.). Academic Press, San Diego, pp. 96-119.

Clarke, A. and N.M. Johnston. 1999. Scaling of metabolic rate with body mass and temperature in teleost fish. *Journal of Animal Ecology* 68: 893-905.

Clarke, R.D. 1992. Effects of microhabitat and metabolic-rate on food-intake, growth and fecundity of 2 competing coral-reef fishes. *Coral Reefs* 11: 199-205.

Clarke, R.D. 1999. Diets and metabolic rates of four Caribbean tube blennies, genus *Acanthemblemaria* (Teleostei: Chaenopsidae). *Bulletin of Marine Science* 65: 185-199.

Crossman, D.J., J.H Choat, K.D. Clements, T. Hardy and J. McConochie. 2001. Detritus as food for grazing fishes on coral reefs. *Limnology and Oceanography* 46: 1596-1605.

Crossman, D.J., J.H. Choat and K.D. Clements. 2005. Nutritional ecology of nominally herbivorous fishes on coral reefs. *Marine Ecology Progress Series* 296: 129-142.

D'Avanzo, C.D., M. Alber and I. Valiela. 1991. Nitrogen assimilation from amorphous detritus by two coastal consumers. *Estuarine, Coastal and Shelf Science* 33: 203-209.

Depczynski, M. and D.R. Bellwood. 2003. The role of cryptobenthic reef fishes in coral reef trophodynamics. *Marine Ecology Progress Series* 256: 183-191.

Floeter, S.R., C.E.L. Ferreira, A. Dominici-Arosemena and I. Zalmon. 2004. Latitudinal gradients in Atlantic reef fish communities: Trophic structure and spatial use patterns. *Journal of Fish Biology* 64: 1680-1699.

Floeter, S.R., M.D. Behrens, C.E.L. Ferreira, M.J. Paddack and M.H. Horn. 2005. Geographical gradients of marine herbivorous fishes: Patterns and processes. *Marine Biology* 147: 1435-1447.

Gibson, R.N. 1968. The food and feeding relationships of littoral fish in the Banylus region. *Vie et Milieu* 19: 447-456.

Gibson, R.N. 1972. The vertical distribution and feeding relationship of intertidal fish on the Atlantic coast of France. *Journal of Animal Ecology* 41: 189-207.

Goldschmid, A. and K. Kotrschal. 1981. Feeding ecology of three populations of *Blennius incognitus* (Pisces: Teleostei: Blenniidae) during the reproductive period and under human influence. *P.S.Z.N.I. Marine Ecology* 2: 1-14.

Goldschmid, A. and K. Kotrschal. 1985. Morphological and functional adaptions in different feeding types of blennies (Perciformes; Teleostei). In: *Vertebrate Morphology*, Duncker and Fleischer (eds.). Fortschritte der Zoologie, Gustav Fischer Verlag, Stuttgart, pp. 241-244.

Goldschmid, A., K. Kotrschal and P. Wirtz. 1984. Food and gut length of 14 Adriatic blenniid fish (Blenniidae; Percomorpha; Teleostei). *Zoologischer Anzeiger* 213: 145-150.

Gonçalves, E.J. and V.C. Almada. 1997. Sex differences in resource utilization by the peacock blenny. *Journal of Fish Biology* 51: 624-633.

Gonçalves, E.J. and V.C. Almada. 1998. A comparative study of territoriality in intertidal and subtidal blennioids (Teleostei, Blennioidei). *Environmental Biology of Fishes* 51: 257-264.

Graham, J.B. 1970. Preliminary studies on the biology of the amphibious clinid *Mnierpes macrocephalus*. *Marine Biology* 5: 136-140.

Grove, D.J. and C. Crawford. 1980. Correlation between digestion rate and feeding frequency in the stomachless teleost, *Blennius pholis* L. *Journal of Fish Biology* 16: 235-247.

Harmelin-Vivien, M.L. 2002. Energetics and fish diversity on coral reefs. In: *Coral Reef Fishes: Dynamics and Diversity in a Complex Ecosystem*, P.F. Sale (ed.). Academic Press, San Diego, pp. 265-274.

Hatcher, B.G. 1981. The interaction between grazing organisms and the epilithic algal community of a coral reef: A quantitative assessment. *Proceedings of the 4th International Coral Reef Symposium* 2: 515-524.

Hay, M.E. 1991. Fish-seaweed interactions on coral reefs: Effects of herbivorous fishes and adaptations of their prey. In: *The Ecology of Fishes on Coral Reefs*, P.F. Sale (ed.). Academic Press, San Diego, pp. 96-119.

Hiatt, R.W. and D.W. Strasburg. 1960. Ecological relationships of the fish fauna on coral reefs of the Marshall Islands. *Ecological Monographs* 30: 65-127.

Hobson, E.S. 1974. Feeding relationships of teleostean fishes on coral reefs in Kona, Hawaii. *Fisheries Bulletin* 72: 915-1031.

Horn, M.H. 1989. Biology of marine herbivorous fishes. *Oceanography and Marine Biology: An Annual Review* 27: 167-272.

Horn, M.H. and R.N. Gibson. 1990. Effects of temperature on the food processing of three species of seaweed-eating fishes from European coastal waters. *Journal of Fish Biology* 37: 237-247.

Horn, M.H., S.N. Murray, M.B. Fris and C. Irelan. 1990. Diurnal feeding periodicity of an herbivorous blenniid fish, *Parablennius sanguinolentus*, in the western Mediterranean. In: *Trophic Relationships in the Marine Environment*, M. Barnes and R.N. Gibson (eds.). *Proceedings of the 24th European Marine Biology Symposium*: Aberdeen University Press, Aberdeen, pp. 170-182.

Illich, I.P. and K. Kotrschal. 1990. Depth distribution and abundance of northern Adriatic littoral rocky reef blennioid fishes (Blenniidae and *Tripterygion*). *P.S.Z.N.I. Marine Ecology* 11: 277-289.

Klumpp, D.W. and N.V.C. Polunin. 1990. Algal production, grazers and habitat partitioning on a coral reef: Positive correlation between grazing rate and food availability. In: *Trophic Relationships in the Marine Environment*, M. Barnes and R.N. Gibson (eds.). *Proceedings of the 24th European Marine Biology Symposium*. Aberdeen University Press, Aberdeen, pp. 372-388.

Kotrschal, K. 1989. Trophic ecomorphology in eastern Pacific blennioid fishes: Character transformation of oral jaws and associated change of their biological roles. *Environmental Biology of Fishes* 24: 199-218.

Kotrschal, K. and D.G. Lindquist. 1986. The feeding apparatus in four Pacific tube blennies (Teleostei: Chaenopsidae): Lack of ecomorphological divergence in syntopic species. *P.S.Z.N.I: Marine Ecology* 8: 241-254.

Kotrschal, K. and D.A. Thomson. 1986. Feeding patterns in eastern tropical Pacific blennioid fishes (Teleostei: Tripterygiidae, Labrisomidae, Chaenopsidae, Blenniidae). *Oecologia* 70: 367-378.

Kotrschal, K., D.G. Lindquist and I.P. Illich. 1991. Opportunistic feeding in Mediterranean blennioid fishes (Blenniidae and *Tripterygion*) as revealed by experimental food provisioning in the field. *P.S.Z.N.I. Marine Ecology* 12: 63-73.

Labelle, M. and J.R. Nursall. 1985. Some aspects of the early life history of the redlip blenny, *Ophioblennius atlanticus*. *Copeia* 1985: 39-49.

Lindquist, D.G. and R.M. Dillaman. 1986. Trophic morphology of 4 western Atlantic blennies (Pisces, Blenniidae). *Copeia* 1986: 207-213.

Lindquist, D.G. and K. Kotrschal. 1987. The diets in four Pacific tube blennies (*Acanthemblemaria*: Chaenopsidae): Lack of ecological divergence in syntopic species. *P.S.Z.N.I. Marine Ecology* 8: 327-335.

Longenecker, K. and R. Langston. 2005. Life history of the Hawaiian blackhead triplefin, *Enneapterygius triceps* (Blennioidei, Tripterygiidae). *Environmental Biology of Fishes* 73: 243-251.

Meekan, M.G. and J.H. Choat. 1997. Latitudinal variation in abundance of herbivorous fishes. *Marine Biology* 128: 373-383.

Milton, P. 1983. Biology of littoral blenniid fishes on the coast of south-west England. *Journal of the Marine Biological Association of the United Kingdom* 63: 223-237.

Mobley, K.B. and W. Fleeger. 1999. Diet of *Scartella cristata*: an artificial habitat-associated blenny (Pisces: Blenniidae). *Vie et Milieu* 49: 221-228.

Moland, E. and G.P. Jones. 2004. Experimental confirmation of aggressive mimicry by a coral reef fish. *Oecologia* 140: 676-683.

Monteiro, N.M., S.M. Quinteira, K. Silva, M.N. Vieira and V.C. Almada. 2005. Diet preference reflects the ontogenetic shift in microhabitat use in *Lipophrys pholis*. *Journal of Fish Biology* 67: 102-113.

Muñoz, A.A. and F.P. Ojeda. 1997. Feeding guild structure of a rocky intertidal fish assemblage in central Chile. *Environmental Biology of Fishes* 49: 471-479.

Muñoz, A.A. and F.P. Ojeda. 1998. Guild structure of carnivorous intertidal fishes of the Chilean coast: Implications of ontogenetic dietary shifts. *Oecologia* 114: 563-573.

Muñoz, A.A. and F.P. Ojeda. 2000. Ontogenetic changes in the diet of the herbivorous *Scartichthys viridis* in a rocky intertidal zone in central Chile. *Journal of Fish Biology* 56: 986-998

Myers, R.F. 1999. *Micronesian Reef Fishes*. Coral Graphics, Barrigada, territory of Guam, USA.

Nieder, J. 1997. Seasonal variation in feeding patterns and food niche overlap in the Mediterranean blennies, *Scartella cristata, Parablennius pili*cornis and *Lipophrys trigloides* (Pisces; Blenniidae). *Marine Ecology* 18: 227-237.

Nieder, J. 2001. Amphibious behaviour and feeding ecology of the four-eyed blenny (*Dialommus fuscus*, Labrisomidae) in the intertidal zone of the island of Santa Cruz (Galapagos, Ecuador). *Journal of Fish Biology* 58: 755-767.

Nursall, J.R. 1981. The activity budget and use of territory by a tropical blenniid fish. *Zoological Journal of the Linnean Society* 72: 69-92.

Odum, H.T. and E.P. Odum. 1955. Trophic structure and productivity of a windward coral reef community on Eniwetok Atoll. *Ecological Monographs* 25: 291-320.

Ojeda, F.P. and A.A. Muñoz. 1999. Feeding selectivity of the herbivorous fish *Scartichthys viridis*: effects on macroalgal community structure in a temperate rocky intertidal coastal zone. *Marine Ecology Progress Series* 184: 219-229.

Patzner, R.A. 1999. Habitat utilization and depth distribution of small cryptobenthic fishes (Blenniidae, Gobiesocidae, Gobiidae, Tripterygiidae) at Ibiza (western Mediterranean Sea). *Environmental Biology of Fishes* 55: 207-214.

Prochazka, K. 1998. Spatial and trophic partitioning in cryptic fish communities of shallow subtidal reefs in false bay, South Africa. *Environmental Biology of Fishes* 51: 201-220.

Prochazka, K. and C.L. Griffiths. 1992. Observations on the distribution patterns, behaviour, diets and reproductive cycles of sand-dwelling clinids (Perciformes: Clinidae) from South Africa. *Environmental Biology of Fishes* 35: 371-379.

Purcell, S.W. and D.R. Bellwood. 2001 Spatial patterns of epilithic algal and detrital resources on a windward coral reef. *Coral Reefs* 20: 117-125.

Randall, J.E. 1967. Food habits of reef fishes of the West Indies. *Studies in Tropical Oceanography* 5: 665-847.

Randall, J.E., G.R. Allen and R.C Steene. 1997. *Fishes of the Great Barrier Reef and Coral Sea*. Crawford House Press, Bathurst, Australia.

Russ, G.R. and J. St John. 1988. Diets, growth rates and secondary production of herbivorous coral reef fishes. *Proceedings of the 6th International Coral Reef Symposium* 2: 37-43.

Sano, M., M. Shimizu and Y. Nose. 1984. Food habits of teleostean reef fishes in Okinawa Island, Japan. *The University Museum of the University of Tokyo* 25: 1-128.

Smith-Vaniz, W.F. 1976. The sabre-toothed blennies, tribe Nemophiini (Pisces: Blenniidae). *Monographs of the Academy of Natural Sciences Philadelphia* 19: 1-196.

Smith-Vaniz, W.F. and V.G. Springer. 1971. Synopsis of the tribe Salariini, with descriptions of five new genera and three new species (Pisces: Blenniidae). *Smithsonian Contributions to Zoology* 73: 1-72.

Springer, V.G. 1968. Osteology and Classification of the Fishes of the Family Blenniidae. *United States National Museum, Bulletin* 284: 1-85.

Springer, V.G. 1988. The Indo-Pacific blenniid fish genus *Ecsenius*. *Smithsonian Contributions to Zoology* 465: 1-134.

Springer, V.G. and W.F. Smith-Vaniz. 1972. Mimetic relationships involving fishes of the family Blenniidae. *Smithsonian Contributions to Zoology* 112: 1-37.

Silberschneider, V. and D.J. Booth. 2001. Resource use by *Enneapterygius rufopileus* and other rockpool fishes. *Environmental Biology of Fishes* 61: 195-204.

Stepien, C.A. 1990. Population structure, diets and biogeographic relationships of a rocky intertidal fish assemblage in central Chile: high levels of herbivory in a temperate system. *Bulletin of Marine Science* 47: 598-612.

Stoner, A.W. and R.J. Livingston. 1980. Distributional ecology and food-habits of the banded blenny *Paraclinus-fasciatus* (Clinidae): a resident in a mobile habitat. *Marine Biology* 56: 239-246.

Taborsky, M. and D. Limberger. 1980. The activity rhythm of *Blennius sanguinolentus* Pallus: an adaptation to its food source? *P.S.Z.N.I. Marine Ecology* 1: 143-153.

Townsend, K.A. and I.R. Tibbetts. 2000. Biomass and distribution of herbivorous blennies in the southern Great Barrier Reef. *Journal of Fish Biology* 56: 774-791.

Townsend, K.A. and I.R. Tibbetts. 2004. The ecological significance of the comb-toothed blenny in a coral reef ecosystem. *Journal of Fish Biology* 65: 77-90.

Vahl, O. and J. Davenport. 1979. Apparent specific dynamic action of food in the fish *Blennius pholis*. *Marine Ecology Progress Series* 1: 109-113.

Vinyoles, D., I.M. Cote and A. de Sostoa. 1999. Egg cannibalism in river blennies: the role of natural prey availability. *Journal of Fish Biology* 55: 1223-1232.

Wallace, J.C. 1973. Observations on relationship between food consumption and metabolic rate of *Blennius pholis*. *Comparative Biochemistry and Physiology* 45: 293-306.

Wilson, S.K. 2000. Trophic status and feeding selectivity of blennies (Blenniidae: Salariini). *Marine Biology* 136: 431-437.

Wilson, S.K. 2001a. Multiscale habitat associations of detritivorous blennies (Blenniidae: Salariini). *Coral Reefs* 20: 245-251.

Wilson, S.K. 2001b. Trophic limitations in small reef fish: an evaluation of herbivory in the Blenniidae. Ph.D. Thesis. James Cook University, Townsville, Queensland, Australia.

Wilson, S.K. 2002. Nutritional value of detritus and algae in blenny territories on the Great Barrier Reef. *Journal of Experimental Marine Biology and Ecology* 271: 155-169.

Wilson, S.K. 2004. Growth, mortality and turnover rates of a detritivorous blenny. *Marine Ecology Progress Series* 284: 253-259.

Wilson, S.K. and D.R. Bellwood. 1997. Cryptic dietary components of territorial Damselfishes (Pomacentridae, Labroidei). *Marine Ecology Progress Series* 153: 299-310.

Wilson, S.K., K. Burns and S. Codi. 2001a. Identifying sources of organic matter in sediments from a detritivorous coral reef fish territory. *Organic Geochemistry* 32: 1257-1269.

Wilson, S.K., K. Burns and S. Codi. 2001b. Sources of dietary lipids in the coral reef blenny *Salarias patzneri*. *Marine Ecology Progress Series* 222: 291-296.

Wilson, S.K., D.R. Bellwood, J.H. Choat and M.J. Furnas. 2003. Detritus in the epilithic algal matrix and its use by coral reef fishes. *Oceanography and Marine Biology Annual Review* 41: 279-309.

Zander, C.D. 1982. Feeding ecology of littoral gobiid and blennioid fishes of the Banyuls area (Mediterranean sea). 1. Main food and trophic dimension of niche and ecotope. *Vie et Milieu* 32: 1-10.

Zander, C.D. and A. Heymer. 1992. Feeding-habits of *Gobius auratus* and other benthic small-sized fish from the French Mediterranean coast under regard of some alternating parameters. *Zoologischer Anzeiger* 228: 220-228.

Zemke-White, W.L., J.H. Choat and K.D. Clements. 2002. A re-evaluation of the diet feeding hypothesis for marine herbivorous fishes. *Marine Biology* 141: 571-579.

Zoufal, R. and M. Taborsky. 1991. Fish foraging periodicity correlates with daily changes of diet quality. *Marine Biology* 108: 193-196.

CHAPTER

3.2

Dentition and Taste Buds in Blennies

Lev Fishelson

INTRODUCTION

Fish employ their oral taste buds (TB) to screen and, subsequently, either reject food particles or accept and then swallow them (Pavlov and Kasumyan, 1998; Kasumyan and Prokopova, 2003; Hansen and Reutter, 2004; Fishelson, 2005a, b). The TB are distributed throughout the skin, but are especially concentrated along various protrusions around the mouth opening and in the oropharyngeal cavity (Whitear, 1971; Kapoor *et al.*, 1975; Gomahr *et al.*, 1992; Hara, 1992; Hansen and Reutter, 2004). As described by Schulte and Holl (1972) and observed during the present study (Fishelson, pers. observ.), some TB are even found on the cirri growing over the nuchal openings as well as supraocularly. Until recently, TB had been studied in catfish (Reutter, 1978; Kapoor and Finger, 2003; Kiyohara and Tsukahara, 2005), rainbow trout (Ezeasor, 1982), minnow (Kiyohara *et al.*, 1980), and lungfish (Reutter, 1991). Several recent

Author's address: Department of Zoology, George S. Wise Faculty of Life Sciences, Tel Aviv University, Tel Aviv 69978, Israel. E-mail: fishelv@post.tau.ac.il

studies compared the TB distribution in groups of species from the same family: e.g., on several flatfishes (Livingstone, 1987), on cardinal fishes (Fishelson and Delarea, 2004), on blennies and gobies (Fishelson *et al.*, 2004), and on various cichlid fishes (Fishelson, 2005a, b). The form and activity of TB have also been reviewed in several books (Hara, 1992; Finger *et al.*, 2000; Doty, 2003), with the most recent one being *Fish Chemosenses*, edited by Reutter and Kapoor (2005). All these studies revealed two important phenomena: first, TB distribution is species-specific and can serve as a specific marker of microadaptation (Fishelson and Delarea, 2004; Reutter and Hansen, 2005; Fishelson, 2005b; Gon *et al.*, 2007); second, this distribution often shows a unique adaptation to the dentition pattern and the type of food and feeding regime of the fish. Various studies have shown that three types of TB are found in the oral cavity (OC) of fish: type I, situated on small papillae or hillocks, 40-60 μm in diameter and 80-100 μm high, rising above the surrounding epithelium; type II, closely resembling type I, but less elevated; and type III, which does not protrude and whose exposed apical ends are level with the surrounding epithelium. Starting with the studies of Murray and Murray (1970), it became evident that all three TB-types possess two kinds of elongated sensory cells, either dark or pale, bearing sensory protrusions of various dimensions at their apical ends (Reutter and Hansen, 2005). From these, the TB signals are transmitted via the sensory parts of the facial (VII), glossopharyngeal (IX), and vagal (X) nerves to the brain centers situated in the medulla oblongata (Fishelson, 2005b; Reutter and Witt, 2005). The present study compares the distribution and numbers of TB relative to dentition in the OC in selected species of the families Labrisomidae, Clinidae, Blenniidae, Tripterygiidae and Dactyloscopidae, together forming a part of the suborder Blennioidei.

MATERIAL AND METHODS

For this study, some of the fish were collected live along the shores of the Israeli Mediterranean and the Red Sea, and from various sites of the Cape Province in South Africa, while others were donated from ichthyological collections in the United States of America, Australia and Israel (see Acknowledgements). The live samples were killed with an overdose of MS 222; all sampled fish were fixed in 10% formaldehyde and 3.5% glutaraldehyde for electron microscopy. For transmission electron microscopy (TEM), the samples were post-fixed in osmium tetra oxide, embedded in Epon, and their sections stained with uranyl acetate and lead citrate. The sections were studied with a JEOL 1200 EX. Samples for scanning electron microscopy (SEM) were dehydrated along ascending solutions of ethanol, critical point dried in CO_2 and gold coated. These samples were studied with a JEOL JSM 840 A. Table 3.2.1 provides data on the studied species and their taxonomic position. A total of 20 species were studied. Measurements were taken using a microcaliber on the samples or their micrographs; and the counting of TB was done under a

Table 3.2.1 Species of blennies studied, locality of origin, and standard length (SL) of the specimens. EM, Eastern Mediterranean; RS, Red Sea; SA, South Africa; AU, Australia.

Species	*Site of origin*	*No. of specimens*	*Size SL (mm)*
Blenniidae			
Cirripectes variolosus	RS	2	44-44
Ecsenius nalolo	RS	2	40-42
Hirculops cornifer	RS	3	48-50
Istiblennius rivulatus	RS	4	49-55
Parablennius sanguinolentus	EM	3	75-83
P. zvonimiri	EM	3	55-61
Salaria pavo	EM	4	67-77
Clinidae			
Cristiceps aurantiacus	AU	3	85-150
Clinus acuminatus	SA	2	90-108
C. arborescens	SA	2	150-184
C. cottoides	SA	4	78-150
C. superciliosus	SA	6	40-118
C. taurus	SA	2	123-134
Heteroclinus perspicillatus	AU	3	62-68
H. heptaeolus	AU	2	62-64
Muraenoclinus dorsalis	SA	1	72.0
Pavoclinus graminis	SA	3	67-79
Dactyloscopidae			
Dactyloscopus lacteus	Galapagos	1	42.0
Labrisomidae			
Labrisomus nuchipinnis	Tabago	2	104-106
L. bucciferus	Bermuda	2	60-64
Malacoctenus macropus	Belize	2	30-32
Starksia lepicoelia	Caribbean	2	22-24
Tripterygiidae			
Tripterygion tripteronotus	EM	3	63-70

dissecting microscope (×70) and on micrographs. Measuring the teeth band enabled calculation of TB/mm^2 in the largest specimens studied (Table 3.2.2).

Here, the oropharyngeal cavity (OC) of *Labrisomus bucciferus* (Labrisomidae) (Fig. 3.2.1A, B) will serve as a model for description of the different sites, dentition and TB numbers, against which the other species will be compared. For simplicity of description, the following sites have been delineated on the jaws: upper jaw (Fig. 3.2.1A): lips, teeth band; upper valve; vomer; palatinum; epipharyngeal bones (EB) that are toothed pharyngobranchial ossicles; lower jaw (Fig. 3.2.1B): lips, teeth band, lower valve, tongue, hypopharyngeal bones (HB), toothed ceratohyale ossicles. Altough in the present studied blennies the EB and HB are not so united and ossified as in labrid and in cichlids (Fishelson, 2005a), I use the similar names for these pharyngeal bones of the studied fish, convinced that here also they will perform the same function-mastication of the food particles. Table 3.2.2 provides data on the numbers and distribution of TB on the various sites in the OC of the studied fishes.

RESULTS

Remarks on the Ontogenesis of TB

The first signs of TB are already visible in 4-mm embryos on the primordial tongue of the oviparous *Salaria pavo*, and 6-mm embryos of the viviparous *Clinus cottoides*. In 8 mm of *S. pavo* and 11 mm *C. cottoides* and *Cristiceps argentatus*, TB protrude on the primordial jaws (Fig. 3.2.1A). In 13 mm embryos, additional TB papillae can now be seen on the vomer, palatinum and pharyngeal ossicles, most of which are premature, small (1.5-2.3 µm in diameter) with few sensory protrusions (Fig. 3.2.1B, inset). In mature juveniles of *S. pavo* and pre-release 18 mm larvae of the viviparous clinids, numerous TB are found on the upper and lower jaws (Fig. 3.2.1C), especially on the tongue, palatinum and pharyngeal bones. In juveniles of 30-40 mm SL, the distribution of TB in the oropharyngeal cavity resembles that observed in adult fish, and the TB cells are already segregated into pale and dark forms (Fig. 3.2.1D), with differentiated protruding sensory endings (Fig. 3.2.1E, F).

Distribution of TB in Adult Fish

In most of the studied blennies, the jaws are surrounded by fleshy lips that often bear type I and type III TB (Fig. 3.2.1E, F). The lips can be roughly

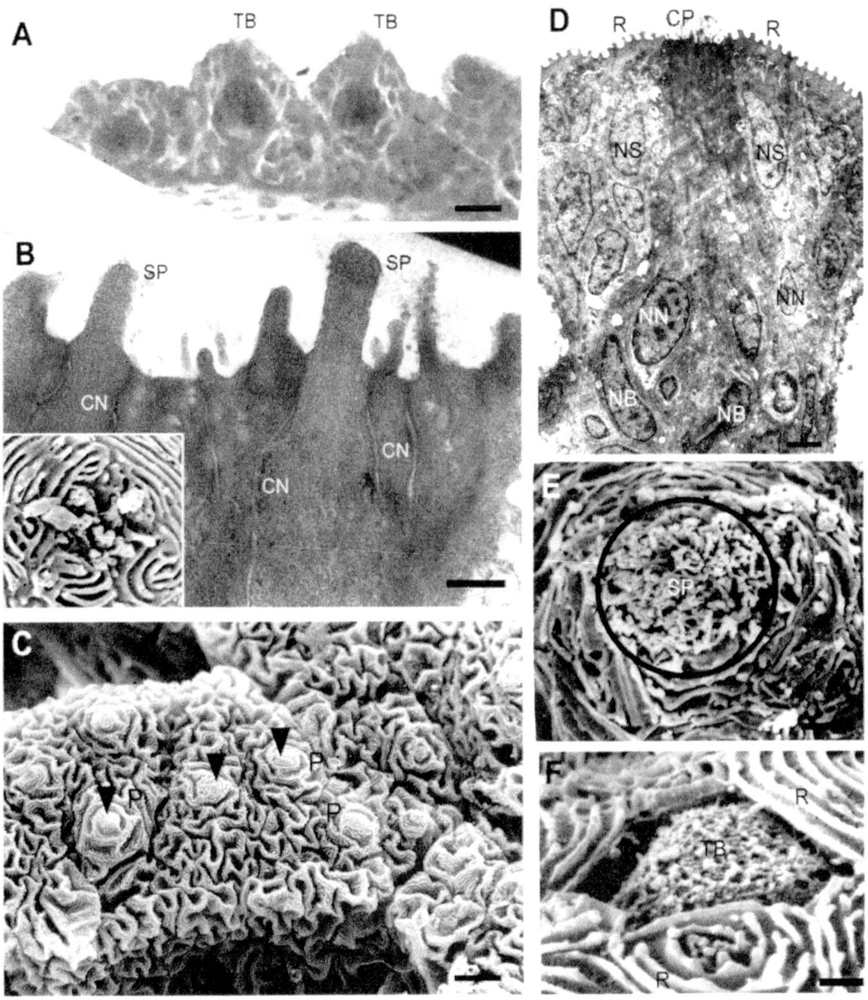

Fig. 3.2.1 Micrographs of taste buds. **A**. Maturing TB on jaw of *Clinus cottoides* embryo. LM. *Scale bar* 20 µm. **B**. Section of TB of *Clinus cottoides*. TEM. *Scale bar* 500 nm (inset, juvenile TB, SEM). **C**. Papillae with TB on lips of *Cristiceps aurantiacus*. SEM. *Scale bar* 10 µm. **D**. Vertical section of a TB of *Heteroclinus perspicillatus*. TEM. *Scale bar* 1 µm. **E**. Sensory protrusions of type I TB of *Clinus cottoides*. SEM. *Scale bar* 1 µm. **F**. Type III TB on HB of *Tripterygion tripteronotus*. SEM. *Scale bar* 1 µm. CN, cell membrane; CP, sensory microvillar protrusions of neuron; NS, supporting cell nuclei; NB, nucleus of basal cell; NN, nucleus of neuron; P, papillae; R, surface cell ridges; SP, sensory protrusions; TB and arrowhead, taste buds.

divided into two areas: the outer, smooth, slightly swollen part; and the inner, marginal part that extends along the band of teeth on the jaws. In some fish varieties, this margin is crenulated and terminates in small lamellae extending above the front line of teeth and often penetrating between them; while in other fish (e.g., Blenniidae) it is smooth. In fishes with crenulations on the lips, their marginal lamellae are largest on the jaw apex, decreasing in size posterior. In *Labrisomus bucciferus* the lamellae are larger on the lower jaw than on the upper (Fig. 3.2.2A, B).

The 'breathing valves', posterior to the teeth band of both jaws, are narrow skin extensions involved in the regulation of water flow during breathing (Bond, 1996). Adjacent and posterior to the valve on the palatal side of the OC is the vomer, shaped differently in the different species. It is frequently prominent in a crescent or V-form line of teeth (Fig. 3.2.2A), but is sometimes toothless and embedded in a group of larger lobules.

Posteriorly, before the entrance into the esophagus, the palate terminates in two small bones, the pharyngobranchiales of the fifth gill arch, which form the epipharyngeal bones (EB). These are covered with teeth and lobules (Fig. 3.2.2A). The palatine ossicles are toothless, except in the clinid *Springeratus xanthosoma* in which they bear small teeth (Shen, 1971).

The lower and upper jaws are usually similar in the form of lips, teeth band and valves (Fig. 3.2.2B). The tongue varies in form among the different species (e.g., it can be short as in *Heteroclinus* spp., narrow as in *Starksia* or wide as in *Labrisoma bucciferus*) (Fig. 3.2.2B). The posterior part of the lower jaw terminates in the toothed fifth ceratohyale plates of the gill arch, which forms the dentate hypopharyngeal bone (HB) (Fig. 3.2.2B). These upper and lower bones form the masticating apparatus at the esophageal entrance, often termed the 'pharyngeal jaws' (Bond, 1996; Fishelson and Delarea, 2004; Fishelson *et al.*, 2004). All these sites feature groups of lobules and papillae bearing various types of TB with their exposed sensory protrusions (Fig. 3.2.1E).

Oral Cavity of *Labrisomus bucciferus (Figure 3.2.2; Table 3.2.2)*

Upper Jaw: In the largest fish studied (104.0 mm S_L), the upper jaw is around 8.0 mm wide. The upper lip (0.6 mm wide) is marginally crenulated by ca. 80 lamellae of various forms, 0.2-0.3 mm long, partly covering the frontline of the teeth. Some of the lamellae bear one to three round papillae, 40 (±5) μm in diameter (totaling 280 (±20) papillae), covered by the typical fingerprint pattern of the exposed epithelial

Table 3.2.2 TB distribution and numbers on lips and other sites of the oropharyngeal cavity in the largest blennioid specimens studied. UL, upper lip; LL, lower lip; UTB, upper teeth band; LTB, lower teeth band; UV, upper valve; LV, lower valve; VOM, vomer; PAL, palatinum; TO, tongue; EP, epipharyngeal bones; HY, hypopharyngeal bones; TO, tongue; TOT, total number of TB; SL, standard length in mm.

Species	*SL*	*UL*	*UTB*	*UV*	*VOM*	*PL*	*EP*	*LL*	*LTB*	*LV*	*TO*	*HY*	*TOT*
Blenniidae													
Cirripectes variolosus	58	0	0	10	14	20	60	0	0	110	10	20	244
Ecsenius nalolo	42	0	0	40	10	120	140	0	0	60	10	80	460
Hirculops cornifer	54	0	12	30	30	60	120	0	0	20	26	60	380
Istiblennius rivulatus	75	10	0	128	16	34	45	18	0	260	30	156	697
Parablennius sanguinolentus	80	0	0	20	15	20	130	0	30	40	20	140	415
Salaria pavo	70	0	18	30	14	60	220	0	20	40	15	120	537
Clinidae													
Clinus acuminatus	108	320	200	150	60	80	340	280	160	160	80	420	2250
C. arborescens	174	160	60	20	48	46	160	140	80	70	35	180	1045
C. cottoides	150	110	240	20	130	64	200	60	120	60	30	260	1454
C. superciliosus	101	380	510	260	160	480	640	320	280	240	260	520	4050
Cristiceps aurantiacus	150	110	170	36	10	72	140	24	78	120	140	230	1130
Heteroclinus perspicillatus	68	90	200	380	80	320	190	120	80	160	120	60	1048
H. heptaeolus	74	216	180	140	80	220	750	56	60	90	160	580	2532
Muraenoclinus dorsalis	54	180	100	48	38	85	180	160	80	30	36	40	977
Pavoclinus graminis	138	80	940	110	240	310	650	300	240	80	200	600	3750
Labrisomidae													
Labrisomus nuchipinnis	100	180	160	120	112	145	220	120	310	20	60	130	1525
L. bucciferus	104	160	110	90	80	110	300	80	160	120	170	300	1680
Malacoctenus macropus	56	12	140	70	12	40	300	0	86	66	30	180	934
Paraclinus nigripinnis	60	10	150	42	80	30	120	0	100	160	42	30	764
Starksia lepicoelia	70	16	30	42	36	46	120	12	14	62	74	160	612
Tripterygiidae													
Tripterygion tripteronotus	72	160	140	125	69	260	150	50	130	115	420	90	1710

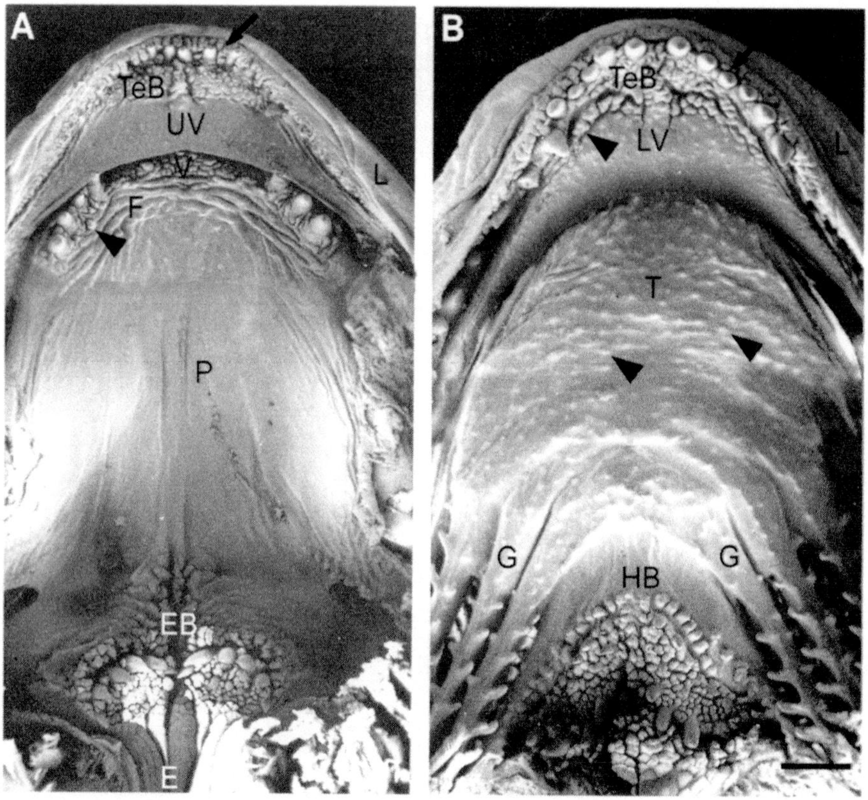

Fig. 3.2.2 Upper (**A**) and lower (**B**) oral surface of *Labrisomus bucciferus* (*Scale bar* 1 mm). E, esophagus; EB, epipharyngeal bones; F, folding of palatinum; G, gill bases; HB, hypopharyngeal bones; L, lips; LV, lower valve; P, palatinum; T, tongue; TeB, teeth band; UV, upper valve; V, vomer; arrow, lip lamellae; arrowhead, taste buds bearing papillae.

membranes. Apical, each of the protruding papillae bears 1-2 type I TB, 5(±1.0) μm in diameter, with exposed sensory protrusions. The teeth band is 1.4 mm wide, with a front row of 30 strong teeth (Fig. 3.2.2A). A second row of 20 teeth extends behind the frontal one, with 4 larger teeth in the middle and a canine-like tooth at each end; among these are ca. 70 papillae bearing 110 type I TB, some situated on elongated ridges. This site features 21 TB/mm^2.

Upper Valve: The valve, 1.8 mm wide, is divided into two parts: the anterior part along the teeth band is smooth, whereas the posterior part bears ca. 80 TB.

Vomer: This bone at the apical end of the palate has the form of a wide crescent, with 2.0 mm long sides, and possesses an irregular row of 7+7 strong teeth separate into two groups of 4 and 3 along each arm; the end teeth are the largest. The teeth are embedded in ca. 60 lobules of various dimensions, bearing around 80 type I TB.

Palate: Predominantly smooth, the anterior part folded, and each fold has several TB; the posterior end possesses ca. 20 folds with type II TB.

Epipharyngeals: The two epipharyngeal bones (EB) are 4.5 mm wide, each with 12 sharp teeth forming two rows. All the teeth are embedded in a total of 200 lobules and papillae of various sizes, forming a sort of cushion and bearing around 300 Type II and III TB, 118/mm^2. Such TB-bearing lobules continue posterior toward the opening of the esophagus.

Lower Jaw: This jaw (Fig. 3.2.2B) is slightly narrower (6.5 mm) than the upper one (Fig. 3.2.1b). Its lip is framed by ca. 50 lamellae that extend to and between the front row of teeth, and bear around 80 type I TB. The teeth band reveal in front a row of 26 strong, slightly flat teeth, behind which is a group of 10 smaller teeth, embedded in the middle of the jaw and divided by a group of lamellae. Within the teeth band are ca. 80 papillae and lobules, bearing around 160 type I TB, 26/mm^2.

Lower Valve: This valve (1.0 mm wide) is wider than the teeth band, and bears ca. 120 lobules at the teeth margin, with ca 160 TB; ca. 120 papillae with TB, 35/mm^2, are dispersed on the valve.

Tongue: The tongue is wide, rounded apical, and bears ca. 170 type II TB, dispersed over the surface and posteriorly passing on the gill bases; the posterior margin carries 30 lobules with ca. 50 type I and II TB.

Hypopharyngeal Bones: Each of the two HB is 2.0 mm wide, oval, with 10 teeth of which the four posterior ones, strong canine teeth, form a row (Fig. 3.2.2B). All the teeth are embedded in ca. 200 lamellae and papillae that bear ca. 300 type I and II TB, 50/mm^2. The distribution of TB in the OC of the studied 108 mm S_L *L. bucciferus* is illustrated in Table 3.2.2. The jaws reveal a total of 1680 TB, and their number is almost identical (850 versus 830) in both jaws. The total 'sensory area (surface occupied by the TB sensory protrusions) in the OC of this fish is approximately 8,350 μm^2.

Comparison of OC and TB Distribution in the Studied Families (Table 3.2.2)

Blenniidae: In general, the studied Blenniidae differ strongly from the other families in having a single row of teeth on both jaws that form a palisade-like 'teeth band', with either a smooth or plicated area behind (Figs. 3.2.3A-C; 3.2.5A-C).In the other fish of the studied families, it is occupied by additional groups or rows of teeth and papillae bearing TB (Figs. 3.2.3E-G; 3.2.5E-G). As observed in *Ecsenius nalolo, Salaria pavo, Cirripectes variolosus* and *Hirculops cornifer*, this site and the adjacent valves are either smooth or folded, both with very few TB (Table 3.2.2). In *E. nalolo* (jaw width of 3.4 mm), this palisade holds ca 96 teeth, and in *C. variolosus* (jaw width 4.0 mm) ca. 180 teeth, all feeble and not attached to the jaw; whereas *S. pavo* has 36 teeth, *H. cornifer* 36, and *P. zvonimiri* 24 strong teeth (on jaw width of 3.8, 2.5, and 3.6, respectively); in all three latter species, the teeth are strongly anchored on the jaw (Fig. 3.2.3B-D). The TB are distributed accordingly (Table 3.2.2). With the exception of *Istiblennius rivulatus*—which has 10 TB on the upper lip—all the other studied blennies lack TB on this site. Similarly, *H. carnifer* and *Salaria pavo* reveal 12 and 18 TB, respectively, on their upper jaw teeth bands, not seen in the other species on this site. *Parablennius zvonimiri* appears to be an exception in this family, as its upper and lower jaws, in addition to the frontal teeth palisade, have some teeth behind them (Fig. 3.2.3D). In this species the vomer, with its strong central teeth, resembles the Labrisomidae (Fig. 3.2.3E) more than the Blenniidae.

Among the blennies, *E. nalolo* and *C. variolosus* differ from the other species in having the tongue and palate entirely covered by papillae and lobules of various dimensions, interspersed with TB (Fig. 3.2.4A, B). These two sites in both species are also cross-divided by a slightly elevated epithelial bar bearing ca. 24 type II TB. In these two species, the EB and HB also strongly differ from the other species: the EB are with two short rows of 4-5 sharp teeth placed parallel to the body axis (Fig. 3.2.4C); the HB is miniature, hides between the gill bases, each with 5 teeth and few lobules with TB (Fig. 3.2.4D).

Labrisomidae: The studied species of this family demonstrate crenulated lips that are wider then their teeth bands. As observed in *Malacoctenus macropus, Starksia lepicoelia* (jaw width1.8 mm), and *Paraclinus nigripinnis* (jaw width 2.2 mm), the lamellae on the margin of the upper and lower lips are narrower on the apical region of the jaws than laterally (Figs. 3.2.3E-G; 3.2.5E). As noted above for *Labrisomus bucciferus* and also seen in *L. nuchipinnis* (Fig. 3.2.3E), both these species possess high numbers of TB in the OC (1680 and 1525, respectively), and in both revealing ca. 300 are on the lips and ca. 400 on the teeth bands (Table

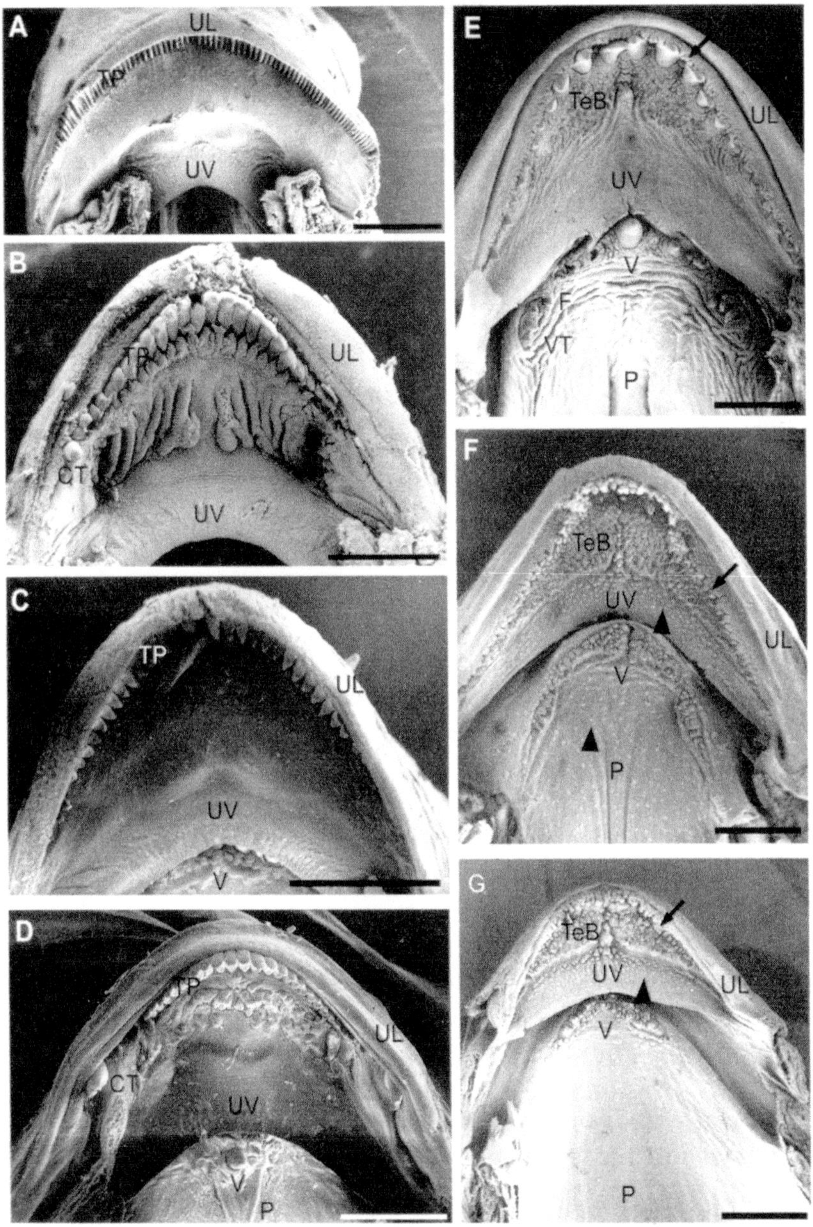

Fig. 3.2.3 Upper jaws of various species of blennioids. **A**. *Ecsenius nalolo*. **B**. *Salaria pavo*. **C**. *Hirculops conifer*. **D**. *Parablennius zvonimiri*. **E**. *Labrisomus nuchipinnis*. **F**. *Tripterygion tripteronotus*. **G**. *Heteroclinus perspicillatus*. *Scale bar* 1 mm. CT, canine teeth; P, palatinum; TeB, teeth band; TP, teeth palisade; V, vomer; VT, lateral teeth of vomer; arrow, marginal lip lamellae; arrowhead, TB bearing papillae.

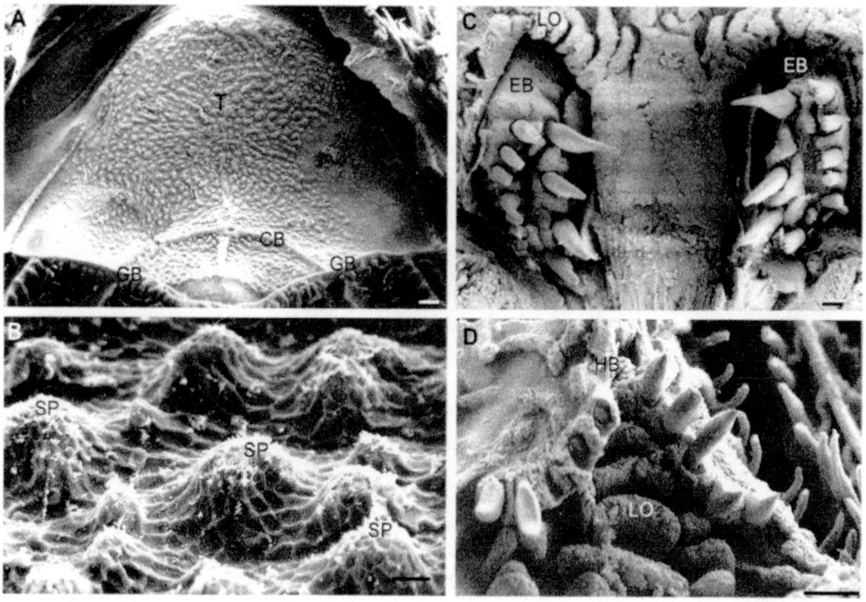

Fig. 3.2.4 Oropharyngeal cavity of *Ecsenius nalolo*. **A**. Tongue and gill base. *Scale bar* 100 μm. **B**. Papillae with type I TB on the tongue. *Scale bar* 10 μm. **C**. Epipharyngeal bones. *Scale bar* 100 μm. **D**. Hypopharyngeal bones. *Scale bar* 100 μm.

3.2.2). In *S. lepicoelia*, *M. macropus* and *P. nigripinnis*, the numbers of TB are much lower, and in some specimens the lower lips lack TB entirely (Table 3.2.2). In contrast to *Labrisomus* spp. in the three latter species, the highest number of TB can be found at the rear of the OC, on the EB and HB, rather than in front of the jaws. The teeth band in the labrosomids is wide; the front row of teeth comprises ca. 26 thick, strong teeth, flat in *Labrisomus* spp. and sharp in *S. lepicoelia* (Fig. 3.2.5C). Beyond this row, the teeth band contains several dispersed small teeth and numerous papillae with type I TB (Table 3.2.2) (Figs. 3.2.3E-G; Fig. 3.2.5F). The high number of TB on the valves of *P. nigripinnis* (80 on the upper and 160 on the lower) is prominent compared to the lower numbers revealed on the much larger specimens of *Labrisomus nuchipinnis* (Table 3.2.2).

Clinidae: The oral cavities of the viviparous Clinidae have been the focus of a dedicated study by Gon *et al.* (2007) and this chapter, therefore, provides certain selected data, in addition to those given in Table 3.2.2. As in the Labrisomidae, also in the studied clinids too the lips are crenulated, and marginal lamellae protrude over the anterior row of teeth; the teeth

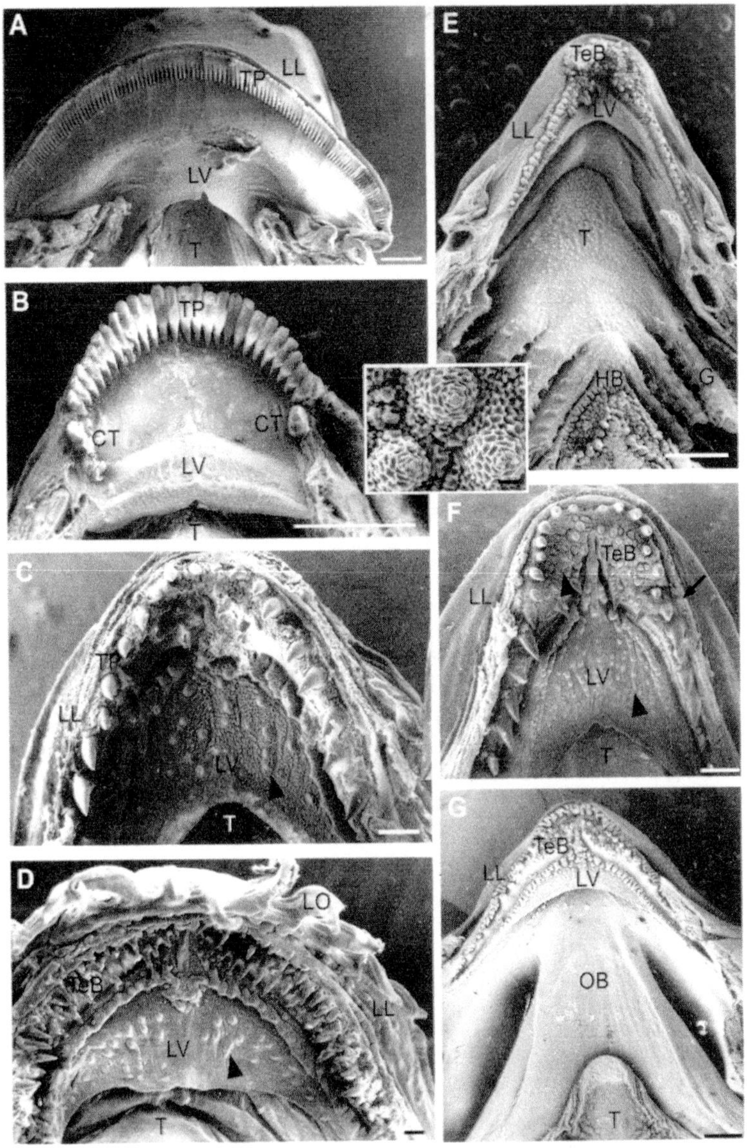

Fig. 3.2.5 Lower jaws of various blennioids. **A**. *Ecsenius nalolo. Scale bar* 1 mm. **B**. *Salaria pavo. Scale bar* 2 mm. **C**. *Starksia lepicoelia. Scale bar* 200 µm. **D**. *Dactyloscopus lacteus. Scale bar* 100 µm. **E**. *Tripterygion tripteronotus. Scale bar* 1 mm (inset, papillae with TB of the tongue, scale bar 20 µm). **F**. *Paraclinus nigripinnis. Scale bar* 1 mm. **G**. *Heteroclinus perspicillatus. Scale bar* 1 mm. CT, canine teeth; G, gill bases; HB, hypopharyngeal bones; LL, lower lip; LO, elongated lip lamellae; LV, lower valve; OB, oral bottom; T, tongue; TeB, teeth band; TP, teeth palisade; arrowhead, papillae with TB; arrow, lamellae on lip margin.

band is wide, with several rows of teeth and numerous TB-bearing papillae (Fig. 3.2.5G). The largest fish studied were of *Clinus arborescens* (174 mm), *C. cottoides* (150 mm), and *Cristiceps aurantiacus* (150 mm). Some of the species of this family, e.g., *Heteroclinus* spp. have very short tongues (Fig. 3.2.5G) that reveal anterior parts of the oral floor; where TB can also be seen. Similarly, in all the species studied, the distribution and number of TB in the specimens of various dimensions correlated with the dimensions of the fish. For example, in *C. superciliosus* of 40 mm the total number of TB on the upper and lower jaws is 720 and 330, respectively; in fish of 68 mm SL the number is 960 and 520, respectively; and in fish of 80 mm S_L it is 1100 and 700, respectively. In *C. superciliosus* of 101 mm and jaw width of 8.0 mm, the number of TB in the upper and lower jaw was 2430 and 1620, respectively (total 4050), with a sensory area of 20,250 μm^2, the highest observed in the studied Blennioidei. In contrast to this, *C. acuminatus* of 108 mm and *C. taurus* of 174 mm had fewer TB (2250 and 1045, respectively) and sensory areas of 11,250 and 5225 μm^2, respectively (Table 3.2.2); while *C. cottoides* of 150 mm possessed 1454 TB and had a 7270 μm^2 sensory area. This latter species differs in having more elongate lamellae (0.2-0.6 mm) along the marginal fringe of the upper and lower lips, each bearing 1-2 type I TB. The upper teeth band is prominent in having three rows of sharp teeth, with the outermost 28 canine-like teeth in the front covered apical by the lamellae, with numerous papillae and type I TB. The number of TB found on the various sites is given in Table 3.2.2.

The teeth band of the lower jaw features 22 sharp front teeth and 28 irregularly placed smaller teeth in the center, divided by a diastem in the middle.

Muraenoclinus dorsalis is the smallest species of the South African fish studied (53 mm S_L), with an upper jaw width of 3.4 mm, and a lower jaw of 2.7 mm. The teeth band of the upper jaw bears a front row of 24 curved strong teeth; the vomer features three rows of short blunt teeth and ca. 30 papillae with type II TB. The division of TB on the various sites of the OC is given in Table 3.2.2. The OC of *Pavoclinus gramistes* generally resembles that of the other South African clinids studied, with numerous TB on the various sites (Table 3.2.2). This genus is unique in the triangular form of the vomer, in which numerous papillae with TB cover the entire space between the arms.

Cristiceps aurantiacus is the largest (150 mm SL) of the Australian clinids studied. It is pale-yellowish in color, with a strongly compressed body and jaw width of 13 mm (upper) and 11 mm (lower). The lips on the inner margin of both jaws are very narrow, with minute lamellae with ca. 120 type I TB. The teeth band has a front row of 30 sharp teeth, a second row of about 26 smaller teeth, and interdispersed with papillae with 220 type I and II TB (Table 3.2.2). The valve is narrow, its basal part features ridges with papillae and TB; the vomer is almost crescent with 8 teeth on each side and 24 type II TB among them. The palate is nearly smooth; the EB has 15+15 teeth, and both are surrounded by 80 lobules with type III TB. Dispersed among the teeth of the EB are 160 TB. The lower jaw too has a narrow teeth band and almost smooth valve. *Heteroclinus perspicillatus* resembles *C. aurantiacus* in its smooth palate and short tongue that exposes a smooth oral floor (Fig. 3.2.5 G). The teeth bands and valves feature numerous papillae and type I TB (Table 3.2.2); the vomer and tongue apex also reveal groups of TB. *Heteroclinus heptaeolus,* with 2906 TB in the OC, especially a high numbers of papillae on the jaws; a special form of the EB and HB (Fig. 3.2.6 D, G), and larger, lobulated vomers. This short-tongued species possesses groups of 40-60 TB on the OC floor (Table 3.2.2). In this species too the tongue is very short, typical to the other studied Australian clinids.

Tripterygiidae: *Tripterygion tripteronotus* (70-72 mm SL) is the most common species of this family along the eastern Mediterranean. The lips along both jaws are crenulated. The upper jaws in the studied fish are 3.5 mm wide and the lower 3.0 mm wide, with numerous papillae and lobulae bearing type I TB (Figs. 3.2.3F; 3.2.5E), especially papillous at the apices. Compared to some other blennies of similar dimensions, this species possesses a very high numbers of TB, i.e., 905 on the upper jaw, and 805 on the lower (Table 3.2.2); the palatinum and tongue as well as the EB and HB display numerous TB-bearing papillae (Fig. 3.2.6E).

Dactylopteridae: *Dactylopterus lacteus* differs from all the other studied fish in having 20-24 relatively large and wide lamellae (0.6-0.8 mm long and 0.2-0.3 mm wide) on the lips, and by having tubular anterior nares of the olfactory organs. On each of the larger lip lamellae are 15-20 type III TB, 3-4 μm in diameter each. Some TB are also on the anterior nares. The teeth-bordering marginal lamellae are very small and bear ca. 60 TB. The teeth band displays two rows of sharp teeth (front 36, back 20) with 120 type I TB. The valves are wider than the teeth band and bear ca. 50 type I TB on small papillae (Fig. 3.2.5).

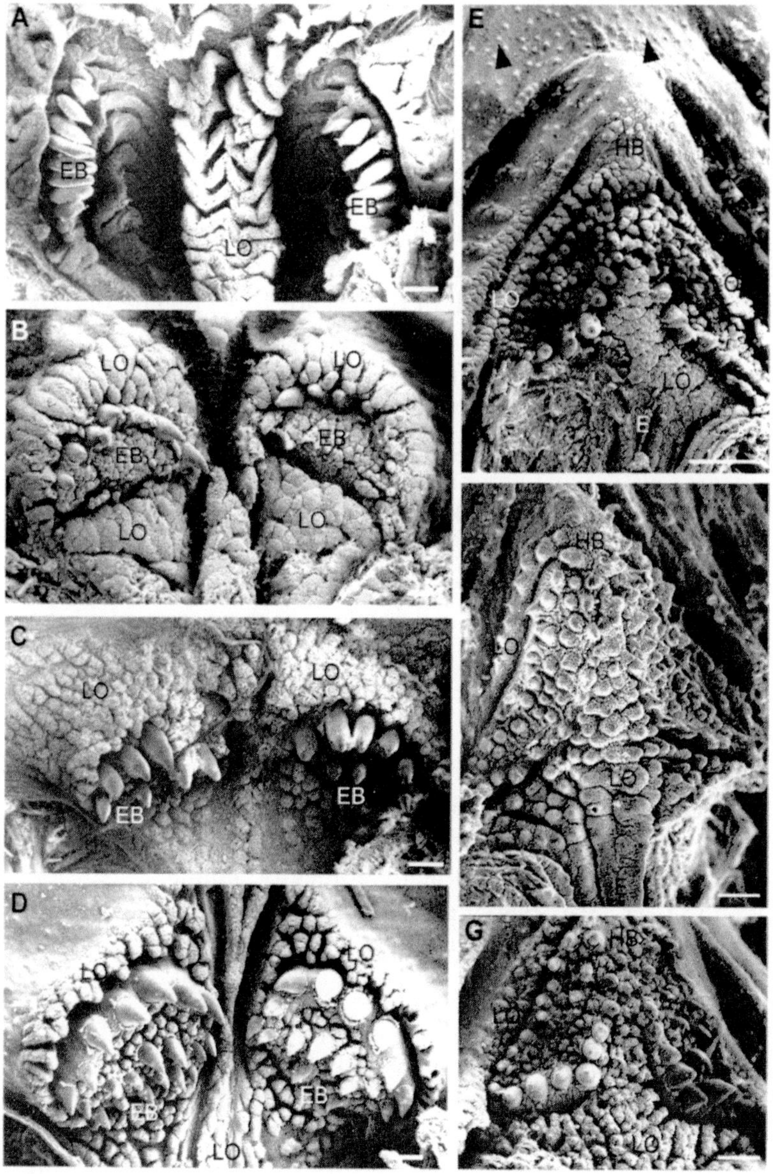

Fig. 3.2.6 Epi- and hypopharyngeal bones of various blennioids. **A**. EB of *Ecsenius nalolo. Scale bar* 150 µm. **B.** EB of *Salaria pavo.* Scale bar 120 µm. **C.** EB of *Malacoctenus macropus. Scale bar* 120 µm. **D**. EB of *Heteroclinus heptaeolus. Scale bar* 100 µm. **E.** HB of *Tripterygion tripteronotus. Scale bar* 1 mm. **F.** HB of *Starksia lepicoelia. Scale bar* 200 µm. **G.** HB of *Heteroclinus heptaeolus. Scale bar* 80 µm. E, esophagus entrance; EB, epipharyngeal bones; G, gill bases; HB, hypopharyngeal bones; LO, lobules with taste buds; T, teeth.

DISCUSSION

The gustatory taste buds (TB) in the oral cavity (OC) of fishes play an important role in feeding and in screening of food items before consumption (Clements and Raubenheimer, 2006 and references therein). These organs have been studied in numerous fish, particularly freshwater species and especially on the oral barbells of silurids (Grover-Johnson and Farbman, 1976; Reutter, 1978). Studies on TB of the OC have recently been summarised, comparing groups of species from a common family (Livingstone, 1987; Fishelson *et al.*, 2004 a, b; 2005a, b; Gon *et al.*, 2007). These studies revealed both a great uniformity in the TB structure of the various fishes, and also that the distribution and number of TB greatly differ among the various taxa studied, a phenomenon that seems to represent species-specific adaptations to food regimes and ecomorphology (Liem, 1986; Linser *et al.*, 1998; Fishelson, 2005b).

This chapter examines the numbers and distribution of TB in the OC of 20 species of fish belonging to several families of the suborder Blennioidei (Springer, 1993), mostly comprising omnivorous or carnivorous, littoral species. As demonstrated in this study, the formation of primordial TB can already be observed in embryos of the bottom-spawning *Salaria pavo* on day 4 following fertilisation, and in 4 mm embryos of the viviparous *Clinus cottoides*. The embryogenesis of TB at such early stages of development, as also documented by Appelbaum *et al.* (1983) and Appelbaum and Riehl (1997) for *Solea solea* and *Cyprinus carpio*, suggests the importance of these organs in the initial life-stages of the fishes. The primordial TB in the OC can be observed in 12 mm post larvae of *S. pavo*, and in 18 mm embryos of *C. superciliosus* and *C. cottoides*, that are still in their ovarian follicles. In juvenile *C. cottoides* (26 mm SL), the TB, 1.6-2.3 μm in diameter, with very few protruding sensory endings, protrude above the surrounding epithelium. As in larger specimens, TB of 4.5-5.0 μm can be found on similar sites, it seems that with age and growth the number of sensory and other cells in the TB increases, as does it the number of sensory protrusions.

As demonstrated by Fishelson and Delarea (2004) and Fishelson (2005a) for gobies, cardinal fish and cichlids, in the studied blennioid families too the same two characters mark the pattern of numbers and distribution of TB in the OC: first, despite intraspecific increase in number of TB with growth (age) of the fish, the final number is species-specific, (Table 3.2.2); and second, the number in the front (Teeth Band + Valves) and the rear part of the OC (EB+HB) varies in the different species (Table 3.2.3). An example of the first character is demonstrated in *Clinus superciliosus* (Clinidae) and *Labrisoma nuchepinnis* (Labrisomidae), of similar body dimensions, ca. 100 mm: in the former the total number of TB is 4050, in the latter it is 1525. *Salaria pavo* of 70 mm has 537 TB in the OC and *Tripterygion tripteronotus* has 1710. These differences have been illustrated in Table 3.2.2. The second characteristic—the differences in TB numbers between the front and rear of the OC—are given in Table 3.2.3. For example, in *T. tripteronotus* the front of the jaw features 770 TB, and the back 1160; in *Labrisoma nuchipinnis* it is 885 and 640, respectively; and in *Heteroclinus heptaeolus* the number is 320 and 1330 (Table 3.2.3). This difference in numbers on the front and rear sites of the OC are very prominent and reversed in the Blenniidae: in *Ecsenius nalolo* the front

Table 3.2.3 Comparison of numbers of TB between the front part of the OC (teeth bands and valves) and rear (EB and HB) parts of the OC in the largest specimens studied

Species	*TeBs+Valves*	*EB+HB*
Blenniidae		
Cirripectes variolosus	10	80
Ecsenius nalolo	100	220
Hirculops conifer	62	180
Istiblennius rivulatus	398	191
Parablennius sanguinolentus	20	270
Salaria pavo	108	340
Clinidae		
Clinus acuminatus	350	760
C. cottoides	260	460
C. superciliosus	770	1160
C. taurus	80	340
Cristiceps aurantiacus	206	370
Heteroclinus heptaeolus	320	1330
Muraenoclinus dorsalis	148	220
Pavoclinus graminis	1050	1250
Labrisomidae		
Labrisomus nuchipinnis	280	350
L. bucciferus	200	600
Malacoctenus macropus	210	480
Paraclinus nigripinnis	192	150
Starksia lepicoelia	72	280
Tripterygiidae		
Tripterygion tripteronotus	265	240

bears 50 TB and the back 220; in *Hirculops cornifer* bears 42 and 180, respectively; and *Salaria pavo* bears 48 and 340. It is possible that the palisade type of teeth on the jaws of these blennies does not provide space for larger numbers of TB. In various gobies that occupy the same habitat as the studied blennies, most of the oral TB are situated at the front of the jaws, with 2000 in *Gobius bucchichi* and 4040 in *Bathygobius fuscus*, compared to 450 and 380 on the pharyngeal bones (Fishelson and Delarea, 2004). Such differences in TB placement have also been observed in cichlid species (Fishelson, 2005a, b): in the more herbivorous cichlid *Tilapia zillii* and in *Deimidiochromis compressus* most TB are on the front of the jaws, whereas in *Astatotilapia flavijosephi* and *Cichlasoma cyanoguttatum*, the majority are on the pharyngeal bones.

The number of available TB in the studied blennies reflects the size of the 'sensory area' occupied by the exposed sensory endings of the TB. Thus, the largest sensory area (20,000 μm^2) is found in *Clinus superciliosus*; 18,750 μm^2 in *Heteroclinus heptaceus*, and 15,750 μm^2 *in Pavoclinus graminis*. The smallest sensory areas are found in the Blenniidae: 2,585 in *Salaria pavo* and 1,220 μm^2 in *Cirripectes variolosus*. These differences in the number of TB and, accordingly, in the dimensions of the sensory or receptor areas in the front and back of the OC, seem to highlight possible differences in localisation of food items for final identification before ingesting or discarding. As revealed from dissecting the fish, the blennies *Escenius nalolo*, *Cirripectus variolosus* and *Hirculops cornifer* are microfeeders, grassing on microalgae and their infauna from hard substrata, whereas most of the clinids as well as some of the labrisomids feed selectively, preying on small mollusks and crustaceans. This may be the reason for the relatively high numbers of oral TB found in all species of clinid fishes (Table 3.2.2), especially in the viviparous species (Prochazka and Griffiths, 1992; Prochazka, 1998).

Table 3.2.4 presents the differences in the numbers of TB on the various sites within the OC of all the species studied. For example, on the upper lip of the various species the number of TB ranges from zero to 380; within the teeth band from 40-940; and on the EB and HB 80-900 and 40-600 respectively. As demonstrated, these differences are species-specific and only partly correlated with fish length.

To understand such differences in the numbers of TB in the OC of the various species from different genera and families, we need to know much more about their diet and feeding behavior. Linser *et al.* (1998), discussing

Table 3.2.4 Lowest and highest numbers of TB on the various sites of the OC in the studied blennioid fishes

	Upper jaw	*Lower jaw*
lips	0-380	0-320
teeth bands	40-940	60-300
valves	20-380	16-300
vomers	12-240	20-260
palates	HB 40-600, EB 80-900	

the taste buds of the bass, *Micropterus salmoides*, found particular importance in the fact that the taste buds occur adjacent to the teeth of the pharyngeal bones, a phenomenon also prominent in all the studied clinid fish. However, it is reasonably clear that over the course of evolution, the distribution and number of TB in the OC of clinids, as well as of other fishes, were shaped—as were other morphological characters—by environmental and nutritional factors in an orchestrated manner. The present investigation of soft structures organisation in the OC cavity of blennioid fishes is the first one of this kind and as our other investigations (Fishelson and Delarea, 2004; Fishelson *et al.*, 2004, 2005a, b) in various fish taxa, provides insights that expose specific structures that can contribute to our understanding of the microevolution within this higher taxon. For example, the adaptive variability is well illustrated at the species level if we compare the pharyngeal bones of *Salaria pavo*, a turf-algae grazer, with this of *Ecsenius nalolo*, a mesoplankton feeder. In the former, the pharyngeal bones are strong, with strong teeth able to grind the passing food particles (Fig. 3.2.3c), whereas in the latter the bones are rudimental and their teeth in form of spicules, are situated in two parallel rows (Fig. 3.2.3a), more adopted to holding the captured pray. Gathering of more information on the feeding and behavior of these, as well as on other fish, will enable to expose the delicate relationships between their chemosensory systems, dentition and ecomorphology.

Acknowledgements

I am most grateful to the following persons who donated the fishes: Ofer Gon, South African Institute of Aquatic Biodiversity Studies, Grahamstown, S. Africa; Glenn Moore, Western Australian Museum, Perth, WA; Victor G. Springer, National Museum of Natural History, Smithsonian Institution, Washington DC, USA; Mark McGrouther,

Australian Museum, Sydney, AU. Thanks are due to Naomi Paz for her editorial help.

References

Appelbaum, S., J.W. Adron, S.G. George, A.M. Mackie and A.M. Pirie. 1983. The development of the olfactory and gustatory organs of the Dover sole, *Solea solea*, during metamorphosis. *Journal of the Marine Biological Association of the United Kingdom* 63: 97-108.

Appelbaum, S. and R. Riehl. 1997. Scanning electron microscopic observations of the chemo- and mechanoreceptors of the carp larvae (*Cyprinus carpio*) and their relation to behaviour. *Aquatic Living Resources* 10: 1-12.

Bond, C.E. 1996. *Biology of Fishes*. Saunders College Publishing, San Diego, 6th Edition.

Clements, K.D. and D. Raubenheimer. 2006. Feeding and nutrition. In: *The Physiology of Fishes*, D.H. Evans and J.B. Claiborne (eds.). Taylor and Francis (CRC), Boca Raton, pp. 47-82, 3rd Edition.

Doty, R.L. 2003. *Handbook of Olfaction and Gustation*. Marcel Dekker, New York, 2nd Edition.

Ezeasor, D.N. 1982. Distribution and ultrastructure of taste buds in the oropharyngeal cavity of the rainbow trout, *Salmo gairdneri* Richardson. *Journal of Fish Biology* 20: 53-68.

Finger, T.E., W.L. Siver and D. Restrepo. 2000. *The Neurobiology of Taste and Smell*. Wiley-Liss, New York.

Fishelson, L. 2005a. Comparison of taste bud types and their distribution on the lips and oropharyngeal cavity, as well as dentition in cichlid fish (Cichlidae, Teleostei). In: *Fish Chemosenses*, K. Reutter and B.G. Kapoor (eds.). Science Publishers, Enfield (NH), USA, pp. 247-275.

Fishelson, L. 2005b. Histogenesis of the oropharyngeal cavity taste buds and the relevant nerves and brain centers in substrate-brooding and mouth-brooding cichlid fish (Cichlidae, Teleostei). *Anatomy and Embryology* 209: 179-192.

Fishelson, L. and Y. Delarea. 2004. Taste buds on lips and mouth of some blennid and gobiid fishes: Comparative distribution and morphology. *Journal of Fish Biology* 65: 651-665.

Fishelson, L., Y. Delarea and A. Zverdling. 2004. Taste buds form and distribution on lips and in the oropharyngeal cavity of cardinal fish species (Apogonidae, Teleostei), with remarks on their dentition. *Journal of Morphology* 259: 316-327.

Gomahr, A., M. Palzenberger and K. Kotrschal. 1992. Density and distribution of external taste buds in cyprinids. *Environmental Biology of Fishes* 33: 125-134.

Gon, O., L. Fishelson and Y. Delarea. 2007. Comparative morphology of the oropharyngeal cavity of clinid fish (Perciformes: Clinidae), with particular attention to the form, number and distribution of taste buds, and dentition. *African Journal of Marine Science* 29: 283-298.

Grover-Johnson, N. and A.I. Farbman. 1976. Fine structure of taste buds in the barbel of the catfish, *Ictalurus punctatus*. *Cell and Tissue Research* 169: 395-403.

Hansen, A. and K. Reutter. 2004. Chemosensory systems in fish: Structural, functional and ecological aspects. In: *The Senses of Fish: Adaptation for the Reception of Natural Stimuli,* G. Vonder Emde, J. Magdans and B.G. Kapoor (eds.). Narosa Publishing House, New Delhi, Kluwer Academic Publishers, Dordrecht, and Springer-Verlag, Heidelberg, pp. 55-89.

Hara, T.J. 1992. *Fish Chemoreception*. Chapman and Hall, London.

Kapoor, B.G. and T.E. Finger. 2003. Taste and solitary chemoreceptor cells. In: *Catfishes,* G. Arratia, B.G. Kapoor, M. Chardon and R. Diogo (eds.). Science Publishers, Inc., Enfield (NH), USA, pp. 753-769.

Kapoor, B.G., H. Evans and I.A. Pevzner. 1975. The gustatory system in fish. *Advances in Marine Biology* 13: 53-108.

Kasumyan, A.O. and O.M. Prokopova. 2003. Taste preference and the dynamics of behavioral taste response in the tench *Tinca tinca* (Cyprinidae). *Journal of Ichthyology* 41: 640-653.

Kiyohara, S., S. Yamashita and J. Kitoh. 1980. Distribution of taste buds on the lips and inside the mouth in the minnow, *Pseudorasbora parva. Physiology and Behaviour* 24: 1143-1147.

Kiyohara, S. and J. Tsukahara. 2005. Barbell taste system in catfish and goatfish. In: *Fish Chemosenses,* K. Reutter and B.G. Kapoor (eds.). Science Publishers, Enfield (NH), USA and Plymouth (UK), pp. 175-209.

Liem, K.F. 1986. The pharyngeal jaw apparatus of the Embiotocidae (Teleostei): A functional and evolutionary perspective. *Copeia* 1986: 311-323.

Linser, P.J., W.E.S. Carr, H.S. Cate, Ch.D. Derby and J.C. Netherton III. 1998. Functional significance of the co-localization of taste buds and teeth in the pharyngeal jaws of the largemouth bass, *Micropterus salmoides. Biological Bulletin* 195: 273-281.

Livingstone, M.E. 1987. Morphological and sensory specialization of five New Zealand flatfish species, in relation to feeding behavior. *Journal of Fish Biology* 31: 775-795.

Murray, R.G. and A. Murray. 1970. The anatomy and ultrastructure of taste endings. In: *Taste and Smell in Vertebrates,* G.E.W. Wolstenhole and J. Knight (eds.). J. Churchill and A. London. pp. 3-30.

Pavlov, D.S. and A.O. Kasumyan. 1998. The structure of the feeding behavior of fish. *Journal of Ichthyology* 38: 116-128.

Prochazka, K. 1998. Spatial and trophic partitioning in cryptic fish communities of shallow subtidal reefs in False Bay, South Africa. *Environmental Biology of Fishes* 51: 201-220.

Prochazka, K. and C.L. Griffiths. 1992. Observations on the distribution patterns, behavior, diets and reproductive cycles of sand dwelling clinids (Perciformes: Clinidae) from South Africa. *Environmental Biology of Fishes* 35: 371-379.

Reutter, K. 1978. Taste organs in the bullhead (Teleostei). *Advances in Anatomy, Embryology and Cell Biology* 55: 1-98.

Reutter, K. 1991. Ultrastructure of taste buds in the Australian lungfish, *Neoceratodes forsteri* (Dipnoi). *Chemical Senses* 16: 404.

Reutter, K. and A. Hansen. 2005. Subtypes of light and dark elongated taste bud cells in fish. In: *Fish Chemosenses,* K. Reutter and B.G. Kapoor (eds.). Science Publishers Inc., Enfield (NH), USA, pp. 211-230.

Reutter, K. and B.G. Kapoor. (eds.) 2005. *Fish Chemosenses.* Science Publishers Inc., Enfield (NH), USA.

Reutter, K. and M. Witt. 2005. Efferent synapses in fish taste buds. In: *Fish Chemosenses,* K. Reutter and B.G. Kapoor (eds.), Science Publishers, Inc., Enfield (NH), USA, pp. 231-245.

Schulte, E. and A. Holl. 1972. Feinbau der Kopftentakel und ihrer Sinnesorgane bei *Blennius tentacularis. Marine Biology* 12: 67-80.

Shen, S.-C. 1971. A new genus of clinid fishes from the Indo-West Pacific, with a redescription of *Clinus nematopterus. Copeia* 1971: 697-707.

Springer, V.G. 1993. Definition of the suborder Blennoidei and its included families (Pisces: Persciformes). *Bulletin of Marine Sciences* 52: 472-495.

Whitear, M. 1971. Cell specialization and sensory function in fish epidermis. *Journal of Zoology (London)* 63: 237-264.

CHAPTER

3.3

Nasal Olfactory Organs and Olfactory Bulbs in Blennies

Lev Fishelson[1] and Ofer Gon[2]

INTRODUCTION

Olfaction has been intensively studied in various fishes, and the accumulated knowledge has revealed both certain common basic structures, as well as various differences connected with specific ecological and foraging adaptations (Kleerkoper, 1969; Hara, 1975, 1992; Hara and Zielinski, 1989; Stabell, 1992; Shipley and Ennis, 1995; Doty, 2003; Hansen and Reutter, 2004). Reviews of these studies, including studies on the neurophysiology of these organs, have been published by Yamamoto (1982), Caprio (1984), Muller and Marc (1984), Zeiske *et al.* (1992), Hara (1994), Laberge and Hara (2001), Zielinski and Hara (2001), Hansen and Zielinski (2005), Valentinčić (2005), and Zeiske and Hansen (2005); and

***Abbreviations*: OB,** olfactory bulb; **OE,** olfactory epithelium; **OL,** olfactory lobes (forebrain); **ON,** olfactory nerves; **OP,** olfactory placodes; **ORN,** olfactory receptor neurons; **RC,** rod cells; **SL,** standard length (in mm)

Authors' addresses: [1]Department of Zoology, George S. Wise Faculty of Life Sciences, Tel Aviv University, Tel Aviv 69978, Israel. E-mail: fishelv@post.tau.ac.il
[2]South African Institute for Aquatic Biodiversity, P.B. 1015, Grahamstown, 6140, South Africa. E-mail: o.gon@ru.ac.za

studies on the role of olfaction in species and gender recognition have been performed on gobies (Tavolga, 1956), and on blennies (Losey, 1969); while major contributions to our understanding of olfaction, especially of the specificity of odor recognition by the olfactory receptor neuron (ORN), have been made by Buck (1996, 2000)and Dynes and Ngai (1998).

The major site of olfactory reception in fish is located in the nasal cavities on the head which, in ditrematous species, open to the outside via the anterior incurrent and posterior excurrent nares (Fig. 3.3.1a). In most fishes, the olfactory epithelium (OE) is spread on lamellae that are grouped into round or elongated rosettes, attached to a central raphe (Bond, 1996). The number and form of the sensory lamellae increase with the growth of the fish, until a final species-specific number develops. Various researchers have suggested that the number and dimensions of the sensory lamellae reflect the olfactory sensitivity of the species (Bond, 1996; Fishelson, 1997).

In all the fishes studied to date, the OE covering the olfactory lamella—in addition to the supporting and basal cells—is composed of two types of ORN, both with exposed dendrites: type I dendrites bear long cilia attached to a swollen, knob-like exposed surface; type II bear cilia and microvilli on the exposed surface. A special sub-type of ciliated neurons, named rod cells, presents an adhesive group of cilia of a single neuron forming a rod-like structure. These rod cells, also called 'olfactory lashes', 'hairs', and 'compound cilia', are, according to Bröuckner *et al.* (1979) derived from fused kinocilia, or, as argued by Muller and Mark (1984), they are type II ORN with cohering cilia. Hansen and Finger (2000) described in detail a type IV ORN from several teleost species, which they termed crypt cells, in which the cilia and microvilli are contained within an invagination of the apical membrane. Such ORN were previously revealed in the Oman shark (Fishelson and Baranes, 1997). In most of the studied teleosts, the number of receptor cells per mm^2 ranges from 50,000-90,000. Yamamoto (1982) reported up to 500,000 cells/mm^2 in some

Fig. 3.3.1 Ontogenesis of olfactory nares in small embryos of blennioid fishes. **a**: Frontal of *Salaria pavo* (SEM). Scale bar 1 mm. **b**: Head of *Cirripectes variolosus* (SEM). Scale bar 1 mm. **c**: Head of *Dactylopterus lacteus* (SEM). Scale bar 8 mm. **d**: Embryo (5 mm) of *Clinus cottoides* (SEM). **e**: OP of the same (SEM). Scale bar 15 μm (inset LM of the OE. Scale bar 15 μm). **f**: Cross-section of the OB in a 4 mm embryo of *C. superciliosus* (LM). Scale bar 30 μm. **g**: Cross-section of the telencephalon of the same (EM). Scale bar 40 μm. **h**: Brain and OP in 6 mm embryo of *Cristiceps aurantiacus* (SEM). AN, anterior nare; C,

Fig. 3.3.1 Contd.

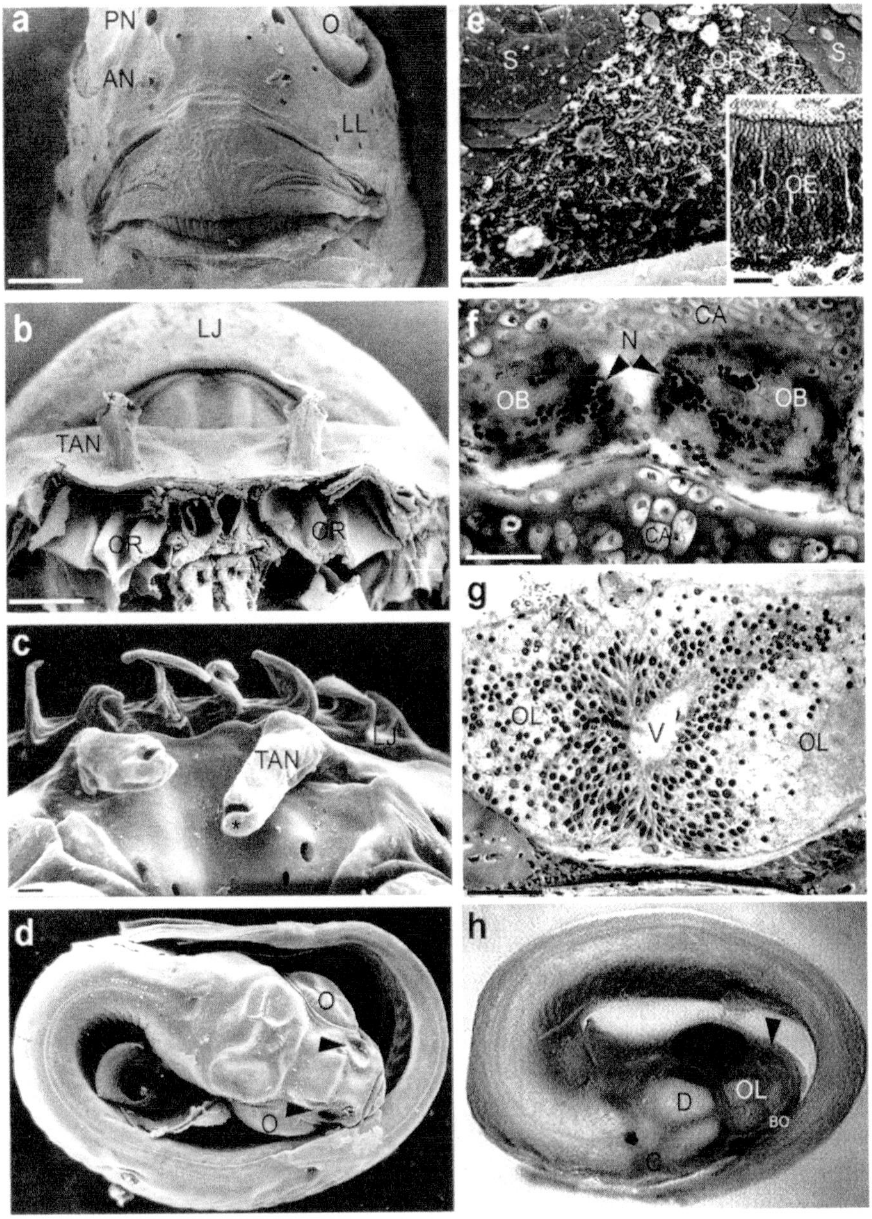

Fig. 3.3.1 Contd.

cerebellum; CA, cartilage; D, diencephalon (optic lobes); LL, lateral line canals; LJ, lower jaw; N, nuclei of mitral cells; O, eye; OB, olfactory bulb; OE, olfactory epithelium; OL, telencephalon (olfactory lobes); OP and arrowhead, olfactory placode; OR, olfactory rosettes; S, surface cells of embryo; PN, posterior nare; TAN, tubular anterior nare; asterisk, "lid" of tubular nare.

species. In addition to the ORN, the nasal epithelium consists of 'non-sensory' cells, bearing long kinocilia that facilitate the inflow and outflow of water across the nasal openings (Zeiske *et al.*, 1992), as well as of supporting cells. The basal cells, from which the other cell types generate, are located along the basal lamina of the OE.

The majority of the primary ORN terminate in the olfactory bulb (OB), from which the neural tract leads to the olfactory lobe (OL) of the forebrain (telencephalon). The major characteristic of OB cytology is revealed in the secondary receptor cells known as mitral cells that mediate between the primary ORN of the OE and the central layer of granular cells. These cells participate in forming the synapses-rich so-called 'olfactory glomeruli', from which signals are carried to the OL and the higher brain centers. According to several authors (e.g., Baier and Korsching, 1994), this pattern of OB organisation is typical for all animals. In various fishes the relationship of the OB to the OL of the brain differs: in *Clarias,* the OB is closely attached to the OE in the nasal cavities and a long neural tract connects between the OB and the OL (Nieuwenhuys, 1982); whereas in other fish, e.g., zebrafish (Byrd and Brunjes, 1995) and salmon (Nieuwenhuys, 1982; Kotrschal *et al.*, 1998), the OB are sessile and closely attached to the OL and no tract is visible. The very early development of these organs during the embryogenesis of fish offers possible evidence of their importance in the early life stages of these organisms (Fishelson 1966, 1997; Appelbaum *et al.*, 1983; Doving and Knutsen, 1993; Dynes and Ngai, 1998; Lindsay and Vogt, 2004).

This chapter focuses on the nasal olfactory organs and OB of species belonging to the fish families Blenniidae, Clinidae, Dactyloscopidae, Labrisomidae and Tripterygiidae, all forming part of the sub-order Blennioidei (Springer, 1993). Representatives of these families are found in all the seas, from the tropics to the ice-covered shores (Bond, 1996), inhabiting various ecological niches and exploiting different food sources. Most species of this sub-order are oviparous, spawning eggs that attach to the substrate, except for a large group of clinids and *Starksia lepicoelia* (Labrisomidae), which are viviparous. Similar to our previous studies (Fishelson and Delarea, 2004; Fishelson, 2005; Fishelson *et al.*, 2006; Gon *et al.*, 2007), this study too is a comparative one, comparing the cytogenesis and adult morphology of the nasal olfactory organs in 16 species of blennioid fishes (Table 3.3.1).

Table 3.3.1 List of species studied, including collection site, standard length of fish (SL in mm) and number of specimens. AU, Australia; SA, South Africa; EM, eastern Mediterranean; RS, Red Sea (Gulf of Aqaba).

Species	*Site of origin*	*No. of specimens*	*Size (SL)*
Blenniidae			
Cirripectes variolosus	RS	2	44, 44
Istiblennius rivulatus	RS	4	49-55
Parablennius sanguinolentus	EM	3	75-83
Salaria pavo	EM	4	67-77
Clinidae			
Clinus acuminatus	SA	2	70, 78
C. arborescens	SA	2	150, 184
C. cottoides	SA	4	88-91
C. superciliosus	SA	4	80-90
C. taurus	SA	2	123, 130
Cristiceps aurantiacus	AU	3	140-150
C. australis	AU	3	62-64
Heteroclinus perspicillatus	AU	3	66-70
Muraenoclinus dorsalis	SA	2	63, 72
Pavoclinus pavo	SA	3	50-54
Dactyloscopidae			
Dactyloscopus lacteus	Galapagos	1	42
Labrisomidae			
Labrisomus nuchipinnis	Tabago	2	104, 106
Malacoctenus macropus	Belize	2	30, 32
Tripterygiidae			
Tripterygion tripteronotus	EM	3	63-70

MATERIALS AND METHODS

The studied fish (Table 3.3.1) were collected from the Red Sea, Eastern Mediterranean Sea, Australia, and the Cape region of South Africa. Samples of fish were also donated by several museums (see Acknowledgements). Organs used for the cytological studies were harvested from freshly collected, decapitated fish, and from specimens fixed in 10% formaldehyde and preserved in 70% ethanol. For light microscopy (LM), the organs from the fresh fish were fixed in Bouin's fluid or 10% formaldehyde, washed in ethanol and embedded in Periplast. These samples were cut serially (8-10 μm) and stained with Ehrlich-hematoxylin, Methylene blue or 0.5% cresyl-violet. For electron microscopy (EM) the samples were fixed in 3.5% glutaraldehyde + 1% formaldehyde, buffered with 0.1M cacodylate at pH 7.2. For transmission electron microscopy (TEM), the samples were post-fixed in osmium tetraoxide and embedded in Epon. Semithin sections of these (2-3 μm) were stained with cresyl-violet and toluidin blue; the microsections were stained with uranyl acetate and lead citrate, and studied with a JEOL 1200 EX. For scanning electron microscopy (SEM), following fixation, the samples were critically point dried with CO_2, gold plated and studied with a JEOL JSM 840A. All measurements were performed on micrographs of TEM and SEM, or with the aid of a light microscope micrometer.

RESULTS

Biology of the Studied Fish

The Blennioidei comprises predominantly small (30-300 mm SL) benthic fish, mainly predatory along the rocky and sandy shores; some are algivorous, found among algae, including kelp, and some dwell among the branches of corals. Their distribution ranges from the intertidal zone to waters over 100 meters deep. Several species, like the Red Sea blennies, *Istiblennius* spp. and especially *Alticus kirkii*, are amphibious in their behavior, dwelling among and on the frequently exposed tidal rocks, whereas only a few species, e.g., *Ecsenius midas*, forage in the open water column. Some species, e.g., *Clinus taurus*, *C. superciliosus* and *C. cottoides* (Clinidae), and *Labrisomus nuchipinnis* (Labrisomidae), are robust, with a sharply pointed or sub obtuse head. Others, like *Tripterygion tripteronotus*, *Parablennius zvonimiri* and *Heteroclinus perspicillatus*, are small, delicate species. The predatory fishes (e.g., Clinidae, Tripterygiidae, Labrisomidae) feed on various crustaceans (isopods, amphipods, shrimps) and on the small patellid and mytilid mollusks of their rocky habitat, or are found on the macroalgae, like kelp and other. Others, like the intertidal Blenniidae, are more algivorous, grazing on turf algae of hard substrata or sieving particles from the soft bottom. The majority of the Blennioidei are oviparous, the males select the nest-site for their territory and the mature females join them there to spawn their adhesive eggs, which remain in the males' care until hatching (Fishelson, 1963; Patzner *et al.*, 1986; Almada *et al.*, 1995). The Clinidae include a large group of viviparous species in which the reproducing males, upon copulating with the receptive females, transfer sperm-bundles (spermatozeugmata) into the females' ovaries

(Fishelson *et al.*, 2006). The embryos develop in their follicles until they are 16-20 mm in length and morphologically mature. At this stage, they are released into the water as juveniles (Veith, 1979; Gunn and Thresher, 1991; Prochazka and Griffiths, 1992).

In the studied fish, the two nasal cavities are situated on the upper side of the head, latero-dorsal or dorsal to the mouth corners, medially separated by the process of the maxillae bone. Each cavity has two nares (nostrils), the anterior one providing the incurrent opening, and the posterior one the excurrent. In numerous species, either all the nares, or only the incurrent pair, are framed by nuchal (nasal) cirri, in the form of a single or branching extension. As these cirri bear taste buds (Fishelson, pers. obs.), it is possible that they serve to taste the inflowing water. In some species, e.g., *Labrisomus nuchipinnis, Salaria pavo, Cirripectes variolosus*, and the sand-dwelling *Dactyloscopus lacteus*, the anterior nare is tubular and provided apically with a lid that prevents the penetration of sand particles (Fig. 3.3.1a, b, c).

Ontogenesis of the Olfactory Organs

The initial indication of olfactory primordia (OP) can be observed in the embryos of *Salaria pavo* (Blenniidae) and *Clinus cottoides* (Clinidae) 24 and 50 hours after fertilisation, respectively. They consist of two groups of neuroectodermal cells, situated near the apical end of the embryo, anterior to and close below the forming neural tube. Neurogenesis of the sensory epithelium in these groups of cells can be first recognised in 2 mm embryos of the oviparous *S. pavo*, and in 4 mm embryos of the viviparous clinids, *C. cottoides* and *Cristiceps aurantiacus*. Shortly after the surface epithelium retreats, these groups of cells become exposed. At this stage, the OP are round and slightly concave, 50-80 μm in diameter (2,500-6,400 μm^2), situated in front of the eyes and sparsely covered with microvilli and cilia (Fig. 3.3.1d). The olfactory epithelium of the OP is pseudo stratified, 30 μm thick, with elongated cells and nuclei situated at various distances from the forming basal lamina (Fig. 3.3.1e and inset). The axons of the ORN begin to protrude across the basal lamina. Concomitantly, organisation and cytological formation of the forebrain progress, with little indication of the OB, sessile on the primordia of the olfactory lobes (OL). At this stage, larger neurons (precursors of mitral cells) are dispersed along the periphery of the OB anlage (Fig. 3.3.1f). Initial stages of cellular segregation can also be seen in the OL (Fig. 3.3.1g).

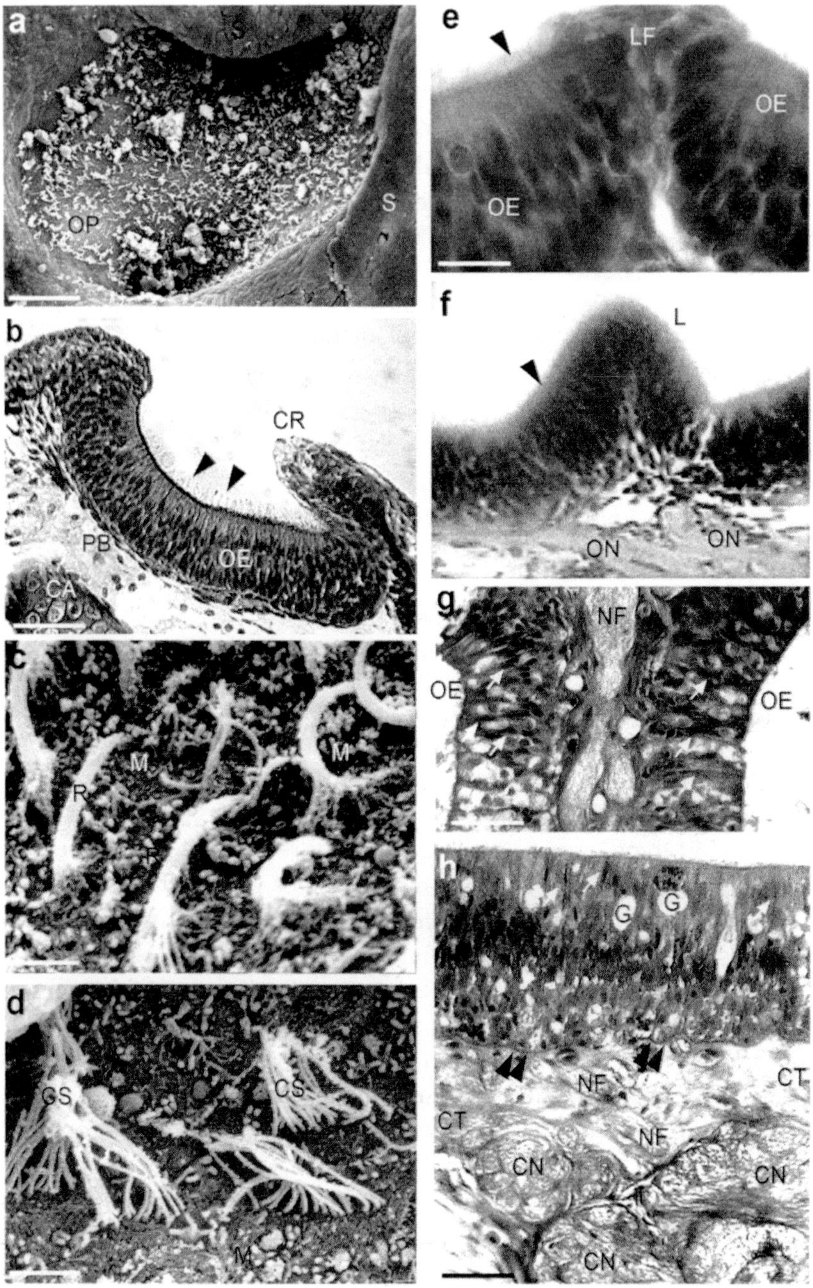

Fig. 3.3.2 Stages in embryogenesis of OE in blennioid fishes. **a**: OP of 6 mm *Clinus superciliosus* (SEM). Scale bar 20 μm. **b**: Section of OP epithelium of *C. cottoides* embryo

Fig. 3.3.2 Contd.

In 6.0 (±06) mm embryos of *C. cottoides*, *C. acuminatus* and *Muraenoclinus dorsalis*, the OP increases in diameter to 160 (± 20) μm (Figs. 3.3.1h, 3.3.2a), measuring ca. 25,400 (±200) μm^2; the density and length of the cilia strongly increase (Fig. 3.3.2b). In some embryos, the first rod-bearing sensory cells can be recognised; each rod is 5-6 μm long (Fig. 3.3.2c). The exposed dendrites of the ORN are evenly spaced, and the differences between the microvillar and ciliated receptors are more pronounced (Fig. 3.3.2d). At this stage, the plication of the OE begins (Fig. 3.3.2e, f), leading toward lamellation and an accompanying increase in size of the sensory epithelium. During this plication, large numbers of sensory axons become enclosed between the OE of the two sides of the lamellae, forming small bundles (Fig. 3.3.2g) that extend into the sub-epithelial connective tissue (Fig. 3.3.2h). Elongation and filia formation of the non-myelinated axons continue below the OE, creating bundles that gradually join to form the future olfactory nerve (ON) (Fig. 3.3.3a and inset) that exit the nasal capsule on its ventral side.

The morphological development of the head in the growing embryos increases the distance between the OE of the nose and the primordial OB, thereby increasing the length of the ON, which is composed of both non-myelinated and myelinated nerve fibers. Interestingly, from the initial stages of the morphogenesis of the trunk somites in the studied viviparous clinids, the caudal part of the embryo's body bends anteriorly and around the head; thus, until hatching, the sensory centers (eyes, brain, oral opening and OP) are covered by the tail and caudal fin and also protected (Fig. 3.3.3b). It remains unknown whether there is a specific function to this positioning, or it is simply the best position for the embryo in the follicle. With additional growth of juveniles of the blennies, *Salaria pavo*

Fig. 3.3.2 Contd.

(6 mm) (LM). Scale bar 25 μm. **c**: Primary rod cells (R1) of *Salaria pavo* embryo (4 mm) (SEM). Scale bar 2 μm. **d**: Ciliate ORC in OE of 7 mm *Clinus acuminatus* embryo (SEM). Scale bar 2 μm. **e**: Primary stage of olfactory lamella formation in *C. cottoides* (7 mm) (LM). Scale bar 8 μm. **f**: Section of lamella at a more advanced stage (8 mm embryo) (LM). Scale bar 15 μm. **g**: Longitudinal section of OE lamella of 6 mm *Cristiceps aurantiacus* (LM). Scale bar 20 μm. **h**: Section of OE and ON-filia of 6 mm *C. superciliosus* (LM). Scale bar 24 μm. CN, cross-section of olfactory nerve bundles; CR, ridges along margins of OP; CS, ciliated dendrites of ORN; CT, connective tissue; G, goblet cells; L, primary olfactory lobules of rosettes; LF, olfactory lobule in formation; M, microvillar cells; NF, nerve filaments; OE, olfactory epithelium; ON, olfactory nerve; OP, olfactory placode; R, rod-type ORN; PB, primordial basal lamina; S, skin surface cells; arrowheads, cilia; white arrow, ORN; double arrowhead, basal lamina.

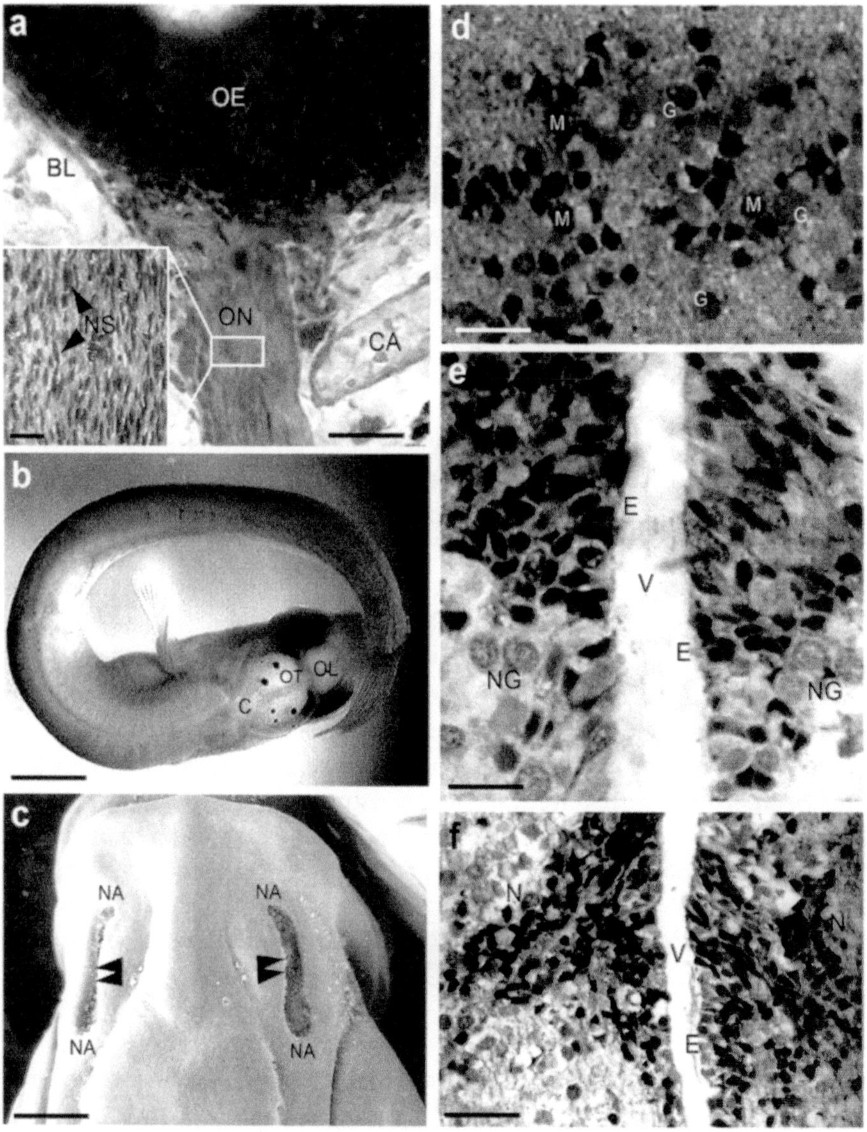

Fig. 3.3.3 Olfactory organ maturation in larger embryos of blennioids. **a**: ON in a maturing embryo (8 mm) of *Clinus cottoides* (LM). Scale bar 15 µm (inset, ON enlarged, Scale bar 5 µm). **b**: 12 mm embryo of *C. superciliosus* (12 mm) (Digital image). **c**: Head of 14 mm embryo of *C. superciliosus* (SEM). Scale bar 12 µm. **d**: Mitral cells in OB of *C. acuminatus* embryo (14 mm) (LM). Scale bar 6 µm. **e**: Ependym formation in the ventricle of the telencephalon (OL) in a 14 mm embryo of *C. cottoides* (LM). Scale bar 20 µm. **f**: Brain-nuclei formation in OL of a 14 mm embryo (LM). Scale bar 25 µm. BL, basal lamina of

Fig. 3.3.3 Contd.

and *Parablennius sanguinolentus*, and of embryos of the viviparous clinids, *Clinus* spp. and *Cristiceps aurantiacus*, three processes become prominent: a pronounced and gradual sinking of the OE, paralleled by closing of the nasal capsules (Fig. 3.3.3c), and elongation and differentiation of the ON. At this stage, the ON incorporate myelin-envelopes around their axons. Concomitantly, cytological maturation of the OB and OL continues. At this stage, again, the length of the OB is ca. 2/3 of that of the OL. In the oviparous blennies, three days after fertilisation, the pericaria of larger neurons, mitral and possible glial cells of the rudimental OB can be seen dispersed on the periphery of the organ, differing in their various forms and nuclei (Fig. 3.3.3d). One day later, the pericaria of the mitral cells gradually take their place along the plexiform layer and internal granular layers of the OB, concomitant with the onset of formation of the ependymal layer along the ventricle of the forebrain (Fig. 3.3.3e). This process is especially prominent in 8 ($\pm$1.0) mm long *Salaria pavo* and 12 ($\pm$0.8) mm embryos of *Clinus cottoides* in which formation of the ependymal layer along the ventricle of the forebrain and the accumulation of neural nuclei, is now more advanced (Fig. 3.3.3f). The axonal endings of the ON fibers and dendrites of the mitral cells begin to establish numerous synaptic junctions, forming the synaptic glomeruli on this site (Fig. 3.3.4a).

In 14 mm SL embryos of *C. cottoides* and *C. superciliosus*, the OP are deeply sunk in the dermis, elongated-oval in shape, ca. 220 μm long and 40 μm wide; the axonal filia are densely distributed within the sub-epithelial connective tissue. At this stage, the forebrain is almost equal in dimensions to the midbrain, with small OB attached anteriorly. In 16 mm SL embryos, the OP are 300 μm long and narrower in the middle than at the apical parts; and in 20 mm SL mature pre-parturition embryos, free in the ovary, the 400 μm long OP are closed in the middle and open at the apices, where the nares of the maturing fish will form. In all the studied blennies, the OE within the nasal capsules continues to plicate (Fig. 3.3.4b), finally resulting in the formation of the olfactory rosettes, each with three lamellae (petals) (Fig. 3.3.4c).

Fig. 3.3.3 Contd.

olfactory epithelium; C, cerebellum; CA, cartilage; D, diencephalon; E, ependym in the first ventricle; G, glial cell nuclei; M, mitral cells; N, aggregates of neurons; NA, nares in formation; NG, cells of forming brain nuclei; NS, nuclei of Schwan cells; OE, olfactory epithelium; OL, olfactory lobes (telencephalon); ON, forming olfactory nerve; V, first ventricle; double arrowhead, olfactory placodes.

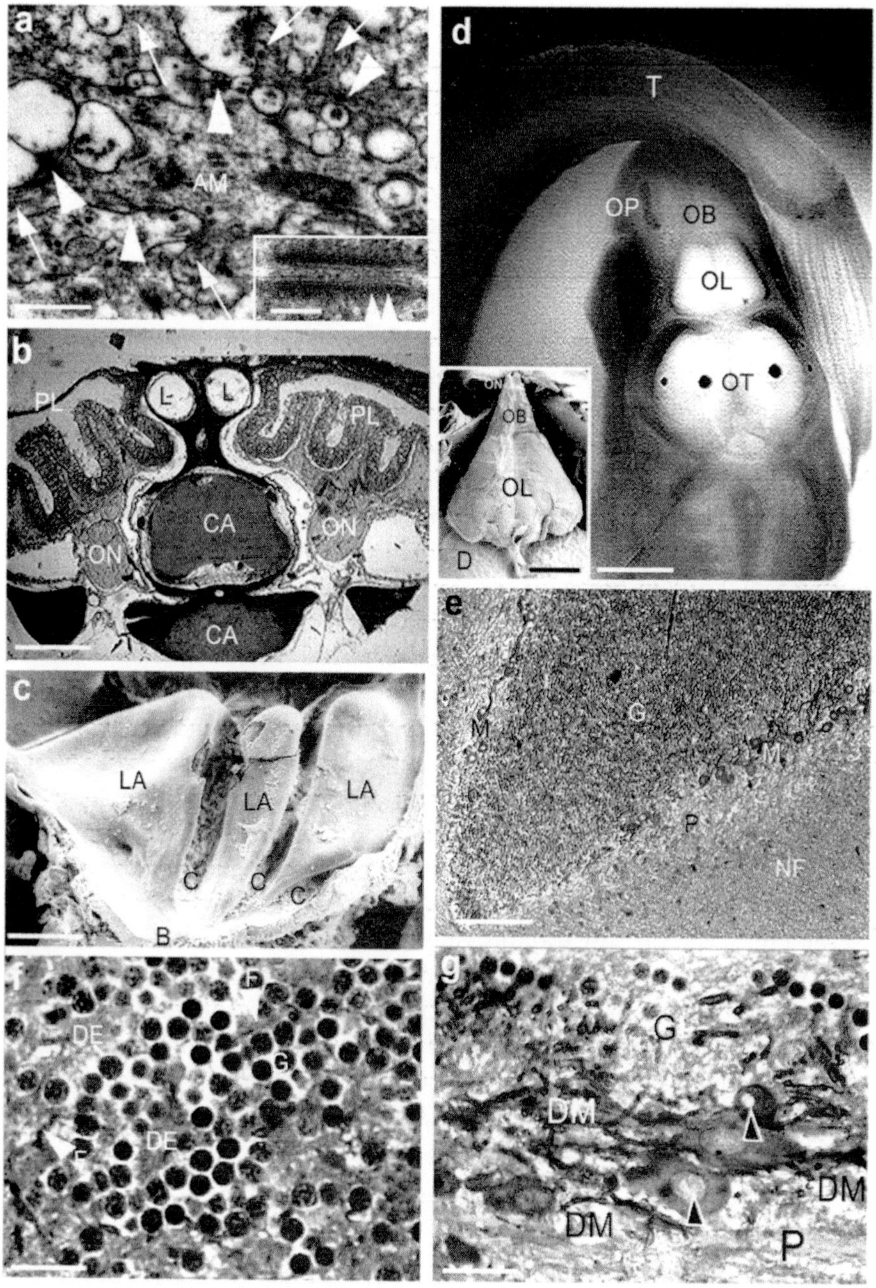

Fig. 3.3.4 Maturation of olfactory rosettes and OB in blennioid fishes. **a**: Apical part of a mitral axon in the OB of *Clinus superciliosus* 12 mm embryo (TEM). Scale bar 0.5 µm (inset,

Fig. 3.3.4 Contd.

In 12 mm SL *S. pavo* and 16 mm embryos of *C. cottoides*, the midbrain has grown larger than the OL (Fig. 3.3.4d). The OB is now also larger and reveals three clearly distinct layers: the inner, granular, core, occupied by numerous smaller anaxonic neurons of 5 μm, surrounded by delicate dendrites. The nuclei of this layer are chromophilic, with an opaque karyosome (Fig. 3.3.4e, f). The second layer, extending around the core, comprises the large mitral cells, prominent in the dimensions of their somata (6-7 μm), rich in granulated endoplasmic reticules, elongated mitochondria, and large and irregular granulated nuclei with a prominent nucleolus (Fig. 3.3.4g). The dendrites of these cells establish many more synaptic connections with axons of the ORN, thereby forming the plexus of the glomerular layer (Fig. 3.3.4d, f). The outermost layer of the juvenile OB is richly occupied by parallel bundles of neural fibers, a few neurons and blood capillaries. Maturation and segregation of the sensory dendrites in the OE continues, their somata are delicate and strongly elongated, ca. 3 μm in diameter at the surface (Fig. 3.3.5a). In these embryos, the ON are often concentrated into three filia.

The surface epithelial cells situated around the margins of the olfactory organ bear microvilli, whereas the more distant cells reveal the well-developed ridge-mazes typical of fish epidermis (Fig. 3.3.5b). The OE is 45 (±6) μm thick, with numerous blood capillaries and axon filia in the sub-epithelial connective tissue. The basal lamina is sieve-like, perforated by numerous axons of the ON. The OB and adjacent parts of the OL are now well advanced in their internal organisation.

As observed in mature clinids and also rainbow trout (Laberge and Hara, 2001), although most of the axons of the ORN terminate in the

---

Fig. 3.3.4 Contd.

tight junction in glomerulus (TEM) Scale bar 4 nm). **b**: Primordial olfactory rosettes of *C. superciliosus* (LM). Scale bar 0.5 mm. **c**: Olfactory rosette of *C. cottoides* (35 mm SL) (SEM). Scale bar 200 μm. **d**: Head of 18 mm *C. cottoides* embryo (Digital image). Scale bar 400 μm (inset OB and OL of the same (SEM). Scale bar 0.6 mm). **e**: Section of the OB of *C. superciliosus* (16 mm embryo) (LM). Scale bar 150 μm. **f**: Granular layer in OB of *C. cottoides* (16 mm) (LM). Scale bar 20 μm. **g**: Mitral cells of the same (LM). Scale bar 6 μm. AM, apical part of a mitral cell axon; B, base of olfactory rosette; C, interlamellar canal of water-flow; CA, cartilage; D, dendrites in granular layer; DE, diencephalon; DM, dendrites of mitral cell; F, neural filaments in granular layer; G, granular layer of OB; L, lateral line canal, cross section; LA, lamella of mature olfactory rosette; OB, olfactory bulb; OL, olfactory lobes; ON, olfactory nerve; OP, olfactory placode; NF, layer of neural filaments; P, plexiform layer in OB; PL, primordial lamellae of rosettes; T, tile of embryo; white arrowhead, synapses in young glomerulus of OB; white arrow, lateral projections of mitral axon head; white-framed arrowhead, nuclei of mitral cells; double arrowhead, tight junction in a glomerulus.

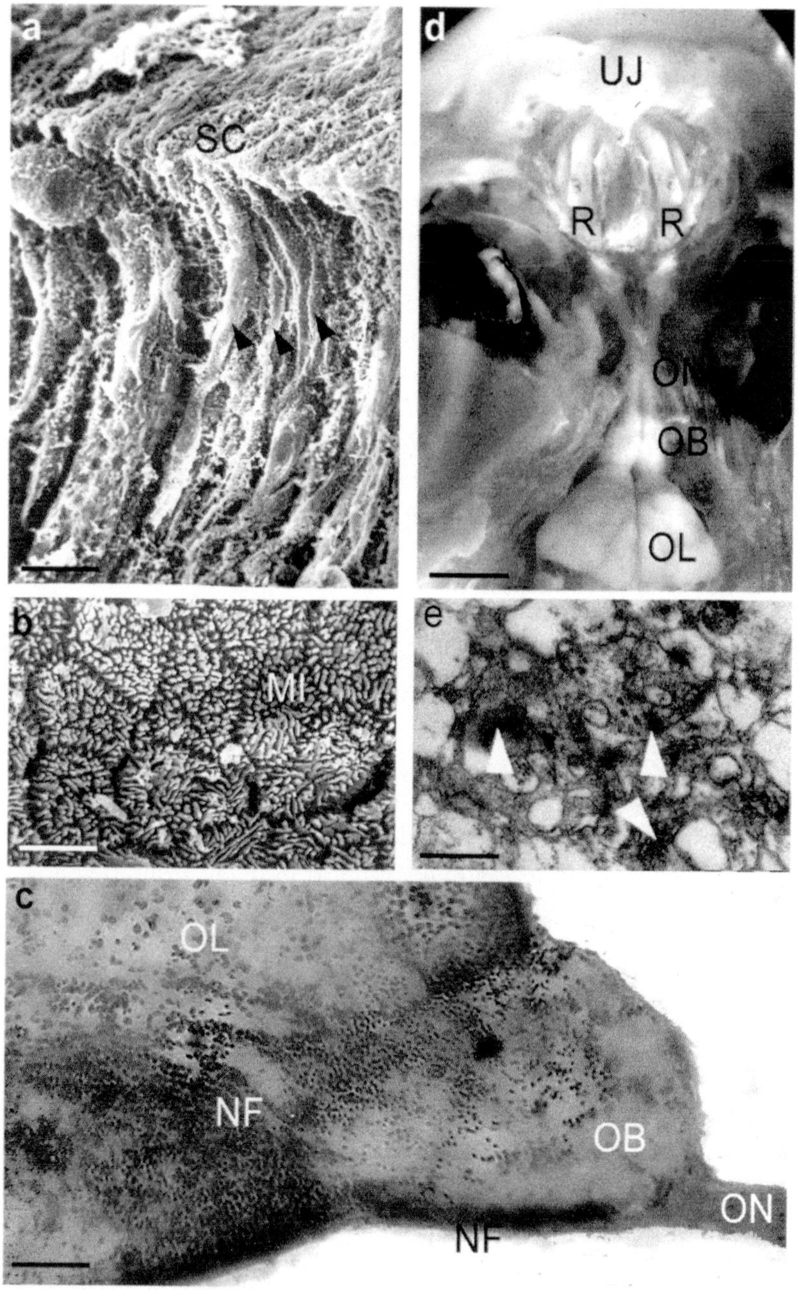

Fig. 3.3.5 Olfactory organs in adult blennioids. **a**: ORN in lamellae of *Heteroclinus perspicillatus* (SEM). Scale bar 6 µm. **b**: Microvillar cells on the margin of olfactory lamella

Fig. 3.3.5 Contd.

plexus of the glomerular layer of the OB, some bypass this site and continue into deeper parts of the forebrain along the ventral part of the OB (Fig. 3.3.5c). At this stage of development, it is already clear that the OB in all the studied blennies remains sessile on the OL. The ON are now 4-8 mm long in the various species, and the olfactory tract, as typical for numerous teleosts (Bond, 1996), integrates here between the OB and OL. In *Salaria pavo* of 12 mm SL and *Clinus* embryos of 20 mm SL the structure of the OB closely resembles that of free-swimming juvenile fish of 30 mm SL, in which both the OB and OL preserve the paired structure observed in embryos (Fig. 3.3.5d). In such embryos, the synaptic networks in the OB are well developed, with numerous globular glomeruli forming a spread of receptor and output neurons—the mitral cells. The numerous branches of the mitral cells make contact with the surrounding dendrites through various types of synapses (Fig. 3.3.5e).

Comparison of Olfactory Organs in Adult Fish

The relative proportions of the various components of the olfactory system in adult specimens of the studied blennioid fish are fairly similar to those observed in the juvenile fish, especially the dimensions of the olfactory rosettes, OB and OL. In all these fishes, the rosettes are composed of three lamellae, each one comprising two layers of OE enclosing a connective tissue sheet, from which the ORC axons and blood vessels exit (Fig. 3.3.4b). The difference lies in the length of the ON and in cytological differentiation of the layers in the OB. In all the studied blennies and clinids, the rosettes are very small compared to the fish body length. For example, in *Parablennius sanguinolentus* of 80 mm SL, they are 1.8 × 1.6 mm in size, in *Clinus taurus* and C. *arborescens* of 150 mm SL they are 2.4 mm × 1.8 mm, and in *Istiblennius rivulatus* and *Salaria pavo* of 60 mm SL they are 1.0 × 0.8 mm. Similar dimensions of the olfactory rosettes have also been found in other studied blennies. The ventral side of the rosettes attaches to the inner membrane of the nasal cavity, while the dorsal side

Fig. 3.3.5 Contd.

of *Parablennius sanguinolentus* (SEM). Scale bar 12 µm. **c**: Vertical section of forebrain of *Salaria pavo* (LM). Scale bar 150 µm. **d**: Olfactory system of *S. pavo*. Scale bar 1.2 mm. **e**: Synaptic center in OB of *Clinus cottoides* (TEM). Scale bar 2 µm. MI, microvilli; NF, neural fibers; OB, olfactory bulb; OL, olfactory lobes; ON, olfactory nerves; R, olfactory rosettes; UJ, upper jaw; black arrowheads, ORN; white arrowheads, synapses.

exposes the lamellae (petals) to the lumen. These lamellae differ in shape among the various species. For example, in *Heteroclinus perspicillatus,* the largest lamella in the rosette is 0.180 mm^2 and bears 55,000 receptor cells (Fig. 3.3.6a). In *Clinus superciliosus,* the two lateral lamellae are larger and triangular, enclosing the narrower and more elongated central one (Fig. 3.3.4c). The largest lamella is 0.136 mm^2, of which the ORN cover 0.108 mm^2, giving a total of ca. 36,000 cells. In *C. cottoides* the rosettes slightly differ, the largest lamella is 0.175 mm^2, and possesses ca. 59,000 ORN. In some species, e.g., *C. taurus, Malacoctenus macropus,* and *Labrisomus nuchipinnis,* the lamellae are robust (Fig. 3.3.6b), whereas in *Tripterygion tripteronotus, I. rivulatus, Cirripectes variolosus, Salaria pavo* and *Pavoclinus graminis* they are narrow and elongated, and in *C. acuminatus* they are leaf-like, delicate, 1.2 × 1.4 mm in dimension, partly overlapping each other. In *Cristiceps aurantiacus* and *C. australis,* the rosettes feature a basal trapezoid plate, and the three lamellae protrude perpendicularly from the plate (Fig. 3.3.6c). The lamellae form three 'channels' at their bases for water flow, covered with kinocilia-bearing cells. In all the other species studied, each lamella attaches separately to a basal point. The OE in mature juveniles is 65 (± 8) μm thick, with a strong basal lamina, with numerous blood capillaries and axon filia below. Within the brain, the OL and optic lobes are almost equal in dimension and the OB forms a prominent apical cone.

Types of ORN in the Studied Species

As already described for various fishes (Zeiske *et al.*, 1992; Morita and Finger, 1998; Zeiske and Hansen, 2005), in the studied juvenile blennioid species too three types of ORN can be distinguished:

Ciliated sensory neurons

Type I: narrow and slender neural cells 20 μm long, with extremely thin (0.3 μm) axons and protruding knob-like dendrites, 3.5 (± 0.5) μm in

Fig. 3.3.6 Olfactory rosettes and ORN of adult blennioid fishes. **a**: Rosette of *Heteroclinus perspicillatus* (SEM). Scale bar 1 mm. **b**: Rosette of *Clinus taurus* (SEM). Scale bar 250 μm. **c**:Rosette of *Cristiceps australis* (SEM). Scale bar 180 μm. **d**: Micrograph of the OE of *Clinus cottoides* (TEM). Scale bar 1 μm. **e**: Ciliated dendrite of ORN of *C. superciliosus* (SEM). Scale bar 2 μm. **f**: OE of *C. cottoides* (SEM). Scale bar 3 μm. **g**: Surface of olfactory lamella of *C. acuminatus* (SEM). Scale bar 8 μm. B, base of rosette; BP, basal plate; C, water-flow canal between lamellae; CE, centriole; D, knob-like dendrite; KC, kinocilia; SC, cilia of sensory neurons.

Fig. 3.3.6 Contd.

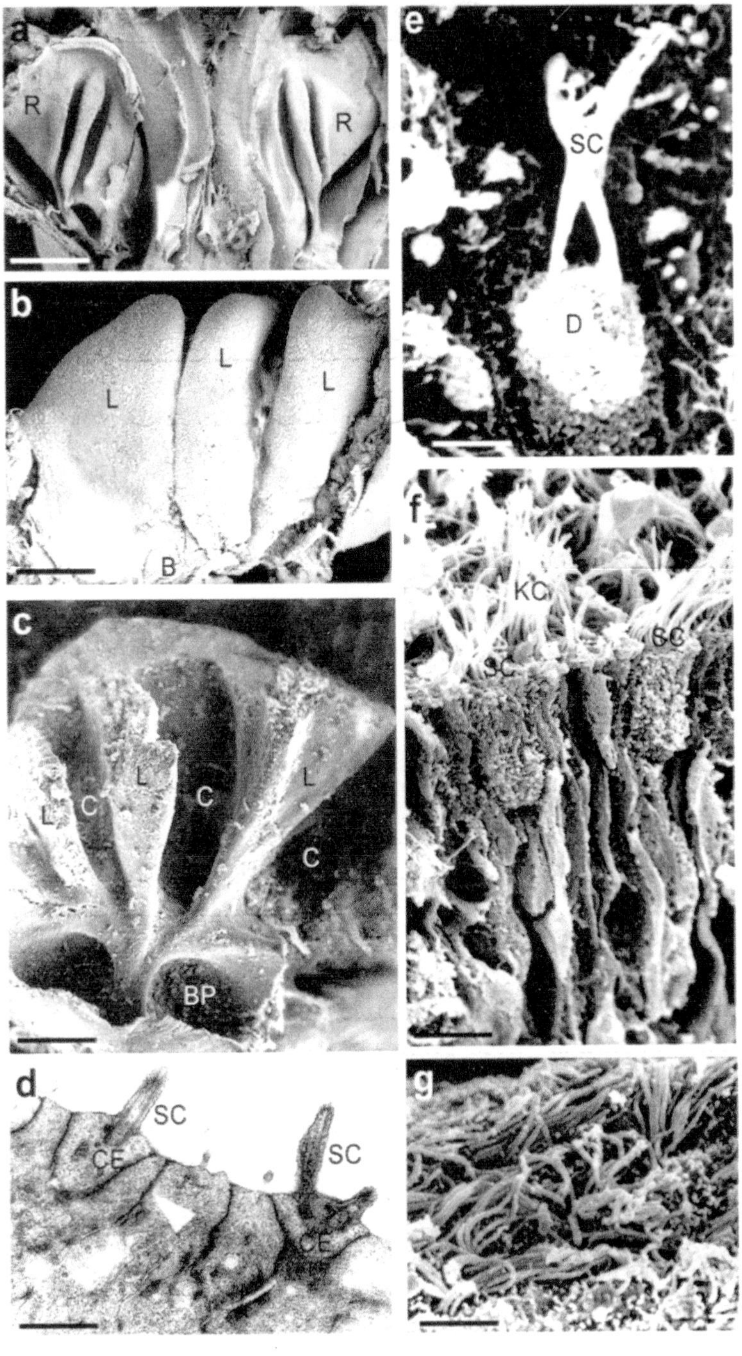

Fig. 3.3.6

diameter. Each of these cells bears 5-6 cilia 6.5 (± 0.5) μm long and 0.3 μm thick (Figs. 3.3.5a, 3.3.6d, e); each cilium is attached to a centriole and supported by 9+2 tubules. Type II: broad, cylindrical in shape at the wider, upper part (4.0 μm) (Fig. 3.3.6f), and narrow-filamentous (0.5 μm) at the lower part; with a flat exposed surface bearing cilia and a few microvilli; their mitochondria are elongate.

Microvillous sensory neurons

Cells with flat, less protruding dendrites, each of which bears numerous microvillii, 2.0 (±0.5) μm long. Ciliated and microvillii-bearing ORN frequently intermingle on the various sites of the rosettes (Fig. 3.3.7a). In numerous instances, the microvillous-type cells also bear a single or a few short cilia (Fig. 3.3.7b).

Rod-like cells

In these cells, a group of cilia from a single ORN adhere together to form a rod-like structure. Some studies have termed such cells true sensory neurons (Zielinski and Hara, 2001), while others consider them to be senescent cells or artifacts, resulting from their being handled for microscopy (Muller and Mark, 1984; Zeiske and Hansen, 2005). The typical rod cell, as observed in the present study (RC1), and described in various publications (Zielinski and Hara, 2001), forms a rod 6.0 (± 0.5) μm long and 1.5 μm thick (Fig. 3.3.7c). This type was observed in all the species studied. In addition, in the studied *Clinus taurus* and *C. arborescens* two different, previously undescribed, rod-type cells were revealed, termed here RC2 and RC3: RC2 cells are 15 (± 2) μm long and bear 24-26 adhering cilia forming curtain-like structures 7.5 (± 1.0) μm long and 3.6 (± 0.6) μm wide (Fig. 3.3.7d). RC3 cells are 25 (± 4) μm long and very slender; each bears a 10 (± 2.0) μm long and 1.0 μm thick rod, with pointed apex (Fig. 3.3.7e). In both the species, RC2 and RC3 are often found together on various sites of the lamellae.

Fig. 3.3.7 SEM micrographs of ORN on the rosettes of blennioid fishes. **a**: Microvillar cells between cells with micro ridges of *Clinus cottoides*. Scale bar 6 μm. **b**: Receptor cells with microvilli and cilia of *Cristiceps aurantiacus*. Scale bar 10 μm. **c**: Rod-like, R1, cells of *C. cottoides*. Scale bar 2 μm (inset, isolated rod, Scale bar 2 μm).**d**: Rod-like R2 cells of *C. taurus*. Scale bar 12 μm. **e**: Elongate R3 rod-like cells of *C. taurus*. Scale bar 10 μm. **f**: Margin of olfactory lamella with ORN and micro ridged surface cells of *Salaria pavo*.

Fig. 3.3.7 Contd.

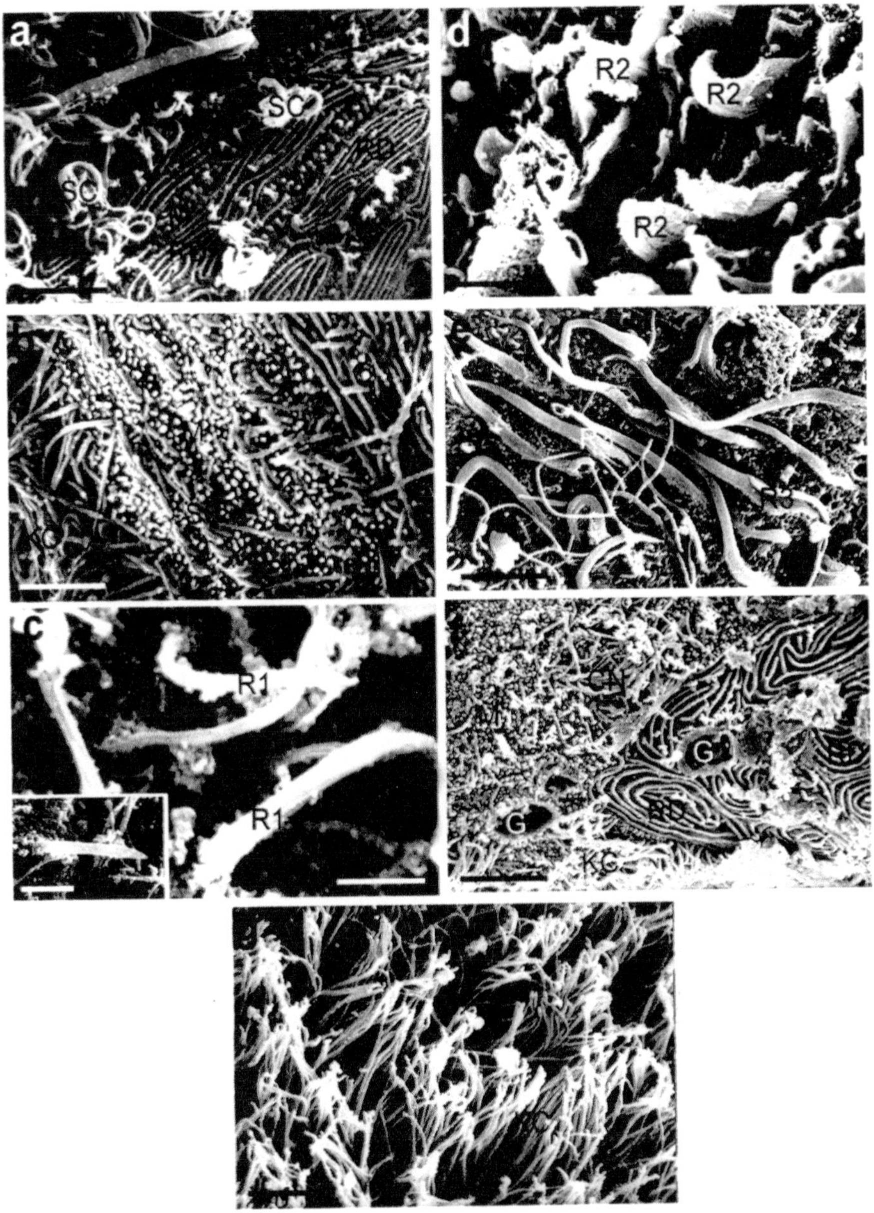

Fig. 3.3.7 Contd.

Scale bar 8 μm.**g**: Kinocilia on olfactory lamella of *C. acuminatus*. Scale bar 20 μm. CN, ciliated ORN; G, goblet mucus cell opening; KC, cells with kinocilia; M, ORN with microvilli and a few cilia; RD, micro ridge-mazes on surface membranes of cells; R1, the 'normal' rod-like cell; R2, curtain type rod cell; R3, elongated rod cell.

Distribution of ORN on the Olfactory Rosettes

In all the species studied, the rims of the lamellae (petals) of the rosettes are covered by surface epithelial cells with their typical ridge mazes and numerous openings of mucus cells (Fig. 3.3.7f). These regular surface cells extend and descend laterally onto the sensory lamellae of the rosettes. On the lamellar surface, they are limited by microvilli cells and cells bearing 9.0 (± 06) μm long and 0.2 μm thick kinocilia of the so-called 'non-sensory' epithelium (Fig. 3.3.7g). The dimensions of the site covered by these neurons differ among the various species. In the studied species of Labrisomidae (*Labrisomus nuchipinnis, Malacoctenus macropus*) and in *Heteroclinus perspicillatus* (Clinidae), the transition between the non-sensory epithelium and ORN is a gradual one, and ciliated and microvillar dendrites can be seen between the ridged cells. In *H. perspicillatus* the upper-central part of the lamellae bears a patch of 0.12 mm^2 of mixed type dendrites, possessing microvilli and some with a few cilia of 5-6 μm long. In *Clinus superciliosus* and *C. cottoides* the upper-exterior parts of the lamellae are covered by kinocilia, while the lower-inner parts of lamella are dominated by ORN with knob-like dendrites, interspersed with RC1 cells. In *Tripterygion tripteronotus, Parablennius sanguinolentus* and *Salaria pavo,* the epithelial cells with ridge mazes also extend along the entire marginal part of the lamellae, while ORN with knobs and RC occupy the inner part of the lamellae. In *C. taurus* and *C. acuminatus,* RC2 and RC3 dominate the rosettes, intermingling with RC1 and kinocilia of non-sensory cells. In all the species studied, the basal channels between the lamellae are covered with kinocillia.

DISCUSSION

Olfaction in fish, based on bipolar neurons with dendrites exposed in the nasal mucosa, has been studied intensively, especially in the second half of the twentieth century (Kleerekoper, 1969; Hasler and Sholtz, 1983, for older literature). More recently, cytological studies of the OE have been followed by physiological investigations, revealing the finely tuned reactions of the olfactory organs to various acids and odors (Buck, 1996, 2000; Laberge and Hara, 2001; Hansen *et al.*, 2003; Hansen and Zielinski, 2005; Hara, 2005 and references therein), including pheromones linked to reproduction (Barata *et al.*, 1998). Moreover, some of these studies have shown that the various ORN are specialised for detection of various multiple odorants (Buck, 2000; Hansen *et al.*, 2005), and the sensitivity of the organ can be compared to that of a retina. Atema (1980) was the first to state that, 'chemical pictures may not be essentially different from visual, mechanical, sound, or electrical pictures in containing specific information' (p. 61).

In most instances, the studies on olfaction have focused on a single species. The present study describes the ontogenesis and morphology of the olfactory organs and OB in 16 species of closely-related blennioid fish, with the aim of revealing the possible common elements as well as structural differences among the various species. As observed in the ontogenesis of other sensory organs, development of the olfactory organs begins very early on in the life history of blennioid fishes, providing the juveniles with a sense of smell at a very early stage of their development, as also observed in other fish species (Fishelson, 1966, 1997; Bröuckner *et al.*, 1979; Appelbaum *et al.*, 1983). According to the results of the present study, the maturation of the OP and the process of their invagination toward the formation of the nasal capsules and nares occur in the oviparous blennies in free-swimming larvae of 4-5 mm, much earlier than in the viviparous species. The latter liberate their offspring at 16-18 mm, at a more mature stage of morphological and neural development. For example, closed olfactory capsules with rosettes of three lamellae are found in free-swimming *S. pavo* of 12 mm SL and in pre-parturition embryos of the viviparous *Clinus cottoides* of 18-20 mm SL, the length at which they are liberated. Such differences in timing could be the result of different ambient temperatures (Hansen and Zielinski, 2005) or an expression of adaptations to micro ecological factors. One example of the second possibility can be seen in the sensory development of cichlids, when comparing the propagules of bottom-spawners to those of mouth-brooders (Fishelson, 1997). In the bottom-spawners (as in the oviparous blennies), the sensory organs develop much earlier than in the mouth-brooding (or viviparous clinids) species, providing the 4-6 mm hatched larvae with sensory ability. The brooding species (oral in cichlids or ovarian follicles in clinids) attend to and protect their propagules for much longer and the sensory organs of these embryos and larvae develop much later, providing an additional example of developmental adaptations.

In the studied blennioid fishes, as described for zebrafish embryos (Dynes and Ngai, 1998), the initial ontogenesis of the olfactory organs begins 24-36 hours following fertilisation, from two OP situated on the head, below the anterior part of the developing brain. It is from this layer of cells that all the cell-types observed in the adult OE develop. The cells that form the OE in blennies are similar to those of other fish (Hansen and Zielinski, 2005 and references therein), and are mainly ciliated or microvillous ORN, with supporting cells between them, and basal cells close to the basal membrane. The ontogenesis and differentiation of the ON and OB take place concomitantly with the differentiation of the OE. With the sinking and closing of the slit of the OP, their ends form a pit, providing a site for formation of the nares. Such a type of inlet and outlet development is termed the 'salmon-type' and is recognised as an ancestral type of organisation.

In the studied blennies *Salaria pavo* and *Parablennius sanguinolentus*, the axons of the OE attach to the forming OB ca. 30 hours after fertilisation, later than that observed in zebrafish embryos (Dynes and Ngai, 1998). Such differences could have been caused by the different water temperatures in which the propagules were held: 26°C in the zebrafish and 22°C in blennies. During this initial differentiation of the OE, the developing axons of the ORN gradually grow and extend, forming the ON that terminate within the OB. The length of these nerves depends on the head-structure of the fish: in species with pointed heads they attain 7-8 mm whereas in those with blunt profiles they reach 4-5 mm. The OB of the studied blennioid fishes is sessile on the forebrain (see also Kotrschal *et al.*, 1998), and cytologically very similar to that described for other fish-species (Laberge and Hara, 2001), especially with respect to the layers of various secondary neurons within the OB.

The major difference between the OE in adults of the studied by us blennies and those of the other studied fish lies in the occurrence in the former of several types of rod-like sensory cells, in addition to the rod cell already described for fishes (Hansen and Zielinski, 2005). In some species of the genus *Clinus*, two additional rod-cell types were revealed, termed here RC2 and RC3. These cells differ in their elongated rods and were not found in the other studied blennioid fishes. In all the species studied, the rods are composed of cohesive groups of cilia, but it remains unclear as to whether these are a special form of ORN, artifacts produced during the handling of the fish, or cells undergoing degeneration. In all the studied blennies, the OE extends over three lamellae that form relatively small rosettes of ca. 1.5 × 2.0 mm. Several authors (see Kleerekoper, 1969; Bond, 1996) have suggested that the dimensions and numbers of lamellae on the rosettes reflect the level of olfactory acuity of the fish. Calculations of the apparent number of ORN on the larger lamellae of fish of similar dimensions revealed that some species, such as *Salaria pavo* possess ca. 55,000/mm^2 on the lamellae, whereas *Hetroclinus perspicillatus* of similar size possess ca.117,000/mm^2. In the larger *Cristiceps aurantiacus* and *Clinus taurus*, there are up to 160,000/mm^2 ORN on the larger lamellae. Despite this apparent richness in olfactory neurons, the studied blennioid fish appear to be macrosmatic, and odor probably does not play a dominant role in their search for food. However, the possibility cannot be excluded that olfaction is involved in reproductive behavior (Barata *et al.*, 1998; Cole and Stacey, 2006).

The OB develops concomitantly with differentiation of the OE and ON, and is sessile on the OL. The first grouping of dendrites toward glomeruli formation and the development of three neural layers within the OB in the OB in *Salaria pavo* and in *Clinus cottoides* was evident in embryos 3 mm length (three days old), and of 4-5 mm (age unknown) respectively. According to Lindsay and Vogt (2004), the larvae of zebrafish four days after fertilisation, and in a similar developmental stage, already respond to chemical stimuli. Li *et al.* (2005), in an elegant study, demonstrated that at this age in zebrafish the spatial map of the synaptic centers in the OB is already established. In the very early stage of OB development, the mitral cells are dispersed on the periphery of the organ; later (in 5-8 mm embryos) these cells become organised as a cell-layer between the external plexus and granular layers of the bulb. At this stage of development, the inner, granular layer of unaxonic neural cells is already clearly organised. Around this core, the dendrites of mitral cells and ON-extensions progress to form the plexus of synaptic glomeruli (Laberge and Hara, 2001; Kosaka and Kosaka, 2005). The axons of the mitral cells project from these synapse centers also into the caudal parts of the forebrain and the hypothalamus, as also observed in brown trout by Anadon *et al.* (1995). In juvenile *Salaria pavo* and in mature embryos of viviparous clinids, some of the olfactory fibers of the OE can clearly be seen to bypass the OB and continue directly into the more-posterior parts of the brain. With continued growth of the fish, the form of the OB changes from conical to more bulbous. In adult fish, its dimension ranges from 0.6 × 0.8 mm and 0.3 mm thick in smaller species (*S. pavo*) to 1.2 × 1.4 mm and 0.5 mm thick in larger ones (*Clinus cottoides*), forming a volume of 0.110 mm^3 to 0.840 mm^3 respectively. Growth of the OB and the OL follows a certain allometry: namely, as the fish grows the OL grows at a greater rate than the

OB. For example, in juveniles, the OB is ca. 1/3 the size of the OL whereas in larger adult specimens it is ca. 1/6 the size. This may suggest that functional and cytological maturation of the OB occurs before that of the OL.

Acknowledgements

The authors are very grateful to the following persons who donated the fishes: Glenn Moore, Western Australian Museum, Perth, WA; Victor G. Springer, National Museum of Natural History, Smithsonian Institution, Washington DC, USA; Mark McGrouther, Australian Museum, Sydney, AU; and Paul Cowley, Monica Mwale and Thembela Bushula, South African Institute for Aquatic Biodiversity, Grahamstown. Thanks are due to Naomi Paz for her editorial help and to Varda Vexler for help with the art work.

References

Almada, V.C., E.J. Gonçalves, R.F. Oliveira and A.J. Santos. 1995. Courting females: Ecological constraints affect sex roles in a natural population of the blenniid fish *Salaria pavo*. *Animal Behavior* 49: 1125-1127.

Anadon, R., M.J. Mamso, I. Rodriguez-Moldes and M. Beccera. 1995. Neurons of the olfactory organ projecting to the caudal telencephalon and hypothalamus: A carbocyanine-dye labeling study in the brown trout (Teleostei). *Neuroscience Letters* 191: 157-160.

Appelbaum, S., J.W. Adron, S.G.George, A.M. Mackie and B.J.S Pirie. 1983. The development of the olfactory and gustatory organs of the Dover sole, *Solea solea*, during metamorphosis. *Journal of the Marine Biological Association of the United Kingdom* 63: 97-108.

Atema, J. 1980. Chemical senses, chemical signals, and feeding behavior in fishes. In: *Fish Behavior and its Use in the Capture and Culture of Fishes*. J.E. Bardach, J.J. Magnuson, R.C. May and J.M. Reinhart (eds.). ICLARM Conference Proceedings 5, Manila, Philippines, pp. 57-101.

Baier, H. and S. Korsching. 1994. Olfactory glomeruli in the zebrafish form an invariant pattern and are identifiable across animals. *Journal of Neurosciences* 14: 219-230.

Barata, E.N., A.V.M. Canario and E. Melo. 1998. The role of olfaction in fish reproduction: its importance for conservation of littoral fish. In: *Behavior and Conservation of Littoral Fishes*, V.C. Almada, R.F. Oliveira and E.J. Gonçalves (eds.), ISBA, Lisboa, pp. 127-148.

Bond, C.E. 1996. *Biology of Fishes*. Saunders College Publishing, San Diego, 6th Edition.

Bröuckner, H., E. Zeiske and R. Melinkat. 1979. Development of the olfactory organ in the rainbow fish, *Nematocentris macculochi* (Atherinoformes, Melanotaeniidae). *Cell and Tissue Research* 200: 53-68.

Byrd, C.A. and P.C. Brunjes. 1995. Organization of the olfactory system in adult zebrafish: histological, immunohistochemical and quantitative analysis. *Journal of Comparative Neurology* 358: 247-259.

Buck, L.B. 1996. Information coding in the vertebrate olfactory system. *Annual Review of Neurosciences* 19: 517-544.

Buck, L.B. 2000. The molecular architecture of odor and pheromone sensing in mammals. *Cell* 100: 611-618.

Caprio, J. 1984. Olfaction and taste in fish. In: *Comparative Physiology of Sensory Systems*. L. Bolis, R.D. Keynes and S.H. Madrell (eds.), Cambridge University Press, Cambridge, pp. 257-284.

Cole, T.B. and N.E. Stacey. 2006. Olfactory responses to steroids in the African mouth-brooding cichlid, *Haplochromis burtoni* (Günther). *Journal of Fish Biology* 68: 661-680.

Doty, R.L. 2003. *Handbook of Olfaction and Gustation*. Marcel Decker, New York, 2nd Edition.

Dynes, J.L. and J. Ngai. 1998. Path finding of olfactory neuron axons to stereotyped glomerular targets revealed by dynamic imaging in living zebrafish embryos. *Neuron* 20: 1081-1091.

Doving, K.B. and L.A. Knutsen. 1993. Chemokinesis in marine fish larvae. In: *Physiological and Biochemical Aspects of Fish Development*. B.T. Walter and K.J. Fyhn (eds.). University of Bergen Press, Norway, pp. 139-145.

Fishelson, L. 1963. Observations on littoral fishes of Israel I. Behavior of *Blennius pavo* Risso (Teleostei, Blenniidae). *Israel Journal of Zoology* 12: 67-80.

Fishelson, L. 1966. Untersuchungen zur vergleichenden Entwicklungsgeschichte der Gattung *Tilapia* (Cichlidae, Teleostei). *Zoologische Jahrbücher (Abt.Anatomie)* 83: 571-656.

Fishelson, L. 1997. Comparative ontogenesis and cytomorphology of the nasal organs in some species of cichlid fish (Cichlidae, Teleostei). *Journal of Zoology (London)* 243: 281-294.

Fishelson, L. 2005. Histogenesis of the oropharyngeal cavity taste buds and the relevant nerves and brain centers in substrate-brooding and mouth-brooding cichlid fish (Cichlidae, Teleostei). *Anatomy and Embryology* 209: 179-192.

Fishelson, L. and A. Baranes. 1997. Ontogenesis and cytomorphology of the nasal olfactory organs in the Oman shark, *Iago omanensis* (Triakidae), in the Gulf of Aqaba, Red Sea. *Anatomical Record* 249: 400-412.

Fishelson, L. and Y. Delarea. 2004. Taste buds on the lips and mouth of some blenniid and gobiid fishes: comparative distribution and morphology. *Journal of Fish Biology* 65: 651-665.

Fishelson, L., O. Gon, V. Holdengreber and Y. Delarea. 2006. Comparative morphology and cytology of the male sperm-transmission organs in viviparous species of clinid fishes (Clinidae: Teleostei, Perciformes). *Journal of Morphology* 267: 1406-1414.

Gon, O., L. Fishelson and Y. Delarea. 2007. Comparative morphology of the oropharyngeal cavity in clinid fishes (Perciformes: Clinidae), with particular attention to the form, number and distribution of taste buds, and dentition. *African Journal of Marine Science* 29: 283-298.

Gunn, J.S. and R.E. Thresher. 1991. Viviparity and reproductive biology of clinid fishes (Clinidae) from temperate waters of Australia. *Environmental Biology of Fishes* 31: 323-344.

Hansen, A. and T.E. Finger. 2000. Phyletic distribution of crypt-type olfactory receptor neurons in fishes. *Brain, Behavior and Evolution* 55: 100-110.

Hansen, A. and K. Reutter. 2004. Chemosensory systems in fish: Structural, functional and ecological aspects. In: *The Senses of Fish: Adaptation for the Reception of Natural Stimuli*, G. Von der Emde, J. Magdans and B.G. Kapoor (eds.). Narosa Publishing House, New Delhi, Kluwer Academic Publishers, Dordrecht and Springer-Verlag, Heidelberg, pp. 55-89.

Hansen, A. and B.S. Zielinski. 2005. Diversity in the olfactory epithelium of bony fishes: Development, lamellar arrangement, sensory neuron cell types and transduction components. *Journal of Neurocytology* 34: 183-208.

Hansen, A., S.H. Rolen, K. Anderson, Y. Morita, J. Caprio and T.E. Finger. 2003. Correlation between olfactory receptor cell type and function in the channel catfish. *Neurosciences* 23: 9328-9339.

Hansen, A., S.H. Rolenc, K. Anderson, Y. Morita, J. Caprio and T.E. Finger. 2005. Olfactory receptor neurons in fish: Structural, molecular and functional correlates. *Chemical Senses* 30 (Supplement 1): 1311.

Hara, T.J. 1975. Olfaction in fish. *Progress in Neurobiology* 5: 271-335.

Hara, T.J. 1992. *Fish Chemoreception*. Chapman and Hall, London.

Hara, T.J. 1994. Olfaction and gustation in fish: An overview. *Acta Physiologica Scandinavica* 152: 207-217.

Hara, T.J. 2005. Olfactory responses to amino acids in rainbow trout: Revisited. In: *Fish Chemoreception*, K. Reutter and B.G. Kapoor (eds.). Science Publishers, Inc., Enfield (NH), USA, pp. 32-64.

Hara, T.J. and B. Zielinski. 1989. Structural and functional development of the olfactory organ in teleosts. *Transaction of the American Fisheries Society* 118: 183-194.

Hasler, A.D. and A.T. Sholtz. 1983. *Olfactory Imprinting and Homing in Salmon. Investigations into the Mechanisms of the Imprinting Process*. Springer-Verlag, Berlin.

Kleerekoper, H. 1969. *Olfaction in Fishes*. Indiana University Press, Bloomingdon.

Kosaka, K. and T. Kosaka. 2005. Synaptic organization of the glomerulus in the main olfactory bulb: Compartments of glomerulus and heterogeneity of the periglomerular cells. *Anatomical Sciences International* 80: 80-90.

Kotrschal, K., M.J. Van Staaden and R. Huber. 1998. Fish brains: Evolution and environmental relationships. *Reviews in Fish Biology and Fishery* 8: 373-408.

Laberge, F. and T.J. Hara. 2001. Neurobiology of fish olfaction: A review. *Brain Research Reviews* 36: 46-59.

Li, J., J.A. Mack, M. Souren, E. Yaksi, S. Higashima, M. Mione, J.R. Fetcha and R.W. Friedrich. 2005. Early development of the functional spatial maps in the zebrafish olfactory bulb. *Journal of Neurosciences* 25: 5784-5795.

Lindsay, J.E. and R.G. Vogt. 2004. Behavioral responses of newly hatched zebrafish (*Danio rerio*) to amino acid chemostimulation. *Chemical Senses* 29: 93-100.

Losey (Jr), G.S. 1969. Sexual pheromone in some fishes of the genus *Hypsoblennius* Gill. *Science* 163: 181-183.

Morita, Y. and T.E. Finger. 1998. Differential projection of ciliated and microvillous olfactory receptor cells in the catfish, *Ictalurus punctatus*. *Comparative Neurology* 7: 539-550.

Muller, J.F. and R.E. Marc. 1984. Three distinct morphological classes of receptors in fish olfactory organs. *Journal Comparative Neurology* 222: 482-495.

Nieuwenhuys, R. 1982. An overview of the organization of the brain of actinopterygian fishes. *American Zoologist* 22: 287-310.

Patzner, R.A., M. Seiwald, M. Adlgasser and G. Kaurin. 1986. The reproduction of *Blennius pavo*. V. Reproductive behavior in natural environment. *Zoologischer Anzeiger* 216: 338-350.

Prochazka, K. and C.L. Griffiths. 1992. Observations on the distribution patterns, behavior, diets and reproductive cycles of sand-dwelling clinids (Perciformes: Clinidae) from South Africa. *Environmental Biology of Fishes* 35: 371-379.

Shipley, M.T. and M. Ennis. 1995. Functional organization of olfactory system. *Journal of Neurobiology* 30: 123-176.

Springer, V.G. 1993. Definition of the suborder Blennioidei and its included families (Pisces: Perciformes). *Bulletin of Marine Sciences* 52: 472-495.

Stabell, O.B. 1992. Olfactory control of homing behavior in salmonids. In: *Fish Chemoreception*, T.J. Hara (ed.). Chapman and Hall, London, pp. 249-270.

Tavolga, W.N. 1956. Visual, chemical and sound stimuli in sex-discriminatory behavior in the gobiid fish *Bathygobius soporator*. *Zoologica* 41: 259-271.

Valentinčić, T. 2005. Olfactory discrimination in fishes. In: *Fish Chemosenses*, K. Reutter and B.G. Kapoor (eds.). Science Publishers Inc., Enfield (NH), (USA), pp. 65-85.

Veith, W.J. 1979. Reproduction of the live-bearing teleost *Clinus superciliosus*. *South African Journal of Zoology* 14: 208-211.

Yamamoto, M. 1982. Comparative morphology of the peripheral olfactory organ in teleosts. In: *Chemoreception in Fishes*, T.J. Hara (ed.). Elsevier, Amsterdam, pp. 39-59.

Zeiske, E., B. Theisen and H. Bröuckner. 1992. Structure, development and evolutionary aspects of the peripheral olfactory system. In: *Fish Chemoreception*, T.J. Hara (ed.). Chapman and Hall, London, pp. 13-39.

Zeiske, E. and A. Hansen. 2005. Development and evolution of the olfactory organ in Gnathostome fish. In: *Fish Chemosenses*, K. Reutter and B.G. Kapoor (eds.) Science Publishers Inc., Enfield (NH), (USA), pp. 1-29.

Zelinski, B.S. and T.J. Hara. 2001. The neurobiology of fish olfaction. In: *The Sensory Biology of Jawed Fishes*, B.G. Kapoor and T.J. Hara (eds.), Science Publishers Inc., Enfield (NH), (USA), pp. 347-366.

SECTION
4

Reproduction and Development

CHAPTER

4.1

Reproductive Organs in Blennies

Robert A. Patzner[1] and Franz Lahnsteiner[2]

INTRODUCTION

The female reproductive organs of blennies (Blenniidae) represent a type which is common in many species of fishes. They consist of a paired ovary which, in most cases, undergoes a yearly reproductive cycle. However, the male reproductive organs differ from those of other groups of fish. The testis is in connection with different accessory organs such as testicular glands, seminal vesicles and blind pouches.

Rathke (1836) described two different parts of the testis in *Parablennius sanguinolentus* and Hyrtl (1850) found in *P. gattorugine* enlarged spermatic ducts which he explained as 'Samenblasen' and the blind pouches as *appendices prostatrices*. Brock (1878) differentiated in *P. sanguinolentus* an outer mature and an inner immature part of the testes. Guitel (1893) studied the male reproductive organs of some Blenniidae and one Clinidae. Courrier (1991, 1922) mentioned interstitial glands between testis and testicular gland in *Lipophrys pholis*. Champy and Gley

Authors' address: Dept. Organismic Biology, University of Salzburg, Hellbrunnerstr. 34, 5020 Salzburg, Austria.
E-mails: [1]robert.patzner@sbg.ac.at; [2]franz.lahnsteiner@sbg.ac.at

(1922) described the testicular glands of *Aidablennius sphynx*, *Blennius ocellaris*, *P. palmicornis* (= *P. sanguinolentus*), *P. tentacularis* and *Salaria pavo*. Eggert (1931) was the first to describe in detail the morphology and the development of sexual organs of several species of Mediterranean and tropical Blenniidae, Tripterygiidae and Clinidae. Sperotti (1940/41) and Coujard (1941) studied the annual cycle of the testicular gland in some species of blennies. Chieffi and Botte (1964) investigated the histochemistry and found 3ß-hydroxysteroid-dehydrogenase positive cells in the testicular glands of Mediterranean blennies. Deliro *et al.* (1965) reported on the testicular glands, too. Tong (1967) made endocrine studies on *L. pholis* and Blüm (1972) and Blüm *et al.* (1972) performed some endocrine experiments and described the effect of the gonadotropic hormones—FSH and LH—upon the male genital tract of *A. sphynx*. Pern and Laumen (1977) carried out a study of the annual cycle of the reproductive organs in *S. pavo*. They studied the endocrine regulation of the organs by application of exogenous hormones. Denoix (1984), Heymer (1985), Singh (1989) and Santos (1995) described the male gonads of *P. pilicornis*, *S. basilisca*, *Ophioblennius atlanticus*, and *P. sanguinolentus parvicornis*, respectively. In a series of publications from the Institute of Zoology at the University of Salzburg, Austria, further descriptions of the morphology of the gonads of Mediterranean blennies were presented (Papitsch *et al.*, 1981; Seiwald, 1986; Lahnsteiner, 1987; Patzner and Seiwald, 1987a; Seiwald and Patzner, 1987a, b; Richtarski and Patzner, 1988; Lahnsteiner *et al.*, 1988; Patzner *et al.*, 1988; Patzner and Seiwald, 1988a, 1988b; Seiwald and Patzner, 1989; Richtarski, 1990; Richtarski and Patzner, 2000). The morphology of the male reproductive organs of two tropical Blenniidae and one Chaenopsidae was described by Patzner (1989, 1991). Only the papers of Lahnsteiner and Patzner (1990a, b, c) and Lahnsteiner *et al.* (1990, 1993a) gave information on the functions of the testicular accessory organs in blennies. De Jonge *et al.* (1989) and de Jonge and Videler (1989) described the testis-testicular gland complex of *Tripterygion* (Tripterygiidae) and Rasotto (1995) described the male reproductive apparatus of some tropical Chaenopsidae, Tripterygiidae, Labrisomidae and Dactyloscopidae. Notes on the development of the testicular organs of *S. pavo* can be found in Kaurin and Patzner (1988) and Patzner and Kaurin (1997).

Blenniid fishes have a specialised mode of sperm production. In their testis, germ cells do not differentiate into spermatozoa but spermatids are released into the testicular gland (Lahnsteiner and Patzner, 1990b). There, and in the spermatic duct, further differentiation processes of spermatids into spermatozoa occur. Studies on reproductive accessory organs of blennies outside the Mediterranean and the Atlantic are rare.

OVARY

Ovary and oogenesis of blenniid fishes were investigated in *Lipophrys* (=*Blennius*) *pholis* (Qasim, 1957; Tong, 1967; Shackley, 1976; Dunne, 1977; Shackley and King, 1977a, b), *Coryphoblennius galerita* (Fives, 1980), *Salaria* (=*Blennius*) *pavo* (Patzner, 1983; Ruchon *et al.*, 1993), *Parablennius cornutus* and *Scartella emarginata* (Eyberg, 1984), *Omobranchus punctatus* (Isamail and Clayton, 1990), *Parablennius parvicornis* (Santos, 1995) and in *Aidablennius sphinx* (Carrasson and Bau, 2003).

The oogenesis of the Mediterranean and East-Atlantic *S. pavo* (Patzner, 1983; Gonçalves and Almada, 1997) is similar to that of other teleost fishes (e.g., Bara, 1960, 1963). The development of the oocytes in the ovary can be divided into seven stages. In September, after the end of the spawning season, the GSI of the female reaches its minimum (Fig. 4.1.1). The ovary of *S. pavo* shows a considerably long resting phase compared to other fish (e.g., Barr, 1963) and even compared to other blennies, as for example *L. pholis* (Dunne, 1977; Shackley and King, 1977a) and *C. galerita* (Fives, 1980), both from the Atlantic.

During the winter, from November to February, no changes can be observed in the ovary. The GSI increases very slightly until the middle of March. At that time, the oocytes have already started their further growth.

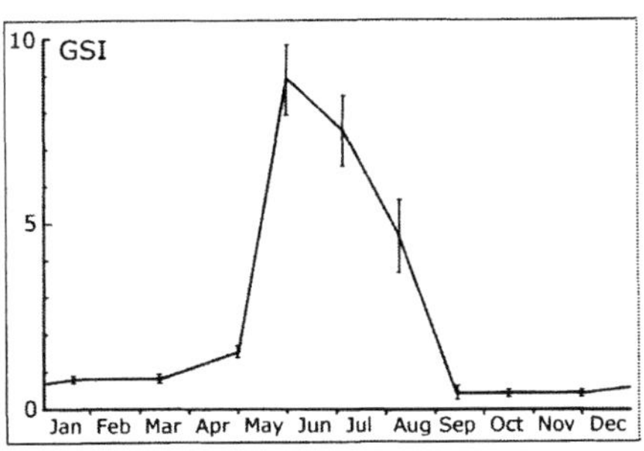

Fig. 4.1.1 Gonadosomatic indices (GSI) of female *Salaria pavo* in the course of one year. Vertical bars represent standard errors (after Patzner, 1983).

At the end of April and beginning of May, when the water temperature is already above 13°C, the GSI has increased to twice its value, and during May there is an enormous increase of the GSI. With the beginning of June, when the water temperature is higher than 18°C, the GSI reaches its maximum (Fig. 4.1.1). At that time, the ovary is nearly filled with oocytes of stage VI and stage VII (Patzner, 1983).

The spawning season of *S. pavo* lasts from the beginning of June to about the middle of August. During this time, a regular decrease of the GSI of the ovary, caused by the repeated spawning of females, can be observed. The ruptured follicle left after discharge of the ripe egg obviously does not form a *corpus luteum*. It shrinks rather quickly and disappears. No histological changes can be seen within the ovary during the spawning season. There is no visible increase in the number of ovulated follicles. All stages of oocytes can still be observed in the middle of August (Patzner, 1983).

TESTIS

All investigated Blenniidae have paired testes. The testes are composed of branching seminiferous tubules which are orientated dorsoventrally and empty into the ducts of the testicular gland (Fig. 4.1.2). The seminiferous

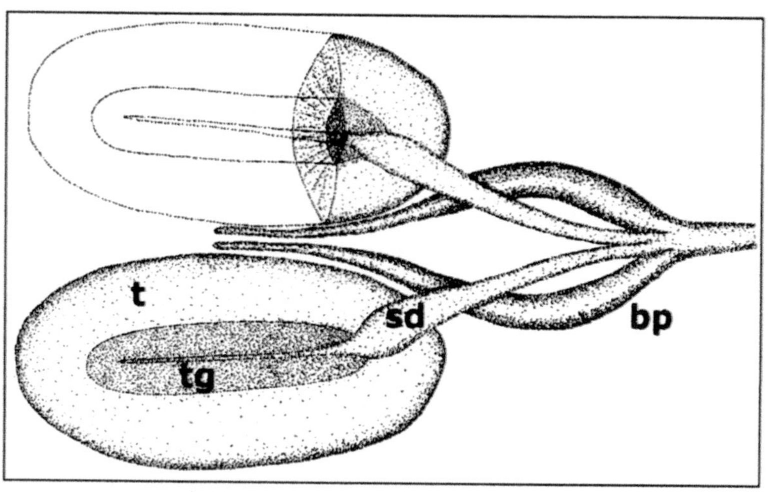

Fig. 4.1.2 Schematic drawing of reproductive system of male *Salaria pavo*, with the posterior side to the right. t = testis; tg = testicular gland, sd = spermatic duct; bp = blind pouch (after Seiwald and Patzner, 1987b).

tubules are embedded in the testicular stroma which consists of connective tissue, blood vessels, nervous tissue and the androgen producing Leydig cells. The Sertoli cells form the walls of the seminiferous tubules where the germ cells are located. Testicular activity can be divided in four phases. Testicular cycle and spermatogenesis have been studied in *Salaria pavo* (Lahnsteiner and Patzner, 1990d), spermiogenesis and sperm differentiation processes in *S. pavo, Aidablennius sphynx, Microlipophrys* (=*Lipophrys*) *adriaticus, Microlipophrys* (=*Lipophrys*) *dalmatinus, Microlipophrys* (=*Lipophrys*) *canevae* and *Scartella cristata* (Lahnsteiner and Patzner, 1990b).

The testis shows an annual cycle. The GSI of nesting males and female-like males reaches the highest values during the spawning season, although in female-like males the GSI is high for a shorter period than for the nesting males. During the interspawing period, the GSI decreases to minimal values (Patzner and Seiwald, 1987; Gonçalves and Almada, 1997). During the interspawning or resting phase, mainly type A spermatogonia are found in the testis. They form a reserve pool of germ cells and are located either singly or in pairs among the Sertoli cells throughout the entire reproductive cycle. During prespawning time, when active spermatogenesis starts, spermatogonia A are surrounded by the Sertoli cell processes and differentiate into spermatogonia B. The number of type B spermatogonia increases considerably by mitotic divisions. Then meiotic divisions follow. After the first meiotic division, type B spermatogonia become the primary spermatocytes. Having terminated the first meiotic division, cells begin the second division which is very fast. This explains why secondary spermatocytes are very rarely seen. After the meiotic divisions are complete, the spermatocytes transform into spermatids which undergo spermiogenesis.

Spermiogenesis is characterised by the condensation of nuclear material, reduction of cytoplasm and cell organelles, polarisation of the cell by formation of an anterior head region containing the nucleus and a posterior midpiece region with the remaining cell organelles (mitochondria). A prominent flagellum develops. Connections (intercellular bridges) between the Sertoli cells and spermatids are broken down during this process. Upon completion of spermiogenesis, the cells transform into spermatozoa. Subsequently, the cyst wall breaks up, whereby the spermatozoa are released into the lumen of the seminiferous tubules.

In contrast to other teleost species, in Blenniidae, final differentiation processes of spermatozoa do not occur in the germinal cysts but in the testicular gland. Therefore, not fully mature spermatozoa but spermatids in late differentiation stages are released into the lumen of the seminiferous tubules. During spawning, the lumen of the seminiferous tubules is filled with spermatids which pass on into the ducts of the testicular gland where the final differentiation processes of spermatids into spermatozoa occur. At this stage, the testis also contains germinal cysts filled with spermatids. After spawning, in the post-spawning period, the Sertoli cells that have formed the walls of the germinal cysts withdraw their processes. Spermatids which did not pass on into the testicular gland but remained in the testis are reabsorbed by phagocytosis by the Sertoli cells. During this process, the vacuoles of the Sertoli cells can increase considerably in size and accumulations of phospholipids (so-called 'testicular clusters') can originate in the testis.

TESTICULAR ACCESSORY ORGANS

The male gonads of Mediterranean blenniid fish consist of testis, testicular gland, spermatic duct and blind pouches (Fig. 4.1.2). The following descriptions are mainly taken from Lahnsteiner and Patzner (1990a, b, c) and Lahnsteiner *et al.* (1990, 1993a). The methods used for fixation and histochemistry are described there.

Testicular Gland and Spermatic Duct

The testicular gland (=testis gland, fat body, 'Hodendrüse', 'Nebenhoden') is located at the ventral side of the testis (Fig. 4.1.2). The seminiferous tubules of the testis change into the glandular ducts which coalescence in order to form the spermatic ducts at the ventral side of the gland. The gland consists of tortuous, irregularly arranged tubules which are formed by a monolayered epithelium of gland cells and of the glandular interstitium, which contains beside fibrocytes, blood capillaries, and nervous tissue from the steroid producing cells. At the beginning of spawning, the ducts of the testicular gland, which have been collapsed during interspawning and prespawning period, open up.

The spermatic duct (= sperm duct, vas deferens, 'Samenblasen') arises at the ventral side of the testicular gland (Fig. 4.1.2). The wall of the spermatic duct is three layered. It consists of: (1) a peritoneum;

(2) a transversally orientated collagen fiber layer intermingled with smooth muscle cells; and (3) a one-layered cylindrical epithelium. The epithelial cells of the spermatic duct have a prominent microvilli layer throughout the entire reproductive season (Fig. 4.1.3).

The fine structural and histochemical characteristics of the testicular gland and spermatic duct of Blenniidae during spawning, the components of the secretion and the fact that final differentiation of spermatids into spermatozoa occurs in the glandular and spermatic ducts in blenniid fishes

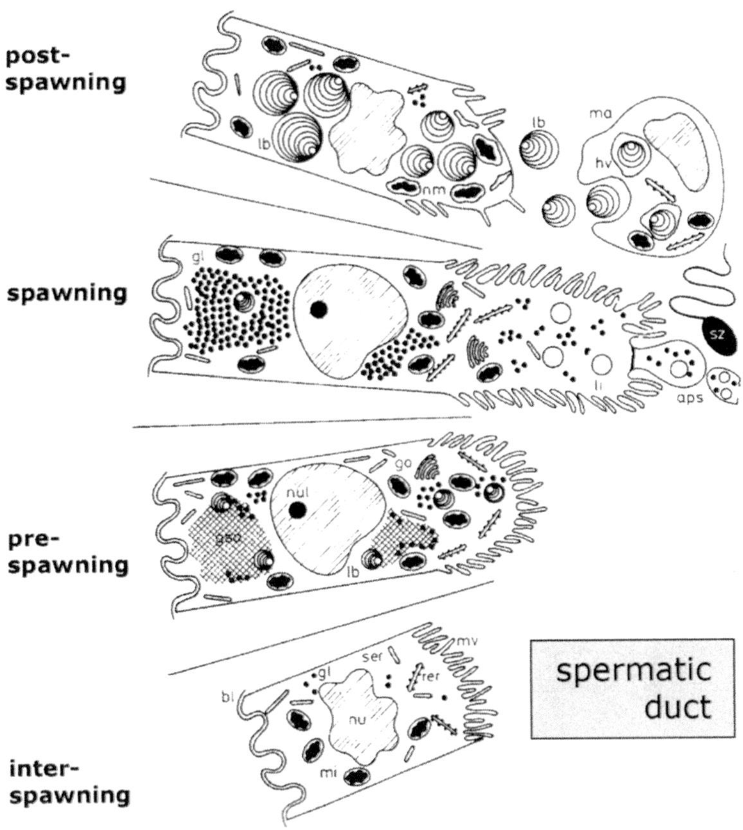

Fig. 4.1.3 Changes in the epithelial cells of the spermatic duct of *Salaria pavo* during the reproductive cycle. Schematic representation. aps = apocrine secretion; bl = basal lamina; gl = glycogen particle; go = Golgi field; gsa = glycogen synthesising area; hv = heterophagic vacuole; lb = lamellar body; li = lipid vacuole; ma = macrophage; mi = mitochondrium; nm = necrotic mitochondrium; mv = microvilli; nu = nucleus; nul = nucleolus; ser = smooth endoplasmatic reticulum; rer = rough endoplasmatic reticulum; sz = spermatozoon (after Lahnsteiner and Patzner, 1990c).

point out to almost similar functions of the two organs concerning: (1) differentiation processes of spermatids and regulation of the ionic concentration of the seminal fluid; (2) nutrition; (3) secretion of mucins; and (4) phagocytosis and storage of lipids and phospholipids during post-spawning and interspawning. Further, the spermatic duct is also involved in storage of differentiated spermatozoa (Figs. 4.1.3, 4.1.4).

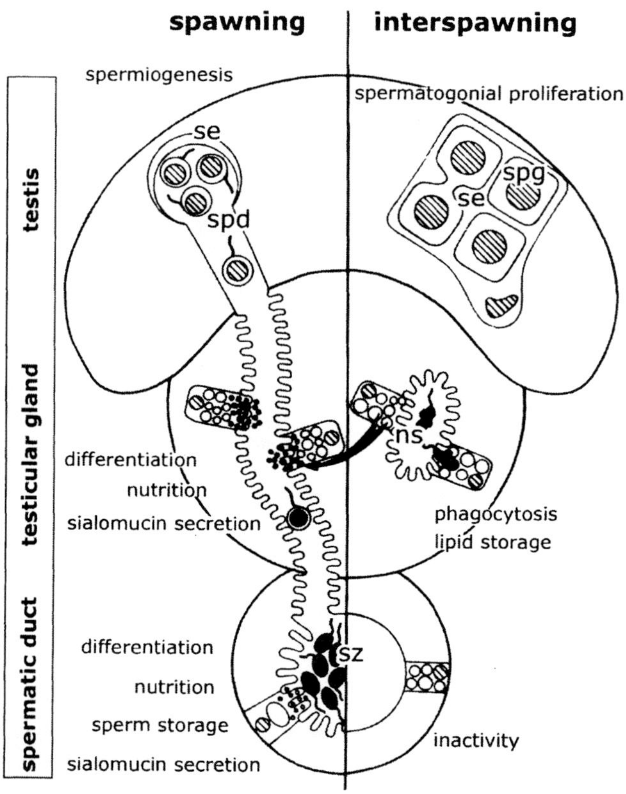

Fig. 4.1.4 The male gonads of blenniid fish and their functions in reproduction during spawning (left) and interspawning period (right). Scheme. Cross section. ns = necrotic spermatid; se = Sertoli cell; spd = spermatid; spg = spermatogonium; sz = spermatozoon (after Lahnsteiner *et al.*, 1990).

Nutrition of germ cells

The occurrence of glycogen depots in the testicular gland and spermatic duct cells, the glycogen secretion into their ducts and the pinocytotic uptake of glycogen granules from the differentiating spermatids and from mature spermatozoa, all reveal the nutritive function of the two organs. Generally, glycolysis is an important energy delivering pathway in spermatids and spermatozoa of teleost fishes (Grier, 1981; Lahnsteiner *et al.*, 1993b). Also, in the testes of teleosts, the developing germ cells use glycogen for energy supply, and there it is synthesised by Sertoli cells (Grier, 1981). In blennies, glycogen deposits occur also in the midpiece region of mature spermatozoa which is also found in spermatozoa of other species with external fertilisation (e.g. Cyprinidae, Cottidae) (Jamieson, 1991). Probably the glycogen is used during the motility period.

Differentiation of spermatids

In blenniid fishes, the sexually mature testes do not contain any spermatozoa, but differentiation processes of spermatids occur in the testicular gland and spermatic ducts. This is a semicystic type of spermiogenesis as also found e.g. in Callichthyidae, some Gobiidae and Scorpaenidae (Jamieson, 1991). The incomplete spermiogenetic cycle in the testes of blenniid fishes and further differentiation processes of spermatids in the testicular gland and spermatic ducts may be interpreted in favor of the production of high amounts of germ cells in the testes in short intervals. Male blennies have a high spawning activity because during the middle of the spawning period, they spawn almost every day. Spawning occurs with a preference to early morning hours which indicates continuous production of spermatids. The testicular gland and spermatic duct may, therefore, be considered to have functions in stabilising and regulating the differentiation processes of spermatids, similar to the Sertoli cells in the testes. In *Microlipophrys* species, the testicular gland is voluminous and the testis is much reduced in comparison to other blenniid species. Those spermatids which are released into the testicular gland in *Microlipophrys* species have an earlier stage of maturation than spermatids of other species, which is pointed out by uncompleted nuclear condensation, an undifferentiated flagellar complex and the presence of smooth endoplasmic reticulum. Thus indicates that in blennies, there is a tendency to reduce the testis in favor of the testicular gland, which takes over the testicular functions in regulating and controlling the processes of spermiogenesis.

Mucin secretion

The fine structural aspects of the synthesis and secretion of secretory vesicles and the histochemical processes of the positive PAS and alcian blue staining in just the same areas are proofs for the production of mucins by the testicular gland and spermatic duct cells. As demonstrated by differing results in the alcian blue staining when varying pH, $MgCl_2$ concentration and after neuraminidase extraction, the testicular glands produce sialomucins, while the spermatic ducts are involved in the secretion of sulfomucins. The specific mixture of sulfo- and sialomucins, added to the seminal fluid of blenniids, is regarded to increase the viscosity of the seminal fluid. In mammals, the mucins secreted by the Cowper's glands give the semen a jelly-like composition, function as a lubricant for the semen, neutralise the acidity of the urine residue in the urethra, help to neutralise the acidity of the vagina and have also roles in immune defense (Mann, 1974; Chughtaib *et al.*, 2005). As eggs of blenniid fish adhere to the substrate with the micropylar region (Patzner, 1984), the mucosubstances of the testicular gland and of the spermatic ducts are regarded to function in the formation of a layer of sperm on which the eggs are deposited. This hypothesis could be ascertained in recent studies in Gobiidae (Marconato *et al.*, 1996; Ota *et al.*, 1996).

Phagocytosis and lipid storage

At the end of spawning and during post-spawning time, the testicular gland and spermatic duct cells have a high phagocytotic activity, related to the elimination of remaining secretion products and of spermatids from the glandular ducts. However, there is no evidence for the reabsorption of remaining spermatozoa in the spermatic duct during the post-spawning period as described for other teleosts (Lahnsteiner *et al.*, 1993c, 1994). In blennies, the reabsorbed material is transformed into phospholipids and neutral lipids in the phagocytotic vacuoles and it is stored in the testicular gland cells during the interspawning period. The close association of rough endoplasmic reticulum and mitochondria to heterophagic vacuoles, their transition stages into secretory products and their drastic decrease during prespawning period (Seiwald and Patzner, 1987b) point out to the hypothesis that lipids and phospholipids may be re-transformed again into secretion products.

Storage of spermatozoa

The spermatic duct also has a role in spermatozoa storage preserving sufficient amounts for fertilisation. This spermatozoa storage role of the spermatic duct is also reported for Salmonidae (Lahnsteiner *et al.*, 1993c),

Esocidae (Lahnsteiner *et al.*, 1993d) and Cyprinidae (Lahnsteiner *et al.*, 1994). To provide the viability and immotility of sperm, the spermatic ducts of blenniid fishes are regarded to function in regulation of the ionic concentrations of the seminal fluid which is established by a high ATPase activity in the supranuclear and apical portions of the spermatic duct epithelium and is similar to what is known from other teleosts (Marshall, 1986; Marshall *et al.*, 1989). There exist great variations in the size of the spermatic ducts in blenniid fishes as they are extended in *Microlipophrys* species in comparison to *Aidablennius sphinx, Parablennius incognitus* and *Salaria pavo*. These differences could be interpreted as different storage capacity for spermatozoa. However, within the Blenniidae family, there are no significant ecological and ethological differences in reproduction and fertilisation modes (Gibson, 1982), which could be related to the necessity for storing different amounts of spermatozoa.

Blind Pouches

The blind pouches (=testicular blind pouches, 'Hodenschläuche', 'Blindschläuche', *appendices prostatrices*) represent paired evaginations of the distal portion of the spermatic ducts (Fig. 4.1.2). They have an annual reproductive cycle similar to the testis, testicular gland and spermatic duct (Lahnsteiner *et al.*, 1990; Lahnsteiner and Patzner, 1990a) with an active secretion phase during spawning, a regression phase with necrotic processes during post-spawning, and an inactive regeneration phase during interspawning and prespawning (Fig. 4.1.5) (Lahnsteiner *et al.*, 1993a).

The fine structural and enzyme-histochemical features of the blind pouches are similar in all blenniid species and indicate the following three functions: (1) synthesis of steroid glucuronides; (2) sialomucin secretion; and (3) a lytic activity. The blind pouches do not function as sperm reservoir as they never contain spermatozoa. They also lack functions in nutrition of spermatozoa since they contain no glycogen granula; a positive PAS reaction is found only in regions which stain also positive for alcian blue; the Best's glycogen detection method is negative; and they have no glucose-6-phosphatase activity. In contrast to the spermatic ducts, the blind pouches also have no adenosine triphosphatase activity and, therefore, they cannot generate an ionic gradient which could influence the motility of spermatozoa.

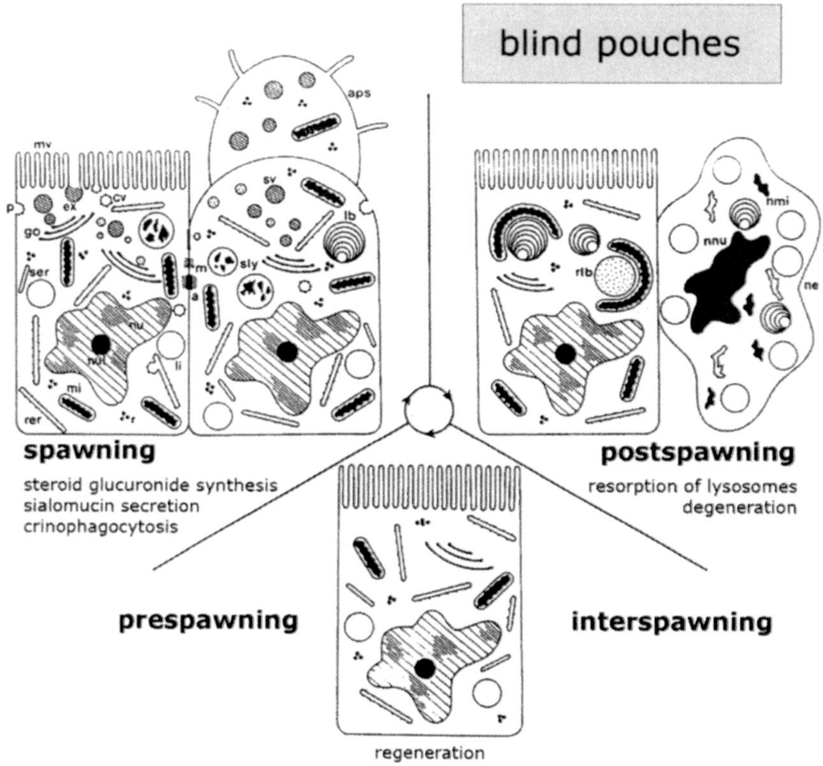

Fig. 4.1.5 Changes in the epithelium of a blind pouch of *Salaria pavo* during the reproductive cycle. Scheme. aps = apocrine secretion; ex = exocrine secretion; go = Golgi field; lb = lamellar body; li = lipid vacuole; mi = mitochondrium; mv = microvilli; nec = necrotic cell; nmi = necotic mitochondrium; nnu = necrotic nucleus; nu = nucleus; nul = nucleolus; p = pinocytosis; ply = primary lysosome; r = ribosomes; rlb = resorbtion of lamellar body; rer = rough endoplasmic reticulum; ser = smooth endoplasmic reticulum; sly = secondary lysosome; sv = secretory vesicle; za = macula adhearens; zo = zonula occludens; mo = macula occludens (after Lahnsteiner *et al.*, 1993a).

As a peculiarity, the blind pouches of *Salaria pavo* have an intensely folded epithelium which is not found in the blind pouches of other species of Blenniidae (Eggert, 1931; Blüm, 1972; Denoix, 1984; Richtarski and Patzner, 2000). These foldings are an enlargement of the internal surface and, subsequently, indicate a higher activity rate. However, by enzyme histochemical features, 3β-steroid dehydrogenase and uridine-5-phosphate dehydrogenase have a lower activity in the blind pouches of *S. pavo* than in the other species and there is also no indication for the production of higher amounts of mucins.

Steroid synthesis

The occurrence of an abundance of tubular rough endoplasmatic reticulum, of mitochondria with tubular, parallelly arranged cristae and of lipid vacuoles as well as activities of 3β-steroid dehydrogenase and glucose-6-phosphate dehydrogenase indicate a function in steroid synthesis in the blind pouches. Moreover, as uridine-diphosphoglucose dehydrogenase activity is also observed, steroids obviously occur as steroid glucuronides. By glucuronidation the generally apolar steroids are rendered water soluble to act as pheromones (Van Weerdt and Richter, 1991). The blind pouches, therefore, must be considered to have functions in pheromone synthesis which is also ascertained by observations of Laumen *et al.* (1974), Barata *et al.* (2008) and Barata and Gonçalves (this book) that mature male *Salaria pavo* are able to attract females by pheromones added into the water. Pheromones are also produced by other accessory reproductive organs like the testicular glands of *Gobius jozo* (= *G. niger*) (Colombo *et al.*, 1977) and the seminal vesicles of *Clarias gariepinus* (Resink *et al.*, 1989).

Sialomucin secretion

Fine structural features of Golgi apparatuses, secretory vesicles as well as positive PAS reactions, alcian blue staining at pH 2.5 and at electrolyte concentrations of 0.2 to 0.6 M indicate the synthesis of sialomucin in the blind pouches of the Blenniidae. These stainings, however, are found only locally and furthermore secretory vesicles often are eliminated by crinophagocytosis. Therefore, we consider that mucin secretion of the blind pouches is very low in comparison to the testicular glands and spermatic ducts and of minor importance.

Lytic capacity

During spawning, the blind pouches have a high crinophagocytotic capacity as the secretory vesicles degenerate to secondary lysosomes and the epithelium shows activities of acid phosphatase, aminopeptidase and β-glucuronidase. Autolysis often occurs in exocrine glands of mammals where it plays a role in the autonomic regulation of the secretion rate (Carr and Toner, 1982). As lysosomes are never cast off from the epithelium and no acid phosphatase, aminopeptidase and β-glucuronidase activities are detected in the lumen of the blind pouches of blennies, these lytic enzymes are not secreted to the seminal fluid.

SPERMIOGENESIS

In blenniid fish, spermiogenesis occurs only until a stage termed spermatid II in the germinal cysts of the testis (Fig. 4.1.6) (Lahnsteiner and Patzner, 1990b). They are released into the seminiferous tubules and designated as spermatids III. During the passage of the testicular gland and in the spermatic ducts, there is further differentiation processes of spermatids III into spermatozoa. These processes are not strictly synchronous in the testicular gland and in the spermatic duct. Only about 80% of the germ

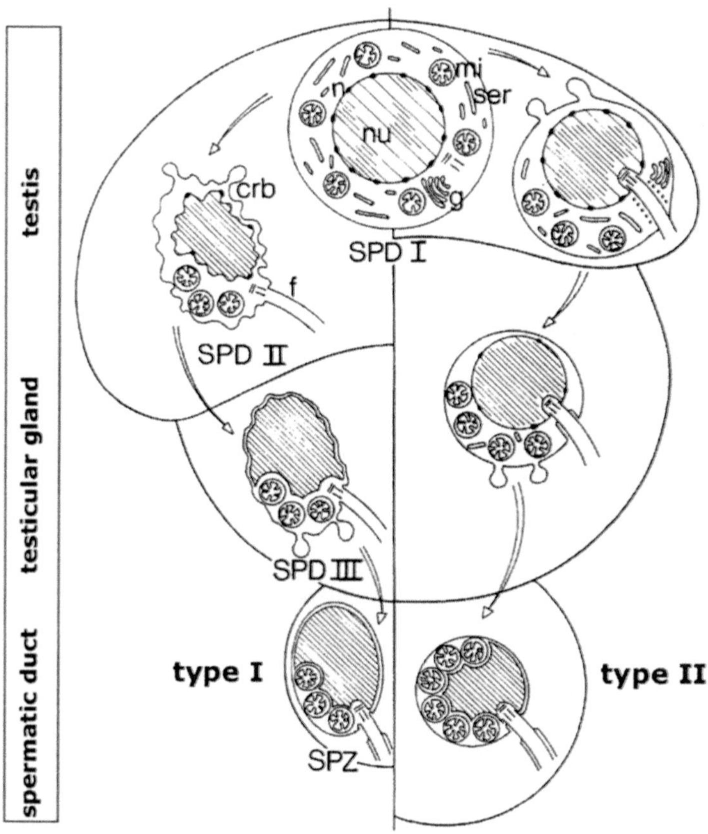

Fig. 4.1.6 Types of spermiogenesis in blenniid fishes. Schematic representation. Left: Spermiogenesis type I (*P. incognitus; P. tentacularis; S. pavo*); right: Spermiogenesis type II (*A. sphynx; M. adriaticus; M. canevae; M. dalmatinus; S. cristata*). crb = casting off of a residual cytoplasmatic bleb; g = Golgi field; mi = mitochondrium; n = nuclear pore; nu = nucleus; ser = smooth endoplasmic reticulum; SPD I-III spermatids stage I-III; SPZ spermatozoon (after Lahnsteiner and Patzner, 1990b).

cells show synchronous development but in the remaining 20% of the germ cells, differentiation occurs more rapidly or lags behind. Subsequently, fully differentiated spermatozoa can be observed in the testicular gland, but spermatids III are also present in the spermatic ducts. There are two different ways of spermiogenesis. Type I spermiogenesis produces elongated spermatozoa and is found, e.g., in *Salaria pavo*, *Parablennius incognitus* and *P. tentacularis*. Type II spermiogenesis generates spherical spermatozoa and is found, e.g., in *Aidablennius sphynx*, *Scartella cristata* and in the *Microlipophrys* species.

Type I Spermiogenesis

In type I spermiogenesis, spermatids I have a similar appearance as in other teleost species. Spermatids II are elongated, ellipsoid. During this stage, the condensation of the nuclear material occurs and only peripheral small, uncondensed areas remain. During the process of nuclear condensation, the nuclear envelope becomes intensively folded, and so does the plasmalemma becomes folded. The cytoplasm is strongly reduced and all cell organelles are concentrated at the caudal cell pole at this stage. The flagellum differentiates and consists of one central and nine peripheral pairs of microtubules. At the end of spermatid II stage, smooth endoplasmic reticulum and Golgi apparatuses are eliminated out of the cytoplasm. Differentiation of spermatids III occurs in the testicular gland (Fig. 4.1.6). The nuclear condensation of the spermatids is completed and the reduction of the cytoplasm proceeds. However, the plasmalemma and the nuclear membranes are still folded. In some species, the mitochondria become located in invaginations of the nucleus during the passage of the testicular gland; in others this differentiation process occurs in the spermatic duct.

In the spermatic ducts, the plasmalemma and the nuclear membranes of spermatids straighten. The mitochondria come in close vicinity to the nucleus as the exterior mitochondrial membrane contacts the exterior nuclear membrane. The centriolar complex also becomes located in close vicinity to the nucleus. At the flagellum, lateral cytoplasmic protrusions differentiate.

Type II Spermiogenesis

In type II spermiogenesis, spermatids retain a spherical shape (Fig. 4.1.6). The chromatin material condenses in the central parts of the nucleus during this differentiation stage but large peripheral areas remain uncondensed. A prominent portion of cytoplasm is found only at the caudal pole of the spermatids, where four mitochondria (in A. *sphynx*) or six mitochondria (in other species) are found. The nucleus gets invaginated at its caudal side and forms a nuclear notch, where the centriolar complex becomes located. The flagellum differentiates and consists of one central and nine peripheral pairs of microtubules.

In spermatids III, during the passage through the testicular gland, the nuclear condensation is completed. The reduction of the cytoplasm proceeds and the differentiation of the flagellum are completed. Cytoplasm reduction is finalised in the spermatic ducts. At the flagellum, lateral cytoplasmatic protrusions differentiate.

TESTICULAR ORGANS IN MEDITERRANEAN BLENNIIDAE

The morphology of the accessory organs of the testes was studied in Mediterranean Blenniidae (Seiwald and Patzner, 1987a; Richtarski and Patzner, 1988; Richtarski, 1990; Richtarski and Patzner 2000). As shown in Table 4.1.1, three main types could be distinguished. Types I and II

Table 4.1.1 Types of accessory organs in Mediterranean Blenniidae

Type	*Species*	*Test. gland*	*Sperm. duct*	*bl. po. thickness*	*bl. po. length*
I a	*M. adriaticus*	+++	+++	+	+
	M. dalmatinus	++++	+++	+	+
I b	*L. trigloides*	+++(+)	+++	+++	+++
	M. canevae	++++	+++	+++	+++
	M. nigriceps	++++	+++	+++	+++
II a	*A. sphynx*	++	++	++	++
	B. ocellaris	++	++	++	++
	C. galerita	++	++	+	++
	P. incognitus	++	++	++	++
	P. rouxi	++	++	+	++
	P. tentacularis	+	++	++	+
	S. cristata	++	++	++	+++
II b	*P. gattorugine*	++	++(+)	+++b	++
	P. zvonimiri	++	++	+++b	++
III	*P. sanguinolentus*	+	+	+	++
	S. pavo	+	+	++	++

bl. po. = blind pouches, + = weakly developed or short, ++++ = very strongly developed or very long, b = branched

could be divided into two subtypes each (Fig. 4.1.7). No clear explanation exists to explain these different types as yet. There is some correlation with taxonomy, e.g., all *Microlipophrys* species and the closely related *Lipophrys trigloides* present type I morphology.

The relation between the volume of the testis and the volume of the testicular gland can vary greatly. In *Salaria pavo,* the testicular gland corresponds to only 15% of the testis. However, in *Microlipophrys canevae,* it corresponds to 3.7 times of the testis volume (Fig. 4.1.6; Table 4.1.2). The spermatic ducts also differ in size. They are very small in *S. pavo* (0.02% of the body volume), whereas in *Microlipophrys adriaticus* and *Microlipophrys dalmatinus* they are voluminous and vesicular-like enlarged (up to 0.3% of the body volume) (Lahnsteiner *et al.*, 1990).

As stated in the previous chapter, there are two types of spermiogenesis in the Mediterranean Blenniidae. There is a relation between these two types and the size of the testicular gland. In species where the testicular gland is smaller than the testis, type I spermiogenesis

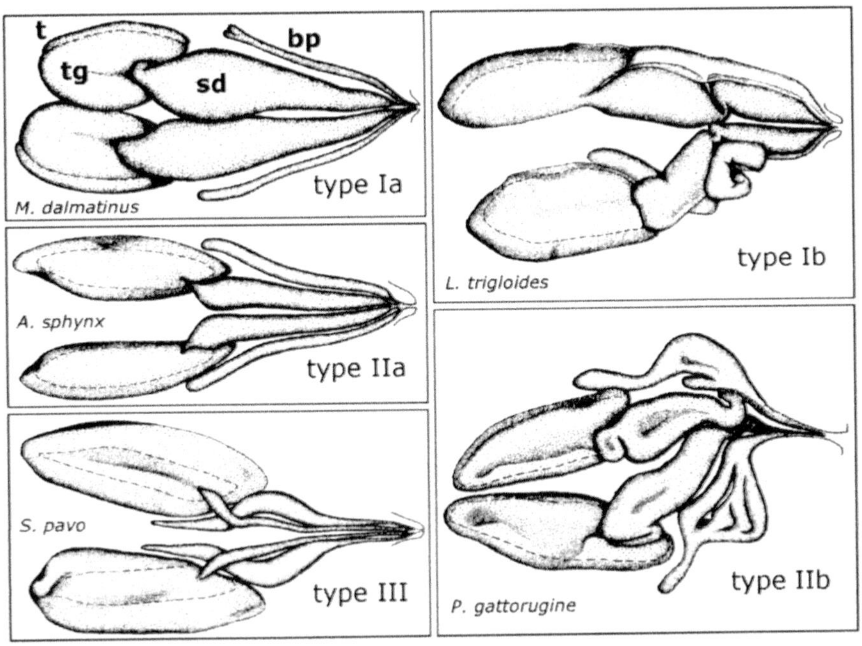

Fig. 4.1.7 Different types of testicular accessory organs of Mediterranean Blenniidae (see text). t = testis; tg = testicular gland; sd = spermatic duct; bp = blind pouch (after Richtarski and Patzner, 2000).

Table 4.1.2 Volume of testicular gland versus testis and the type of spermiogenesis in some Mediterranean Blenniidae

Species	*Volume of test. gland in % to testis*	*Type of spermiogenesis*
S. pavo	15%	I
P. incognitus	60%	I
P. tentacularis	60%	I
M. adriaticus	130%	II
M. dalmatinus	220%	II
M. canevae	370%	II

is present and in species with enlarged testicular glands type II of spermiogenesis is found (Table 4.1.2).

TESTICULAR ORGANS IN TROPICAL BLENNIIDAE

Knowledge on testicular organs of tropical Blenniidae is either very poor or nonexistent. The testicular organs of *Alticus* (= *Salarias*) *kirkii*, *Blenniella* (= *Salarias*) *bilotonensis*, *Entomacrodus striatus* (=*Salarias frenatus*), three species of *Istiblennius* (= *Salarias*): *I. dussumieri*, *I. edentulus* (= *quadricornis*), *I. unicolor*, two species of *Omobranchus* (= *Petroscirtes*): *O. punctatus*, *O. elegans* and two species of *Salarias*: *S. alboguttatus*, *S. luctuosus* (= *andersoni*) have been described by Eggert (1931). Later, *Salarias fasciatus* and *Ecsenius bicolor* were studied by Patzner (1989).

The testicular organs of *Omobranchus* spp. resemble the organs described for the Mediterranean and Atlantic species of this family (type II a, see above). However, the organs of the other tropical Blenniidae investigated till date show distinct differences to those. The following descriptions were made after Eggert (1931) and Patzner (1989).

Testis

The testes of all investigated species are paired elongated organs (Fig. 4.1.8). In *E. bicolor*, the testis is comparatively thin and extremely elongated.

Testicular Gland

The testicular glands of all investigated species except *E. bicolor* are organized, as in the Mediterranean species, interposed between the seminiferous tubules and the spermatic duct (Table 4.1.3). In contrast to

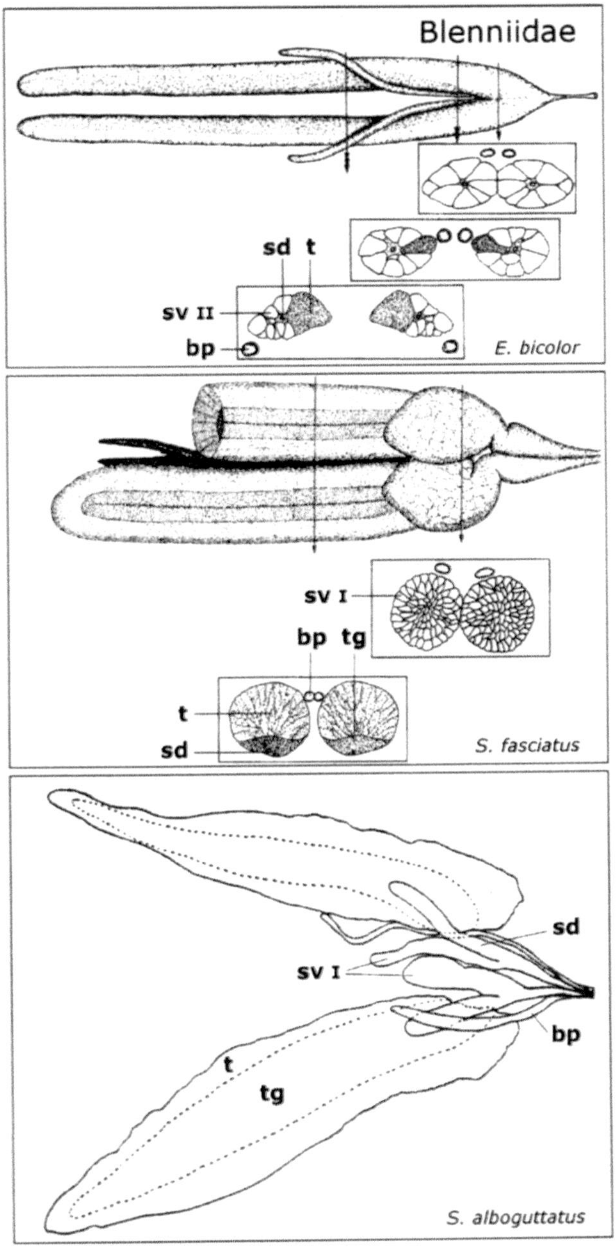

Fig. 4.1.8 Ventral views and cross sections of testis and testicular organs of the tropical Blenniidae *Ecsenius bicolor, Salarias fasciatus* and *S. alboguttatus*. t = testis; tg = testicular gland; sd = spermatic duct; bp = blind pouch; sv I = seminal vesicles, type I; sv II = seminal vesicles, type II (after Patzner, 1989; Eggert, 1931).

Table 4.1.3 Male accessory organs in tropical and Mediterranean species of the family Blenniidae

Species	*Gland*	*Pouch*	*s.v. I*	*s.v. II*	*Sperm.d. enl.*	*Authors*
Alticus kirkii	+	+	–	–	+	1
Blenniella bilotonensis	+	+	–	–	+	1
Ecsenius bicolor	–	+	–	+	–	2
Entomacrodus striatus	+	+	+	–	= s.v. I	1
Istiblennius dussumieri	+	+	+	–	= s.v. I	1
I. edentulus	+	+	–	–	+	1
I. unicolor	+	+	+	–	= s.v. I	1
Omobranchus punctatus	+	+	–	–	+	1
O. elegans	+	+	–	–	+	1
Salarias alboguttatus	+	+	+	–	= s.v. I	1
S. luctuosus	+	+	+	–	= s.v. I	1
S. fasciatus	+	+	+	–	= s.v. I	2
Microlipophrys spp.	++	+	–	–	++	3
other Mediterr. spec.	+	+	–	–	+	3

+ = present, – = absent, ++ = enlarged. Testicular gland, blind pouches, seminal vesicle of types I and II, enlarged spermatic duct, after 1 = Eggert (1931), 2 = Patzner (1989), 3 = Richtarski and Patzner (2000)

this, no testicular gland can be found in *E. bicolor* (Fig. 4.1.8). It appears to be the only species of the Blenniidae without a testicular gland. Maybe the unusually thick epithelium of the spermatic duct takes over at least part of the function of the testicular gland of the other blennies.

Spermatic Duct

In the investigated species except for *E. bicolor*, the spermatic duct is at its beginning either a single or double canal at the ventral side of the testicular gland. In *A. kirkii*, *B. bilitonensis* and *I. edentulus* the duct resembles that of the Mediterranean blennies. In the other *Salarias* and *Istiblennius* species, they fuse with the seminal vesicles (see below). The spermatic duct of *E. bicolor* is always a single canal with a high epithelium. It runs along the border between the elongated seminal vesicles and the testis. Unlike in other species, it does not fuse with the seminal vesicles when leaving the testis but lies between them, embedded in a capsule of connective tissue.

Seminal Vesicles, Type I

In *Entomacrodus striatus*, *I. unicolor*, *S. alboguttatus* and *S. luctuosus*, the seminal vesicles are unchambered sacks, attached to the spermatic duct

(Fig. 4.1.8). During spawning, they are filled with ripe spermatozoa. In *I. dussumieri* and *S. fasciatus*, the seminal vesicles are protrusions of the spermatic duct and consist of bladder-like chambers with a flat monolayered epithelium (Fig. 4.1.8). The single chambers are separated from each other by a thin layer of connective tissue. In *A. kirkii*, *B. bilitonensis*, *I. edentulus* and *Omobranchus* spp. no seminal vesicles are formed.

Seminal Vesicles, Type II

The seminal vesicles of *E. bicolor* run along the entire testis and widen posterior to this organ (Fig. 4.1.8). They also consist of chambers formed by a monolayered epithelium and a layer of connective tissue. The number of the chambers in *E. bicolor* is small compared to the diameter of the whole seminal vesicle and the number of chambers in *S. fasciatus*. In mature *E. bicolor*, all the vesicles are filled with a strong-staining (Azan) secretion. No spermatozoa were present within the seminal vesicles of type II. In other species of Blenniidae no similar structure was found until now. It may be speculated that there are parallels with the seminal vesicles of gobiid fishes (Lahnsteiner *et al.*, 1992).

Blind Pouches

In all investigated species of tropical Blenniidae, the blind pouches are paired tubes (Fig. 4.1.8). They are round to oval in cross-section and show some villous formations towards the lumen. No spermatozoa were found within the blind pouches. The length of the pouches is about half of the testis length or even less. In *E. bicolor*, they flow into the posterior part of the seminal vesicles. There is no connection to the testis and the spermatic duct.

TESTICULAR ORGANS IN OTHER BLENNIOIDEI

Tripterygiidae

Morphological descriptions of testicular organs of tripterygiid fishes are scarce. The Mediterranean *Tripterygium tripteronotus* (= *nasus*) was studied by light microscopy (Eggert, 1931) and electron microscopy (de Jonge *et al.*, 1989; Richtarski, 1990). De Jonge *et al.* (1989) also used histochemistry and investigated a second Mediterranean species,

T. xanthosoma (=*delaisi*). Rasotto (1995) gave coarse descriptions of the testicular organs of the Pacific *Enneanectes boelkei* and *E. pectoralis*.

Testis and testicular gland

The testes are similar to those of the Blenniidae. The structure of the testicular gland is slightly different to that of the Blenniidae described above (Fig. 4.1.9). Radiating seminiferous tubules drain into an efferent duct system situated against the testicular gland and penetrate it at several sites. The glandular mass consists of radially oriented strands of polyhedral cells surrounded with capillaries or sinusoids. The glandular cells contain smooth endoplasmic reticulum, numerous lipid droplets and mitochondria with large cristae. No steroidogenic activity was found within the testicular gland.

Spermatic duct

The efferent ducts passing through the testicular gland connect to a longitudinal main duct, located ventrally and centrally in the testicular gland. At the posterior end of the gland, the sperm duct is dilated and leaves the gland caudally. The spermatic ducts are slightly enlarged (like in most Blenniidae) and fuse together before entering the genital papilla. No blind pouches are formed in either Tripterygiidae (Fig. 4.1.9; Table 4.1.4).

Table 4.1.4 Accessory organs in different groups of Blennioidei: testicular glands, enlarged spermatic ducts, blind pouches, seminal vesicles of type I (containing spermatozoa) and type II (containing mucus, no spermatozoa)

Group	*Test. gland*	*Sperm. duct*	*Blind pouch*	*Sem. ves. I*	*Sem. ves. II*
Mediterr. Blenniidae	+	+	+	–	–
Blenniidae	+/–	+/–	+	+/–	+/–
Tripterygiidae	+[1]	+	–	–	–
Labrisomidae	+	+/–	–	–	–
Clinidae	+	(+)?[2]	–	(+)?[2]	–
Chaenopsidae	– (+[3])	+/–	–	–	+[4]
Dactyloscopidae	+	–	–	–	–

[1]different from that of Blenniidae; [2]not clear, see text; [3]only in *Neoclinus*; [4]unpaired, not in *Neoclinus*
+ = present, – = absent

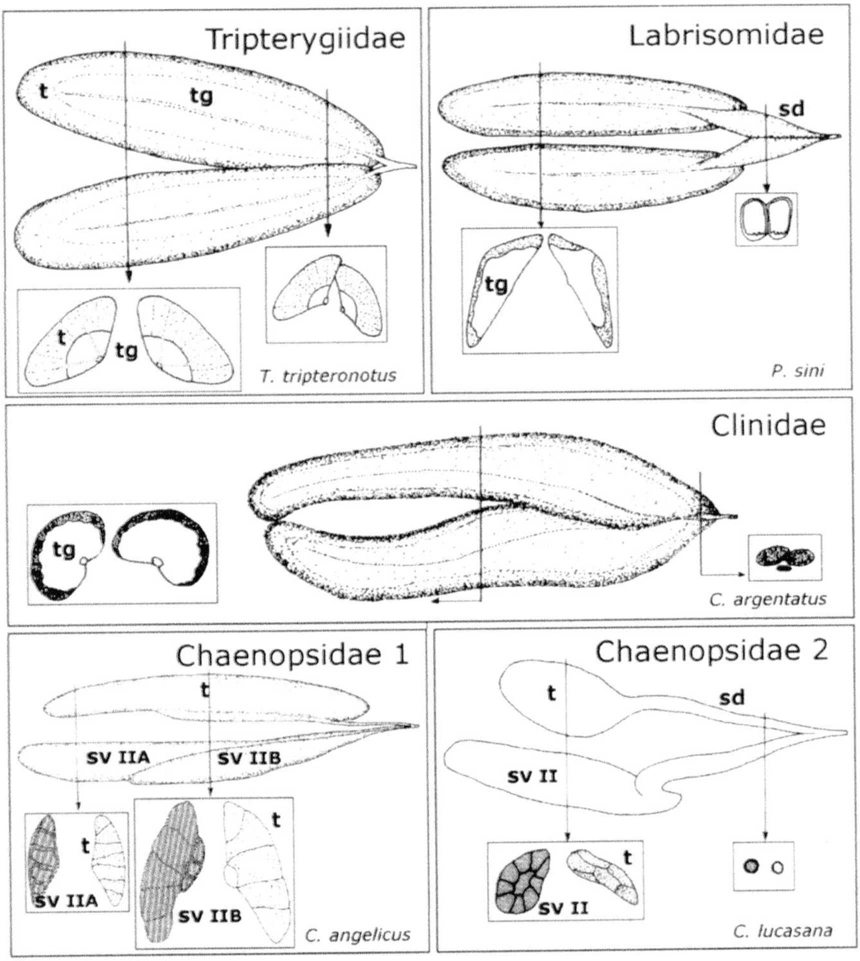

Fig. 4.1.9 Ventral views and cross sections of testis and testicular organs of *Tripterygion tripteronotus* (Tripterygiidae), *Paraclinus sini* (Labrisomidae), *Cristiceps argentatus* (Clinidae), *Coralliozetus angelicus* (Chaenopsidae) and *Cirriemblemaria lucasana* (Chaenopsidae). t = testis; tg = testicular gland; sd = spermatic duct; sv II = seminal vesicle type II (after Patzner (1991) and Patzner and Hastings, unpublished data).

Labrisomidae

The male testicular accessory organs were investigated in *Labrisomus bucciferus*, *L. haitensis*, *L. nuchipinnis*, *Malacoctenus macropus*, *M. triangulatus* and *Paraclinus nigripinnis* by Rasotto (1995) and in *L. xanti* and *P. sini* by Patzner and Hastings (unpublished data).

Testis and testicular gland

The testis-testicular gland complex is the same as described for the Mediterranean Blenniidae (see above). The gland is highly developed, its volume ranging between less than or up to about 100% (in *Labrisomus bucciferus, L. xanti, P. sini*) and more than 200% (in *L. haitensis, M. versicolor*) that of the testis (Fig. 4.1.9).

Spermatic duct

In *P. sini* the duct is enlarged but rather short. In *Labrisomus* spp. and *Malacoctenus* spp. the ducts are longer and convoluted. No blind pouches were found (Table 4.1.4).

Clinidae

Till date, the testicular organs were studied only in the Mediterranean *Clinitrachus* (=*Cristiceps*, =*Clinus*) *argentatus* (Guitel, 1893; Eggert, 1931, Patzner, unpublished).

Testis and testicular gland

In coarse morphology, the testis-testicular gland complex of *Clinitrachus* resembles that of the most Blenniidae (Fig. 4.1.9). However, as more thorough studies are missing, a similarity to that of Tripterygiidae (see above) cannot be excluded.

Spermatic duct

In contrast to most Blenniidae and Tripterygiidae, here the ducts are not enlarged. After Eggert (1931) they fuse into a 'Samenblase' which would correspond to the seminal vesicle, type I of some tropical Blenniidae (see above). This structure was neither found by Guitel (1893) nor by Patzner (unpublished). No blind pouches are present (Fig. 4.1.9; Table 4.1.4).

Chaenopsidae

The following descriptions were made after studies on *Coralliozetus angelicus* (Patzner, 1991), *Acanthemblemaria spinosa* (Rasotto, 1995) and *A. crockeri, A. macrospilus, Chaenopsis alepidota, Cirriemblemaria lucasana, Coralliozetus micropes, Ekemblemaria myersi, Emblemaria hypacanthus, Mccosterichthys sandae, Neoclinus stephensae, N. uninotatus, Protemblemaria bicirris* and *Stathmonotus sinuscalifornici* (Patzner and Hastings, unpublished data).

Testis

In all species of Chaenopsidae studied up to now—except *M. sandae* and *Neoclinus* ssp.—the testis is unpaired and situated in the left part of the abdomen. It is oblong in most species and rounded in *A. spinosa and C. lucasana* (Fig. 4.1.9). The testicular tubules are very short and have a disproportionately large diameter, so that they should be termed 'chambers' instead of 'tubules'. Not all of these chambers open directly into the spermatic duct, but some of them flow into each other. In *M. sandae*, the testis is paired and very small (3% of the length of the fish).

There are no accessory organs in *M. sandae*. In *N. stephensae* and *N. uninotatus* the testis-testicular gland complex is the same as in Labrisomidae (see above). The volume of the gland ranges between less or up to about 100% (in *N. uninotatus*) and more than 200% (in *N. stephensae*) that of the testis.

Spermatic duct

In *C. angelicus*, the single spermatic duct does not run along the entire testis but starts at the anterior third of the testicular lobe (Fig. 4.1.9). In the middle region of the testis, the duct has a relatively large oval diameter. The epithelium is usually flat but can also show villous formations. Along the duct, the separate chambers of the testis discharge directly into it. Towards the caudal end of the testis, the diameter of the spermatic duct decreases. At the place where it leaves the testis the diameter is 100 to 200 μm and in the region proximal to the genital papilla, where it comes in contact with the ducts of the seminal vesicles, the duct is only 20 to 25 μm in diameter. The spermatic duct fuses with the seminal vesicle ducts just before entering the genital papilla. In *S. sinuscalifornici* and *Neoclinus* spp. the spermatic ducts are quite short and not enlarged.

Seminal vesicles

On the right side of the abdomen there are unpaired seminal vesicles (Table 4.1.4). Three types can be distinguished: (1) *C. angelicus* has two kinds of unpaired vesicles, 'seminal vesicle IIA' and 'seminal vesicle IIB' (Fig. 4.1.9). They both correspond to seminal vesicle type II (see above) and lay partly one behind the other and partly side by side but have no connection with each other. The vesicles consist of chambers very similar to the chambers of the testis. In all the investigated specimens, the epithelium of the seminal vesicle IIA shows a very high secretory activity. Seminal vesicle IIA is drained by a single duct which exits the body at the end of the genital papilla. Seminal vesicle IIB, on the other hand, possesses up to ten or more efferent ducts. These ducts do not fuse with each other or with the spermatic duct until the region just before the genital papilla. (2) In *Acanthemblemaria crockeri*, *A. macrospilus*, *Chaenopsis alepidota*, *Cirriemblemaria lucasana*, *Coralliozetus micropes* and *Protemblemaria bicirris*, one cambered seminal vesicle can be found, similar to the above seminal vesicle IIA (Fig. 4.1.9). (3) In *A. spinosa*, *Ekemblemaria myersi* and *Emblemaria hypacanthus*, a single seminal vesicle consisting of only one chamber is observed. No spermatozoa were found within either type of seminal vesicles. They have a similar structure and a similar staining property to the seminal vesicles (type II) of another blenniid fish, *Ecsenius bicolor* (Patzner, 1989, see above) and the gobiid fishes (Lahnsteiner *et al.*, 1992). In the latter, the chambers of the seminal vesicles open directly into the spermatic duct (Patzner *et al.*, 1991) and have no efferent duct like the blennies *E. bicolor* and *C. angelicus*. There is no explanation for the presence of two types of seminal vesicles found in *C. angelicus* and the reason as towhy they both have efferent ducts which are separated from each other.

In *M. sandae* as well as in *Neoclinus* spp. no seminal vesicles are developed. The testicular organs of the *Neoclinus* species investigated up to now did not show any similarities with those of other Chaenopsidae but represent in detail the type of the Labrisomidae (see above and Table 4.1.4).

Dactyloscopidae

The accessory organs were investigated only in *Platygillellus* (=*Gillellus*) *rubrocinctus* (Rasotto, 1995). They show great similarities to the organs of the Labrisomidae (Table 4.1.4). The volume of the testicular gland is more than 90% of the testis volume.

CONCLUSIONS

Blenniidae have a very specialised reproductive system in comparison to other teleost fish. Testicular glands and spermatic ducts take over testicular functions during spawning as final differentiation and maturation of spermatids into spermatozoa occur in these organs. This process enables the production of high amounts of germ cells in short time intervals which could increase the spawning activity of males. The testicular gland and spermatic duct produce seminal fluid of a very specific composition containing considerable amounts of mucosubstances. These mucosubstances are considered to increase the viscosity and stickiness of the spawned semen that is adhered to the substrate in form of a sperm layer on which eggs are deposited. This type of spawning can increase the efficiency of fertilisation. Testicular blind pouches are another type of accessory reproductive organs in the Blenniidae which main functionis the synthesis of steroid glucuronides (= pherormones). Territorial males may attract the females by releasing pheromones into the water.

Acknowledgements

We especially want to thank the co-workers in our group, Margit Seiwald, Ute Richtarski and Barbara Nussbaumer. Further thanks for valuable discussion and providing material are due to Phil Hastings, Kurt Kotrschal, Ricardo Serrão Santos, Peter Wirtz and Alfred Goldschmid. We are grateful to Anette Galabov for technical assistance. Most of the studies were supported by the Austrian 'Fonds zur Förderung der wissenschaftlichen Forschung' (grant no. 5338), by the Austrian National Bank (project no. 1568) and the 'Stiftungs- und Förderungsgesellschaft der Paris-Lodon-Universität, Salzburg' (Austria). Working facilities were provided at the marine biological laboratory of 'Sorgenti di Aurisina' (University of Trieste, Italy).

References

Bara, G. 1960. Histological and cytological changes in the ovaries of the mackerel '*Scomber scomber* L.', during the annual cycle. *Revue de la Faculté des Sciences, de l'Université d'Istanbul* B25: 49-91.

Barata, E.N., R.M. Serrano, A. Miranda, R. Nogueira, P.C. Hubbard and A.V.M. Canário. 2008. Putative pheromones from the anal glands of male blennies attract females and enhance male reproductive success. *Animal Behaviour* 75: 379-389.

Barr, W.A. 1963. The endocrine control of the sexual cycle in the plaice, *Pleuronectes platessa* (L.). I. Cyclical change in the normal ovary. *General and Comparative Endocrinology* 3: 197-204.

Blüm, V. 1972. The influence of ovine follicle stimulating hormone (FSH) and luteinizing hormone (LH) on the male reproductive system and the skin of the Mediterranean blenniid fish *Blennius sphinx* (Valenciennes). *Journal of Experimental Zoology* 181: 203-216.

Blüm, V., L. Machemer and W.D. Bartmann. 1972. Die Wirkung von TSH und LtH auf das Verhalten und auf die Histologie der Haut, der Schilddrüse und der Gonaden beim Pfauenschleimfisch *Blennius pavo*. *Zoologisches Jahrbuch Physiologie* 76: 340-355.

Brock, B. 1878. Beiträge zur Anatomie und Histologie der Geschlechtsorgane der Knochenfische. *Morphologisches Jahrbuch* 4: 505-572.

Carr, E.K. and P.C. Toner. 1982. *Cell Structure. An Introduction and Biomedical Electron Microscopy*. Churchill & Livingstone, Edinburgh, London.

Carrasson, M. and M. Bau. 2003. Reproduction and gonad histology of *Aidablennius sphinx* (Pisces: Blenniidae) of the Catalan Sea (northwestern Mediterranean). *Scientia Marina* 67: 461-469.

Champy, C. and P. Gley. 1922. La glande du testicule des blennies et sa signification. *Bulletin de la Société de France* 47: 199-208.

Chieffi, G. and V. Botte. 1964. Osservazioni sul significato funzionale della ghiandola annessa del testicolo dei Blennidii. *Bollettino di Zoologica* 31: 471-477.

Chughtaib, B., A. Sawas, R.L. O'Malley, R.R. Naik, S.A. Khan and S. Pentyala. 2005. A neglected gland: A review of Cowper's gland. *International Journal of Andrology* 28: 74–77.

Colombo, L., P.C. Belvedere and A. Pilati. 1977. Biosynthesis of free and conjugated 5ß-reduced androgens by the testis of the black goby, *Gobius jozo* L. *Bulletin of Zoology* 44: 131-134.

Coujard, R. 1941. L'évolution de la glande testiculaire des blenniides. *Comptes Rendus des Séances de la Société de Biologie et des ses filiales* 155: 371-373.

Courrier, R. 1921. Sur l'existence d'une glande interstitielle dans le testicule des poissons. *Comptes Rendus des Séances de la Société de Biologie et des ses filiales* 85: 939-941.

Courrier, R. 1922. Sur l'existence d'une glande interstitielle dans le testicule des blennies. *Bulletin de la Société de Zoologie de France* 47: 458-462.

de Jonge, J. and J.J Videler. 1989. Differences between the biologies of *Tripterygion tripteronotus* and *T. delaisi* (Pisces, Perciformes, Tripterygiidae): The adaptive significance of an alternative mating strategy and a red instead of a yellow nuptial colour. *Marine Biology* 100: 431-437.

de Jonge, J., A.J.H. de Ruiter and R. van den Hurk. 1989. Testis-testicular gland complex of two *Tripterygion* species (Blennioidei, Teleostei): Differences between territorial and non-territorial males. *Journal of Fish Biology* 35: 497-508.

Deliro, G., V. Botte and G. Chieffi. 1965. Ulteriori osservazioni sul tessuto interstitiale del testiculo die Teleostei. *Bollettino di Zoologica* 32: 199-205.

Denoix, M. 1984. *Zur Biologie des Schleimfisches* Parablennius pilicornis *Cuvier, 1829 (Blenniidae, Perciformes) unter besonderer Berücksichtigung der sekundären Geschlechtsmerkmale des Männchens nach Hormonbehandlung.* Thesis, University of Tübingen, Tübingen.

Dunne, J. 1977. Littoral and benthic investigations on the west coast of Ireland. VII. (Section A: Faunistic and ecological studies) The biology of the shanny, *Blennius pholis* L. (Pisces) at Carna, Connemara. *Proceedings of the Royal Irish Academy* B77: 207-226.

Eggert, B. 1931. Die Geschlechtsorgane der Gobiiformes und Blenniiformes. *Zeitschrift für wissenschaftliche Zoologie* 139: 249-558.

Eyberg, I. 1984. The biology of *Parablennius cornutus* (L.) and *Scartella emarginata* (Guenter) (Teleostei: Blenniidae) on a Natal Reef. *South African Association for Marine Biological Research, Oceanographic Research Institute. Investigation Report* 54: 1-16.

Fives, J.M. 1980. Littoral and benthic investigations on the west coast of Ireland. XI. The biology of Montagu's blenny, *Coryphoblennius galerita* L. (Pisces), on the Connemara coast. *Proceedings of the Royal Irish Academy* B80: 61-77.

Gibson, R.N. 1982. Recent studies on the biology of intertidal fishes. *Oceanographical and Marine Biology Annual Review* 7: 376-410.

Gonçalves E.J. and V.C. Almada. 1997. Sex differences in resource utilization by the peacock blenny. *Journal of Fish Biology* 51: 624-633.

Grier, H.J. 1981. Cellular organization of the testis and spermatogenesis in fishes. *American Zoologist* 21: 345-357.

Guitel, F. 1893. Description des orifices génito-urinaires de quelques *Blennius*. *Archives de Zoologie expérimentale et générale*, Série 3 1: 611-658.

Heymer, A. 1985. Morphologie, coloration, intersexualite et anatomie des organes reproducteurs chez *Blennius basiliscus*. *Revue françaises de Aquariologie* 12: 39-52.

Hyrtl, J. 1850. Beiträge zur Morphologie der Urogenitalorgane der Fische. *Denkschriften der Akademie der Wissenschaften in Wien* 1: 391-411.

Isamail, W.A. and D.A. Clayton. 1990. Biology of *Omobranchus punctatus* (Blenniidae) on rocky shores in Kuwait. *Cybium* 14: 285-293.

Jamieson, B.G.M. 1991. *Fish Evolution and Systematics: Evidence from Spermatozoa*. Cambridge University Press, Cambridge.

Kaurin, G. and R.A. Patzner. 1988. Gonadal differentiation in *Blennius pavo* (abstract). *VI. Congress of European Ichthyologists, Budapest, 1988*, p. 122.

Lahnsteiner, F. 1987. *Die Vaskularisation der Gonaden des Pfauenschleimfisches* Blennius pavo *(Pisces: Teleostei: Blenniidae)*. M.Sc. Thesis, University of Salzburg, Salzburg, Austria.

Lahnsteiner, F. and R.A. Patzner. 1990a. Functions of the testicular gland of blenniid fish: Structural and histochemical investigations. *Experientia* 46: 1005-1007.

Lahnsteiner, F. and R.A. Patzner. 1990b. Spermiogenesis and structure of mature spermatozoa in blenniid fishes (Pisces, Blenniidae). *Journal of Submicroscopical Cytology and Pathology* 22: 565-576.

Lahnsteiner, F. and R.A. Patzner. 1990c. The spermatic duct of blenniid fish (Teleostei, Blenniidae): Fine structure, histochemistry and function. *Zoomorphology* 110: 63-73.

Lahnsteiner, F. and R.A. Patzner. 1990d. The mode of germ cell reneweal and fine structure of early spermatogenesis in *Blennius pavo* (Teleostei, Blenniidae). *Zoologischer Anzeiger* 224: 129-131.

Lahnsteiner, F., A. Lametschwantdner, R.A. Patzner and H. Adam. 1988. Vascularization of male gonads in *Blennius pavo* (Teleostei, Blenniidae) as revealed by scanning electron microscopy of vascular corrosion casts. *Scanning Microscopy* 2: 2077-2086.

Lahnsteiner, F., U. Richtarski and R.A. Patzner. 1990. Function of the testicular gland in two blenniid fishes, *Salaria* (=*Blennius*) *pavo* and *Lipophrys* (=*Blennius*) *dalmatinus*

(Blenniidae, Teleostei) as revealed by electron microscopy and enzyme histochemistry. *Journal of Fish Biology* 37: 85-97.

Lahnsteiner, F., M. Seiwald, R.A. Patzner and E.A. Ferrero. 1992. The seminal vesicles of the male grass goby, *Zosterisessor ophiocephalus*. Fine structure and histochemistry. *Zoomorphology* 111: 239-248.

Lahnsteiner, F., B. Nussbaumer and R.A. Patzner. 1993a. Unusual testicular accessory organs, the testicular blind pouches of blennies (Teleostei, Blenniidae). Fine structure, (enzyme-) histochemistry and possible functions. *Journal of Fish Biology* 42: 227-241.

Lahnsteiner, F., R.A. Patzner and T. Weismann. 1993b. Energy resources of spermatozoa of the rainbow trout (*Oncorhynchus mykiss*) (Pisces, Teleostei). *Reproduction, Nutrition and Development* 33: 349-360.

Lahnsteiner, F., R.A. Patzner and T. Weismann. 1993c. The spermatic ducts of salmonid fishes (Salmonidae, Teleostei). Morphology, histochemistry and composition of the secretion. *Journal of Fish Biology* 42: 79-93.

Lahnsteiner, F., R.A. Patzner and T. Weismann. 1993d. The efferent duct system of the male gonads of the European pike (*Esox lucius*): Testicular efferent ducts, testicular main ducts and spermatic ducts. *Journal of Submicroscopical Cytology and Pathology* 25: 487-498.

Lahnsteiner, F., R.A. Patzner and T. Weismann. 1994. The testicular main ducts and spermatic ducts in cyprinid fishes I. Morphology, fine structure and histochemistry. *Journal of Fish Biology* 44: 937-951.

Laumen, J., U. Pern and V. Blüm. 1974. Investigations on the functions and hormonal regulation of the anal appendices in *Blennius pavo* (Risso). *Journal of Experimental Biology* 190: 47-56.

Mann, T. 1974. *The Biochemistry of Semen and of the Male Reproductive Tract*. Methuen and Co. Ltd., London.

Marconato, A., C. Rasotto and M.B. Mazzoldi. 1996. On the mechanism of sperm release in three gobiid fishes (Teleostei: Gobiidae). *Environmental Biology of Fishes* 46: 321-327.

Marshall, W.S. 1986. Sperm duct epithelium of the brook trout (*Salvelinus alpinus*): Na^+ transport and seminal plasma composition. *Canadian Journal of Zoology* 64: 1827-1830.

Marshall, W.S., S.E. Bryson and R.D. Idler. 1989. Gonadotropin stimulation of K^+ secretion and Na^+ absorption by sperm duct epithelium. *General and Comparative Endocrinology* 75: 118-128.

Ota, D., M. Marchesan and E.A. Ferrero. 1996. Sperm release behaviour and fertilization in the grass goby. *Journal of Fish Biology* 49: 246-256.

Papitsch, M., R.A. Patzner and H. Adam. 1981. Effects of photoperiod and temperature on gonadal maturation of *Blennius* (= *Salaria*) *pavo* (Teleostei, Blenniidae). *Vie et Milieu* 31: 215-219.

Patzner, R.A. 1983. The reproduction of *Blennius pavo* (Teleostei, Blenniidae). I. Ovarian cycle, environmental factors and feeding. *Helgoländer Meeresuntersuchungen* 36: 105-114.

Patzner, R.A. 1984. The reproduction of *Blennius pavo* (Teleostei, Blenniidae) II. Surface structures of the ripe egg. *Zoologischer Anzeiger, Jena* 213: 44-50.

Patzner, R.A. 1989. Morphology of the reproductive systems of two IndoPacific blenniid fishes, *Salarias fasciatus* and *Ecsenius bicolor* (Blenniidae, Teleostei). *Zeitschrift für zoologische Systematik und Evolutionsforschung* 27: 135-141.

Patzner, R.A. 1991. Morphology of the male reproductive system of *Coralliozetus angelicus* (Pisces, Blennioidei, Chaenopsidae). *Journal of Fish Biology* 39: 867-872.

Patzner, R.A. and G. Kaurin. 1997. Sexual differentiation in *Salaria* (= *Blennius*) *pavo*. *Journal of Fish Biology* 50: 887-894.

Patzner, R.A. and M. Seiwald. 1987a. The reproduction of *Blennius pavo* Risso (Teleostei, Blenniidae). VII. Secondary sexual organs and accessory glands of the testis during the reproductive cycle. *Proceedings of the V. Congress of European Ichthyologists, Stockholm, 1985*, pp. 293-298.

Patzner, R.A. and M. Seiwald. 1987b. The reproduction of *Blennius pavo* (Teleostei, Blennidae). VI. Testicular cycle. *Zoologischer Anzeiger, Jena* 219: 265-273.

Patzner, R.A. and M. Seiwald. 1988a. Morphological and histochemical differences in the testicular glands of blennioid and gobioid fishes (abstract). *14. Conference of European Comparative Endocrinologists, Salzburg, 1988*.

Patzner, R.A. and M. Seiwald. 1988b. Different types of testicular accessory organs in blennioid fishes (abstract). *VI. Congress of European Ichthyologists*, Budapest, 1988, p. 156.

Patzner, R.A., F. Lahnsteiner and B. Nussbaumer. 1988. Functions of the testicular accessory organs in blenniid fishes (Blenniidae). Morphological, histochemical and biochemical studies (abstract). *VII. Congress of European Ichthyologists, Den Haag, 1988*.

Patzner, R.A., M. Seiwald, S. Angerer, E. Ferrero and P. Giulianini. 1991. Genital system and reproductive cycle of the male grass goby, *Zosterisessor ophiocephalus* (Teleostei, Gobiidae), in the northern Adriatic Sea. *Zoologischer Anzeiger, Jena* 226: 205-219.

Pern, U. and J. Laumen. 1977. *Untersuchungen zur Fortpflanzungsbiologie des mediterranen Schleimfisches* Blennius pavo *(Risso)*. Ph.D. Thesis, University of Bochum, Germany.

Qasim, S.Z. 1957. The biology of *Blennius pholis* L. (Teleostei). *Proceedings of the Zoological Society of London* 128: 161-208.

Rasotto, M.B. 1995. Male reproductive apparatus of some Blennioidei (Pisces: Teleostei). *Copeia* 1995: 907-914.

Rathke, H. 1836. Beitäge zur Anatomie der Fische. *Archiv für Anatomie und Physiologie* 3: 170-186.

Resink, J.W., W.G.E.J. Schoonen, P.C.H. Albers, D.M Filé, C.D. Notenboom, R. Van den Hurk and P.G.W.J. Van Oordt. 1989. The chemical nature of sex attracting pheromones from the seminal vesicle of the African catfish, *Clarias gariepinus*. *Aquaculture* 83: 137-151.

Richtarski, U. 1990. *Morphologie der akzessorischen Geschlechtsorgane männlicher Mittelmeerschleimfische (Pisces, Teleostei, Blenniidae)*. M.Sc. Thesis, University of Salzburg, Salzburg, Austria.

Richtarski, U. and R.A. Patzner. 1988. Morphology of male reproductive organs of four Mediterranean *Lipophrys* species (Blenniidae) (abstract). *VI. Proceedings of the Congress of European Ichthyologists, Budapest, 1988*, p. 170.

Richtarski, U. and R.A. Patzner. 2000. Comparative morphology of male reproductive systems in Mediterranean blennies (Blenniidae). *Journal of Fish Biology* 56: 22-36.

Ruchon, F., T. Laughier and J.-P. Quingard. 1993. Variation saisonnière de la taille des oefs et de la fécondité par acte de ponte de *Lipophrys pavo* (Teleostei, Blenniidae) dans une lagune Nord-Mediterraneenne (France, Mauguio). *Cybium* 17: 197-214.

Santos, R.S. 1995. Anatomy and histology of secondary sexual characters, gonads and liver of the rock-pool blenny, *Parablennius sangiunolentus parvicornis* (Pisces: Blenniidae) of the Azores. *Arquipelago* 13A: 21-38.

Schoonen, W.G.E.J., J.C.M. Granneman, J.G.D. Lambert and P.G.W.J. Van-Oordt. 1987. Steroidogenesis in the testes and seminal vesicles of spawning and non-spawning African catfish, *Clarias gariepinus*. *Aquaculture* 63: 77-88.

Seiwald, M. 1986. *Morphologie und Funktion der Hodendrüse des Pfauenschleimfisches* Blennius pavo *(Pisces: Teleostei: Blenniidae)*. Ph.D. Thesis, University of Salzburg, Salzbuerg, Austria.

Seiwald, M. and R.A. Patzner. 1987a. The reproduction of *Blennius pavo* (Teleostei, Blenniidae). VIII. The testicular gland—Preliminary results. *Proceedings of the V. Congress of European Ichthyologists, Stockholm, 1985*, pp. 299-304.

Seiwald, M. and R.A. Patzner. 1987b. Ultrastructure of the testicular gland of *Blennius pavo* (Pisces, Teleostei). *Zoomorphology* 107: 26-32.

Seiwald, M. and R.A. Patzner. 1989. Histological, fine-structural and histochemical differences in the testicular glands of gobiid and blenniid fishes. *Journal of Fish Biology* 35: 631-640.

Shackley, S.E. 1976. Aspects of reproduction in a marine teleost. Ph.D. Thesis, University of Swansea, UK.

Shackley, S.E. and P.E. King. 1977a. The reproductive cycle and its control. Frequency of spawning and fecundity in *Blennius pholis* L. *Journal of Experimental Marine Biology and Ecology* 30: 73-83.

Shackley, S.E. and P.E. King. 1977b. Oogenesis in a marine teleost, *Blennius pholis* L. *Cell and Tissue Research* 181: 105-128.

Singh, B. 1989. *The Morphology, Endocrine Mediation of Development and Correlation with Reproductive Activity of the Anal-fin-fold, a Secondary Sexual Character in Male* Ophioblennius atlanticus *(Pisces: Blenniidae)*. Ph.D. Thesis, University of Alberta, Edmonton, Canada.

Sperotti, G. 1940/41. Ricerche sull'apparato riproduttivo maschile del *Blennius fluviatilis*. *Atti dell'Instituto veneto di scienze lettere ed arti* 100: 117-147.

Tong, L.J. 1967. Endocrine studies on *Blennius pholis* L. Ph.D. Thesis, University of North Wales, Bangor, UK.

Van Weerdt, J.H. and C.J.J. Richter. 1991. Sex pheromones and ovarian development in teleost fish. *Comparative Biochemistry and Physiology* A100: 517-527.

CHAPTER

4.2

Sexual Selection in Blennies

Francis Neat[1] and Wouter Lengkeek[2]

INTRODUCTION

Several features of blennies immediately catch the naturalist's eye: the ornaments, the colour patterns, the occupancy of a nest site, the threat displays, the courtships and the sometimes extreme degree of sexual dimorphism. These are, of course, behaviours and morphological adaptations associated with reproduction. The theory of sexual selection provides a framework for understanding the unusual nature of the traits and the often bewildering variation between species. In this chapter, we will review the literature on sexual selection in blennies and our own experience of studying the reproductive habits of blennies with the aim of drawing it together under the current sexual selection theory.

As early as 1686, blennies were being named and illustrated with reference to their secondary sexual ornaments (Willughby, 1686; Fig. 4.2.1). However, it was of course Darwin (1872) who hypothesised that

Authors' addresses: [1]Fisheries Research Services, Marine Laboratory, P.O. Box 101, 375 Victoria Road, Aberdeen AB11 9DB, Scotland. E-mail: f.neat@marlab.ac.uk

[2]Bureau Waardenburg bv consultants for environment & ecology, marine ecology group P.O. Box 3654100 AJ Culemborg, The Netherlands.
E-mail: w.lengkeek@buwa.nl

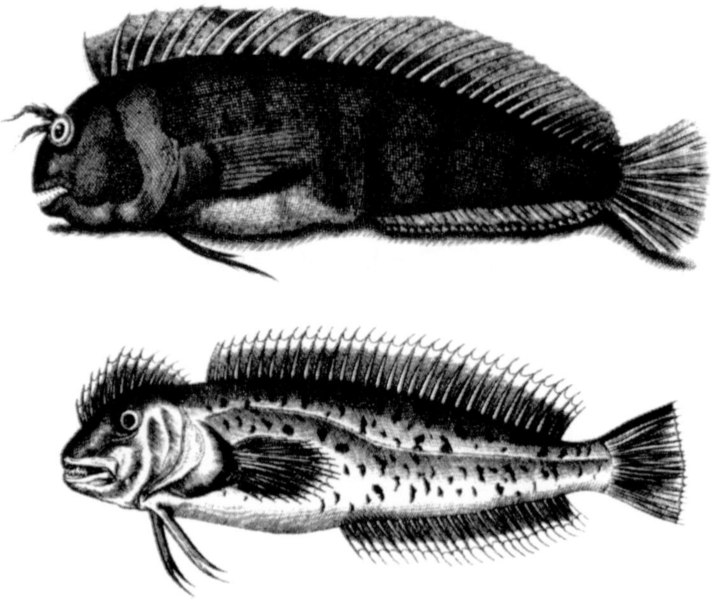

Fig. 4.2.1 An illustration from Francis Willughby's 1686 edition of 'De Historia Piscium' commissioned by the Royal Society of London. The blennies illustrated are *Parablennius gattorugine* (above) and *Scartella cristata* (below) and referenced with respect to their sexually dimorphic head cirri (or tentacles).

such traits were sexually selected traits. He referred to blennies to illustrate how variable the expression of sexually selected traits could be within genera. He wrote of blennies on p. 12 of *The Descent of Man Vol. II*:

> 'a crest is developed on the head of the male only during the breeding season, and their bodies at the same time become more brightly coloured. There can be little doubt that this crest serves as a temporary sexual ornament, for the female does not exhibit a trace of it. In other species of the same genus both sexes possess a crest, and in at least one species neither sex is thus provided.'

The empirical study of sexual selection in blennies, however, remained dormant until the late 1970s. Blennies lend themselves particularly well to the study of sexual selection and reproductive behaviour for several reasons: they are found in shallow waters and easily observed and captured for individual marking (due to their territorial habits and nest occupancy) and also measurement of their reproductive success is easily quantified due to their readiness to accept artificial nests such as glass vials which facilitate direct counting of eggs received (Hastings, 1988a, b; Kraak and Videler, 1991; Neat and Locatello, 2002). Since then, more than 50 studies on sexual selection of blennies have been published. The early studies were pioneering in testing predictions of sexual selection theory and, as the field has developed, blennies have continued to act as model systems (key studies are summarised in Table 4.2.1). It should be noted that studies of sexual selection in the sub-order is strongly biased toward the Blenniidae with some data available for the Chaenopsidae and Tripterygiidae, but very limited information is available for the Labrisomidae, Clinidae and Dactyloscopidae. Therefore, our conclusions should not be extended to all Blenniodei and, since blennies have several unique reproductive habits—certainly not to all fishes—for which recent comprehensive reviews of sexual selection exist (Sargent *et al.*, 1998; Forsgren *et al.*, 2002; Amundsen, 2003).

THE THEORY OF SEXUAL SELECTION

Darwin (1872) proposed the process of sexual selection to account for the evolution of elaborate morphologies and behaviours, usually in males that function solely in the acquisition of mates and enhancing reproductive success. He defined sexual selection as competition within one sex (more often males) for members of the opposite sex (more often females), and the differential choice by members of one sex (more often females) for members of the opposite sex (more often males). In other words, it is the component of overall fitness attributable to traits associated with variance in mating success and fertilisation. Sexual selection is generally of greater intensity in males because, prior to mating, they tend to have lower reproductive investment as a consequence of their smaller gamete size relative to females (Trivers, 1972). This means that they have higher potential reproductive rates which lead to a male-biased operational sex-ratio (Emlen and Oring, 1976; Kvarnemo and Ahnesjo, 1996, 2002) and more intense male-male competition (Parker and Simmons, 1996). The study of sexual selection has flourished in recent years (Andersson, 1994; Arnquist and Rowe, 2005) and it is now widely accepted to be a pervasive and particularly diversifying component of evolution.

Sexual selection underlies many of the most remarkable and aesthetically striking morphological traits and behaviours in the animal kingdom. There are three processes by which sexual selection operates, intrasexual competition, intersexual competition and sexual conflict. We

Table 4.2.1 Key blenny studies on sexual selection

Family	*Species*	*Brief summary*	*References*
Blenniidae	*Aidablennius sphynx*	Female choice: preference for males with eggs Constraints on male success: consequences for female choice Female reproductive allocation	Kraak and Videler (1991) Neat and Locatello (2002) Locatello and Neat (2005)
Blenniidae	*Salaria pavo*	Male choice: courting females and female competition Female choice: male traits and mating success Male traits: functional significance and direct benefits	Almada *et al.* (1995) Oliveira *et al.* (1999) Giacomello *et al.* (2006)
Blenniidae	*Salaria fluviatilis*	Mutual mate choice and plastic sex roles Male mating success: nest site preference by females	Lengkeek and Didderen (2006) Vinyoles *et al.* (2003)
Blenniidae	*Parablennius tentacularis*	Female choice: male ornaments and mating success	Giacomello and Rasotto (2005)
Blenniidae	*Parablennius sanguinolentus parvicornis*	Female choice: preference for large male and territory Male mating success/female preference: alternative tactics	Oliveira *et al.* (2000) Oliveira *et al.* (2002)
Blenniidae	*Petroscirtes breviceps*	Mate choice: mutual, female competition	Shibita and Kohda (2006)
Blenniidae	*Ophioblennius atlanticus*	Female choice: preference for large males Female choice: costs of being choosy Male traits: functional significance and direct benefits	Côté and Hunte (1989) Reyonolds and Côté (1995) Giacomello *et al.* (2006)
Chaenopsidae	*Coralliozetus angelica*	Female choice: preference for large males	Hastings (1988b)
Chaenopsidae	*Acanthemblemaria crockeri*	Female choice: preference for large males Nest size constraints of male reproductive success	Hastings (1988b) Hastings (1992)
Labrisomidae	*Malacoctenus macropus*	Male mating success: large males preferred by females	Petersen (1988)
Labrisomidae	*Malacoctenus hubbsi*	Male mating success: no large male advantage	Petersen (1988)
Tripterygiidae	*Axoclinus carminalis*	Female choice: preference for nests with eggs	Petersen (1989)
Tripterygiidae	*Fosterygion varium*	Female choice: preference for safe territories	Thompson (1986)

will consider each of these processes in turn with respect to blennies and assess how the predictions match field based correlations, the outcome of experiments and comparative patterns within the sub-order. Before doing so, however, it is necessary to review the background on sexual ecology of blennies and gain some appreciation of the morphological and behavioural diversity associated with reproduction.

SEXUAL ECOLOGY OF THE BLENNIOIDEI AND THE OPPORTUNITY FOR SEXUAL SELECTION

The Blenniidae, Chaenopsidae, Tripterygiidae and some Labrisomidae share several reproductive and ecological traits that predispose them to the action of sexual selection: (1) A promiscuous mating system, i.e., males and females mate with multiple partners, as evidenced by multiple behavioural studies as well as genetic parentage studies (Mackiewicz *et al.*, 2005). Sexual selection is predicted to be strong in promiscuous species because variance in reproductive success will be greater. (2) A demersal spawning strategy (external fertilisation) with males establishing a breeding territory. Territoriality usually means there is opportunity for intrasexual competition between males and an increased likelihood that alternative reproductive tactics will arise (Oliveira *et al.*, this book). (3) The occurrence of paternal care of eggs which means that males invest relatively heavily in reproduction. This has two important consequences: females stand to gain direct benefits from choosing high quality caring males, and male mate choice is predicted. (4) A shallow water habitat with plenty of light for mate choice to be based on visual cues and display behaviours. (5) The habit of spawning in holes, cavities and biogenic features such as empty shells or bore-holes means that the males are relatively protected from predators and, therefore, the costs of ornamentation may be less curbed by the action of natural selection.

Sexually Selected Ornaments and Behaviours in Blennies

Although we should bear in mind that female traits and behaviours may also be subject to sexual selection in blennies (Almada *et al.*, 1995; Lengkeek and Didderen, 2006), we will mainly be concerned with traits and behaviours that are present or exaggerated in mature males. Sexual size dimorphism is the most common trait that is likely to have been shaped by sexual selection (Lengkeek *et al.*, 2008). Throughout the sub-order, a wide range of sexual size differences occurs (Table 4.2.2), although there is clearly a prevalence of male-biased sexual dimorphism. The degree of sexual size dimorphism can be ecologically driven and evolve through natural selection (Shine, 1989), for example, when it reflects sex-specific feeding strategies. Or in cases where natural selection favours a reduction in male body size, reversed sexual size dimorphism may arise as a consequence of the constraints imposed by a lower limit of body size necessary for producing eggs. More often, however, sexual size dimorphism is often associated with the intensity of sexual selection (Selander, 1972; Andersson, 1994). In blenniids there is evidence of both female mate choice and male-male competition favouring large male body size and in the majority of species, males are the larger sex. A few exceptions occur, *Lipophrys* (= *Paralipophrys*) *trigloides* is sexually monomorphic, and there a few species where females are larger than the males, for example, the Tripterygiid, *Enneanectes* (= *Axoclinus*) *carminalis* (Neat, 2001) and the Labrisomid *Malacoctenus hubbsi* (Petersen, 1988). It should also be noted that Lengkeek *et al.* (2008) demonstrated that there is often significant variability in sexual size dimorphism between populations of the same species.

Sexual dichromatism is widely spread in the sub-order with males being typically brighter, and more varied in colour than females (Zander, 1979). Intense male colouration is usually associated with territoriality. Bright colourations are generally assumed to be attractive to the opposite

Table 4.2.2 Sexual size dimorphism (SSD) in studies of Blennioidea genera. + = male-biased SSD, 0 = no SSD, – = female-biased SSD.

Family	*Genus*	*SSD*
Blenniidae	*Aidablennius*	+
	Coryphoblennius	+
	Lipophrys	+ 0
	Parablennius	+ 0
	Scartella	+
	Ophioblennius	0
	Alticus	+
	Paralipophrys	0
	Salaria	+
Chaenopsidae	*Coralliozetus*	+
	Acanthemblemaria	+
Labrisomidae	*Malacoctenus*	+ –
Tripterygiidae	*Tripterygion*	+
	Axoclinus	+ –

sex (Andersson, 1994), but are also associated with negative fitness consequences through increased predation risk (Wiens, 2001). The trade-off between mate-attraction and predation risk has important consequences for sexual selection and the design of signals. For example, *Microlipophrys* (= *Lipophrys*) *canevae* are strikingly sexually dichromatic, with males displaying bright yellow cheeks, but a dark cryptic frontal part of the head (Zander, 1979). Although it remains to be demonstrated, males are likely to be highly conspicuous to females when they position their front part of the body outside of the breeding hole during courtship, but inconspicuous when they withdraw into their hole in the presence of predators. In a number of species, such as *Salaria pavo* and *Aidablennius sphynx*, breeding males exhibit striking sky-blue stripes or spots (Fig. 4.2.2a, b). Females also exhibit distinctive colourations prior to spawning, for example, female *Aidablennius sphynx*, *Salaria fluviatilis* and *Salaria pavo* show a distinct change in colour from cryptic colours to a contrasting dark and light banding pattern across their head and body (Fig. 4.2.2b).

Male blennies possess multiple secondary sexual ornaments. The Blenniidae appear to be the only family within the order that has evolved a fleshy head crest. It is present in at least 10 genera and is particularly well developed in the genera *Alticus*, *Istiblennius*, *Omobranchus*, *Praealticus* and *Salaria*. The crest is a feature of territorial males, being absent in non-territorial 'sneaker' males and females. In *Salaria pavo*, the crest develops in height with the onset of the spawning season and reaches a maximum height mid-way through the season (Oliveira *et al.*, 1999). The majority of blennioids have sensory cirri or tentacles on their head region which are thought be involved in chemoreception (Von Bartheld and Meyer, 1985). Blenniids are known to produce sexual pheromones (Losey, 1969; Barata *et al.*, 2008) and it is reasonable to hypothesise that the often highly bifurcated morphology of the cirri provides a large surface area for promoting sensitivity to pheromones. In many male blenniids (at least 19 genera) and chaenopsids, the cirri are larger and more elaborate in males and it is likely that they function to attract females (Zander, 1979; Giacomello and Rasotto, 2005). In *Parablennius tentacularis*, the length of the cirri was greater during the spawning season than outside it and found to positively correlate with male mating success, indicating that the trait is sexually selected (Giacomello and Rasotto, 2005). The presence of glandular structures on the fins of mature territorial males is also frequently reported (Zander, 1979; Northcott and Bullock, 1991; Neat *et al.*, 2003a; Giacomello and Rasotto, 2005). The role of these glands

Fig. 4.2.2 **a.** A brightly coloured male *Aidablennius sphynx* from the Mediterranean displays to a female by erecting his dorsal fin and head nodding (photo: W. Lengkeek). **b.** Nesting male and female *Salaria pavo* in a laboratory set-up. Note colouration of female. (photo: D. Gonçalves). **c.** A nesting male *Coryphoblennius galerita* exhibiting enlarged, coloured lips and head cirri (photo: J.J. Videler). **d.** A male *Tripterygion xanthosoma* displaying contrasting head and body colours (photo: R.A. Patzner).

appears to be in pheromone and mucous production (Losey, 1969; Laumen *et al.*, 1974; Neat *et al.*, 2003a; Barata *et al.*, 2008) and these authors speculated they functioned in mate attraction. Gonçalves *et al.* (2002b), however, found that mate choice by female *Salaria pavo* was unaffected by exposure to male chemical cues. Giacomello *et al.* (2006) tested the hypothesis that the glands also serve to protecting the eggs from microbial attack. In *Ophioblennius atlanticus* and *Salaria pavo,* they demonstrated that the protein fraction of the secretion has lysozyme-like activity. The dorsal fin of males appears to have been co-opted into a sexual display signal. It is exaggerated in a number of the Chaenopsidae, e.g., *Emblemaria hypacanthus* (Thompson *et al.*, 2000) and Blenniidae, e.g., *Aidablennius sphynx* where it is noted to show extended growth during the spawning season (Zander, 1979). The dorsal fin is erected during intrasexual competition and courtship (Fig. 4.2.2a). In many species, an eye-spot is present toward the front of the dorsal fin, e.g., *Blennius ocellatus* and *Parablennius tentacularis* (Giacomello and Rasotto, 2005), although this feature is also exhibited by females and it is questionable whether it is a sexually selected trait (see later). Enlarged canine teeth are found in a number of male blenniids (Kotrschal and Goldschmid, 1992) and in *P. incognitus* the males have been observed to open their mouths, displaying canine teeth during territorial disputes (FN pers. obs.). Males of the family Chaenopsidae have a more robust and musculature jaw and mouth morphology than females (Hastings, 1991, 2002) and the lips of some blennies, e.g., *Coryphoblennius galerita* (Fig. 4.2.2c), are larger and more distinctive in breeding males. Only one study, on *Parablennius parvicornis*, has demonstrated that breeding males make acoustic signals to attract females (De Jong *et al.*, 2008).

The Blenniidae and Chaenopsidae appear to show the highest degree of sexual dimorphism. The Tripterygiidae tend to be less morphologically diverse, although often exhibit striking dichromatism. The few data on the Labrisomidae suggest the family lacks secondary sexual ornaments and is generally cryptically coloured, although exceptions occur, e.g., the redside blenny, *Malacoctenus hubbsi* (Thompson *et al.*, 2000). Too little is known about the Clinidae or Dactyloscopidae to draw firm conclusions, although many are also cryptically coloured and appear to be sexually monomorphic. Many of the Blenniidae and most of the Chaenopsidae are hole-dwelling species and this may underlie the high degree of elaboration of head ornaments in these two families because the head is the most visible part of the fish. The Tripterygiidae, on the other hand, spawn in more open areas and this may underlie the often dramatic nuptial body colourations of males (de Jonge and Videler, 1989).

The Blenniidae are the best studied family with respect to sexual selection and associated traits and behaviours. We undertook a review of potential male ornaments within the family. The degree of exaggeration of five ornaments in 94 species from 48 genera was assessed based on information from FishBase (Froese and Pauly, 2007). We used a maximum of four species per genera, to ensure that each genus contributed equally to the overall patterns. The five focal morphological traits were: (1) presence and degree of development of a head crest; (2) presence and degree of development of head cirri or 'tentacles'; (3) height of dorsal fin relative to body size; (4) presence and degree of development of an 'eye-spot' on the dorsal fin; and (5) presence of a blue-stripe colouration. The degree of development of the traits 1, 2 and 4, as well as the height of the dorsal fin was assessed using scores on an ordinal scale (0, 1, 2 or 3; 0 for absence of the trait to 3 for maximum development). For the presence of a blue stripe colouration a score of 1 or 0 (present or not) was assigned. Our aim was to identify broad patterns within the family, controlling for body size but without controlling for phylogenetic effects. To this end we undertook a partial correlation analysis of the degree of development of the five traits. Significant correlations were detected in the following comparisons: height of the dorsal fin with head crest ($r = 0.35$, $P < 0.001$) and tentacle exaggeration ($r = 0.27$, $P = 0.009$). Blue stripes were pronounced in species with exaggerated crests ($r = 0.23$, $P = 0.023$) and extended dorsal fins ($r = 0.25$, $P = 0.014$), but not in genera with highly exaggerated tentacles. Dorsal fin 'eye-spots' seemed unrelated to any of the other traits and was the only trait that seemed to occur independently of other sexual traits. Thus, four of the five most distinct and widely occurring sexually dimorphic traits, crests, tentacles, blue stripes and dorsal fin height tended to positively correlate with one another. This indicates that these four traits have been or are subject to sexual selection, and are part of a suite of ornaments important for mate attraction. and/or male/male competition (these traits can be used to signal male competitive abilities to other males).

Courtship behaviours of the blennies include 'bob and weaving' described for *Coraliozetus angelicus* (Hastings, 1986), body extensions or 'lunges' in *Acanthemblemaria crockeri* (Hastings, 1988b), 'head-nodding' in *Aidablennius sphynx* (Kraak and Videler, 1991; Kraak, 1996; Fig. 4.2.2a), *Salaria fluviatilis* (Neat *et al.*, 2003b; Lengkeek and Didderen, 2006) and *Meiacanthus nigrolineatus* (Fishelson, 1974), 'quivering', 'shaking' or 'shivering' in *Hypsoblennius* sp. (Losey, 1969), *Aidablennius sphynx* (Kraak, 1996) and *Scartella cristata* (pers. obs.), 'loop-swimming' in Mediterranean *Microlipophrys* (= *Lipophrys*) species (pers. obs.), 'figure-of-eight' swimming in *Tripterygion* sp. (Wirtz, 1978) and *Axoclinus* sp. (pers. obs.), 'nudge and biting' in *Labrisomus* sp. (Gibran *et al.*, 2004) and 'circling and leading' in *Parablennius parvicornis* (Oliveira *et al.*, 2000). Although qualitatively diverse, these behaviours appear to attract attention to the male, lead toward the location of his nest site, expose some aspect of the male's body to the female or perhaps demonstrate his readiness to spawn.

Female courtship has been observed in the genus *Salaria* in the form of 'head-nodding' (Almada *et al.*, 1995; Lengkeek and Didderen, 2006), rapid respiratory movements and display of the gravid belly (Patzner *et al.*, 1986). Similarly, female *Aidablennius sphynx* and *Petroscirtes breviceps* expose their gravid belly when they approach a nesting male (Kraak, 1996; Shibata and Kohda, 2006; pers. obs.). Oliveira *et al.* (2000) report that female courtship in *Parablennius parvicornis* occurs at least as frequently as male courtship, especially in response to the preferred larger males. It is suspected that female courtship is far more common in blennies than has been reported.

Intrasexual Selection: Reproductive Territoriality and Male-Male Competition

Intrasexual selection occurs through competition between same-sex individuals for access to mates, either directly such as male-male competition for females, or indirectly through competition for essential resources such as nest sites (Andersson, 1994). Typically, it does not involve active participation by the other sex. The mechanism by which traits such as a large body size or weaponry evolve under intrasexual selection is fairly straightforward, i.e., direct selection on traits that accrue higher fitness to their bearer when in competition over mates or resources that attract mates. It is also generally assumed that females stand to benefit from mating with those males that are successful in competition, although this is not always the case and this issue will be dealt with later, under sexual conflict. Generally, however, intrasexual sexual selection is relatively uncontroversial and accounts for numerous male traits that are used in agonistic displays and interaction. In blennies, traits such as body size, canine teeth and jaw musculature clearly function in the context of intrasexual competition. All male blennioids establish, maintain and defend a territory during spawning (Nursall, 1977; Gonçalves and Almada, 1998). Spawning sites range from open spaces on algal turfs and rocks (tripterygiids), cracks and crevices in rocks (many blenniids, labrisomids), biogenic holes (blenniids and chaenopsids), barnacle tests and sand burrows (Chaenopsids), mollusc shells and even human artefacts such as bottles (blenniids). Variation in nest sites is a key determinant of male and female reproductive success and males fight over access to the best sites. Competition is probably most intense in species that spawn in biogenic holes or shells which are patchily distributed in space and can be a limiting factor to the number of breeding individuals (Buchheim and Hixon, 1992; Almada *et al.*, 1994). Disputes over territories involve displays such as erection of the dorsal fin, a raising-up of the head, e.g., blenniids (Kraak, 1996), chaenopsids (Hastings, 1988a, b) and tripterygiids (Wirtz, 1978), an upward curving of the entire body (*Axoclinus* sp.) and a gaping of the mouth (Blenniidae, pers. obs.). If the dispute continues, escalation to mouth-wrestling and direct attacks by biting has been observed in blenniids such as *Aidablennius sphynx* and *Parablennius incognitus*.

As with most animals, body size is a critical factor influencing dominance and the outcome of aggressive interactions, with larger contestants generally defeating smaller opponents (Hastings, 1988a; Buchheim and Hixon, 1992; Lengkeek, 2006). Larger males are also involved in more territorial disputes and attract more intrusions from sneaker males (Oliveira *et al.*, 2000, author's unpublished data). The habit of hole dwelling introduces unusual dynamics to competition because the optimal hole size is relative to the size of the individual. This means that small individuals can occupy a good nest hole, simply because the nest hole is too small for the larger, dominant individuals in the population. This may be part of the reason that in hole-dwelling species, small males attempt to breed unhindered by larger individuals (Neat and Locatello, 2002) and it may act to constrain selection on body size.

It is important to appreciate that many blennies share their reproductive habitat with other blennies and that many are similar sized, especially the hole-dwelling species. This leads to males of different species competing over the same breeding holes. The concept of the ecological niche and competitive equilibrium has perhaps led to interspecific competition being largely overlooked, but recently, Lengkeek (2006) demonstrated that interspecific male-male competition occurs over breeding holes in an assemblage of Mediterranean blennies. This potentially intensifies selection on traits that are used to settle territorial disputes. Lengkeek (2006) tested for an effect of body size on the outcome of interspecific contests for breeding holes. They found that which species were competing was as important as relative size in determining eventual occupancy. Thus, the overall level of nest site competition is an important factor in considering the evolution of intrasexually selected male traits.

Intersexual Selection: Female and Male Choice

Intersexual selection, or more generally 'mate choice' (Andersson, 1994) is the process of selecting among potential mating partners that vary: (a) in the genetic quality they will contribute to the next generation or (b) in some resource such as a nest site that affects the viability of the next generation or (c) in some resource that directly enhances the reproductive output or survival. The mechanistic basis of mate choice can be traced to Fisher (1930), who developed Darwin's original argument by proposing a genetic mechanism whereby the evolution of a female preference and the male trait for which the preference exists become correlated. In order for the process to get started all that is needed is that females evolve a preference for a trait that is characteristic of slightly fitter males. Later, Kirkpatrick (1982) showed that even a female preference for a trait that confers no advantage to the male can also spread by the same mechanism. In each generation, the male trait and the female preference for that trait exaggerate further or 'run-away' until a point is reached at which the negative consequences for survival outweigh the advantages for reproducing. In other words, males carry the genes that determine their sexual attractiveness that, in turn, enhance the sexual attractiveness of their sons. Selection on the female's part is indirect, i.e., her preference is selected solely because it is correlated with the male trait. The male trait, on the other hand, is under direct selection by females. The Fisher process assumes no costs associated with being choosy; all that is needed is genetic covariance between female choice and male trait. However, costs of mate choice are apparent and any cost implies direct selection because being choosy affects the female's fitness and fitness is always directly selected for. Although some models of the Fisher process can accommodate costly female preferences (Day, 2000), the introduction of female costs to the original Fisher model cause the process to grind to a halt. A series of models known under various titles such as 'good genes', 'viability', 'handicap' or 'indicator' models proposed that secondary sexual traits are expensive and only the fittest males will be selected to fully express them (Zahavi, 1977; Johnstone, 1995; Kokko *et al.*, 2002). Once again, a genetic correlation between female preference and male trait is assumed, but now male trait expression is an increasing function of male condition and his overall genetic quality or fitness. Net fitness is always subject to direct selection and as a consequence the cost of a female's choosiness can be borne up to the value of the indirect benefits she stands to gain. Thus, expensive female preferences can be accounted for and the good genes models extend Fisher's argument. Fisherian and good genes models have recently been harmonised into one conceptual process of indirect selection, rather than interpreted as alternative hypotheses (Kokko *et al.*, 2002, 2003). Females may, of course, also receive direct benefits as a result of being choosy. For example, in cases where a male's traits are correlated with the level of paternal care, there are reduced mate search costs, nutritional benefits or reduced parasite load.

Evidence for non-random mating by females has been found without exception in studies of mate choice in blennioids. The evidence for female choice of male partner in blennies suggest that three criteria are important to females: (1) traits of the male; (2) characteristics of his nest site; and (3) the male's recent and current reproductive success. While there are a large number of field studies that demonstrate that male body size correlates with reproductive success, i.e., number of eggs received (Thompson, 1986; Petersen, 1988; Hastings, 1988b, 1992; Côté and Hunte, 1989; Oliveira *et al.*, 2000; Gonçalves *et al.*, 2002a; Neat and Locatello, 2002), which indicates that females prefer large males, rather fewer studies have actually demonstrated the female choice of larger males. Those studies that have include the laboratory studies of Hastings (1988a, b) who studied the Chaenopsids, *Acanthemblemaria crockeri* and *Coralliozetus angelicus*. Sixty percent of female *A. crockeri* chose to spawn with the larger male and 66% of *C. angelicus* did likewise. The most direct field evidence for female choice for male size is the study of Côté and Hunte (1989) on the redlip blenny, *Ophioblennius atlanticus*. Based on courtship and spawning observations, they demonstrated that the probability of a female entering a nest and spawning increased with the size of the male. A similar pattern was observed in the Azorean rock-pool blenny (*Parablennius parvicornis*) (Oliveira *et al.*, 2000), and in *Parablennius zvonimiri* (Lengkeek, 2006).

Several studies suggest that females gain direct benefits from mating with males of larger than average body size. Côté and Hunte (1989) reported that by mating with a larger male, a female has a decreased chance that her batch of eggs will be lost as larger males are more attentive of eggs and more likely to retain their territory until the eggs hatch. In *P. sanguinolentus parvicornis*, Oliveira *et al.* (2000) showed that larger males undertook more cleaning behaviour of their nest and possibly also fanned the eggs at a higher rate. Neat and Locatello (2002) also suggested that larger males *Aidablennius sphynx* care better based on a positive correlation between the rate of pectoral fin fanning of eggs in the nest and body size. In the same study, these authors also measured sperm release just after ejaculation (characterised by a quivering behaviour) by collecting a water sample from inside the nest through a tube attached to the nest. There was a positive exponential relationship between the number of sperm released and male body size. In a follow-up study, Locatello and Neat (2005) found evidence that when forced to pair with small males, females slow the rate at which they release eggs. This is congruent with the hypothesis that small males are sperm limited and, therefore, sub-optimal partners.

Body size and age and are often confounded and only a single study has been able to separate the two (Côté and Hunte, 1993). In this study, it was concluded that in *Ophioblennius atlanticus*, once the female had entered the nest she then had some mechanism of assessing male age and preferred older males over younger males. Older males guarded their broods for longer durations and, therefore, female choice accrues direct benefits. Theoretical studies suggest that females may benefit by mating with older males because older males reveal more information in their displays as a consequence of their limited future reproductive potential (Proulx *et al.*, 2002). Body size is not, however, always associated with the reproductive success of the male and presumably not female choice. Several studies did not detect an advantage to larger males either in the number of matings (*Malacoctenus hubbsi*, Petersen, 1988 and *Enneanectes* (=*Axoclinus*) *carminalis*, Petersen, 1989) or the number of eggs spawned (*Aidablennius sphynx*: Kraak and Videler, 1991).

As we have seen, multiple male ornaments appear to have evolved among blennies. Several evolutionary scenarios have been hypothesised to explain the evolution of multiple ornaments (Brooks and Couldridge, 1999; Van Doorn and Weissing, 2004). Each ornament may reflect a different component of overall quality or several ornaments may serve to reinforce one another regarding the overall quality of the male. Alternatively, the ornament may have become abandoned by females because it was an inaccurate indicator of quality, but persisted because it bears little cost and is, therefore, not selected against. It is noteworthy in our study that none of the correlations were particularly strong, reflecting the high variability within the family and suggesting that particular ornaments are not stable over evolutionary time. The evolutionary loss of sexually selected traits (and the preferences for them) is reported to be relatively common (Wiens, 2001) and it may be that ornaments are in a continual evolutionary flux of gain and loss. When it comes to empirical evidence, however, there is rather little evidence for ornaments being correlated with reproductive success of males or female choice. In *Parablennius tentacularis*, Giacomello and Rasotto (2005) found a significant correlation between the relative length of the head cirri (tentacles) of this species and male reproductive success. A field study on *Salaria pavo* suggested that neither the blue eye-spots of males nor the crest size was different between males that successfully attracted females and those that failed (Oliveira *et al.*, 1999). In the laboratory, however, a female preference was detected for males with more developed crests (Gonçalves *et al.*, 2002a, b). Similarly, despite the often extensive and elaborate courtship behaviours exhibited by blennies, there is little evidence to suggest that females prefer males that court more intensively. Hastings (1988a) found no correlation between courtship intensity and male reproductive success in *Acanthemblemaria crockeri*. Similarly, our own unpublished observations on *Salaria fluviatilis* and *Aidablennius sphynx* suggest no relationship between the frequency of male courtship and mating success. We have only been able to find one study in *Parablennius parvicornis sanguinolentus* that suggested that males that courted more, received more eggs (Oliveira *et al.*, 2000). Courtship behaviours, nevertheless, clearly serve a purpose in the initial attraction of females. Notably, many of the behaviours such as the 'looping' of *Lipophrys* species and 'figure of eight' display of the Tripterygiidae indicate to the female the location of the nest hole. Other behaviours such as the exaggerated quiver may signal capacity for sperm release or might be associated with sperm and pheromone release; quivering is associated with ejaculation during actual spawning (Neat and Locatello, 2002). However, when it comes to potential indicators of male quality, the lack of female preference suggest that courtship is unreliable. Courtship in blennies is a complex process, involving many qualitatively different behaviours and it may be that a detailed analysis of the individual components is needed to tease out the functional significance for mate choice.

Nest site characteristics can be an important component of female preference. In the Tripterygiid, *Forsterygion varium*, Thompson (1986) found that females preferred to mate with males whose territories were surrounded by rocks and hiding places. This may reflect upon the trait of females selecting mates in places where the risk of predation is lower and is, therefore, perhaps better understood in terms of female costs of mate choice. Nest site depth and nest opening area have been reported as important aspects of female choice in *P. sanguinolentus parvicornis* (Oliveira *et al.*, 2000). Vinyoles *et al.* (2003) studied nest orientation in the river blenny, *Salaria fluviatilis* and found that males whose nest contained the most eggs tended to be orientated to the southeast and that males adapt their nest orientation relative to the current direction in such a way that localised current speed directly at the nest entrance is always similar. Similarly, Lengkeek (2006) documented that territorial male river blennies settled nearer to upstream rapids as the river current speed decreased over the season. The choice of nest may also be determined by the availability of space inside the nest (Matsumoto *et al.*, 2008): in *Rhadoblennius elipes*, females prefer nests in which they can deposit eggs deep inside rather than near the entrance.

Several studies that did not find a preference for or an advantage of large males, discovered that females preferred to spawn with males that had a high previous number of spawning (Petersen, 1989). This was confirmed by Kraak and Videler (1991), who demonstrated that females prefer to mate with males whose nests contained more eggs. This trait has also been observed in *S. pavo* (Fagundes *et al.*, 2007). Females may gain direct benefits as a result; depositing eggs among other eggs dilutes the probability of one's own eggs being eaten by other females or the male himself and egg survival may be increased in a nest containing many eggs as the male has a greater vested interest in caring (Côté and Hunte, 1989). It may also be the case that females are using the presence of eggs as a proxy for a good territory or nest site (Thompson, 1986). The study of Kraak and van der Berghe (1992) that showed that females often lay a single or a few 'test' eggs in a nest and then return to that particular nest suggests that females assess the male's dedication to parental care and are, thereby, seeking direct benefits.

Female choice may also be influenced by the benefits to be gained from hedging one's bets (Yasui, 1998). Hastings (1988a) noted in his choice experiments that for *A. crockeri*, 22% of females spawned with both males. Our own observations suggest that females frequently distribute a single clutch among multiple nests (e.g., unpublished data on *Lipophrys* sp.). Therefore, while a female may prefer one male over another, there may be something to be said for 'not putting all your eggs in one basket', both in terms of increasing the genetic diversity of offspring and reducing the risk of entire clutch loss. From a different perspective, alternative reproductive tactics are common in blennies (Oliveira *et al.*, this book) and sneaker males are morphologically equipped to compete at a spermatic level (Neat *et al.*, 2003a). In *Scartella cristata,* parasitic males fertilised on an average 12% of the brood of territorial males (Mackiwicz *et al.*, 2005). Females might gain indirect benefits from having some fraction of their clutch fertilised by highly competitive 'sexy sperm' (Harvey and May, 1989). Interestingly, Oliveira *et al.* (2002) found that territorial males that associated with a satellite male were upto five times more successful than lone males and, in *S. pavo,* the most successful territorial males had more sneakers associated with their nest (Gonçalves *et al.*, 2003).

Female choice is clearly based on multiple criteria. It is often the case that the largest males occupy the best nest sites and secure the most matings, with preferences becoming reinforced and mating success highly skewed (Oliveira *et al.*, 2000). However, this is not always the case, as was demonstrated by Hastings (1988a), who documented an example of how body size and nest site quality negatively interact to determine female preference in the barnacle nest-dwelling chaenopsid, *Coralliozetus angelica.* Larger males have higher reproductive success and are preferred by females. However, the nests vary in the degree to which they are covered in algae (fouled) and this increases with size of the nest and, hence, also the size of the male. The more fouled the nest, the lower the reproductive success of the male and less attractive he is. The outcome is that there are many combinations of male quality and nest quality and optimal female mate choice is likely to represent a trade-off between indirect and direct benefits.

The idea that mate choice is expensive has important implications for the process of sexual selection. Costs associated with mate choice include the time and energy spent searching for optimal mates or travelling to the best nest sites, reduced feeding opportunities and simply delaying reproducing. There is also the likelihood that mate searching will increased the risk of predation. In blennies, there is evidence that females often have feeding territories that can be at significant distances from male nest sites (Petersen, 1988; Reynolds and Côté, 1995). To date, one study has explicitly addressed the costs of female choice in blennies (Reynolds and Côté, 1995). By choosing larger and older males, female redlip blennies travel further than they would have to if they mated randomly. This entails a cost in terms of the distance travelled and, more importantly, in terms of injuries inflicted by attacks from damselfish as the females are forced to cross their feeding territories.

Male blennies should, like females, be expected to be selective in their choice of mating partner. Limited nest site or hole availability, limited space for eggs within the nest-hole and extensive uniparental effort limit male reproductive rates and, under certain circumstances, may cause males to have lower reproductive rates than females. Almada *et al.* (1995) documented male choosiness, female courtship and female competition for males in a population of *Salaria pavo*. This was caused by shortage of good nest sites and territorial males being outnumbered by females. Lengkeek and Didderen (2006) demonstrated the manner in which male choosiness occurred and varied in intensity under the influence of fluctuating female availability for spawning in *Salaria fluviatilis*. In addition, as the number of females increased, female courtship of males began to be observed, a behaviour that was previously absent when few females were around and males were less choosy. Similarly, Shibata and Kohda (2006) showed in the blenniid, *Petroscirtes breviceps* that during the peak of the breeding period when nest sites are scarce males become highly selective in the females they allow to enter their nests. Although examples are still relatively scarce, male mate choice may be much more common in blennies because of the limited reproductive rate of males. What exactly males prefer and what the consequences are for sexual selection on female size or ornamentation remains poorly understood. It is likely that males will select females on the basis of traits that directly indicate fecundity such as the degree to which the belly is swollen with eggs (Gonçalves and Oliveira, 2003; Lengkeek and Didderen, 2006). If the reason that males are choosy about females is to obtain direct fecundity benefits, they should avoid females that invest too heavily in the trait at the expense of fecundity. However, if females do not invest enough in the trait, the male may be unable to assess the potential fecundity of the female at all. Thus, female traits are predicted to stabilise at intermediate levels of exaggeration (Fitzpatrick *et al.*, 1995; Chenoweth *et al.*, 2006). Surprisingly, costs of male choosiness have not been studied at all. Our own anecdotal evidence suggests that some males fail to sire large broods of eggs, not for lack of females, but for being too choosy about the females they allow to enter the nest. This is a study area that would be relatively easy to acquire data and may shed much light into the area of male choice.

The relative importance of male and female choice may vary within species across the reproductive season, depending on the operational sex-ratio (OSR) or mate encounter rate. A shifting OSR and variable mate quality can lead to sex role plasticity, i.e., males and females differentially assuming the role of the more competitive sex. Traditionally, the sex role of males is to be the more competitive sex (Reynolds, 1996), but when males become the limited sex, females may assume the more competitive role and sex roles are referred to as being reversed (Kvarnemo and Ahnesjö, 1996). In many studies of sexual selection in blennies, there is evidence of sex-role plasticity, for example, female courtship of males; *Salaria pavo* (Almada *et al.*, 1994), *Salaria fluviatilis* (Lengkeek and Didderen, 2006) as well as examples of female-female competition over males *Aidablennius sphynx* (Neat and Locatello, 2002) and *Petroscirtes breviceps* (Shibata and Kohda, 2006). It is clear that mate choice in blennies is better described as mutual (Johnstone *et al.*, 1996). This is because the reproductive rates of males and females are relatively equal due to paternal care, the restricted availability of nest sites and the limited nest area to receive eggs. Such mutual mate choice is likely to lead to what is termed assortative mating where large males tend to mate with large females and there are examples of this in blennies (Côté and Hunte, 1989; Locatello and Neat, 2005).

Females may be unable to mate with their preferred choice due to a number of circumstances: (1) Preferred males cannot currently mate because their reproductive rate is constrained, e.g., by nest size (Hastings, 1992; Neat and Locatello, 2002). (2) Females are actively excluded from mating with their preferred choice by other females (Shibata and Kohda, 2006) or by the male himself (Neat and Locatello, 2002). (3) Females have their eggs fertilised by parasitic males/sneakers (Neat *et al.*, 2003; Mackiewicz *et al.*, 2005). Faced with such variance in mate quality, females might adjust their reproductive effort according to male quality and allocate a greater quota of their reproduction to a mate that is, on an average, higher in quality than they expect to find in the future (Sheldon, 2000). As we have seen, male quality is highly variable in blennies as a consequence of body size, nest site and genetic quality and females might be expected to differentially allocate their reproductive effort according to male quality. Locatello and Neat (2005) tested the differential allocation hypothesis in *Aidablennius sphynx*. While females prefer large males that confer direct benefits through fanning their eggs more and releasing more sperm, they frequently spawn with small males due to the reproductive rate of large males being constrained by nest size and the operational sex-ratio being female biased. When paired to large males, females spawned more eggs per unit of time than when paired with smaller males, which is consistent with the theory of differential allocation.

Sexual Conflict

In sexual reproduction, two genetically unrelated partners share genetic interests in their offspring. However, because reproduction is expensive and one sex usually invests more in a given reproductive event than the other, their overall interests tend not to be equal (Trivers, 1972; Parker, 1979; Arnqvist and Rowe, 2005). This inequality is the root of sexual conflict. The type of sexual conflict that is of particular relevance to sexual selection is termed 'interlocus' sexual conflict and arises from the selective advantage that will be conferred on an individual that exploits its mate to its own advantage (Arnqvist and Rowe, 2005). In other words, genes that increase the reproductive success of one sex can be selected for despite them coding for traits that are detrimental to genes expressed in the other sex. For example, in species with parental care, each partner stands to benefit if it can somehow cause the other partner to invest more. Such sexual conflict or sexually antagonistic co-evolution can occur over any form of reproductive investment, be it pre-zygotic when it is largely concerned with conflicts over fertilisation such as sperm competition, or post-zygotic when it is mostly concerned with conflicts over investment into caring for offspring (Chapman *et al.*, 2003). To illustrate how sexual conflict extends the sexual selection theory, consider a hypothetical example; a male trait evolves that benefits males in gaining fertilisations or attracting females, but results in a direct cost to females. Even though females may gain indirectly through their successful sons, if the direct cost outweighs this, selection will act to favour females that resist or become insensitive to the trait. This leads to unpredictable patterns of co-evolution (Chapman *et al.*, 2003). It could lead to further exaggeration of the male trait and, ultimately, if the trait then becomes expensive to bear, it will become condition dependent and, thus, an indicator of genetic quality. This is reminiscent of Fisherian run-away sexual selection and it has become known as 'chase-away' selection. Alternatively, however, it could happen that the male trait and the preference are both lost or just the female preference is lost and the male trait is retained if it poses no cost to the male.

The theory of sexual conflict has proven very insightful in explaining many aspects of reproductive behaviour and morphology (Arnquist and Rowe, 2005) and it is worth exploring whether the evidence so far reviewed can be better interpreted in the light of sexual conflict theory. Male aggression toward females or rejection from the nest site is commonly observed in blennies (Almada *et al.*, 1995; Neat and Locatello, 2002; Shibata and Kohda, 2006). Such overt conflict between the sexes suggests costs to both the male (being aggressive is expensive) and the female (being the recipient of aggression is costly). On the one hand, such observations may simply arise as a consequence of females competing for an undisturbed access to the limited space in nests, or males choosing females above a certain threshold of potential fecundity, especially if there is a risk that what appears to be female may, in fact, be a female-mimic male (Gonçalves and Alamada, 1996). On the other hand, the theory of sexual selection struggles to explain the existence of directly imposed costs by the other partner, whereas sexual conflict theory predicts exactly such dynamics. Consider the finding of two recent studies that suggest male choosiness and aggression toward females varies according to the operational sex ratio (Lengkeek and Didderen, 2006; Shibata and Kohda, 2006). Now consider how males assess the number of females available for mating. A simple rule of thumb might be by keeping track of the number of female visits he receives (rather than try to estimate the actual number of females available), i.e., he monitors the mate encounter rate, rather than the actual OSR (Kokko and Johnstone, 2002). If this is the case, sexual conflict theory might predict that a female manipulates a male's perception of available mates by preventing other females from coming close to the nest. By doing so, the male may be more likely to accept her because he assesses the OSR to be male biased and, furthermore, he is less likely to eat her eggs because he perceives his chances of future reproduction as being low. Guarding of the nest site by females, female-female and male-female aggression have all been observed during courtship in *Salaria pavo* (Almada *et al.*, 1995), *A. sphynx* (Neat and Locatello, 2002), and *P. breviceps* (Shibata and Kohda, 2006). Thus sexual conflict may explain why females are aggressive to one another and why males may be aggressive toward females. Male-female aggression and female-female interaction during reproduction is most certainly an area that deserves further study.

Filial cannibalism is widespread among brood-guarding fish, including blennies, for example, *S. fluviatilis* (Vinyoles *et al.*, 1999) and *S. pavo* (Gonçalves and Almada, 1997). Our own unpublished observations suggest that this also occurs in *A. sphynx*; 8 out 13 nesting males and 3 out of 6 females sampled were found to contain eggs in their stomachs. Filial cannibalism can be explained by the trade-off between current and future reproduction (Fitzgerald, 1992), but there is clearly the potential for sexual conflict (Lindstrom, 2000). Females will always stand to gain by eating the eggs of others while in a nest, whereas the males will always stand to lose if a female eats any of the eggs he has fertilised. Females will obviously stand to lose if a male eats their eggs (unless by eating a few eggs he is provided with the energy to successfully care for the remainder of her clutch). From the male's perspective, he must ensure that a female will lay more eggs than she will eat. From the female's perspective, she must minimise the risk of the male eating her eggs and take any opportunity to eat eggs. Thus, this could explain the common finding that females prefer to lay eggs in a nest that already contains eggs; by doing so she reduces the probability of her eggs being eaten while providing herself with the opportunity to eat the eggs of others.

A final potential aspect of blenny reproductive systems that may be subject to sexual conflict is sperm competition and alternative reproductive tactics (Oliveira et al., present book). The majority of studies on sperm competition and alternative reproductive tactics in blennies deal with the topic from the male perspective. One exception has been Oliveira *et al.* (2002), who suggested that female *Parablennius sanguinolentus parvicornis* prefer to mate with territorial males that have an association with a satellite male. However, the advantage to the female was argued in terms of the direct benefits she would gain via better parenting on the part of the territorial male. No study has considered whether females may actually prefer to have their clutch fertilised by multiple males from the point of view of indirect benefits. In Blenniids, the male testicular anatomy is such that sperm production is highly economised (Giacomello *et al.*, 2008; Patzner and Lahnsteiner, present book) and, thus, sperm concentration could be a constraint on fertilisation success as has been reported in labrid species (Warner *et al.*, 1996). Female polyandry and cryptic female choice via sperm competition has been shown to be an important component of female mating strategies in numerous invertebrate and avian species (Eberhard, 1996; Yasui, 1998). If we are to entertain the consequences of such female mating preferences, it must be framed in terms of sexual conflict because nesting males can only stand to lose by being involved in sperm competition as they end up investing in genetically unrelated offspring. Females, on the other hand, may stand to benefit from sperm competition either directly by increasing fertilisation success or indirectly through her sons carrying the genes for being effective sperm competitors.

CONCLUSIONS

Sexual selection has and continues to play a pivotal role in the evolution of blennies and is probably crucial to understanding their diversity. Blennies exhibit an array of ornaments, behaviours and colour patterns that defy explanation by natural selection and call upon the theory of sexual selection. Sexual selection is predicted to be a significant evolutionary force since territoriality, promiscuity and paternal care, all combine to create a high variance in male mating success and sexual competition. There is clear evidence for choosiness in females and the most important mate criteria appear to be body size, nest site and the past success of the male. Surprisingly, there is less evidence that the plentiful ornaments, colours and dramatic courtship behaviours are linked with mate choice decisions. There is also evidence that males are choosy, although male mate choice criteria are not at well understood and this represents a serious gap in our understanding. A salient point of this chapter is that mating strategies need to be understood from both the male and female perspective; mate selection is a mutual process in blennies due to the equity of investment in reproduction by both sexes. While there is clear evidence for the costs of being choosy for females, surprisingly, there are not many studies that persuasively demonstrate the direct benefits of being choosy. This is a central tenet of sexual selection theory and a lack of convincing evidence suggests either that sexual selection theory is inadequate or that conclusive studies are yet to be carried out. The field of sexual conflict has only recently begun to flourish. Many observations during spawning—such as male-female aggression—make more sense in the light of what we now know about sexual conflict. All this adds up to a complex picture of the underlying causes of variance in mating success. There is little doubt that some factors such as body size can be considered principle factors, but there are many other secondary factors that have lesser but nevertheless significant effects on mating success. Just as we now take for granted the occurrence of alternative reproductive tactics in a intrasexual context, it is perhaps time to consider that there exist a multitude of alternative reproductive tactics in the more general context of female-male, male-female, female-female and male-male interactions.

Studies on blennies provided some of the first examples of mate-choice criteria in the field and blennies remain an excellent model group to study new questions concerning the mechanisms involved with sexual selection. Perhaps the most valuable asset of blennies in studying sexual selection is the common occurrence of plastic sex roles and the frequency with which reversals in courtship and competition are observed between the sexes. Such variation is unusual and must be taken advantage of; it is the raw material needed for experimental approaches to sexual selection.

Acknowledgements

We would like to dedicate this chapter to Professor John J. Videler who helped pioneer the study of sexual selection in blennies and who inspired us both to consider in detail the reproductive habits of blennies. The authors wish to thank the EC Marie Curie Fellowship Program (FN) and the Fisheries Society of the British Isles (WL) for funding our blenny research.

References

Almada, V.C., E.J. Gonçalves, A.J. Santos and C. Baptista. 1994. Breeding ecology in a population of *Salaria pavo* (Pisces: Blenniidae) in an area where nest sites are very scarce. *Journal of Fish Biology* 45: 819-830.

Almada, V.C., E.J. Gonçalves, R.F. Oliveira and A.J. Santos. 1995. Courting females: Ecological constraints affect sex roles in a natural population of the blenniid fish *Salaria pavo*. *Animal Behaviour* 49: 1125-1127.

Amundsen, T. 2003. Fishes as models in studies of sexual selection and parental care. *Journal of Fish Biology* 63 Supplement A: 17-52.

Andersson, M. 1994. *Sexual Selection*. Princeton University Press, Princeton.

Arnquist, M. and L. Rowe. 2005. *Sexual Conflict*. Princeton University Press, Princeton.

Barata, E.N., R.M. Serrano, A. Miranda, R. Nogueira, P.C. Hubbard and A.V.M. Canario. 2009. Putative pheromones from the anal glands of male blennies attract females and enhance male reproductive success. *Animal Behaviour* 75: 379-389.

Brooks, R. and V. Couldridge. 1999. Multiple sexual ornaments co-evolve with multiple mating preferences. *American Naturalist* 154: 37-45.

Buchheim, J.R. and M.A. Hixon. 1992. Competition for shelter holes in the coral-reef fish, *Acanthemblemaria spinosa* Metzelaar. *Journal of Experimental Marine Biology and Ecology* 164: 45-54.

Chapman, T., G. Anquist, J. Bangham and L. Rowe 2003. Sexual Conflict. *Trends in Ecology and Evolution* 18: 41-47.

Chenoweth, S.F., P. Doughty and H. Kokko. 2006. Can non-directional male mating preferences facilitate honest female ornamentation? *Ecology Letters* 9: 179-184.

Côté, I.M. and W. Hunte. 1989. Male and female mate choice in the redlip blenny: Why bigger is better. *Animal Behaviour* 38: 78-88.

Côté, I.M. and W. Hunte. 1993. Female blennies prefer older males. *Animal Behaviour* 46: 203-205.

Darwin, C.R. 1872. *The Descent of Man, and Selection in Relation to Sex*. John Murray, London.

Day, T. 2000. Sexual selection and the evolution of costly female preferences: Spatial effects. *Evolution* 54: 715-730.

de Jonge, J. and J.J. Videler. 1989. Differences between the reproductive biologies of *Tripterygion tripteronotus* and *T. delaisi* (Pisces, Perciformes, Tripterygiidae): The adaptive significance of red instead of yellow nuptial colouration. *Marine Biology* 100: 431-437.

Emlen, S.T. and L.W. Oring. 1977. Ecology, sexual selection, and the evolution of mating systems. *Science* 197: 215-223.

Endler, J.A. and A.L. Basolo. 1998. Sensory ecology, receiver biases and sexual selection. *Trends in Ecology & Evolution* 13: 415-420.

Eberhard, W.G. 1996. *Female Control: Sexual Selection by Cryptic Female Choice*. Princeton University Press, Princeton.

Fisher, R.A. 1930. *The Genetical Theory of Natural Selection*. Clarendon Press, Oxford.

Fishelson, L. 1974. Observations on the behaviour of the fish *Meiacanthus nigrolineatus* Smith-Vaniz (Blenniidae) in Nature (Red Sea) and in captivity. *Australian Journal of Marine and Freshwater Research* 26: 329-341.

Fitzgerald, G.J. 1992. Filial cannibalism in fishes: Why do parents eat their offspring? *Trends in Ecology & Evolution* 7: 7-10.

Fitzpatrick, S., A. Berglund and G. Rosenquist. 1995. Ornaments or offspring? Costs to reproductive success restrict sexual selection processes. *Biological Journal of the Linnean Society* 55: 251-260.

Forsgren, E., J.D. Reynolds and A. Berglund. 2002. Behavioural ecology of reproduction in fish. In: *Handbook of Fish Biology and Fisheries*, P.J.B. Hart and J.D. Reynolds (eds.), Blackwell Scientific Publishing, Oxford, Vol. 1, *Fish Biology*, pp. 225-247.

Giacomello, E. and M.B. Rasotto. 2005. Sexual dimorphism and male mating success in the tentacle blenny *Parablennius tentacularis* (Teleostei: Blenniidae). *Marine Biology* 147: 1221-1228.

Giacomello, E., D. Marchini and M.B. Rasotto. 2006. A male sexually dimorphic trait provides antimicrobials to eggs in blenny fish. *Biology Letters* 2: 330-333.

Giacomello, E., F. Neat and M.B. Rasotto. 2008. Mechanisms enabling sperm economy in blenniid fishes. *Behavioral Ecology and Sociobiology* 62: 671-680.

Gibran, F.G., F.B. Santos, H.F. dos Santos and J. Sabino. 2004. Courtship behavior and spawning in the hairy blenny *Labrisomus nuchipinnis* (Labrisomidae) in southeastern Brazil. *Neotropical Ichthyology* 2: 163-166.

Gonçalves, E.J. and V.C. Almada. 1997. Sex differences in resource utilization by the peacock blenny. *Journal of Fish Biology* 51: 624-633.

Gonçalves, E.J. and V.C. Almada. 1998. A comparative study of territoriality in intertidal and subtidal blennioids (Teleostei: Blennioidei). *Environmental Biology of Fishes* 51: 257-264.

Gonçalves, D.M. and R.F. Oliveira. 2003. Time spent close to a sexual partner as a measure of female mate preference in a sex-role-reversed population of the blenny *Salaria pavo* (Risso) (Pisces: Blenniidae). *Acta Ethologica* 6: 1-5.

Gonçalves, D., T. Fagundes and R. Oliveira. 2003. Reproductive behaviour of sneaker males of the peacock blenny. *Journal of Fish Biology* 63: 528-532.

Gonçalves, E.J., V.C. Almada, R.F. Oliveira and R.S. Santos. 1996. Female mimicry as a mating tactic in males of the blenniid fish *Salaria pavo*. *Journal of the Marine Biological Association of the United Kingdom* 76: 529-538.

Gonçalves, D.M., P. C. Simões, A.C. Chumbinho, M.J. Correia, T. Fagundes and R.F. Oliveira. 2002a. Fluctuating asymmetries and reproductive success in the peacock blenny. *Journal of Fish Biology* 60: 810-820.

Gonçalves, D.M., E.N. Barata, R.F. Oliveira and A.V.M. Canário. 2002b.The role of male visual and chemical cues on the activation of female courtship behaviour in the sex-role reversed peacock blenny. *Journal of Fish Biology* 61: 96-105.

Harvey, P.H. and R.M. May. 1989. Out for the sperm count. *Nature (London)* 337: 508-509.

Hastings, P.A. 1986. Habitat selection, sex ratio and sexual selection in *Coralliozetus angelica* (Blennioidae: Chaenopsidae). In: *Indo-Pacific Fish Biology: Proceedings of the Second International Conference on Indo-Pacific Fishes*, T. Uyeno, R. Arai, T. Taniuchi and K. Matsuura (eds.), Ichthylogical Society of Japan, pp. 785-793.

Hastings, P.A. 1988a. Female choice and male reproductive success in the angel blenny, *Coralliozetus angelica* (Teleostei: Chaenopsidae). *Animal Behaviour* 38: 115-124.

Hastings, P.A. 1988b. Correlates of male reproductive success in the brown cheeked blenny, *Acanthemblemaria crockeri* (Blennioidae: Chaenopsidae). *Behavioral Ecology and Sociobiology* 22: 95-102.

Hastings, P.A. 1991. Ontogeny of sexual dimorphism in the angel blenny, *Coralliozetus angelica* (Blennioidei: Chaenopsidae). *Copeia* 1991: 969-978.

Hastings, P.A. 1992. Nest-site size as a short-term constraint on the reproductive success of paternal fishes. *Environmental Biology of Fishes* 34: 213-218.

Hastings, P. A. 2002. Evolution of morphological and behavioral ontogenies in females of a highly dimorphic clade of blennioid fishes. *Evolution* 56: 1644-1654.

Johnstone, R.A. 1995. Sexual selection, honest advertisement and the handicap principal: Reviewing the evidence. *Biological Reviews* 70: 1-65.

Johnstone, R.A., J.D. Reynolds and J.C. Deutsch. 1996. Mutual mate choice and sex differences in choosiness. *Evolution* 50: 1382-1390.

Kirkpatrick, M. 1982. Sexual selection and the evolution of female choice. *Evolution* 36: 1-12.

Kokko, H. and R.A. Johnstone. 2002. Why is mutual mate choice not the norm? Operational sex ratios, sex roles and the evolution of sexually dimorphic and monomorphic signalling. *Philosophical Transactions of the Royal Society of London* B357: 319-330.

Kokko, H. and J. Mappes. 2005. Sexual selection when fertilization is not guaranteed. *Evolution* 59: 1876-1885.

Kokko, H., R. Brooks, J.M. McNamara and A.I. Houston. 2002. The sexual selection continuum. *Proceedings of the Royal Society of London* B269: 1331-1340.

Kokko, H., R. Brooks, M. Jennions and J. Morely. 2003. The evolution of mate choice and mating biases. *Proceedings of the Royal Society of London* B270: 653-664.

Kotrschal, K. and A. Goldschmid. 1992. Morphological evidence for the biological role of caniniform teeth in combtooth blennies (Blenniidae, Teleostei). *Journal of Fish Biology* 41: 983-991.

Kraak, S.B.M. 1996. A quantitative description of the reproductive biology of the Mediterranean blenny *Aidablennius sphynx* (Teleostei, Blenniidae) in its natural habitat. *Environmental Biology of Fishes* 46: 329-342.

Kraak, S.B.M. and J.J. Videler. 1991. Mate choice in *Aidablennius sphynx* (Teleostei, Blenniidae); Females prefer nests containing more eggs. *Behaviour* 119: 243-266.

Kraak, S.B.M. and E.P. van der Berghe. 1992. Do female fish assess paternal quality by means of test eggs? *Animal Behaviour* 43: 865-867.

Kvarnemo, C. and I. Ahnesjo. 1996. The dynamics of operational sex ratios and competition for mates. *Trends in Ecology & Evolution* 11: 404-408.

Kvarnemo, C. and I. Ahnesjo. 2002. Operational sex ratios and mating competition. In: *Sex Ratios: Concepts and Research Methods*, I.C.W. Hardy (ed.). Cambridge University Press, Cambridge, pp. 366-382.

Lauman, J., U. Pern and V. Blum. 1974. Investigations on the function and hormonal regulations of the anal appendages in *Blennius pavo* (Risso). *Journal of Experimental Zoology* 190: 47-56.

Lengkeek, W. 2006. Dynamic reproductive behaviour and the evolution of sexual size dimorphism in Mediterranean blennies. Ph.D. Thesis. University of East Anglia, Norwich.

Lengkeek, W. and K. Didderen. 2006 Breeding cycles and reproductive behaviour in the river blenny (*Salaria fluviatilis*). *Journal of Fish Biology* 69: 1837-1844.

Lengkeek, W., K. Didderen, I.M. Cote, E.M. van der Zee, R.C. Snoek and J.D. Reynolds. 2008. Plasticity in sexual size dimorphism and Rensch's rule in Mediterranean blennies (Blennidae). *Canadian Journal of Zoology* 86: 1173-1178.

Lindström, K. 2000. The evolution of filial cannibalism and female mate choice strategies as resolutions to sexual conflict in fishes. *Evolution* 54: 617-627.

Locatello, L. and F.C. Neat. 2005. Reproductive allocation in *Aidablennius sphynx* (Teleostei: Blenniidae): Females lay more eggs faster when paired with larger males. *Journal of Experimental Zoology* A303: 922-926.

Losey, G.S. 1969. Sexual pheromone in some fishes of the genus *Hypsoblennius* Gill. *Science* 163: 181-183.

Mackiewicz, M., B.A. Porter, E.E. Dakin and J.C. Avise. 2005. Cuckoldry rates in the molly millar (*Scartella cristata*, Blenniidae) a hole nesting marine fish with alternative reproductive tactics. *Marine Biology* 148: 213-221.

Matsumoto, Y., T. Takegaki, A. Tawa and Y. Natsukari. 2008. Female within-site-spawning-preference and its affect on female mate choice. *Journal of Zoology* 276: 48-53.

Neat, F.C. 2001. Parasitic spawning in two species of triplefin blenny: Contrasts in demography, behaviour and gonadal characteristics. *Environmental Biology of Fishes* 55: 57-64.

Neat, F.C. and L. Locatello. 2002. No reason to Sneak: Why males of all sizes can breed in the hole-dwelling blenny, *Aidablennius sphynx* (Teleostei: Blenniidae). *Behavioral Ecology and Sociobiology* 52: 66-73.

Neat, F.C., L. Locatello and M.B. Rasotto. 2003a. Reproductive morphology in relation to alternative reproductive tactics in *Scartella cristata*. *Journal of Fish Biology* 62: 1381-1391.

Neat, F.C., W. Lengkeek, P. Westerbeek, B. Laarhoven and J.J. Videler. 2003b. Behavioural and morphological differences between lake and river populations of *Salaria fluviatilis*. *Journal of Fish Biology* 63: 374-387.

Northcott, S.J. and A.M. Bullock. 1991. The morphology of the club glands on the dorsal fin of mature male shannies, *Lipophrys pholis* (L.) (Bleniidae: Teleostei). *Journal of Fish Biology* 39: 795-806.

Nursall, J.R. 1977. Territoriality in the redlip blenny *Ophioblennius atlanticus* (Pisces, Blenniidae). *Journal of Zoology* 182: 205-223.

Oliveira, R.F., V.C. Almada, E. Forsgren and E.J. Gonçalves. 1999. Temporal variation in male traits, nesting aggregations and mating success in the peacock blenny. *Journal of Fish Biology* 53: 499-512.

Oliveira, R.F., J.A. Miranda, N. Carvalho, E.J. Gonçalves, M.S. Grober and R.S. Santos. 2000. Male mating success in the Azorean rock-pool blenny: The effects of body size, male behaviour and nest characteristics. *Journal of Fish Biology* 57: 1416-1428.

Oliveira, R.F., N. Carvalho, J. Miranda, E.J. Gonçalves, M. Grober and R.S. Santos. 2002. The relationship between the presence of satellite males and nest-holders' mating success in the Azorean rock-pool blenny *Parablennius sanguinolentus parvicornis*. *Ethology* 108: 223-235.

Parker, G.A. 1979. Sexual selection and sexual conflict. In: *Sexual Selection and Reproductive Competition in Insects*, M.S. Blum and N.A. Blum (eds.), Academic Press, London, pp. 123-166.

Parker, G.A. and L.W. Simmons. 1996. Parental investment and the control of sexual selection: Predicting the direction of sexual competition. *Proceedings of the Royal Society of London* B263: 315-321.

Patzner, R.A., M. Seiwald, M. Adlgasser and G. Kaurin. 1986. The reproduction of *Blennius pavo* (Teleostei, Blenniidae) V. Reproductive behavior in natural environment. *Zoologischer Anzeiger* 216: 338-350.

Petersen, C.W. 1988. Male mating success, sexual size dimorphism, and site fidelity in two species of *Malacoctenus* (Labrisomidae). *Environmental Biology of Fishes* 21: 3-183.

Petersen, C.W. 1989. Females prefer mating males in the carmine triplefin, *Axoclinus carminalis*, a paternal brood-guarder. *Environmental Biology of Fishes* 26: 213-221.

Proulx, S.R., T. Day and L. Rowe. 2002. Old males signal more reliably. *Proceedings of the Royal Society of London* B269: 2291-2299.

Reynolds, J.D. 1996. Animal breeding systems. *Trends in Ecology and Evolution* 11: 68-72.

Reynolds, J.D. and I.M. Côté. 1995. Direct selection on mate choice: Female redlip blennies pay more for better mates. *Behavioral Ecology* 6: 175-181.

Sargent, R.C., V.N. Rush, B.D. Wisenden and Y.Y. Hong. 1998. Courtship and mate choice in fishes: Integrating behavioural and sensory ecology. *American Zoologist* 38: 82-96.

Selander, R.K. 1972. Sexual selection and dimorphism in birds. In: *Sexual Selection and the Decent of Man: 1871-1971*, B. Campbell (ed.). Aldine Publishing Company, Chicago, pp. 180-230.

Sheldon, B.C. 2000. Differential allocation: tests, mechanisms and implications. *Trends in Ecology and Evolution* 15: 397-402.

Shibata, J. and M. Kohda. 2006. Seasonal sex role changes in the blenniid *Petroscirtes breviceps*, a nest brooder with paternal care. *Journal of Fish Biology* 69: 203-214.

Shine, R. 1989. Ecological causes for the evolution of sexual dimorphism: A review of the evidence. *Quarterly Review of Biology* 64: 418-461.

Thompson, S. 1986. Male spawning success and female choice in the mottled triplefin, *Forsterygion varium* (Pisces: Tripterygiidae). *Animal Behaviour* 34: 580-589.

Thompson, D.A., L.T. Findlay and A.N. Kerstitch. 2000. *Reef Fishes of the Sea of Cortez*. University of Texas Press, Austin.

Trivers, R.L. 1972. Parental investment and sexual selection. In: *Sexual Selection and the Decent of Man: 1871-1971*, B. Campbell (ed.). Aldine Publishing Company, Chicago, pp. 136-169.

Vinyoles, D., I.M. Côté and A. De Sostoa. 1999. Egg cannibalism in river blennies: The role of natural prey availability. *Journal of Fish Biology* 55: 1223-1230.

Vinyoles, D., I.M. Côté and A. De Sostoa. 2002. Nest orientation patterns in *Salaria fluviatilis*. *Journal of Fish Biology* 61: 405-416.

Van Doorn, S. and F.J Weissing. 2004. The evolution of female preferences for multiple indicators of quality. *American Naturalist* 164: 173-186.

Von Bartheld, C.S. and D.L. Meyer. 1985. Trigeminal and facial innervation of cirri in three teleost species. *Cell and Tissue Research* 241: 615-622.

Warner, R.R., D.Y., Shapiro, A. Marconato and C.W. Petersen. 1996. Sexual conflict—Males with the highest mating success convey the lowest fertilization benefits to females. *Proceedings of the Royal Society of London* B262: 135-139.

Wiens, J.J. 2001. Widespread loss of sexually selected traits: How the peacock lost its spots. *Trends in Ecology & Evolution* 16: 517-523.

Willoughby, F. 1686. *De Historia Piscium*. The Royal Society, London.

Wirtz, P. 1978. The behaviour of the Mediterranean *Tripterygion* species (Pisces, Blennioidei). *Zeitschrift fur Tierpsychologie* 48: 142-174.

Yasui, Y. 1998. The 'genetic benefits' of female multiple mating revisited. *Trends in Ecology and Evolution* 13: 246-250.

Zahavi, A. 1975. Mate selection—A selection for a handicap. *Journal of Theoretical Biology* 53: 205-214.

Zander, C.D. 1975. Secondary sex characteristics of Blennioid fishes (Perciformes). *Pubblicazioni della Stazioni Zoologica di Napoli* 39: 717-727.

CHAPTER

4.3

Alternative Reproductive Tactics in Blennies

Rui F. Oliveira[1], David M. Gonçalves[2] and Albert Ros[3]

INTRODUCTION

WHAT ARE ALTERNATIVE REPRODUCTIVE TACTICS?

In some species, individuals of the same sex adopt alternative tactics to compete for mates. Instead of basing their intra-sexual competition in the investment of resources towards the attraction and monopolisation of sex partners (e.g. expression of visual, chemical or acoustic courtship signals and/or the defense of breeding territories), individuals using the alternative tactic exploit the investment made by conventional individuals (e.g., by sneaking into the breeding territories during spawning episodes). Therefore, the terms bourgeois and parasitic have been proposed to describe these two tactics (Taborsky, 1997) and the traits selected in the two male types are usually divergent. In bourgeois males,

Authors' address: Unidade de Investigação em Eco-Etologia, Instituto Superior de Psicologia Aplicada, Rua Jardim do Tabaco 34, 1149-041 Lisboa, Portugal.
E-mails: [1]ruiol@ispa.pt [2]davidg@ispa.pt [3]aros@ispa.pt

traits related with mate attraction and monopolisation will be selected, while in parasitic males those traits that increase the probability of stealing fertilisations from bourgeois males will prevail. This disruptive selection, acting on a constellation of phenotypic traits, may result in the evolution of polyphenisms within one sex, usually the males, where the expression of male reproductive behaviour and male secondary sex characters in parasitic males becomes dissociated from the differentiation of a functional male gonad. In summary, alternative reproductive tactics (ART) are discontinuous behavioural and morphological traits selected to maximise fitness in two or more mutually exclusive ways in the context of reproductive competition. Based on the descriptive patterns of observed behaviour, alternative reproductive phenotypes can be categorised as fixed or plastic (Moore, 1991; Brockmann, 2001). In fixed alternative phenotypes, the individuals adopt one of the tactics for their entire lifetime. In plastic (or flexible) alternative phenotypes, the individuals may change tactics during their lifetime. Within plastic ART, again, two sub-categories can be further distinguished: irreversible sequential patterns, when individuals switch from one tactic to another at a particular moment in their lifetime (developmental switches); and reversible patterns, when individuals can change back and forth between patterns (Moore, 1991; Moore *et al.*, 1998; Brockmann, 2001; Fig. 4.3.1).

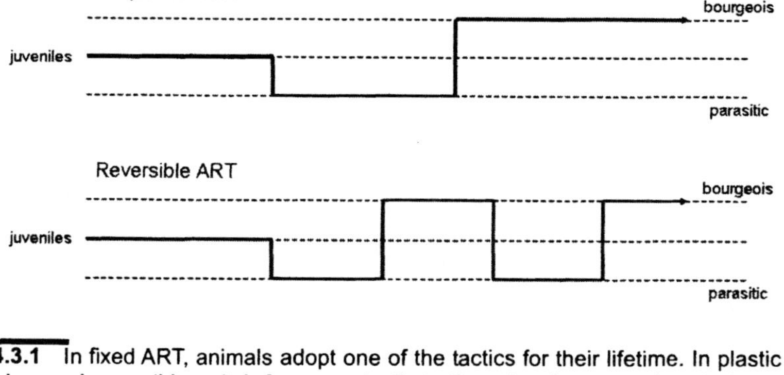

Fig. 4.3.1 In fixed ART, animals adopt one of the tactics for their lifetime. In plastic ART, animals may irreversibly switch from one tactic to the other (sequential tactics) or switch back and forth between tactics (reversible tactics).

Other schemes of classification of alternative reproductive phenotypes have been proposed in the literature based on the underlying evolutionary processes such as isogignous (i.e., with equal fitness outcomes) versus allogignous (i.e., not equally successful strategies, Austad, 1984), alternative versus conditional strategies (Gross, 1996), Mendelian strategies versus developmental strategies versus behavioural strategies (Shuster and Wade, 2003). However, these approaches require a great deal of information to be collected before the alternative phenotypes can be classified. For example, the classic distinction between frequency-dependent and condition-dependent strategies requires the knowledge about the lifetime reproductive success of the alternative phenotypes, since in the former they are expected to be isogignous while in the latter this is not expected (in this case, alternative phenotypes were seen as 'making the best of a bad job', Dawkins, 1980). Moreover, the simple dichotomy between alternative versus conditional strategies requires the investigation of the genetic mechanisms underlying the sexual polymorphism in the population (see below). Therefore, we have adopted a descriptive classification of ART.

ORIGIN OF ALTERNATIVE REPRODUCTIVE TACTICS: EVOLUTIONARY AND DEVELOPMENTAL MECHANISMS

Alternative reproductive tactics do not necessarily imply the occurrence of genetic polymorphisms. They may emerge both from monomorphic and polymorphic genotypes (Austad, 1984; Gross, 1996; Shuster and Wade, 2003). When there is no genetic variation underlying the expression of alternative phenotypes (genetic monomorphism), these may be conditionally triggered by current conditions or by developmental switches. Therefore, alternative tactics emerge among individuals with the same genetic architecture due to the diverging conditions that they experience. Usually, thresholds exist that determine the switching point between conditional tactics. For example, the decision to adopt one or the other tactic may depend on a size threshold so that above a certain size, individuals would do better by continuing to grow and postponing reproduction (as a bourgeois male), while below the threshold it may pay to reproduce early as a parasitic breeder (i.e., condition-dependent selection, Fig. 4.3.2a; Taborsky *et al.*, 2008). Alternative reproductive tactics may also result from polymorphic genotypes, in which case the frequencies of the each tactic-based genotype should be balanced by

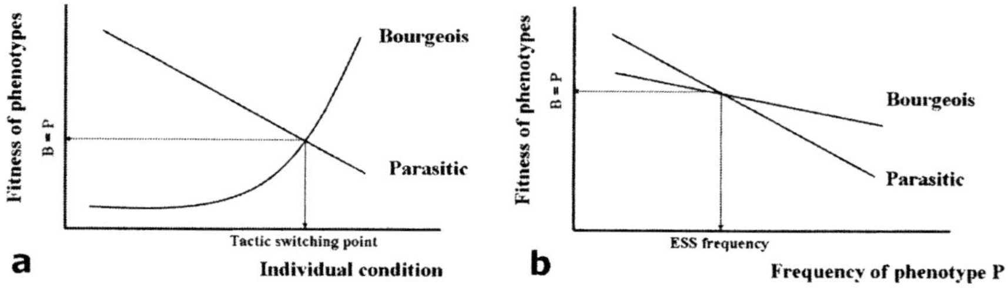

Fig. 4.3.2 **a.** Condition-dependent ART are based on genetic monomorphism, where the individual condition determines the best tactic to adopt. For example, males may gain by adopting a parasitic tactic while they are small and switch into a bourgeois tactic when they grow. **b.** Frequency-dependent ART are based on polymorphic genotypes. In this case, the lifetime fitness of animals using the different tactics is expected to be equal and the frequencies of each tactic-based genotype are balanced by frequency-dependent selection.

frequency dependent selection (Fig. 4.3.2b), leading to equal lifetime fitness of animals using the different tactics (Shuster and Wade, 1991; Ryan *et al.*, 1992; Repka and Gross, 1995). There has been considerable debate in the existing literature on the relative importance of conditional tactics as opposed to genetic polymorphism for the evolution of alternative reproductive tactics (Gross, 1996; Shuster and Wade, 2003). Since conditional responses have been demonstrated for the majority of the cases of alternative reproductive tactics described so far and since conditional choices have been associated with genetic monomorphism, it has been argued that genetic polymorphisms play a minor role in the causation of alternative tactics (Gross, 1996; but see also Shuster and Wade, 2003). This dichotomy has led to a terminological distinction often made in evolutionary biology where the term strategy is used to classify a 'genetically based program', while the tactic describes the rules that are part of a strategy (i.e., the phenotype; Shuster and Wade, 2003). Despite this ongoing debate, it is too reductionist to assume that alternative tactics could origin purely from either genetic or environmental factors (Caro and Bateson, 1986). In fact, most dimorphic traits seem to be threshold traits influenced by quantitative trait loci, which means that tactic switch points can have a genetic basis, implying that trait expression is both conditional and heritable (Tomkins *et al.*, 2004). This scenario allows alternative phenotypes to evolve independently from each other (i.e., in a non-frequency dependent fashion), which greatly increases the scope for the evolution of alternative tactics (Taborsky *et al.*, 2008). In this chapter, we have opted to use the term 'tactic' because we will mainly deal with phenotypes and not with the functional implications of the term 'strategy'.

PHYSIOLOGICAL MECHANISM UNDERLYING THE EXPRESSION OF ALTERNATIVE REPRODUCTIVE TACTICS

Due to their role in sexual differentiation and in the control of male reproduction in vertebrates (e.g., Dixon, 1998; Wilson *et al.*, 2002; Nelson, 2005), sex steroids are prime candidates to play a key role in the causal mechanisms underlying the differentiation and the maintenance of alternative reproductive morphs. In particular, androgens participate in the differentiation of primary and secondary sex characters, in the expression of reproductive behaviours, in the feedback regulation of the hypothalamus and pituitary, and in spermatogenesis (Nelson, 2005), which makes them a preferential target for studies of endocrine correlates of male ARTs. A review of the available data on androgen levels in species with ART indicates that alternative phenotypes in teleost species consistently differ in their circulating 11-ketotestosterone (KT) levels, with bourgeois males having significantly higher levels of circulating KT than parasitic males, but that there was no clear pattern regarding testosterone (T) (Brantley *et al.*, 1993a; Oliveira, 2006). Interestingly, the exceptions to this rule occur in species where the ART do not involve morphological specialisations (i.e., polyphenisms), suggesting that androgens are particularly relevant for the expression of tactic-dependent morphological traits but not as much for the expression of tactic-specific behavioural traits (Oliveira, 2006). Therefore, the development of male ARTs is likely to be influenced by other neuroendocrine systems in addition to gonadal steroids.

Another neuroendocrine candidate that has received growing attention in recent years is the neuropeptide arginine vasotocin (AVT). This decapeptide is the teleost homologue to the mammalian arginine vasopressin (AVP) and, together, AVP/AVT have been demonstrated to influence the expression of social behaviours—including courtship behaviour—in a wide range of vertebrate species (Goodson and Bass, 2001). In all the teleost species with ART studied so far, an association has been found between the expression of alternative phenotypes and forebrain AVT, either in terms of soma size or number of AVT-ir neurons, or of AVT mRNA expression on a per cell basis (see Oliveira, 2006 for a review). Furthermore, in studies that have manipulated AVT by exogenous administration in fish species with ART, AVT has been effective in eliciting tactic-specific behaviours (Goodson and Bass, 2000; Carneiro *et al.*, 2003; Semsar and Godwin, 2003).

Finally, glucocorticoids have also been suggested as potential players in ART differentiation (Moore *et al.*, 1998). Recent studies in fish suggest that glucocorticoid profiles are associated with distinctive behavioural styles and that the interactions between the hypothalamic-pituitary-interrenal axis and the hypotalamic-pituitary-gonadal axis could be the causal link to explain the glucocorticoid modulation of alternative reproductive traits (Knapp, 2004).

To summarise, in recent years, we have moved from a simplistic view of the causal mechanisms of ART being based in pleiotropic-like effects of one endocrine agent on a constellation of tactic-specific traits, to a more complex model in which different neuroendocrine mechanisms interact in the control of the different traits that together make-up the specific profiles of alternative phenotypes (Fig. 4.3.3).

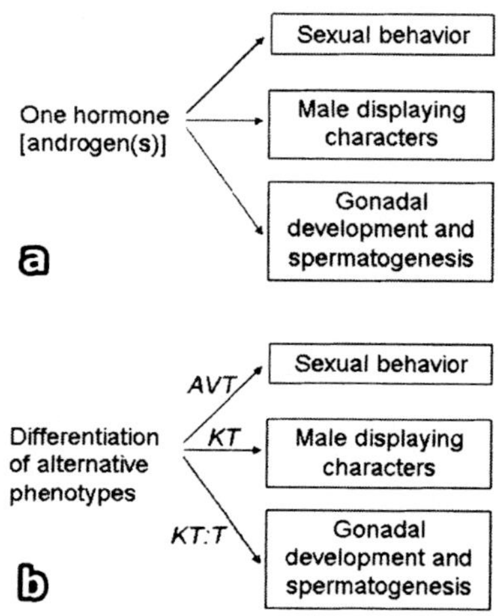

Fig. 4.3.3 **a.** Classically, the causal mechanisms of ART were explained by one endocrine agent acting on all tactic-specific traits. For example, androgens have been suggested to be causally linked to the behavioural, morphological, anatomical and gonadal traits of bourgeois males. **b.** Recently, the interaction between different neuroendocrine mechanisms has been suggested to be necessary to induce and maintain alternative phenotypes.

OCCURRENCE OF ALTERNATIVE REPRODUCTIVE TACTICS IN BLENNIIDS

Teleosts are by far the vertebrate taxon with the highest incidence of species with ART (for comparative data on ART in different vertebrate taxa see Oliveira *et al.*, 2008). The last published count of ART in fish identified 170 species from 32 different families (Taborsky, 2008), and new examples of species with ART are described annually, even among common temperate species studied on a regular basis. Three main factors have been suggested to explain this ubiquity of ART among teleosts: (1) the occurrence of indeterminate growth in fish that promotes large intra-sexual size differences; (2) external fertilisation; and (3) male parental care (Taborsky, 1999).

Among blennioids, ART have been described in 10 species from the families Blenniidae and Tripterygidae (Table 4.3.1). No published records that we are aware of have described ART in the other 4 Blennioid families (Chaenopsidae, Clinidae, Dactyloscopidae and Labrisomidae). A phylogenetic analysis of ART in blenniids suggests that this trait evolved several times and that it never become a permanent feature of a deeper clade, comprising many genera (Almada and Robalo, 2008). Two Northeastern Atlantic blenniids have been studied in great detail regarding both the mechanisms and the ecology of ART and, therefore, we will concentrate on these two species in the following section.

Table 4.3.1 Blenniioid species with alternative reproductive tactics

Family	*Species*	*References*
Tripterygidae	*Axoclinus nigricaudus*	Neat (2001)
	Axoclinus carminalis	Thresher (1984); Neat (2001)
	Tripterygion tripteronotus	Wirtz (1978); Mohr (1986); de Jonge and Videler (1989)
	Tripterygion delaisi	Wirtz (1978); Mohr (1986); de Jonge and Videler (1989)
	Tripterygion melanurus	Mohr (1986)
Blenniidae	*Scartella cristata*	Neat *et al.* (2003a); MacKiewicz *et al.* (2005)
	Parablennius parvicornis	Santos (1985b); Oliveira *et al.* (2001b)
	Salaria pavo	Ruchon *et al.* (1995); Gonçalves *et al.* (1996)
	Salaria fluviatilis	Neat *et al.* (2003b)

TWO TALES FROM THE INTERTIDAL: SATELLITES AND SNEAKERS

'Satellites of Love': Satellite Males in the Azorean Rock-pool Blenny

The mating system of the Azorean rock-pool blenny *Parablennius parvicornis* has been well characterised in a population occurring at the south coast of Faial, one of the islands in the central group of the Azores.

In the main studied site, Feteira (38° 31′N; 28° 40′W), some horizontal enclosures in a large basaltic platform have collapsed giving rise to large inter-tidal enclosures that remain with shallow water during low tide (Fig. 4.3.4a). These pools accommodate dense aggregations of *P. parvicornis* and offer perfect conditions for behavioural observations and field experiments (Fig. 4.3.4b, c). Male sequential polymorphism associated with alternative reproductive tactics has been described in this population (Santos, 1985a, b; Oliveira *et al.*, 2001b). These include a male morph without secondary sexual character expression (M–) and involved in parasitic tactics: sneakers and satellites; and a morph characterised by the presence of secondary sexual characters (M+) that may reproduce as nest-holders or act as non-territorial floaters (Santos and Almada, 1988; Fig. 4.3.4d).

At the end of May, males start to compete aggressively for natural cavities in the pools and under boulders (Santos, 1985b). Only the larger males establish territories and many smaller males are seen in the pools and adjacent deeper littoral (Santos *et al.*, 1995; Oliveira *et al.*, 2005). Territorial males change to a distinct brown-dark coloration, and possess well-developed secondary sexual characters, including a distinct head hump and an anal gland in the first two rays of the anal fin (Santos, 1985b). They clean out the crevice or an area under the boulder from sand and debris, and this serves as their nest. Nest-holding males court females by signaling the location of the nest, circling, and leading the females (Santos and Barreiros, 1993). Bigger males are more active in all these behaviours, probably signaling their higher competitive ability and larger resources (Oliveira *et al.*, 2000). The anal glands are likely to play a role in attracting females, as in other blennies they have been shown to release pheromones (see Chapter 5.1 by Barata and Gonçalves, this book). Spawning occurs inside the nest and after laying demersal adhesive eggs the female leaves the nest. Over the reproductive season, the nest-holding males receive spawns of multiple females and their nests may hold more than 30000 eggs (Santos, 1989; Santos and Barreiros, 1993). The male takes care of the eggs until hatching and egg care behaviours include cleaning out the nest of debris, territorial defense against egg predators and replacing water in the nest with a fanning movement of the tail in order to provide the eggs with oxygen.

Although males and females remain in the same pool area for several years, the females are less site-attached than males (Santos and Almada, 1988; Carneiro *et al.*, 2001). Females may be seen in different pools during high tide and often leave again with low tide (Santos and Almada, 1988). During their passage, the nest-holding males actively court females. Some females stay in the pools during low tide and often aggregate in groups. Fecund females overlap in body size with the males (total length = 11 ± 2.0 cm) and between June and July, they lay sequential clutches of eggs (Santos, 1985b; Santos and Almada, 1988). Females show a preference to spawn in the relatively large nests of the bigger males (Oliveira *et al.*, 2000). During the reproductive season, all potential nests are occupied and many aggressive interactions can be observed near the nests (Santos and Barreiros, 1993). Although only the larger males manage to establish and maintain a nest-holding territory, nest defense is difficult, either because the nests are exposed or have a large opening (for example, under boulders), or because the nest has two or more openings that cannot be simultaneously monitored (Santos, 1985a; Oliveira *et al.*, 2000). This allows ample opportunity for smaller sub-dominant males to sneak into the nest and get access to the eggs, allowing male alternative reproductive tactics to evolve. *P. parvicornis* males mature after their first winter when they are about 7 cm in length. These low-competitive males have developed testes and sperm in their *vas deferens* (Santos, 1985b). Testes weight corrected for body weight (gonadosomatic index) is significantly higher in these young males than in the nest-holding males (Santos, 1995; Fig. 4.3.4e). They further differ from the larger M+ males in that they do not express secondary sexual characters (e.g., anal glands, Fig. 4.3.4f; the M+ versus M– morph; Santos, 1985b). The smaller M– males roam around several pools, trying to enter different nests in order to fertilise eggs, and are named 'sneakers'. Larger M– males associate with one or two nests in the same pool and are named 'satellite' males. These males defend the nests they associate with, displaying aggressive behaviours towards nearly all the conspecifics or heterospecifics that enter the territory, with the exception of females and the nest-holder. Satellites are too small to take over and defend a nest, as becomes evident when occasionally a nest-holder disappears due to predation or exhaustion. Initially, the satellite might defend the nest opening against intruders, but soon a larger male will enter the nest and establish himself as a new nest-holder. Santos (1985b) found that associations between a parental male and one or two satellites are often stable over time. These observations have put forward the following question: Why do satellites defend nests? And why do nest-holders allow these fertile males to associate with their nests? To answer these questions, a brief description of the possible costs and benefits associated with each of the tactics is presented.

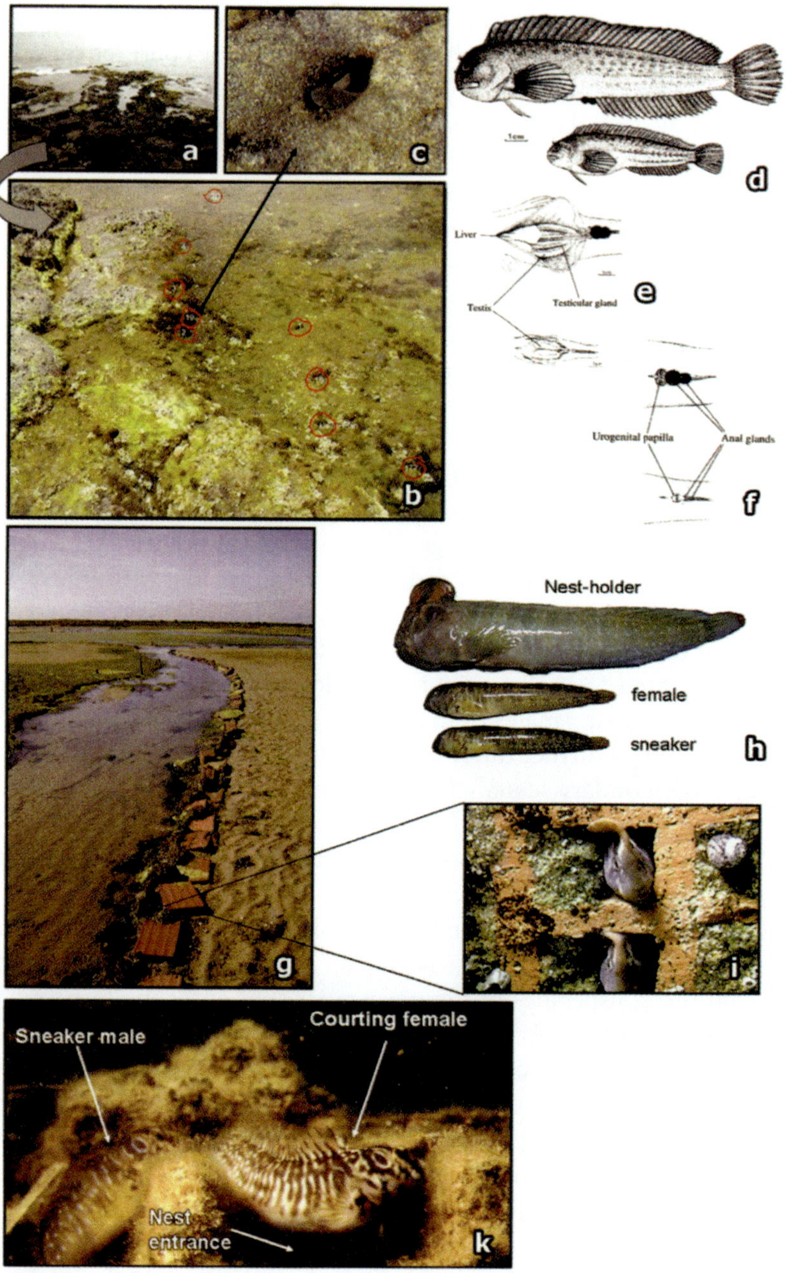

Fig. 4.3.4 **A-J** Alternative reproductive tactics in the rock-pool blenny *P. parvicornis*. **a.** A general view of the intertidal enclosures in Faial (Azores islands) where the studied

Fig. 4.3.4 Contd.

Nest-holding males can be found nearly all the time in or near (within 20 cm) the nest and they respond aggressively towards conspecifics and other species that come close to the nest (Santos, 1985a; Ros *et al.*, 2004). This aggression is highly effective to protect the eggs from predation, as is illustrated by capturing the nest-holder from the nest whereafter many fish home in and remove the eggs within minutes (Santos and Almada, 1988). The high investment of nest-holder males in territorial defense and nest care is costly not only in terms of the allocation of metabolic resources but also because this double investment impedes the male from leaving the nest to forage (Santos and Almada, 1988; Ros *et al.*, 2004). This results in a substantial decrease in body condition over the reproductive period (up to 30% of bodyweight, Santos *et al.*, 1996). Nest-holders need to trade-off vigilance and territorial activities with egg care and mating behaviour. This, together with the fact that many nests have a complex morphology, leaves many opportunities for small males to sneak fertilisations. Larger M– males might have developed the territorial satellite tactic in order to monopolise these sneaking opportunities. Satellites may build up knowledge about the properties of the nest and its surroundings increasing their success in escaping the aggressive advances of the nest-holder and in detecting the best opportunities for parasitic fertilisations of the eggs.

Satellites have been reported to be much more successful to enter the nest during the mating phase of the parental males (observations up to 60 sec) than non-attached sneakers (Santos, 1985b). Nest-holders, thus, loose part of their reproductive potential to the satellites, and respond accordingly. Fifty percent of their aggressive behaviour is directed to the satellite (Santos, 1985b). Still, if corrected for the time present on the

---

Fig. 4.3.4 Contd.

population occurs. **b.** A detailed view of one intertidal enclosure. The nests of nine bourgeois males are marked with a red circle. **c.** A bourgeois male at its nest entrance. **d.** Alternative reproductive phenotypes include a larger male morph with secondary sexual characters (M+) that may reproduce as nest-holders or act as non-territorial floaters and a smaller male morph without secondary sexual character expression (M-) involved in sneaker and satellite tactics. **e.** When compared with M+ males (top image), M- males have larger testis and lack a testicular gland (bottom image). **f.** M+ males have an anal gland in the first two rays of the anal fin (top image) that M- males lack (bottom image); (photo credits: a, Rui Oliveira, b and c, Albert Ros; drawings d, e and f adapted from Santos, 1995). **g.** General view of the habitat of the peacock blenny *S. pavo* at the Ria Formosa (southern Portugal). **h.** A female-like sneaker male, a female and a nesting male of *S. pavo*. **i.** Two nesting males in adjacent holes of a brick (photos g and i by Rui Oliveira; photo h by David Gonçalves). **k.** A courting female is trying to enter a nest of a bourgeois male while a female-like sneaker male is in close proximity (photo by Rui Oliveira).

nest-holder territory and compared to other conspecifics, it becomes clear that the nest-holders direct proportionally less aggressive acts towards satellites (Oliveira *et al.*, 2002). This suggests that nest-holders benefit from the presence of satellites, being nest-holders and satellites involved in a reciprocal altruistic relationship (Santos and Almada, 1988). The possible beneficial effect of satellites for nest-holders has been investigated by comparing nests with and without an associated satellite male (Oliveira *et al.*, 2002). Satellites have attacked or chased sneakers and other conspecifics that might compete for fertilisations in the nest (neighboring males and floaters) (see also Santos, 1985b). This shared territoriality was negatively correlated with the time nest-holders invested per challenge on their territory (Oliveira *et al.*, 2002). Moreover, satellites never attack females and have even been seen to court females (loop swimming; Santos, 1985b), which may increase female attraction and also benefit the nest-holder. Satellite removal and replacement experiments have not been carried out yet but are critical in showing whether satellites really accrue benefits in terms of shared territoriality and attracting females. In agreement with the hypothesis that satellites trade-off this investment by exploring the reproductive potential of males, satellites preferred bigger nests and nests with two openings over smaller nests, and were found to associate more with more successful nest-holders. However, paternity data demonstrating that satellites gain an advantage in parasitic fertilisations of the eggs when compared with other parasitic males is still lacking.

From the female perspective, it might be beneficial to select a male associated with a satellite because they signal high quality males, and/or because the territorial behaviour of the satellite reduces the amount of sneaking attempts which might interrupt the spawning bout (discussed by Oliveira *et al.*, 2002). Furthermore, females might benefit from sperm diversity, which increases the genetic variability of the offspring. It is, thus, conceivable that a trade-off could have evolved, in which parental males have to pay some costs in lost fertilisations compensated for by lower costs in nest defense and possibly higher attractiveness (Santos and Almada, 1988).

Change in polymorphism in the Azorean rock-pool blenny is sequential (Santos, 1995; Oliveira *et al.*, 2001b). Males of one to two years of age show the M– morph and may switch between sneaker and satellite tactics. Males older than two years develop secondary sexual characters and compete for nests to become nest-holders or are non-territorial floaters. The morphological changes associated with the switch from M– to M+ males seem to be mediated by androgens, specifically by KT. During the reproductive season, KT and T are elevated in M+ males, while in M– males only T is elevated (Oliveira *et al.*, 2001b; Ros *et al.*, 2006b). Moreover, free circulating levels of KT but not T show a positive correlation with the size of the primary secondary sexual character, the anal glands (Oliveira *et al.*, 2001b), and treatment of M- males with slow release implants of KT but not T induces the growth of these glands (Oliveira *et al.*, 2001c; Ros *et al.*, 2006a).

The relationship between androgen levels and behavioural tactics in *P. parvicornis* seems less straightforward than the just described relation between androgens and morphotype (Oliveira *et al.*, 2001b). Despite the conspicuous within-morphotype variation in behavioural traits (nest-holder versus floater; sneaker versus satellite), no significant differences have been found within the morphotype for both T and KT levels in two different studies (Oliveira *et al.*, 2001b; Ros *et al.*, 2006a). Also, Oliveira *et al.* (2001b) found in nest-holders some weakly negative correlations between androgen levels and several reproductive behaviours.

In a field study where nest-holders were treated with slow release implants filled with KT in oil or with oil only, sexual behaviour was not facilitated by the KT treatment (Ros *et al.*, 2004). However, KT males increased territorial aggression, responding to conspecifics at longer distances from the nest and chasing them until they were further from the nest than the control males (Ros *et al.*, 2004). In aquaria experiments, KT implanted satellites did not show any difference in the frequency of aggressive displays towards a mirror or of courtship displays towards females when compared with control animals (Oliveira *et al.*, 2001c). These results, in combination with the basal KT levels measured in satellites in the field, seems a paradox in the light of the positive effect of KT on territorial behaviour in nest-holders. Thus, variation in androgen levels seems to be more associated with morphological rather than behavioural differences between male morphotypes. Possibly, other neuroendocrine mechanisms may be more relevant in the mediation of behavioural variation between male tactics.

'A Fish Called Tootsie': Female-like Males in the Peacock Blenny

Male alternative reproductive tactics have been well characterised in the peacock blenny *Salaria pavo*. Ruchon *et al.* (1995) first described the occurrence of male alternative phenotypes in this species at a brackish lagoon in Maugio, France, and soon after, female-mimicry was identified as an alternative male mating tactic in a population at Ria Formosa, Portugal (Gonçalves *et al.*, 1996; Fig. 4.3.4g). The Ria Formosa population has since then been extensively studied and a number of interesting features have been described. Most data presented below on ART, thus, concerns this population, but first a general review on the species reproductive biology and sexual behaviour is presented.

Sexual dimorphism is intense in *S. pavo*, with males being larger than females and having a well-developed head crest and an anal gland in the first two rays of the anal fin (Fishelson, 1963; Patzner *et al.*, 1986; Fig. 4.3.4h). The species reproductive biology follows the general pattern for blennies: the males defend a nest in a rock crevice or hole from which they attract the females; the females attach demersal eggs in a monolayer to the cavity walls; and parental care is provided exclusively by the male until egg hatching.

The patterns of sexual displays are very different between males and females. When approaching a nest, females court the males by beating the pectoral fins and opening and closing the mouth in synchrony while displaying a typical nuptial coloration that consists of contrasting dark and light bands across the anterior portion of the body (Fishelson, 1963; Patzner *et al.*, 1986). Males court the females usually from the nest by increasing the contrast between yellow and brown bands in the throat and head crest while performing intense lateral body jerkins. More rarely, the male leaves the nest and swims around the female while displaying body jerkins and then returns to the nest. In the nest, the male moves back and forth, alternating between protruding its head from the nest and going completely inside the nest. This behaviour is usually followed by the female further approaching the nest while courting and entering the nest (Fishelson, 1963; Patzner *et al.*, 1986; pers. obs.).

In the Ria Formosa, ecological constraints seem to promote both sex-role reversal and the occurrence of male alternative reproductive tactics. The species occurs in a mudflat area where the only hard substrates available for nesting are artificial debris, such as bricks, used by clam culturists to delimit concessions (Fig. 4.3.4g). Most males nest in brick holes and do not defend any area around the nest. Several males are usually found reproducing in different holes of the same brick (range: 1-6 nesting males per brick, Almada *et al.*, 1994; Fig. 4.3.4i). Because nest availability is low, male-male competition for nests is high. Accordingly, at the peak of the breeding season there are many non-nesting floater males in reproductive conditions who are, on an average, smaller than the nesting males (Almada *et al.*, 1994; Oliveira *et al.*, 1999). The scarcity of nests is likely to bias the operational sex ratio towards females (see Almada *et al.*, 1995; Oliveira *et al.*, 1999). Accordingly, sex-roles are reversed and females take the initiative in courtship and exhibit courtship displays more often than males (Almada *et al.*, 1995; Fig. 4.3.4k).

Small males are unable to compete for the few available nests and adopt an alternative reproductive tactic. These males are morphologically similar to females and mimic the complex female reproductive displays and nuptial coloration in order to approach the nesting males and achieve parasitic fertilisations of eggs (Gonçalves *et al.*, 1996; Fig. 4.3.4k). Nesting males are deceived by these parasitic males, as is suggested by the fact that they attack and court matched for size parasitic males and females with the same probability (Gonçalves *et al.*, 2005). However, as body size increases, parasitic males are more attacked and less courted by nesting males when compared with matched for size females, suggesting larger parasitic males are poorer female-mimics (Gonçalves *et al.*, 2005). The cues used by nesting males for better discrimination at larger sizes are still unclear.

Evidence that both sex-role reversal and male alternative reproductive tactics are promoted by nest shortage comes from a comparative study between this population and a population from the Adriatic. The Adriatic population occurs in a rocky shore area where nest availability is high. Males reproduce in rock crevices and actively defend the area around the nest. Sex-roles are not reversed, as males take the initiative in courtship and perform more courtship displays than females and, in comparison with the Ria Formosa population, males leave the nest to court females more often (J. Saraiva and R.F. Oliveira, unpublished data). These differences suggest that nest availability modulates the frequency of male and female sexual displays. Because nest availability is high in the Adriatic, small males can acquire nests and try to attract females. In accordance, the average size of nesting males is smaller in the Adriatic and some nesting males are as small as parasitic males in the Ria Formosa (J. Saraiva, and R.F. Oliveira, unpublished data). Nevertheless, some small males lacking secondary sexual characters with enlarged testis occur in a small proportion, suggesting male ART are not completely absent in the Adriatic (J. Saraiva, and R.F. Oliveira, unpublished data). These data suggest that ART in *S. pavo* are condition-dependent.

In the Ria Formosa, parasitic males show a strong site fidelity, usually associating with only one brick that contains one or more nesting males throughout the breeding season. Bricks that on an average receive more eggs throughout the breeding season have more and larger parasitic males attached (Gonçalves *et al.*, 2003). This suggests that either the parasitic males compete between them for access to the most successful breeding sites or that females prefer to spawn in nests that have more and larger parasitic males attached.

The parasitic reproductive tactic is only adopted during the males' first breeding season, as has been revealed by an ongoing mark-recapture study. All parasitic males recaptured in their second breeding season had well-developed male secondary sexual characters and some successfully established nests and guarded eggs (T. Fagundes, J. Saraiva, D. Gonçalves and R.F. Oliveira, unpublished data). Alternatively to this ontogenetic reproductive route, some males apparently skip reproduction during the first year, having poorly developed testis during the breeding season (Gonçalves *et al.*, 1996). These males also develop male characters and may succeed in establishing a nest in the following breeding season (T. Fagundes, J. Saraiva, D. Gonçalves and R.F. Oliveira, unpublished data). The mechanisms influencing the young males' decision to reproduce or not during their first breeding season remains unknown.

Males adopting sequential alternative reproductive tactics undergo major morphologic, physiologic and behavioural changes. Morphologically, body size increases, the head crest and the anal gland develop and a gland also develops in the testis. This testicular gland has been shown to be the main source of gonadal androgens (Oliveira *et al.*, 2001a) and, accordingly, nesting males have higher circulating levels of both T and KT when compared to parasitic males (Gonçalves *et al.*, 2008). The increase in circulating androgens may be responsible for some of the morphological changes taking place during tactic transition, as suggested above for *P. parvicornis*. The administration of KT implants to sneakers promoted the development of the genital papilla and anal gland within one week (Oliveira *et al.*, 2001d), while T was less effective in promoting the development of these male characters (Gonçalves *et al.*, 2007). In accordance, the KT/T ratio gradually increases from sneakers to transitional males to nest-holders (Gonçalves *et al.*, 2008).

Androgens have also been shown to mediate behavioural changes in *S. pavo*. Parasitic males switch from courting nesting males with female-like displays during their first breeding season into courting females with typical male behaviours during the remaining reproductive periods. Thus, both a defeminisation and a masculinisation of the males' reproductive behaviour occurs during tactic switching. The administration of both KT and T to parasitic males decreases the frequency of female-like displays (Oliveira *et al.*, 2001d; Gonçalves *et al.*, 2007). However, the mechanism through which androgens exert this action remains unknown. One hypothesis is that androgens modulate AVT circuits, as shown across vertebrates (e.g., De Vries and Panzica, 2006). Both parasitic males and females express more AVT mRNA in a per-cell basis in the pre-optic area (Grober *et al.*, 2002) and AVT administration promotes female displays in these two morphs (Carneiro *et al.*, 2003). Thus, a down regulation of AVT as a consequence of an increase in circulating androgens during tactic switching could be a possible mechanism underlying the parasitic males' defeminisation.

The influence of androgens on behavioural defeminisation contrasts with the results found for behavioural masculinisation where androgen implants failed to promote male behaviors in parasitic males (Oliveira *et al.*, 2001d; Gonçalves *et al.*, 2007). There are three possible interpretations for these results: (1) the time length of the endocrine manipulations (8 days) was insufficient to behaviourally masculinise parasitic males; (2) other neuroendocrine mechanisms underlie behavioural masculinisation; (3) changes in other components of the steroid pathway (e.g., steroid receptors, steroidogenic enzymes) are necessary for androgens to exert a masculinising effect. Further experiments should allow testing these hypotheses.

TOWARDS A UNIFYING THEORY FOR THE INTRA-SEXUAL VARIATION IN MALE REPRODUCTIVE BEHAVIOUR IN FISH: LESSONS FROM BLENNIES

The comparative approach to the understanding of the proximate mechanisms of intra-sexual variation in fish is a valuable tool for various reasons. First, it promotes the development of a conceptual framework to explain these phenomena that is not species-centered. One major problem in this area is that a lot of research effort has been invested in only a reduced number of species, so that the information available for these few species abounds in detail but it tends to be extrapolated, as is valid to the whole group of teleosts. Therefore, the collection of data on different species exhibiting alternative tactics contributes to the awareness that similar functional phenomena may have different underlying mechanisms and promotes the search for commonalities among species. In turn, these prompt the generation of hypotheses that organise the observed variation and, thus, contribute to the development of a framework that explains the evolution of proximate mechanisms underlying alternative tactics. In the following paragraphs, we will try to develop such a framework based on the comparative results summarised in this chapter.

An important distinction that should be made—when studying the proximate mechanisms of intra-sexual variation in male reproduction—is whether the alternative phenotypes diverge only in terms of behavioural traits or if differences in the expression of morphological and anatomical traits are also present.

Empirical evidence suggests that major changes in reproductive behaviour may occur without significant variation in circulating hormone levels. On the contrary, morphological and anatomical modifications in reproductive structures are thought to be more dependent on hormonal effects (see below). Thus, behaviour is expected to be more labile than morphology and anatomy and the mechanisms underlying the expression of behavioural variation should be more flexible than those underlying morphological and anatomical variations. It further follows that alternative reproductive tactics that only involve differences in behaviour are expected to differ in the activation of different neural substrates but not necessarily to display different hormone profiles. On the other hand, alternative reproductive phenotypes that diverge in morphological traits (i.e., intra-sexual polymorphisms) are expected to also have different hormone profiles. It could be argued that differences in hormone levels should only be present at the period of the differentiation of the tactic, if their effects were to be organisational. However, there are several pieces of evidence suggesting that androgen-dependent traits, typical of bourgeois males, need the continuous exposure to androgens in order to be maintained. For example, in species that seasonally reproduce the development of male secondary sexual characters is maximum during the breeding season and coincident with an increase in circulating androgen levels (e.g., Mayer *et al.*, 1990). Also, it has long been demonstrated that castration induces the regression and exogenous administration of androgens restores the development of male secondary sexual characters (e.g., sonic muscles in vocalising males of *Porichthys notatus*, Brantley *et al.*, 1993b; but see also Modesto and Canário, 2003).

Below, the hypothesis that the differences between male morphs among different phenotypic traits (i.e., behavioural, morphological and gonadal) may be ruled by different proximate mechanisms will be evaluated.

In order to compare the magnitude of the differences between alternative male types in androgen levels and in the forebrain AVT system, we have calculated the effect sizes for each of the variables for the two species studied (Thalheimer and Cook, 2002; Fig. 4.3.5). We have used Cohen's d as a measure of effect size since it has become the standard measure in the literature, enabling the comparison of the data presented here with other published studies. Moreover, the fact that Cohen's d values give per se an indication of the magnitude of the effect size (i.e., 0.2 = small; 0.5 = medium; 0.8 = large) allows us to interpret the values obtained by comparing them to these known benchmarks (Rosnow and Rosenthal, 1996; Rosnow *et al.*, 2000; Thalheimer and Cook, 2002). To facilitate the reading of the following sections, we have also compiled an overview of the behavioural, morphological and gonadal differences between the alternative morphs for the species reported here (Table 4.3.2).

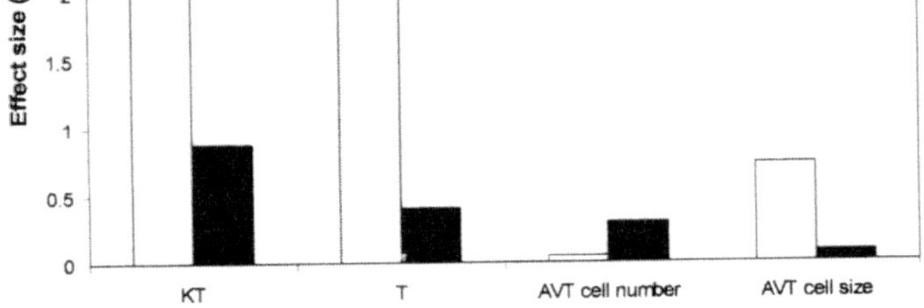

Fig. 4.3.5 The magnitude of the differences between alternative male types of *P. parvicornis* and *S. pavo* in androgens (KT and T) and in forebrain AVT (cell number and cell size) was calculated by computing the Cohen's d effect size measure.

Table 4.3.2 Behavioural, morphological and gonadal characteristics of mature males in *Parablennius parvicornis* and *Salaria pavo*

	Parablennius parvicornis				*Salaria pavo*	
	M–		M+			
	Sneakers	*Satellites*	*Floaters*	*Nesting males*	*Sneakers*	*Nesting males*
Behavioural	Roam around several pools with a cryptic behaviour in order to enter nests and fertilise the eggs.	Help in territory defense and female attraction to attain a privileged position for parasitic fertilisations.	Non-territorial, are looking for a vacant territory to establish a nest.	Actively defend the territory and court females. Provide parental care to the eggs.	Assume a female-like behaviour in order to approach the nests of nesting males and parasitically fertilise the eggs.	Adopt a passive role during courtship, rarely court the females. Do not defend a territory around the nest. Provide parental care to the eggs.
Morphological	The smallest of all male classes. Do not express male secondary sexual characters.	Do not express male secondary sexual characters.	Do not display the nesting male nuptial coloration. Smaller than nesting males. Intermediate pattern of development of male secondary sexual characters.	Have a distinct dark-brown coloration, a distinct head hump and an anal gland in the first two rays of the anal fin.	Much smaller than nesting males. Do not display male secondary sexual characters.	Have a well-developed head-crest and an anal gland in the first two rays of the anal fin.
Gonadal	Enlarged testis in relation to M+ males. The testicular gland is either absent or vestigial.	Enlarged testis in relation to M+ males. The testicular gland is absent or vestigial.	Smaller testis than M– males. The testicular gland is present.	Smaller testis than M– males. The testicular gland is present.	Enlarged testis in relation to nesting males. The testicular gland is absent or vestigial.	Smaller testes than sneakers. The testicular gland is present.

Hormones and Differences between Alternative Phenotypes in Secondary Sex Characters

Ketotestosterone (KT) has been demonstrated to be the most potent androgen in the induction of secondary sex characters in male teleosts (Borg, 1994). Thus, differences in KT levels may be less important in species with alternative tactics that lack major tactic-specific morphological specialisations (e.g., the expression of male secondary sex characters in bourgeois males). Between the two species studied here *S. pavo,* that displays a higher degree of phenotypic specialisation of the alternative tactics, also exhibits a larger effect size in androgen levels between alternative male types (Fig. 4.3.5). These results suggest that differential KT levels are necessary in order to induce or to maintain tactic-specific morphological characters but may not be needed when the alternative tactics only differ in behavioural traits. If this hypothesis is correct an association between the degree of dimorphism between the alternative male phenotypes and the differential variation in KT levels between alternative morphs is to be expected. In our two study species, major differences in KT levels occur between morphotypes that differ in the development of secondary sexual characters (i.e., M+ versus M– males in *P. parvicornis*, and sneakers versus nesting males in *S. pavo*) while in morphotypes that only differ in behaviour (i.e., sneakers versus satellites and floaters versus nesting males in *P. parvicornis*) KT levels are similar (Oliveira *et al.*, 2001a, b and unpublished data). Additionally, from all species for which androgen levels are known when the ART involves a morphological intra-sexual dimorphism (apart from differences in body size), the KT (but not T) levels are higher in the bourgeois than in the parasitic male, and irrespective of the ART they display (Oliveira, 2006). This suggests a parallel to the androgen correlates of sex-change in fish in that androgens are thought to play a major role in the morphological differentiation during sex-change but are not essential for the occurrence of behavioural sex-change (Godwin *et al.*, 1996; Grober, 1998; Reavis and Grober, 1999).

Hormones and Differences between Alternative Phenotypes in Reproductive Behaviour

Several studies suggest that reproductive behaviour may change dramatically without major variations in circulating hormone levels. For example, in the protogynous blue-head wrasse, *Thalassoma bifasciatum*, the experimental removal of a territorial male allows large females to attain social dominance and to change sex. In this experimental setting, ovariectomised females do not differ from sham-operated females in the sense that they also change their behaviour and display the full suite of male reproductive behaviours (Godwin *et al.*, 1996). This suggests that the regulation of reproductive behaviours is more dependent on other neuroendocrine mechanisms and less on gonadal hormones. Brain neuropeptides systems are likely candidates for this regulation. In the above example of *T. bifasciatum*, female-to-male behavioural switch is closely followed by major changes in hypothalamic AVT mRNA levels (Godwin *et al.*, 2000) and blocking AVT action (with an AVP V1 receptor antagonist) inhibits females from behaviourally changing into males (Semsar and Godwin, 2004). The AVT system has also been studied in our two blenniid species in relation to reproductive behaviour. The asymmetry in the expression of courtship behaviour between alternative morphs is larger in *S. pavo* than in *P. parvicornis* (Table 4.3.2). In the Ria Formosa population of *S. pavo*, nest-holders lack male courtship behaviour (or display it in a low frequency), while sneaker males display female-like courtship behaviours. In *P. parvicornis,* there is only a quantitative differentiation in reproductive behaviour between satellites and nesting males, as satellites also participate in territory defense and female attraction (Santos, 1985b; Oliveira *et al.*, 2002). Interestingly, the effect size is largest for AVT cell size in *S. pavo* (Fig. 4.3.5) while in *P. parvicornis* AVT cell size presents a negligible effect size and only a small effect size is detected for AVT cell number (Fig. 4.3.5). These results, together with the fact that in some species AVT promotes female displays (e.g., *S. pavo*) and in others male displays (e.g., *T. bifasciatum*), suggest that AVT is closely associated with the general expression of courtship behaviour rather than specifically with male or female sexual displays.

Nevertheless, sex steroids have also been shown to significantly influence reproductive displays in species with alternative reproductive tactics. For example, androgens (both T and KT) inhibit the expression of female-like displays in sneakers of *S. pavo* (Oliveira *et al.*, 2001d; Gonçalves *et al.*, 2007). As AVT promotes these displays (Carneiro *et al.*, 2003), it seems possible that androgens act on these behaviours through the suppression of the AVT system. Similar results have been reported for the midshipman *Porichthys notatus* (Goodson and Bass, 2000). Thus, sex steroids do not seem to be critical to induce or maintain differences in reproductive behaviour between males displaying alternative tactics but are likely to act as modulators of the neural substrates underlying the expression of these behaviours.

Hormones and Differences between Alternative Phenotypes in Gonadal Allocation

Apart from gonadotrophins, androgens also participate in the regulation of spermatogenesis, with KT stimulating later stages of the process and T being involved in the negative feedback mechanisms needed to control KT-dependent spermatogenesis. Thus, a balance between T and KT is needed for the control of spermatogenesis (Schulz and Miura, 2002). Therefore, the ratio between KT and T should be more informative than absolute KT or T levels. If the KT:T ratio is computed for the species for which both androgen levels and GSI values are available ($n = 9$ species) a trend for the magnitude of the difference in the KT:T to be negatively correlated with the magnitude of the difference in GSI values between alternative male types is found (Oliveira, 2006). In other words, in species for which the magnitude of the difference in KT levels between bourgeois and parasitic is larger, there is a smaller difference in relative gonad size. This also means that among parasitic males, a higher GSI is associated with a lower KT:T ratio, which probably allows them to have larger testes without a linked expression of the secondary sex characters and of bourgeois male behaviour.

To summarise, the relative importance of different physiological mechanisms for the differentiation of tactic-specific traits might vary among behavioural, morphological and gonadal traits and, therefore, the classic paradigm of androgens controlling the expression of the whole set of reproductive characters that characterise each tactic should be questioned (Fig. 4.3.3).

Finally, it should also be mentioned that the decoupling of different male traits in parasitic males may be achieved by other mechanisms apart from differences in hormone levels, namely by varying the local micro-environments in the different target tissues, due to differential expression of receptors or to differential levels of activity of catabolic enzymes that modulate the availability of the active hormone to specific targets (e.g., 11-β-hydroxilase and 11-β-HSD, that metabolise T into KT, are key steps in the expression of male secondary sex characters, spermatogenesis and the modulation of the expression of reproductive behaviour in male teleosts, Borg, 1994). This focus on target tissues has rarely been used when studying the mechanisms of intra-sexual variation in reproduction. One of the rare examples of such an approach is a study on the relative levels of brain steroid receptors between alternative reproductive phenotypes in the protogynous wrasse, *Halichoeres trimaculatus*, where it was found—using competitive RT-PCR—that the levels of AR transcripts were significantly higher in the brain of terminal phase males than in initial phase males (Kim *et al.*, 2002). No other significant differences in gene expression were observed either for AR in the gonads or for ER both in the brain and in the gonads. Thus, by varying the expression of AR in specific tissues (brain versus gonad), bourgeois males (in this case, terminal phase males) can both increase their sensitivity to circulating androgen levels in specific targets (the brain) and, simultaneously, allow a compartmentalisation of the effects of androgens, so that the potential deleterious effects of androgens can be avoided by varying AR densities in different tissues (Ketterson and Nolan Jr., 1999). This mechanism hypothetically makes it possible to activate the expression of an androgen-dependent reproductive behaviour in bourgeois males without having the associated costs of increasing spermatogenesis or the expression of a sex character, since the androgen action can be independently modulated at each compartment (brain versus gonad versus morphological secondary sex character). Studies focusing on target tissues are, thus, a major avenue for future research in this area.

Acknowledgements

We would like to thank all the colleagues that, over the years, have worked with us in the different projects on the challenging world of fish sexuality, in particular Vitor Almada, Adelino Canário, Emanuel Gonçalves, Matthew Grober and Ricardo Serrão Santos, with whom various discussions over the years have helped to shape the current view we have of these phenomena. We would also like to thank Kurt Kotrschal and Robert Patzner for their constructive comments on an early version of this manuscript. The writing of the current manuscript has been funded by two research grants from Fundação para a Ciência e a Tecnologia (POCTI/BSE/38395/2001 and UI&D/331/2001) and the EU Program FEDER.

References

Almada, V.C. and J.I. Robalo. 2008. Phylogenetic analysis of alternative reproductive tactics: problems and possibilities. In: *Alternative Reproductive Tactics: An Integrative Approach*, R.F. Oliveira, M. Taborsky and H.J. Brockmann (eds.). Cambridge University Press, Cambridge, pp. 52-62.

Almada, V.C., E.J. Gonçalves, A.J. Santos and C. Baptista. 1994. Breeding ecology and nest aggregations in a population of *Salaria pavo* (Pisces, Blenniidae) in an area where nest sites are very scarce. *Journal of Fish Biology* 45: 819-830.

Almada, V.C., E.J. Gonçalves, R.F. Oliveira and A.J. Santos. 1995. Courting females—Ecological constraints affect sex-roles in a natural population of the blenniid fish *Salaria pavo*. *Animal Behaviour* 49: 1125-1127.

Austad, S.N. 1984. A classification of alternative reproductive behaviors and methods for field-testing ESS models. *American Zoologist* 24: 309-319.

Borg, B. 1994. Androgens in teleost fishes. *Comparative Biochemistry and Physiology* C109: 219-245.

Brantley, R.K., J.C. Wingfield and A.H. Bass. 1993a. Sex steroid levels in *Porichthys notatus*, a fish with alternative reproductive tactics, and a review of the hormonal bases for male dimorphism among teleost fishes. *Hormones and Behavior* 27: 332-347.

Brantley, R.K., M.A. Marchaterre and A.H. Bass. 1993b. Androgen effects on vocal muscle structure in a teleost fish with inter- and intra-sexual dimorphism. *Journal of Morphology* 216: 305-318.

Brockmann, H.J. 2001. The evolution of alternative strategies and tactics. *Advances in the Study of Behavior* 30: 1-51.

Carneiro, L., R.P. Andrade, R.F. Oliveira and K. Kotrschal. 2001. Sex differences in home range and dorso-lateral telencephalon in the Azorean rock-pool blenny, Society for Neuroscience Abstracts, Volume 27, Program No. 535.4.

Carneiro, L., R. Oliveira, A.V.M. Canário and M. Grober. 2003. The effect of arginine vasotocin on courtship behaviour in a blenniid fish with alternative reproductive tactics. *Fish Physiology and Biochemistry* 28: 241-243.

Caro, T.M. and P. Bateson. 1986. Organization and ontogeny of alternative tactics. *Animal Behaviour* 34: 1483-1499.

Dawkins, R. 1980. Good strategy or evolutionary stable strategy? In: *Sociobiology: Beyond Nature/Nurture*, G.W. Barlow and S. Silverberg (eds.). Westview Press for the American Association for the Advancement of Science, Boulder, Colorado, pp. 331-367.

de Jonge, J. and J.J. Videler. 1989. Differences between the reproductive biologies of *Tripterygion tripteronotus* and *T. delaisi* (Pisces, Perciformes, Tripterygiidae): The adaptive significance of an alternative mating strategy and a red instead of a yellow nuptial colour. *Marine Biology* 100: 431-437.

De Vries, G.J. and G.C. Panzica. 2006. Sexual differentiation of central vasopressin and vasotocin systems in vertebrates: Different mechanisms, similar end-points. *Neuroscience* 138: 947-955.

Dixon, A.F. 1998. *Primate Sexuality. Comparative Studies of the Prosimians, Monkeys, Apes, and Human Beings*. Oxford University Press, Oxford.

Fishelson, L. 1963. Observations on littoral fishes of Israel I. Behaviour of *Blennius pavo* Risso (Teleostei, Blenniidae). *Israel Journal of Zoology* 12: 67-91.

Godwin, J., D. Crews and R.R. Warner. 1996. Behavioural sex change in the absence of gonads in a coral reef fish. *Proceedings of the Royal Society of London* B263: 1683-1688.

Godwin, J., R. Sawby, R.R. Warner, D. Crews and M.S. Grober. 2000. Hypothalamic arginine vasotocin mRNA abundance variation across sexes and with sex change in a coral reef fish. *Brain, Behavior and Evolution* 55: 77-84.

Gonçalves, D., T. Fagundes and R. Oliveira. 2003. Reproductive behaviour of sneaker males of the peacock blenny. *Journal of Fish Biology* 63: 528-532.

Gonçalves, D., R. Matos, T. Fagundes and R. Oliveira. 2005. Bourgeois males of the peacock blenny, *Salaria pavo*, discriminate female mimics from females? *Ethology* 111: 559-572.

Gonçalves, D., J. Alpedrinha, M. Teles and R.F. Oliveira. 2007. Endocrine control of sexual behavior in sneaker males of the peacock blenny *Salaria pavo*: Effects of castration, aromatase inhibition, testosterone and estradiol. *Hormones and Behavior* 51: 534-541.

Gonçalves, D., M. Teles, J. Alpedrinha and R.F. Oliveira. 2008. Brain and gonadal aromatase activity and steroid hormone levels in female and polymorphic males of the peacock blenny *Salaria pavo*. *Hormones and Behavior* 54: 717-725.

Gonçalves, E.J., V.C. Almada, R.F. Oliveira and A.J. Santos. 1996. Female mimicry as a mating tactic in males of the blenniid fish *Salaria pavo*. *Journal of the Marine Biological Association of the United Kingdom* 76: 529-538.

Goodson, J.L. and A.H. Bass. 2000. Vasotocin innervation and modulation of vocal-acoustic circuitry in the teleost *Porichthys notatus*. *Journal of Comparative Neurology* 422: 363-379.

Goodson, J.L. and A.H. Bass. 2001. Social behavior functions and related anatomical characteristics of vasotocin/vasopressin systems in vertebrates. *Brain Research Reviews* 35: 246-265.

Grober, M.S. 1998. Socially controlled sex change: Integrating ultimate and proximate levels of analysis. *Acta Ethologica* 1: 3-17.

Grober, M.S., A.A. George, K.K. Watkins, L.A. Carneiro and R.F. Oliveira. 2002. Forebrain AVT and courtship in a fish with male alternative reproductive tactics. *Brain Research Bulletin* 57: 423-425.

Gross, M.R. 1996. Alternative reproductive strategies and tactics: Diversity within sexes. *Tree* 11: 92-98.

Ketterson, E.D. and V. Nolan Jr. 1999. Adaptation, exaptation, and constraint: A hormonal perspective. *American Naturalist* 154: S4-S25.

Kim, S.J., K. Ogasawara, J.G. Park, A. Takemura and M. Nakamura. 2002. Sequence and expression of androgen receptor and estrogen receptor gene in the sex types of protogynous wrasse, *Halichoeres trimaculatus*. *General and Comparative Endocrinology* 127: 165-173.

Knapp, R. 2004. Endocrine mediation of vertebrate male alternative reproductive tactics: the next generation of studies. *Integrative Comparative Biology* 43: 658-668.

MacKiewicz, M., B.A. Porter, E.E. Dakin and J.C. Avise. 2005. Cuckoldry rates in the molly miller (*Scartella cristata*; Blenniidae) a hole-nesting marine fish with alternative reproductive tactics. *Marine Biology* 148: 213-221.

Mayer, I., B. Borg and R. Schulz. 1990. Seasonal changes in and effect of castration/androgen replacement on the plasma levels of five androgens in the male three-spined stickleback, *Gasterosteus aculeatus* L. *General and Comparative Endocrinology* 79: 23-30.

Modesto, T. and A.V.M. Canário. 2003. Hormonal control of swimbladder sonic muscle dimorphism in the Lusitanian toadfish. *Journal of Experimental Biology* 206: 3467-3477.

Mohr, H.P. 1986. Kleptogamie: Vergleich dreier mediterraner Arten der Tripterygidae (Perciformes, Teleostei). Diploma Thesis, Albert Ludwig Universität, Freiburg, Germany.

Moore, M.C. 1991. Application of organization-activation theory to alternative male reproductive strategies: A review. *Hormones and Behavior* 25: 154-179.

Moore, M.C., D.K. Hews and R. Knapp. 1998. Hormonal control and evolution of alternative male phenotypes: Generalizations of models for sexual differentiation. *Integrative and Comparative Biology* 38: 133-151.

Neat, F.C. 2001. Male parasitic spawning in two species of triplefin blenny (Tripterigiidae): Contrasts in demography, behaviour and gonadal characteristics. *Environmental Biology of Fishes* 61: 57-64.

Neat, F.C., L. Locatello and M.B. Rasotto. 2003a. Reproductive morphology in relation to alternative male reproductive tactics in *Scartella cristata*. *Journal of Fish Biology* 62: 1381-1391.

Neat, F.C., W. Lengkeek, E.P. Westerbeek, B. Laarhoven and J.J. Videler. 2003b. Behavioural and morphological differences between lake and river populations of *Salaria fluviatilis*. *Journal of Fish Biology* 63: 374-387.

Nelson, R.J. 2005. *An Introduction to Behavioral Endocrinology*. Sinauer Associates Inc., Sunderland MA, 3rd Edition.

Oliveira, R.F. 2006. Neuroendocrine mechanisms of alternative reproductive tactics in fish. *Behaviour and Physiology of Fish* 24: 297-357.

Oliveira, R.F., V.C. Almada, E. Forsgren and E.J. Gonçalves. 1999. Temporal variation in male traits, nesting aggregations and mating success in the peacock blenny. *Journal of Fish Biology* 54: 499-512.

Oliveira, R.F., J.A. Miranda, N. Carvalho, E.J. Goncalves, M.S. Grober and R.S. Santos. 2000. Male mating success in the Azorean rock-pool blenny: The effects of body size, male behaviour and nest characteristics. *Journal of Fish Biology* 57: 1416-1428.

Oliveira, R.F., A.V. Canário and M.S. Grober. 2001a. Male sexual polymorphism, alternative reproductive tactics, and androgens in combtooth blennies (Pisces: Blenniidae). *Hormones and Behavior* 40: 266-275.

Oliveira, R.F., A.V. Canário, M.S. Grober and R.S. Santos. 2001b. Endocrine correlates of male polymorphism and alternative reproductive tactics in the Azorean rock-pool blenny, *Parablennius sanguinolentus parvicornis*. *General and Comparative Endocrinology* 121: 278-288.

Oliveira, R.F., L.A. Carneiro, A.V. Canário and M.S. Grober. 2001c. Effects of androgens on social behavior and morphology of alternative reproductive males of the Azorean rock-pool blenny. *Hormones and Behavior* 39: 157-166.

Oliveira, R.F., L.A. Carneiro, D.M. Gonçalves, A.V. Canário and M.S. Grober. 2001d. 11-Ketotestosterone inhibits the alternative mating tactic in sneaker males of the peacock blenny, *Salaria pavo*. *Brain, Behavior and Evolution* 58: 28-37.

Oliveira, R.F., N. Carvalho, J. Miranda, E.J. Gonçalves, M. Grober and R.S. Santos. 2002. The relationship between the presence of satellite males and nest-holders' mating success in the Azorean rock-pool blenny *Parablennius sanguinolentus parvicornis*. *Ethology* 108: 223-235.

Oliveira, R.F., A.F. Ros and D.M. Goncalves. 2005. Intra-sexual variation in male reproduction in teleost fish: A comparative approach. *Hormones and Behavior* 48: 430-439.

Oliveira, R.F., A.V.M. Canário and A.F.H. Ros. 2008. Hormones and alternative reproductive tactics in vertebrates. In: *Alternative Reproductive Tactics: An Integrative Approach*, R.F. Oliveira, M. Taborsky and H.J. Brockmann (eds.). Cambridge University Press, Cambridge, pp. 132-174.

Patzner, R.A., M. Seiwald, M. Adlgasser and G. Kaurin. 1986. The reproduction of *Blennius pavo* (Teleostei, Blenniidae) V. Reproductive behavior in natural environment. *Zoologischer Anzeiger* 216: 338-350.

Reavis, R.H. and M. Grober. 1999. An integrative approach to sex change: social, behavioural and neurochemical changes in *Lythrypnus dalli* (Pisces). *Acta Ethologica* 2: 51-60.

Repka, J. and M.R. Gross. 1995. The evolutionarily stable strategy under individual condition and tactic frequency. *Journal of Theoretical Biology* 176: 27-31.

Ros, A.F., R. Bruintjes, R.S. Santos, A.V. Canario and R.F. Oliveira. 2004. The role of androgens in the trade-off between territorial and parental behavior in the Azorean rock-pool blenny, *Parablennius parvicornis*. *Hormones and Behavior* 46: 491-497.

Ros, A.F., C. Ferreira, R.S. Santos and R.F. Oliveira. 2006a. Regulation of immunocompetence by different androgen metabolites in a blenny with alternative reproductive tactics. *Journal of Experimental Zoology* A305: 986-994.

Ros, A.F., N. Bouton, R.S. Santos and R. F. oliveira. 2006b. Alternative male reproductive tactics and the immunocompetence handicap in the Azorean rock-pool blenny, *Parablennius parvicornis*. *Proceedings of the Royal Society of London* B273: 901-909.

Rosnow, R.L. and R. Rosenthal. 1996. Computing contrasts, effect sizes, and counternulls on other people's published data: General procedures for research consumers. *Psychological Methods* 1: 331-340.

Rosnow, R.L., R. Rosenthal and D.B. Rubin. 2000. Contrasts and correlations in effect size estimation. *Psychological Science* 11: 446-453.

Ruchon, F., T. Laugier and J.P. Quignard. 1995. Alternative male reproductive strategies in the peacock blenny. *Journal of Fish Biology* 47: 826-840.

Ryan, M.J., C.M. Pease and M.R. Morris. 1992. A genetic polymorphism in the swordtail *Xiphophorus nigrensis*: testing the prediction of equal fitnesses. *American Naturalist* 139: 21-31.

Santos, R.S. 1985a. Estrutura e função dos territórios em machos parentais de *Blennius sanguinolentus* Pallas (Pisces: Blenniidae). *Memórias do Museu do Mar-Série Zoológica* 3: 1-46.

Santos, R.S. 1985b. Parentais e satélites: tácticas alternativas de acasalamento nos machos de *Blennius sanguinolentus* Pallas (Pisces: Blenniidae). *Arquipélago-Série Ciências da Natureza* 6: 119-146.

Santos, R.S. 1989. Observações sobre os intervalos de desenvolvimento de *Parablennius sanguinolentus* (Pallas) (Pisces: Blenniidae) dos Açores: Períodos embrionário, larvar e juvenil. *Arquivos do Museu Bocage*, Nova série 1: 293-310.

Santos, R.S. 1995. Anatomy and histology of secondary sexual characters, gonads and liver of rock-pool blenny, *Parablennius sanguinolentus parvicornis* (Pisces: Blenniidae) of the Azores. *Arquioélago*-Série Ciências da Natureza A13: 27-38.

Santos, R.S. and V. Almada. 1988. Intraspecific variations in reproductive tactics in males of the rocky intertidal fish *Blennius sanguinolentus* in the Azores. In: *Behavioral Adaptation to Intertidal Life*, G. Chelazzi, R. Hartnoll and M. Vannini (eds.). Plenum Publishing Corporation, New York, pp. 421-447.

Santos, R.S. and J.P. Barreiros. 1993. The ethogram of *Parablennius sanguinolentus parvicornis* (Valenciennes in Cuvier & Valenciennes, 1836) (Pisces: Blenniidae) from the Azores. *Arquipélago*-Série Ciências da Natureza A11: 73-90.

Santos, R.S., R.D.M. Nash and S.J. Hawkins. 1995. Age, growth and sex ratio of the azorean rock-pool blenny, *Parablennius sanguinolentus parvicornis*. *Journal of the Marine Biological Association of the United Kingdom* 75: 751-754.

Santos, R.S., S.J. Hawkins and R.D.M. Nash. 1996. Reproductive phenology of the Azorean rock pool blenny a fish with alternative mating tactics. *Journal of Fish Biology* 48: 842-858.

Schulz, R.W. and T. Miura. 2002. Spermatogenesis and its endocrine regulation. *Fish Physiology and Biochemistry* 26: 43-56.

Semsar, K. and J. Godwin. 2003. Social influences on the arginine vasotocin system are independent of gonads in a sex-changing fish. *Journal of Neuroscience* 23: 4386-4393.

Semsar, K. and J. Godwin. 2004. Multiple mechanisms of phenotype development in the bluehead wrasse. *Hormones and Behavior* 45: 345-353.

Shuster, S.M. and M.J. Wade. 1991. Equal mating success among male reproductive strategies in a marine isopod. *Nature (London)* 350: 608-610.

Shuster, S.M. and M.J. Wade. 2003. *Mating Systems and Strategies*. Princeton University Press, Princeton.

Taborsky, M. 1997. Bourgeois and parasitic tactics: Do we need collective, functional terms for alternative reproductive behaviours? *Behavioral Ecology and Sociobiology* 41: 361-362.

Taborsky, M. 1999. Conflict or cooperation: What determines optimal solutions to competition in fish reproduction? In: *Behaviour and Conservation of Littoral Fishes*, V.C. Almada, R.F. Oliveira and E.J. Gonçalves (eds.). ISPA, Lisboa, pp. 301-349.

Taborsky, M. 2008. Alternative reproductive tactics in fish. In: *Alternative Reproductive Tactics: An Integrative Approach*, R.F. Oliveira, M. Taborsky and H.J. Brockmann (eds.). Cambridge University Press, Cambridge, pp. 251-299.

Taborsky, M., R.F. Oliveira and H.J. Brockmann. 2008. The evolution of alternative reproductive tactics: concepts and questions. In: *Alternative Reproductive Tactics: An Integrative Approach*, R.F. Oliveira, M. Taborsky, and H.J. Brockmann (eds.). Cambridge University Press, Cambridge, pp. 1-22.

Thalheimer, W. and S. Cook. 2002. *How to calculate effect sizes from published research articles: A simplified methodology*. Available at: http://work-learning.com/effect_sizes.htm.

Thresher, R.E. 1984. Combtooth Blennies (Blenniidae). In: *Reproduction in Reef Fishes*, R.E. Thresher (ed.). T.H.F. Publications, Inc., Neptune City, N.J., pp. 314-333.

Tomkins, J.L., N.R. LeBas, J. Unrug and J. Radwan. 2004. Testing the status-dependent ESS: Population variation in fighter expression in the mite *Sancassania berlesei*. *Journal of Evolutionary Biology* 17: 1377-1388.

Wilson, J.D., M.W. Leihy, G. Shaw and M.B. Renfree. 2002. Androgen physiology: unsolved problems at the millennium. *Molecular and Cellular Endocrinology* 198: 1-5.

Wirtz, P. 1978. The behaviour of the Mediterranean *Tripterygion* species (Pisces, Blennioidei). *Zeitschrift für Tierpsychologie* 48: 142-174.

CHAPTER

4.4

Larval Development in Blennies

William Watson

INTRODUCTION

Early life history information is available for about 40% of the 151 blennioid genera (Table 4.4.1), but this is less than 20% of the 875 species. Blennioid larvae are planktonic. At hatching, the oviparous species are about 1.5-6.5 mm long (Table 4.4.2), preflexion-stage larvae that are usually ready to begin feeding, with fully pigmented eyes, functional jaws, and only a small yolk reserve. Viviparous species typically are larger and further developed at birth: clinins, for example, complete notochord flexion prior to birth. Blennioids generally undergo notochord flexion at about 4-12 mm and settle from the plankton at about 10-25 mm, although some clinids and blenniids are much larger at settlement.

Blennioid larvae are somewhat compressed and elongate, with body depth commonly ≤ 20% of the body length (Table 4.4.3). Most have a short gut (pre-anal length typically ≤ 50% of body length) which is coiled

Author's address: NOAA Southwest Fisheries Science Center, 8604 La Jolla Shores Drive, La Jolla, CA 92037-1508 USA.
E-mail: william.watson@noaa.gov

Table 4.4.1 Literature sources for descriptions of eggs and larvae of blennioid species.

Family	*Genus*	*Egg description*	*Larval description*
Tripterygiidae	*Axoclinus*		Beltrán-León and Ríos Herrera (2000)
	Enneanectes		Brogan (1992); Watson (1996a); Cavalluzzi and Olney (2006a)
	Enneapterygius	Shiogaki and Dotsu (1973)	Shiogaki and Dotsu (1973); Miller *et al.* (1979); Matarese *et al.* (1984)
	Forsterygion	Graham (1939); Ruck (1980)	Graham (1939); Ruck (1980, 1998)
	Gilloblennius	Ruck (1980)	Ruck (1980)
	Grahamina	Ruck (1973)	Ruck (1973)
	Helcogrammoides		Mongdard (1979)
	Ruanoho	Ruck (1980)	Ruck (1980)
	Springerichthys	Shiogaki and Dotsu (1988a)	Shiogaki and Dotsu (1988a)
	Tripterygion		Padoa (1956b); Gordina and Duka (1979)
Dactyloscopidae	*Dactylagnus*	Dawson (1976)	Brogan (1992)
	Dactyloscopus	Dawson (1975); Watson (1996b)	Brogan (1992); Watson (1996b); Cavalluzzi and Olney (2006b)
	Gillellus	Böhlke (1968); Dawson (1977)	Watson (1996b); Cavalluzzi (1997); Cavalluzzi and Olney (2006b)
	Leurochilus	Böhlke (1968)	
	Myxodagnus	Dawson (1976)	Brogan (1992); Watson (1996b)
	Sindoscopus	Dawson (1977)	
Labrisomidae	*Alloclinus*		Watson (1996c)
	Cryptotrema		Watson (1996c)
	Dialommus	Howe (1991)	Herrera and Lavenberg (1999)
	Exerpes		Brogan (1992)
	Labrisomus		Brogan (1992); Watson (1996c);. Herrera and Lavenberg (1999)
	Malacoctenus		Brogan (1992); Herrera and Lavenberg (1999)
	Paraclinus	Breder (1939); Watson (1996c)	Breder (1939, 1941); Matarese *et al.* (1984); Brogan (1992); Watson (1996c); Cavalluzzi and Olney (2006c)
	Starksia		Herrera and Lavenberg (1999)
Clinidae	*Clinitrachus*		Padoa (1956b)
	Gibbonsia		Wang (1981); Watson (1996d)
	Heterostichus	Barnhart (1932); Feder *et al.* (1974); Wang (1981); Matarese *et al.* (1984,	Wang (1981); Matarese *et al.* (1984, 1989); Stepien (1986); Watson (1996d)

Table 4.4.1 Contd.

Table 4.4.1 Contd.

		1989); Stepien (1986); Watson (1996d)	
	Heteroclinus		Gunn and Thresher (1991)
	Xenomedea	Rosenblatt and Taylor (1971)	
Chaenopsidae	*Acanthemblemaria*		Stephens *et al.* (1966); Brogan (1992)
	Chaenopsis		Böhlke (1957); Brogan (1992)
	Coralliozetus		Brogan (1992)
	Emblemaria		Brogan (1992)
	Neoclinus	Shiogaki and Dotsu (1972); Lindquist (1981); Wang (1981); Watson (1996e)	Shiogaki and Dotsu (1972, 1988b); Lindquist (1981); Wang (1981); Watson (1996e)
	Stathmonotus		Brogan (1992); Cavalluzzi (1997); Cavalluzzi and Olney (2006d)
Blenniidae	*Aidablennius*	Kalinina (1976)	Padoa (1956a); Dekhnik (1973); Kalinina (1976); Sabates (1994)
	Atrosalarias		Watson (2000)
	Blenniella		Springer and Williams (1994)
	Blennius	Cunningham (1889); Lebour (1927); Padoa (1956a); Fives (1970); Russell (1976)	Ehrenbaum (1905); Ford (1922); Lebour (1927); Padoa (1956a); Fives (1970); Russell (1976)
	Chasmodes	Hildebrand and Cable (1938); Fritzsche (1978); Peters (1981); Ditty *et al.* (2006); Fahay (2007)	Hildebrand and Cable (1938); Lipson and Moran (1974); Fritzsche (1978); Peters (1981, 1985); Ditty *et al.* (2006); Fahay (2007)
	Cirripectes		Kojima and Shiogaki (1988)
	Coryphoblennius	Fives (1980)	Padoa (1956a); Fives (1970, 1980); Sabates (1994)
	Enchelyurus		Miller *et al.* (1979); Matarese *et al.* (1984); Watson (1987, 2001); Kubo and Sasaki (2000b)
	Entomacrodus		Kojima and Shiogaki (1988); Brogan (1992); Watson (1996f)
	Exallias		Watson (1983, 2000); Kojima and Shiogaki (1988)
	Haptogenys	Dutt and Visweswara Rao (1960)	Dutt and Visweswara Rao (1960)
	Hypleurochilus	Hildebrand and Cable (1938); Catalano (1978); Ditty *et al.* (2006); Fahay (2007)	Hildebrand and Cable (1938); Catalano (1978); Ditty *et al.* (2005, 2006); Fahay (2007)

Table 4.4.1 Contd.

Table 4.4.1 Contd.

	Hypsoblennius	Hildebrand and Cable (1938); Fritzsche (1978); Balbontín and Pérez (1979); Watson (1996f); Ditty *et al.* (2006); Fahay (2007)	Hildebrand and Cable (1938); Lipson and Moran (1974); Fritzsche (1978); Balbontín and Pérez (1979); Stevens and Moser (1982); Ninos (1984); Peters (1985); Brogan (1992); Watson (1996f); Able and Fahay (1998); Herrera (2000); Beltrán-León and Herrera (2000); Ditty *et al.* (2005, 2006); Fahay (2007)
	Istiblennius	Mito (1954)	Mito (1954); Miller *et al.* (1979); Matarese *et al.* (1984); Kojima and Shiogaki (1988); Springer and Williams (1994)
	Laiphognathus		Kubo and Sasaki (2000a)
	Lipophrys	M'Intosh (1905); Hefford (1910); Lebour (1927); Eggert (1932); Cipria (1936); Qasim (1956); Padoa(1956a); Fives (1970); Faria *et al.* (2002, 2005)	Garstang (1900); Hefford (1910); Ford (1922); Lebour (1927); Cipria (1936); Padoa (1956a); Qasim (1956); Fives (1970); Faria *et al.* (2002, 2005)
	Lupinoblennius	Ditty *et al.* (2006)	Peters (1985); Ditty *et al.* (2006)
	Meiacanthus	Fishelson (1976); Tellock and Alig (1998); Suzuki and Nakayama (2005); Suzuki *et al.* (2006)	Fishelson (1976); Kojima (1988); Tellock and Alig (1998); Suzuki and Nakayama (2005); Suzuki *et al.* (2006)
	Microliopophrys		Sabates (1994)
	Omobranchus	Thomson and Bennett (1953); Visweswara Rao (1970); Dotsu and Oota (1973)	Seale (1935); Thomson and Bennett (1953); Mito (1966); Visweswara Rao (1970); Dotsu and Oota (1973); Houde *et al.* (1986); Kojima (1988); Sufei and Weibing (1989); Watson and Miskiewicz (1998); Kawaguchi *et al.* (1999)
	Ophioblennius	Ré and Almeida (1981); Labell and Nursall (1985); Ditty *et al.* (2006)	Ré and Almeida (1981); Labell and Nursall (1985); Olivar and Fortuño (1991); Brogan (1992); Watson (1996f); Ditty *et al.* (2006)
	Parablennius	Lebour (1927); Cipria (1934); Padoa (1956a); Fives (1970); Kalinina (1976); Russell (1976); Dotsu and Moriuchi (1980); Chesalina (1998)	Ford (1922); Lebour (1927); Cipria (1934, 1938); Padoa (1956a); Fives (1970); Kalinina (1976); Russell (1976); Dotsu and Moriuchi (1980); Watson (1983, 2000); Olivar (1986); Kojima and Shiogaki (1988); Olivar and Fortuño (1991); Chesalina (1998); Watson and Miskiewicz (1998); Ditty *et al.* (2005, 2006); Fahay (2007)

Table 4.4.1 Contd.

Table 4.4.1 Contd.

	Paralipophrys		Padoa (1956a)
	Petroscirtes	Munro (1955); Dotsu (1982)	Munro (1955); Mito (1966); Dotsu (1982); Watson (1983, 2000); Kojima and Shiogaki (1988); Watson and Miskiewicz (1998);
	Plagiotremus		Watson (1983, 1996f, 2000); Kojima and Shiogaki (1988); Beltrán-León and Herrera (2000)
	Salaria	Padoa (1956a); Wickler (1957); Fishelson (1963); Kalinina (1976); Patzner (1984); Chesalina (1998)	Cipria (1936); Padoa (1956a); Wickler (1957); Fishelson (1963); Kalinina (1976); Chesalina (1998)
	Scartella	De Leo *et al.* (1976); Suzuki (2003)	De Leo *et al.* (1976); Kojima and Shiogaki (1988); Olivar and Fortuño (1991); Suzuki (2003); Ditty *et al.* (2005, 2006)
	Stanulus		Kojima and Shiogaki (1988)
	Xiphasia		Houde *et al.* (1986); Kojima and Shiogaki (1988)

Table 4.4.2 Size ranges of eggs, and of larvae at selected developmental stages, for the blennioid families. Size at hatching for Clinidae includes the size at birth for the clinins.

		Approximate size (mm BL) at:		
Family	*Egg diameter (mm)*	*Hatching*	*Notochord flexion*	*Settlement*
Tripterygiidae	0.9-1.4	3-6	4-9	10-26
Dactyloscopidae	0.9-1.2	≤ 4	4-8	10-20
Labrisomidae	0.6-1.5	2.5-5	5-10	10-25
Clinidae	1.2-1.4	5-14	6-10	15-45
Chaenopsidae	0.9-1.5	3.5-6.5	6-12	15-30
Blenniidae	0.4-1.6	1.5-5.5	4-11	6.5-66

at hatching (dactyloscopids, many blenniids) or coils before settlement (all others currently known except *Stathmonotus*). The gas bladder initially is anterior but shifts to the posterior with development, and usually is prominent, except in dactyloscopids and blenniids where it is often inconspicuous and no longer visible after the flexion stage.

Spines are lacking on the pectoral girdle in all larval blennioids, and on the head in most varieties except blenniids. Some dactyloscopids and labrisomids have a few small preopercular spines, while blenniids often have large and/or numerous preopercular spines and, less commonly, small supraocular spines or serrations.

Table 4.4.3 Selected morphometric characters of blennioid larvae. Values given for body depth, pre-anal length, and head length are percentages of body length; values for eye diameter and snout length are percentages of head length; Pr = pre-flexion stage, Po = post-flexion stage.

Family	*Body depth*	*Pre-anal length*	*Head length*	*Eye diameter*	*Snout length*
Tripterygiidae	12-22	30-55	15-27	Pr 38-45 Po 25-35	Pr 16-24 Po 22-29
Dactyloscopidae	10-20	25-35	15-25	Pr 35-55 Po 20-30	Pr 20-30 Po 25-30
Labrisomidae	10-20	35-45	18-27	Pr 30-50 Po 25-35	Pr 10-30 Po 20-35
Clinindae	10-20	40-50	17-27	Pr 30-45 Po 20-35	Pr 10-20 Po 20-30
Chaenopsidae	10-16	25-53	18-25	Pr 30-50 Po 25-37	Pr 10-15 Po 20-30
Blenniidae	5-40	25-57	10-35	Pr 30-50 Po 25-45	Pr 15-30 Po 10-30

The caudal-fin rays are the first fin rays to begin forming, usually late in the pre-flexion stage, except in blenniids where the pectoral-fin rays often begin to form before, or concurrently with the caudal-fin rays. In tripterygiids, clinids, chaenopsids, most labrisomids, and also many blenniids, the soft rays of the dorsal and anal fins form next, usually beginning near the flexion-post-flexion stage transition. In some blenniids, the pelvic-fin rays form simultaneously with the dorsal- and anal-fin soft rays, and in dactyloscopids both pectoral- and pelvic-fin rays form before the dorsal and anal soft rays. Spines of the dorsal and anal fins typically form late, before the pelvic-fin rays in labrisomids, clinids, and some blenniids, or after the pelvic rays in tripterygiids, dactyloscopids, chaenopsids, and some blenniids.

Melanistic pigment in blennioid larvae ranges from being nearly absent to extensive, but commonly is light (Table 4.4.4). Melanophores typically form dorsally, at least on the head during the pre-flexion or flexion stage in blenniids and during the post-flexion stage in most others. Pigment is present dorsally on the gas bladder and gut; on the gut, this is typically only on the hindgut, at least initially, except in blenniids where it is more extensive. Ventral pigmentation on the isthmus and gut is typically sparse or absent; in many species, melanophores form a 'V'-shape pattern, with the apex near the cleithral symphysis.

Most blennioid larvae are pigmented on the ventral margin of the tail, usually with a single row of melanophores along much of its length, but

Table 4.4.4 Main areas of melanistic pigmentation in larval blennioid fishes, pre-flexion through late post-flexion stages (does not include transformation-settlement stages when pigmentation typically increases substantially, or fins, which typically are unpigmented in all except Blenniidae). Abbreviations are: Pr = pre-flexion stage; Fl = notochord flexion stage; Po = post-flexion stage; melano = melanophore(s); A = anal fin; D = dorsal fin.

Family	*General appearance*	*Head*			*Gut area*		*Trunk and tail*			
		Dorsal	*Jaws*	*Opercular area*	*Dorsal*	*Ventral*	*Dorsal*	*Lateral*	*Ventral*	*Hypural margin*
Tripterygiidae	Lightly to heavily pigmented, usually light	Usually none initially, on mid- and/or hindbrain areas in Po	Usually none	Usually none; some present in some species	On gas bladder and hindgut	Usually none initially, at cleithral symphysis starting late Pr to Po	None to heavy; commonly few melano, on margin posteriorly	Internal over notochord in some, beginning during Fl	Single row on margin of tail; "Y"-shaped melano at fin-ray bases	Present during Po in some
Dactyloscopidae	Lightly pigmented	Usually none; a little in some species	Usually none	Usually none	On gas bladder and most or all of gut	Usually present on gut and at cleithral symphysis	None to 1-2 melano on margin posteriorly, may increase	None	Single row on margin of tail; "Y"-shaped melano at fin-ray bases	Present only during Po in some
Labrisomidae	Lightly pigmented	Usually none before Po	Usually none	Usually none	On gas bladder and hindgut; on foregut in some	Usually present on gut and at cleithral symphysis	Usually none; at D ray bases in some species during Po	Internal over notochord in some, beginning during Po	Usually single row on margin of tail; "Y"-shaped melano at fin-ray bases	Commonly none; present in some by Fl or Po

Table 4.4.4 Contd.

Table 4.4.4 Contd.

Clinidae	Lightly pigmented through Fl	Usually none initially, on mid- and hindbrain in Po	Usually none	Usually none	On gas bladder and hindgut	Usually present, at least at cleithral symphysis	Usually none initially, some on margin posteriorly in Po	Internal over notochord in some, beginning during Po	Usually single row on margin of tail; "Y"-shaped melano at fin-ray bases	Little or none
Chaenopsidae	Lightly pigmented	Commonly none before Po	Usually none before Po	Commonly none before Po	On gas bladder and hindgut	Usually present near cleithral symphysis and on hindgut; on midgut in some species	On margin in some species, beginning posteriorly in Fl or Po	Internal over notochord in some, beginning during Fl or Po	Usually single row on margin of tail; "Y"-shaped melano at fin-ray bases	Little or none
Blenniidae	Lightly to heavily pigmented, usually light	Usually some, light to heavy	Often some in Fl or Po	Often some in Fl or Po	On gas bladder and most or all of gut	Usually some, at least at hindgut, often elsewhere	Usually none before late Po; commonly light to heavy as early as Pr in Nemophini	Internal over notochord in some, beginning during Fl or Po; commonly external as early as Pr in Nemophini	Usually single row on margin tail; "Y"-shaped melano at fin-ray bases	Commonly some by Po

initially on only a few myomeres in some species. This pigment increases or decreases in some species, and may be lost in some blenniids. As the anal fin develops, the melanophores at the fin-ray bases typically acquire a characteristic 'Y' shape with the base of the 'Y' extending along the ray.

Pigmentation dorsally and laterally on the trunk and tail typically is sparse or absent before the post-flexion stage. In some tripterygiids and nemophin blenniids, the dorsum is heavily pigmented and, in some nemophins, moderate lateral pigmentation is present as early as the pre-flexion stage.

The fins are typically unpigmented until late in the post-flexion stage, except that pectoral-fin pigmentation is common in blenniids other than nemophins. Many blennioid species develop melanophores along the hypural margin in the flexion stage, or, more commonly, the post-flexion stage.

During transformation and settlement, blennioids typically become more heavily pigmented, beginning at the head and spreading to the posterior. Barred and mottled patterns commonly form by, or soon after settlement.

Descriptions of larval development within each family follow. These generalised descriptions of patterns common at the level of family or genus do not include all species-level variations. Descriptions of pigmentation refer only to melanistic pigment. The nomenclature of developmental stages (pre-flexion, flexion, post-flexion, transformation) and body parts follow definitions in Moser (1996) and Leis and Carson-Ewart (2000). Larval length is body length (BL = notochord length prior to completion of notochord flexion, and standard length thereafter), and in the following descriptions the 'tail' refers to the post-anal region of the body, as defined by Moser (1996).

TRIPTERYGIIDAE

The planktonic larvae occur primarily in shallow, near-shore waters (Brogan, 1994). Egg and larval stages have been described for just over 30% of the 31 genera (Table 4.4.1) but fewer than 10% of the approximately 166 species.

Tripterygiid eggs are nearly spherical to somewhat flattened, ca. 0.9-1.4 mm in diameter, with homogeneous, colorless to red-orange yolk containing small pigment inclusions and 1-100 or more colorless to orange oil globules. The eggs are attached to the nest by adhesive filaments that arise more or less evenly around the chorion, or from its lower half, or in groups from several loci, or in a tuft from the basal pole.

Following an incubation of about 2-3 weeks (ca. 12-18°C: Shiogaki and Dotsu, 1973; Ruck, 1973, 1980), the larvae hatch at about 3-6 mm with pigmented eyes, functional mouth, and a small yolk sac (Fig. 4.4.1). Notochord flexion typically occurs within about the 4-9 mm range, and larvae transform and settle at about 10-26 mm. Larvae are moderately elongate (body depth usually < 20% BL) and somewhat compressed. At hatching, the head is rounded, with a blunt snout and large eyes. The snout becomes more acute early in the pre-flexion stage and the eyes become relatively smaller by the post-flexion stage (Table 4.4.3). The short gut (pre-anal length commonly ~40-50% BL) coils during notochord flexion. The gas bladder initially is located anteriorly, but shifts to mid-gut during the pre-flexion stage and to the posterior part of the abdominal cavity during the post-flexion stage. Larvae have about 30-43 myomeres, depending on species; commonly there are 10-12 pre-anal myomeres.

Caudal-fin rays begin forming late in the pre-flexion stage or early in the flexion stage, followed by the soft rays of the dorsal and anal fins late in the flexion stage to early post-flexion stage. Pectoral-fin rays begin to form simultaneously with, or just after, the first dorsal- and anal-fin rays, followed by the pelvic-fin rays and, finally, the dorsal- and anal-fin spines during the post-flexion stage. Larval tripterygiids have no spines on the head or pectoral girdle, but may have ocular and nasal cirri that begin to form shortly before settlement.

Larval pigmentation is sparse to heavy, but commonly is sparse before transformation (Table 4.4.4). The typical pattern includes melanophores dorsally on the gas bladder and hindgut, in a single row on the ventral margin of the tail (commonly one melanophore per myomere), and in a single row on the dorsal margin of the tail (most often one to a few melanophores posteriorly). Some species (and some individuals of others) lack dorsal pigment on the trunk and tail through at least the early part of the post-flexion stage, and at least one species is heavily pigmented on much of the dorsum (Ruck, 1980). When the anal-fin rays form, a Y-shaped melanophore each is commonly present at the bases of several to most soft rays. Melanophores form internally at the nape and posterior part of the notochord in some species beginning late in the pre-flexion stage or during notochord flexion. A few melanophores may form over the

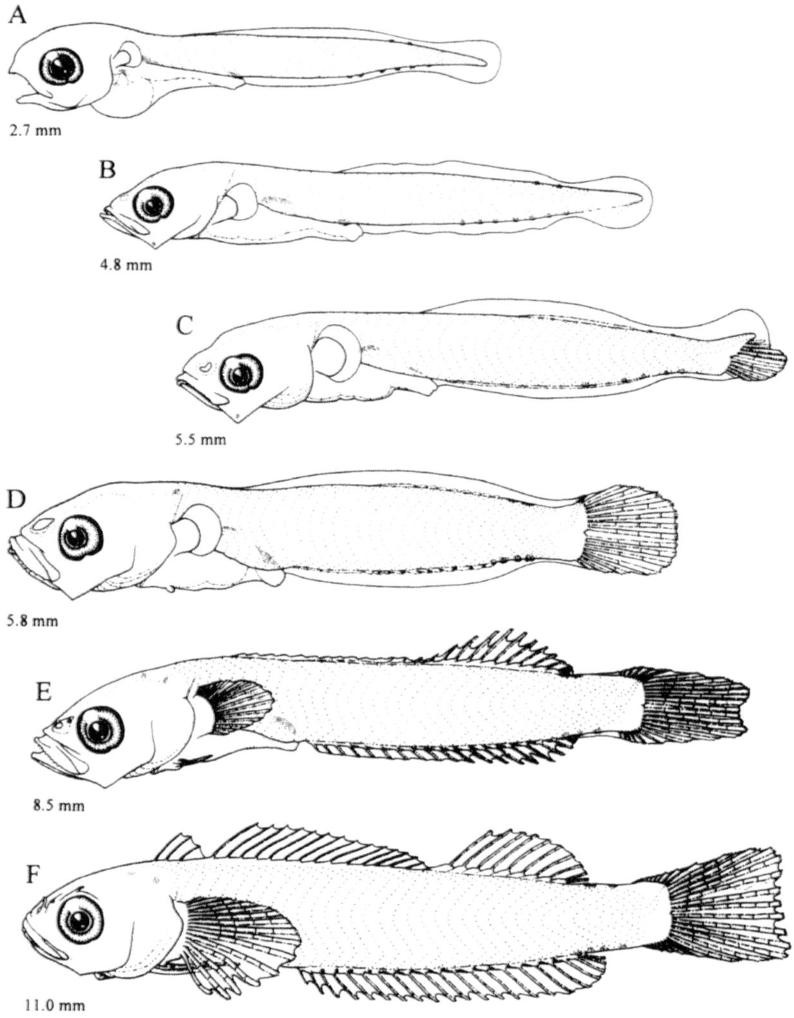

Fig. 4.4.1 Larvae of the Hawaiian tripterygiid, *Enneapterygius atriceps*. A: 2.7 mm, recently hatched (Kaneohe Bay, Oahu, August 1971); B: 4.8 mm, pre-flexion stage (Kahe Point, Oahu, July 1974); C: 5.5 mm, flexion stage (Kahe Point, Oahu, July 1974); D: 5.8 mm, post-flexion stage (Kaneohe Bay, Oahu, August 1971); E: 8.5 mm, post-flexion stage (Kaneohe Bay, Oahu, August 1971); F: 11.0 mm, late post-flexion stage (Kahe Point, Oahu, March 1974).

midbrain and on the hindbrain during the post-flexion stage. A V-shaped (in ventral view) melanophore commonly forms at the cleithral symphysis by the post-flexion stage. Some species have melanophores on the hypural margin in the post-flexion stage.

DACTYLOSCOPIDAE

The planktonic larvae occur primarily in shallow coastal waters (Brogan, 1994). Some information is available for ovarian eggs of about a dozen species representing six of the nine genera, and larvae have been described for seven of the 43 species, representing four genera (Table 4.4.1).

Nearly mature ovarian eggs are spherical, about 0.9-1.2 mm in diameter, and have a granular yolk, which may lack oil globules. In at least the egg-carrying genera (*Dactylagnus*, *Dactyloscopus*, *Myxodagnus*), small conical projections are scattered over the surface of the chorion and a cluster of long filaments is attached at the basal pole (Watson, 1996b). The filaments intertwine to attach the eggs to one another.

Larvae hatch at a length of 4 mm or less, with pigmented eyes and a functional mouth. Notochord flexion occurs within about the 4-8 mm range and larvae transform and settle at about 10-20 mm. Larvae are elongate throughout development, with a relatively small head, short snout, and initially large eyes (Table 4.4.3). The gut is coiled and short (Fig. 4.4.2) and the small, usually inconspicuous gas bladder is located anteriorly. The eyes become relatively smaller (Table 4.4.3), and shift from a lateral to a dorsolateral position late in the post-flexion stage. In *Myxodagnus*, the snout becomes acute as the lower jaw projection develops in the post-flexion stage. Larvae have about 33-53 myomeres, depending on species; there are usually 6-10 pre-anal myomeres.

Caudal-fin rays begin to form late in the pre-flexion stage, followed by pectoral- and then pelvic-fin rays during notochord flexion. Soft rays of the dorsal and anal fins apparently form simultaneously early in the post-flexion stage, followed by the fin spines. Most dactyloscopid larvae lack spines on the head and the pectoral girdle, but at least one *Gillellus* species has a few small preopercular spines from late pre-flexion to post-flexion stage (Fig. 4.4.2). The characteristic opercular fimbriae begin to form late in the post-flexion stage and the interopercular and labial fimbriae form during transformation.

Larval pigmentation is light (Table 4.4.4). The typical pattern initially includes melanophores dorsally on the gas bladder, dorsally and ventrally on the gut, one or two at the cleithral symphysis or on the isthmus near the symphysis, a single row along the ventral margin of the tail, and in some species, a melanophore or two posteriorly on the dorsal margin of the tail and/or dorsally on the head. Some species also have internal pigment

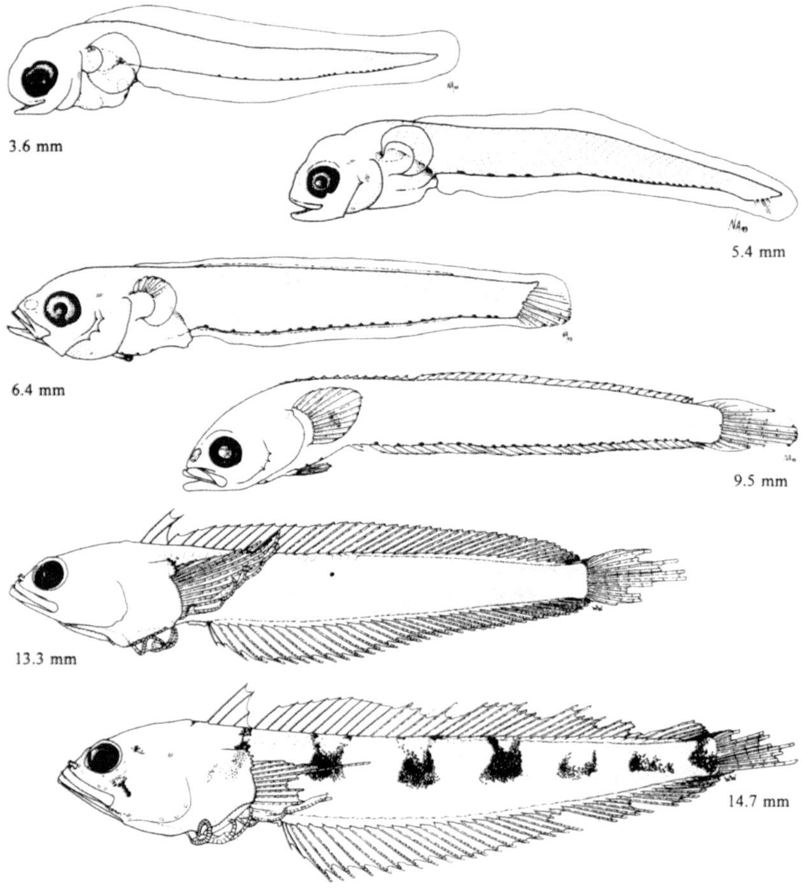

Fig. 4.4.2 Larvae of the dactyloscopid *Gillellus* sp. (probably *G. semicinctus*). Illustrations from Watson (1996b).

near the anterior margin of the liver. Pigmentation changes relatively little in larvae. When anal-fin rays form, a Y-shaped melanophore each is present at the bases of most or all the soft rays. In *Myxodagnus*, the number of melanophores dorsally on the tail increases to extend nearly the full length of the body by the post-flexion stage, and *Dactyloscopus* also may develop dorsal melanophores posteriorly in the post-flexion stage (Brogan, 1992; Watson, 1996b). *Myxodagnus* develops melanophores on the hypural margin in the flexion stage.

LABRISOMIDAE

The planktonic larvae occur primarily in shallow, near-shore waters (Brogan, 1994). Eggs have been described for four species, and larvae for 13 of the approximately 106 species; however, the 13 species represent just over half of the 14 genera (Table 4.4.1).

Labrisomid eggs are slightly elliptical to spherical, ca. 0.6-1.5 mm in diameter, with a granular yolk containing one or more small oil globules. In at least one species, the yolk and oil globules are amber colored (Breder, 1939). Melanophores form on the yolk and embryo during the latter part of embryonic development. Egg attachment is via long, adhesive filaments that arise from the basal pole. *Dialommus fuscus* may also have a network of smaller filaments surrounding the chorion (Howe, 1991), and *Malacoctenus* eggs have two filament clusters at opposing poles (Brogan, pers. comm. in Watson, 1996c). In at least one *Paraclinus* species and *D. fuscus*, the eggs resemble dactyloscopid eggs in the sense that small, triangular or spike-like projections are scattered on the chorion (Fig. 4.4.3; Howe, 1991; Watson, 1996c).

Larvae of the oviparous labrisomids hatch at about 2.5-4 mm with pigmented eyes, a functional mouth, and a small to moderate yolk sac. Starksiin larvae are unknown but are presumably at least as well developed; *Xenomedea* embryos reach at least 5 mm before birth (Rosenblatt and Taylor, 1971). Notochord flexion occurs within about the 5-10 mm range and larvae transform and settle at about 10-25 mm.

Larvae are compressed and elongate (body depth commonly < 15% BL), with a short gut (pre-anal length commonly ~ 40% BL) that coils during the preflexion stage (Fig. 4.4.3). The gas bladder shifts from the anterior half of the abdominal cavity before notochord flexion to the posterior half after flexion. At hatching, the head is rounded, with a short, blunt snout that lengthens and becomes more acute during the pre-flexion stage (Table 4.4.3). The larvae usually have about 8-13 preanal myomeres and, depending on species, 31-49 total myomeres.

Caudal-fin rays begin to form near the beginning of notochord flexion, followed early in the post-flexion stage by soft rays of the dorsal and anal fins. The middle rays apparently form first in the dorsal and anal fins. In at least one *Paraclinus* species, pectoral-fin rays begin to form before the dorsal- and anal-fin rays (Watson, 1996c). Dorsal- and anal-fin spines and pelvic-fin rays typically are the last to develop; the pelvic rays form just

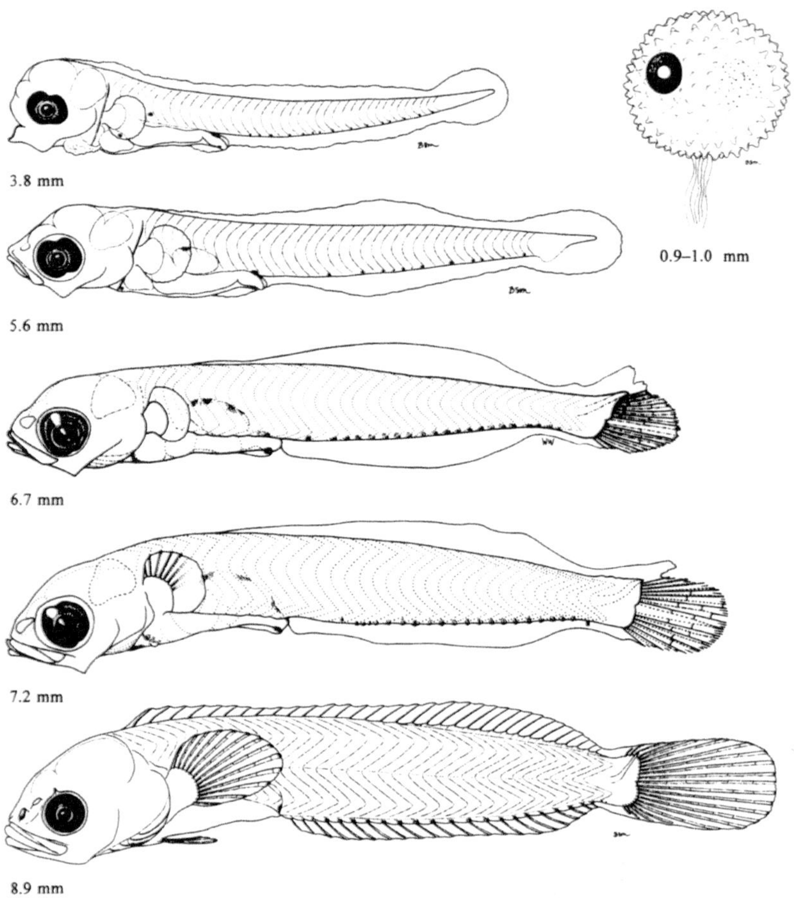

Fig. 4.4.3 Egg and larvae of the labrisomid *Paraclinus integripinnis*. Illustrations from Watson (1996c).

before to just after the dorsal and anal spines, depending on species. Dorsal-fin spines are apparently added from posterior to anterior; most may form about simultaneously in *Paraclinus*. Most labrisomids lack spines on the head and pectoral girdle, but some *Labrisomus* and *Malacoctenus* species have a few, small spines on the preopercular margin from the preflexion or flexion stage until the latter part of the post-flexion stage (Brogan, 1992; Watson, 1996c; Herrera and Lavenberg, 1999). Nasal, ocular, and nucchal cirri form late in the post-flexion stage.

The pigmentation is typically light, with melanophores largely limited to the gas bladder, gut, and ventral margin of the tail before mid- to late post-flexion stage (Table 4.4.4). Pigment commonly is present anteriorly or anteroventrally on or adjacent to the liver. Ventral melanophores on the gut may form a 'V' pattern anteriorly, but posteriorly they are typically on or near the longitudinal midline. Later in the post-flexion stage, only the pigment near the pelvic-fin base usually remains, except in *Labrisomus* and *Malacoctenus*, which often retain much of the ventral pigment.

Ventral pigmentation on the tail commonly consists of a melanophore per myomere along most of its length, with the melanophores at the fin-ray bases usually Y-shaped. In *Labrisomus*, the ventral series is sparser through early post-flexion stage and one or two melanophores are distinctly larger; *L. xanti* has a single large melanophore through the flexion stage (Brogan, 1992; Watson, 1996c). Larval *Labrisomus* develop the Y-shape melanophore at each anal-fin ray base during the post-flexion stage.

Dorsal pigmentation on the trunk and the tail may be present at hatching in some species, but commonly it is absent until late in the post-flexion stage or transformation stage when a melanophore each forms at the bases of some or all of the dorsal-fin rays. Caudal pigment typically is lacking until late in the post-flexion stage, except in *Labrisomus*, where it is commonly present through much or all of the larval period.

Pigmentation may be present on the midbrain and/or nape areas at hatching in some species, but more commonly is absent before the post-flexion stage. Internal pigment may form in the otic capsule in some species. Pigmentation increases on the head late in the post-flexion stage, but remains largely dorsal until settlement, when it rapidly increases over the body and fins, commonly forming a barred or mottled pattern in early juveniles.

CLINIDAE

Eggs have been described for only one (~ 10%), and larvae for four (~ 40%) myxodin species, representing three of the five myxodin genera (Table 4.4.1). Larvae have not been described for any of the ophiclinins, and for only one clinin species (~ 2%), although clinin reproduction has been well studied (e.g., Gunn and Thresher, 1991; Prochazka and Griffiths, 1992) and the larvae may be relatively common in the plankton near shore (Neira and Sporcic, 2002). Myxodin larvae, likewise, occur primarily near shore (Barnett *et al.*, 1984).

Eggs of *Heterostichus rostratus* are spherical, 1.2-1.4 mm in diameter, with a single oil globule ~ 0.3 mm in diameter, and with pink to purple, brown, or greenish yolk (Fig. 4.4.4). Feder *et al.* (1974) noted that *Gibbonsia* eggs are white. Egg attachment in *H. rostratus* is via long filaments that arise in a cluster from the basal pole of each egg.

Larval myxodins hatch at about 5-6 mm (incubation period ~ 2 weeks in *H. rostratus*: Stepien, 1986), undergo notochord flexion at about 6-10 mm, and settle at about 20-45 mm (~ 8 weeks after hatching in *H. rostratus*). Clinins are about 7-14 mm at birth, having already completed notochord flexion, and settle at about 15-20 mm about 3-7 weeks later (Gunn and Thresher, 1991).

Development through notochord flexion is similar in clinins and myxodins. Larvae are compressed and elongate (body depth commonly ≤ 15% BL), with preanal length commonly ~ 45% BL (Table 4.4.3; Fig. 4.4.4). Myxodins hatch with a small to moderate yolk sac, which is absorbed within the first day or two (Stepien, 1986); clinins complete the yolk sac stage before birth (Gunn and Thresher, 1991). The myxodin gut coils during the post-flexion stage; in clinins it coils before birth. The gas bladder initially is anterior, but shifts to near the terminus of the gut by the post-flexion stage in myxodins. At hatching, myxodins have a rounded head with a short, blunt snout that becomes more acute during the pre-flexion stage and gradually lengthens (Table 4.4.3). Relative eye diameter gradually decreases through the larval phase. Larvae have about 15-20 pre-anal and 38-75 total myomeres, depending on species (Clinini and Myxodini).

Caudal-fin rays begin forming near the beginning of notochord flexion. Soft rays of the dorsal and anal fins form next, beginning about simultaneously near the end of notochord flexion. Addition of dorsal rays is anteriorly from the posterior-most rays. In *Gibbonsia* and *Heterostichus* the posterior anal-fin rays also form first, with rays initially added in an anterior direction, but then the anterior-most rays form, with subsequent additions progressing toward the middle from both sites. Pectoral-fin rays form next, with rays added ventrad from the upper-most rays. Dorsal- and anal-fin spines follow; in *Gibbonsia* and *Heterostichus* the dorsal spines are initially added anteriorly from the posterior-most spines, with the anterior-most spines forming before completion of the fin and the final spines added from both anterior and posterior. Pelvic-fin rays form last. Clinids lack spines on the head and pectoral girdle.

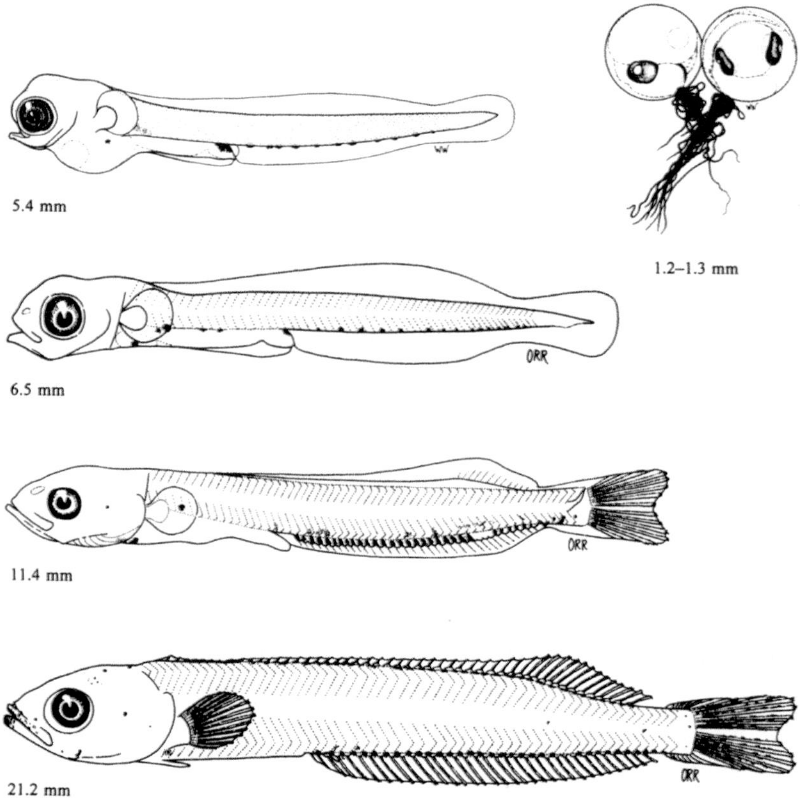

Fig. 4.4.4 Eggs and larvae of the myxodin clinid *Heterostichus rostratus*. Illustrations from Watson (1996d).

Clinid larvae are lightly pigmented through at least notochord flexion, with melanophores largely limited to the dorsum of the gas bladder and hindgut, the vicinity of the cleithral symphysis, and the ventral margin of the tail (Table 4.4.4). Ventral pigment on the tail may be sparse initially, but increases to about one melanophore per myomere or anal-fin ray base (with the typical blennioid Y-shape at the fin-ray bases) in the post-flexion stage. Melanophores may form internally on the mid- and/or hindbrain during the post-flexion stage, followed by pigment above the vertebral column and externally on the head (primarily dorsally) and dorsal margin (at least posteriorly). Melanophores may form on the hypural margin late in the post-flexion stage, and at settlement clinids become nearly completely pigmented, often with a barred pattern.

CHAENOPSIDAE

Larvae are planktonic in shallow, near-shore waters (Brogan, 1992, 1994). Although spawning is well known and nests are often observed, eggs have been described for fewer than 5% of the species, all in the genus *Neoclinus*. Descriptions are available for larvae of fewer than 20% of the 90 species, representing six of the 14 genera (Table 4.4.1).

Neoclinus eggs are slightly flattened, oval to nearly spherical, 0.9-1.5 mm in diameter, with a pale yellow to orange, granular yolk and one to several yellowish to orange oil globules (commonly one, ~ 0.2-0.3 mm, plus a few scattered globules < 0.1 mm). Attachment of chaenopsid eggs is via a cluster of long, adhesive filaments, or an adhesive pad (Brogan, pers. com. in Watson, 1996e). Melanophores form on the embryo, and may form on the yolk adjacent to the embryo, during the latter part of embryonic development. The chaenopsid incubation period ranges from just under one week (Hastings, 1988) to over one month (Shiogaki and Dotsu, 1972).

Larvae hatch at about 3.5-6.5 mm with pigmented eyes, a functional mouth, and a small yolk sac. The notochord flexes in about the 6-12 mm range, and larvae settle at about 15-30 mm. Larvae are compressed and elongate (body depth commonly ~ 12-14% BL). The gut coils during the flexion or post-flexion stage, or does not coil in *Stathmonotus* (Brogan, pers. com. in Watson, 1996e). Pre-anal length ranges from short in *Neoclinus* (~ 25-40% BL; Fig. 4.4.5) to moderate in *Stathmonotus* (~ 50-53% BL). The gas bladder, often prominent through the beginning of the post-flexion stage, shifts posteriorly with development. At hatching, the head is rounded, with a short, blunt snout that lengthens and becomes more acute by the flexion stage. The eyes initially are large but become relatively smaller (Table 4.4.3). Larvae have about 9-25 pre-anal (~ 11-15 in most species) and about 35-60 total myomeres, depending on the species.

The first caudal-fin rays form near the beginning of notochord flexion, and the first dorsal- and anal-fin soft rays (the middle rays, at least in *Neoclinus*) follow, near the end of flexion. Pectoral-fin rays begin to form at the same time or soon after the dorsal and anal rays. The first pelvic-fin rays and posteriormost dorsal-fin spine(s) form next, with dorsal spines sequentially added anteriorly. Chaenopsids lack spines on the head and pectoral girdle, but most develop nasal and ocular cirri late in the post-flexion stage.

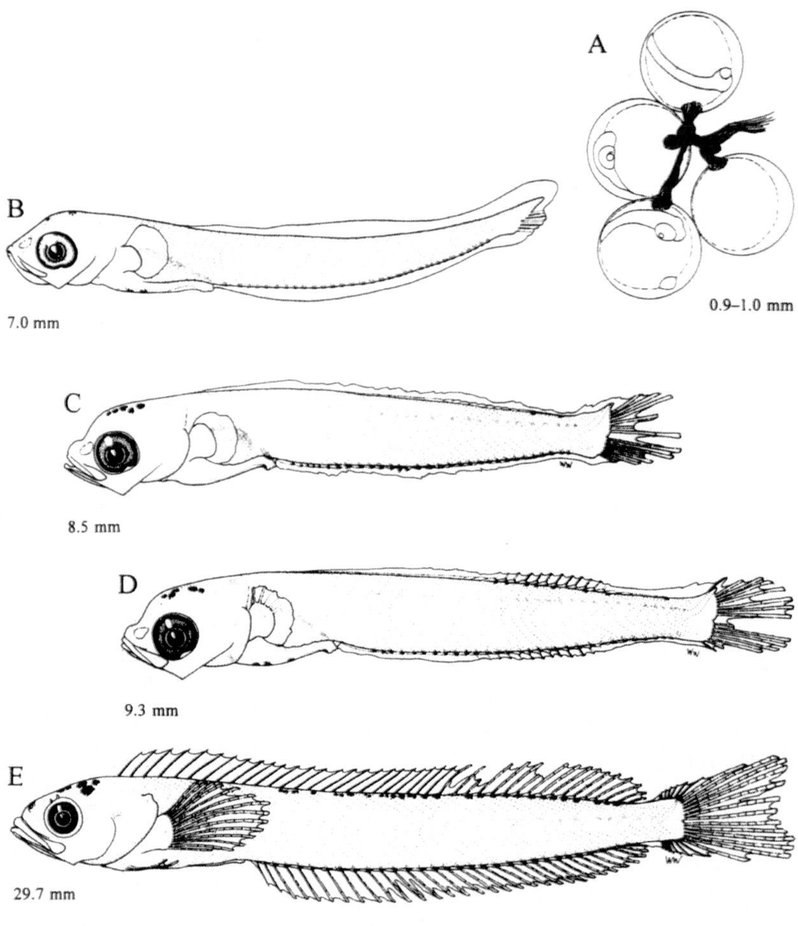

Fig. 4.4.5 Eggs and larvae of the chaenopsid *Neoclinus blanchardi*. Illustrations A, C-E from Watson (1996e); B 7.0 mm, late pre-flexion stage (CalCOFI 9210 NH, station 83.3.42, October 1992).

Chaenopsid larvae are lightly pigmented, with melanophores primarily on the gas bladder, dorsally and ventrally on the gut, and on the ventral margin of the tail (Table 4.4.4). Some species also have pigment in one or more of the following locations, at least by the post-flexion stage: dorsally on the head and body margin, internally over the notochord/vertebral column, and in some interneural and interhaemal spaces. Ventral pigment on the gut typically includes melanophores on the longitudinal midline of the hindgut and often at mid-gut, and ventrolaterally near the liver which may coalesce in the pre-flexion stage to form a V-shaped pattern, then condense to a streak between the pelvic-fin bases in the post-flexion stage. Pigment on the ventral margin of the tail typically consists of a melanophore or two (usually one) at most post-anal myomeres or at the bases of most or all of the anal-fin rays (typically one per ray base, usually Y-shaped).

External pigment may form dorsally on the head during the pre-flexion stage, at least in some *Neoclinus* species, but commonly is absent before the post-flexion stage. A melanophore or two may form internally over the hindbrain during the pre-flexion stage, and in the otic capsule. In some species, melanophores form on the jaws and opercular area late in the post-flexion stage.

Melanophores form on the dorsal margin in some species beginning late in the flexion stage or during the post-flexion stage, and may extend most of the length of the body by late in the post-flexion stage, with a melanophore or two at or adjacent to the bases of a few to most dorsal-fin rays. In *Neoclinus*, internal melanophores form posteriorly over the notochord during the flexion stage and spread anteriorly to mid-tail or beyond during the post-flexion stage. Pigment may be present in a few middle interneural and interhaemal spaces and some may be present proximally on the caudal fin in post-flexion stage *Stathmonotus hemphilli* (Cavalluzzi, 1997). Other chaenopsid larvae lack fin pigmentation apart from the melanophores at the bases of the anal- and dorsal-fin (in some species) rays.

BLENNIIDAE

Larvae are planktonic and most varieties occur in shallow coastal waters; pelagic pre-juveniles (ophioblennius stage) of some salariin species (Williams, 1990; Springer and Williams, 1994) may range far offshore (e.g., Leis and Miller, 1976; Clarke, 1993; McGowen, 1993; Brogan, 1994; Riginos and Victor, 2001). Blenniid eggs and larvae are the best known among blennioid fishes, with some early life history stage(s) described for more species than for all other blennioids combined. Nevertheless, these represent just under half the 57 blenniid genera (Table 4.4.1) and fewer than 25% of the 385 known species.

Blenniid eggs are nearly spherical to hemispherical, about 0.4-1.6 mm in diameter (commonly, vertical axis ~ 0.5-0.8 mm, horizontal axis ~ 0.8-1.0 mm). The yolk is usually granular, with none to many small oil globules (usually $\geq$ 1). Oil globules range from about 0.2-0.3 mm in diameter when a single globule is present, to < 0.1 mm when there are many. The yolk is commonly pale yellow to bright red-orange, but may be pink, purple, or brown in some species and commonly fades or becomes more brownish during development. Some species have orange and/or violet yolk inclusions, which fade with development. When pigmented, oil globules are commonly yellow. In late embryos, the eyes become pigmented and larval pigment patterns form, including melanophores on the gut and ventral margin of the tail, often dorsally on the head, and on the pectoral fins in some species. Eggs are attached via a filamentous, adhesive pad or pedestal (Fig. 4.4.6). The incubation period is about 5-21 days (commonly ≤ 10 days at temperatures > 20°C).

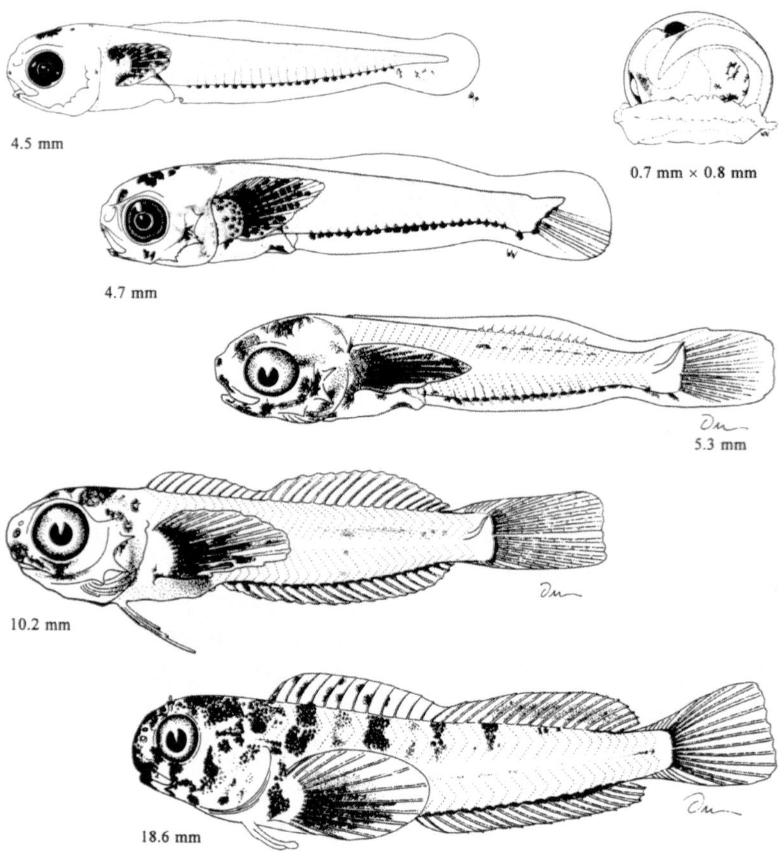

Fig. 4.4.6 Egg and larvae of the parablenniin blenniid *Hypsoblennius jenkinsi*. Illustrations from Watson (1996f).

Larvae hatch at about 1.5-5.5 mm (commonly 2.5-5 mm) and undergo notochord flexion at about 4-11 mm, beginning about 1-3 weeks after hatching. They settle at about 10-20 mm about 3-7 weeks after hatching, except for the ophioblennius stage, when the individuals are more typically > 20 mm at settlement (up to at least 66 mm in *Ophioblennius*; Fig. 4.4.7).

At hatching, the larvae have a functional mouth in most species, a small to moderate yolk sac, and are slightly to moderately compressed and elongate (body depth commonly ≤ 20% BL). The gut is short and coiled (pre-anal length commonly ~ 35-45% BL) in most, but is initially longer (~45-55% BL) and uncoiled in some salariins (e.g., *Entomacrodus*, *Exallias*,

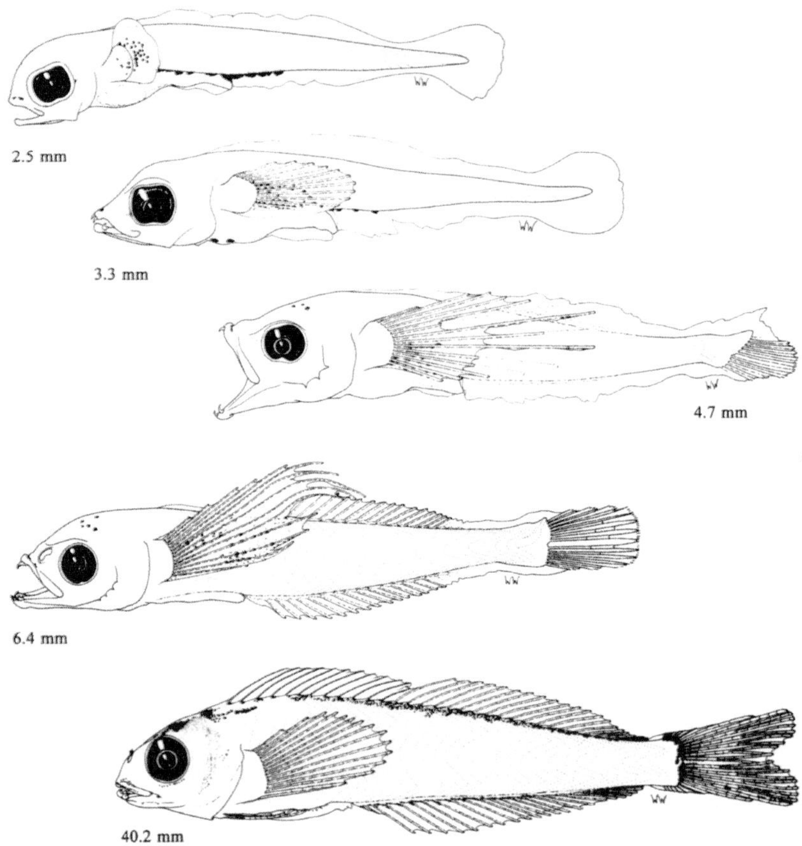

Fig. 4.4.7 Larvae and pelagic prejuvenile ("ophioblennius" stage) of the salariin blenniid *Ophioblennius steindachneri*. Illustrations from Watson (1996f).

Ophioblennius: Watson, 1996f, 2000). Later-stage larvae range from somewhat deep-bodied (e.g., *Blennius*, *Stanulus*) to very elongate (e.g., *Xiphasia*, some *Plagiotremus* species; Ford, 1922; Kojima and Shiogaki, 1988; Watson, 2000); most remain moderately elongate. The eyes are moderate to large (Table 4.4.3). The snout is initially short (typically ≤ 20% head length) and commonly remains so, but it may lengthen to nearly one-third of head length in some salariins. The gas bladder is small and anterior, but commonly is obscure and not visible after notochord flexion. Larval blenniids have about 8-12 pre-anal and about 28-135 total myomeres (commonly ~ 30-40).

The order of fin development is variable. Caudal- and/or pectoral-fin rays usually begin forming first, but in *Meiacanthus* pelvic-fin rays form first (Watson, 2000), during the pre-flexion stage. In many other blenniids pectoral-fin rays are first, beginning just before hatching in some omobranchins (e.g., some *Omobranchus*: Dotsu and Oota, 1973; Watson and Miskiewicz, 1998), but more commonly during the pre-flexion stage. In some nemophins (e.g., some *Petroscirtes* and *Plagiotremus*) and salariines (e.g., some *Parablennius*), the caudal-fin rays begin forming just before the pectoral-fin rays late in the pre-flexion stage (e.g., Watson and Miskiewicz, 1998; Watson, 2000). Soft rays of the dorsal and anal fins begin to form during notochord flexion: the anterior several rays form about simultaneously (Watson, 1987, 2001), or the middle rays form first (Ditty *et al.*, 2005). The dorsal- and then the anal-fin spines typically form after flexion. The posterior dorsal spines commonly form first, and in the anal fin either the second spine is first or both spines form simultaneously. Pelvic-fin rays often form last in the post-flexion stage; in some species (mainly parablenniins), they form during the flexion stage simultaneously with, or before, the dorsal and anal rays.

Spines are lacking on the head in larval nemophins and some salariins, but they are present on the preopercle in other blenniids, and on the supraocular ridge in some. Preopercular spines form early in the pre-flexion stage in omobranchins and many salariines (especially parablenniins). The first spine typically is at the angle of the posterior margin, usually with others added on the margin during the pre-flexion and/or flexion stages. In omobranchins, the spine at the angle commonly becomes long and broad, and is ornamented with small spinules in some species, while the other spines remain small, unornamented, and may disappear during the flexion or post-flexion stages (Fig. 4.4.8). In larval

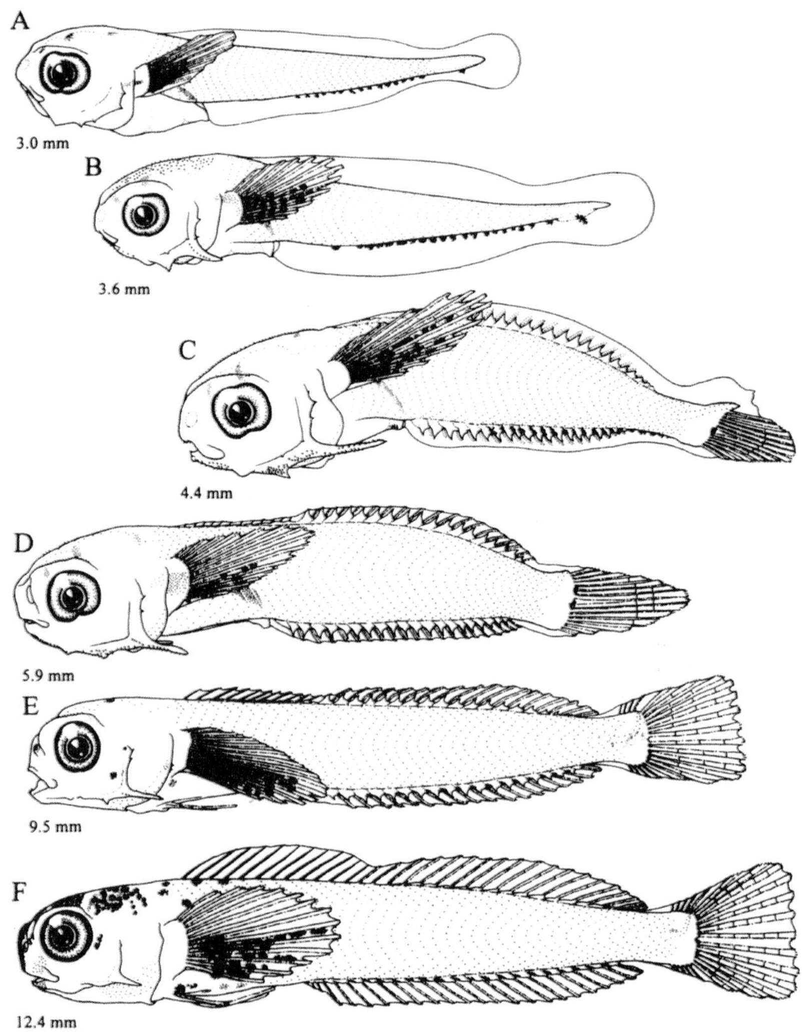

Fig. 4.4.8 Larvae of the omobranchin blenniid *Omobranchus rotundiceps obliquus*, from Kaneohe Bay, Oahu, Hawaii. A: 3.0 mm (January 1971); B: 3.6 mm (March 1971); C: 4.4 mm, specimen is slightly shrunken (March 1971); D: 5.9 mm (March 1971); E: 9.5 mm (October 1971); F: 12.4 mm (March 1967). Accessory spinules on the head are slightly exaggerated for clarity, and are not shown on the dorsum except along the dorsal edge in C-E.

parablenniins, the preopercular spines are often numerous, with the spine at the angle commonly longest but less enlarged than in the omobranchins (Fig. 4.4.6). Among the salariins, when the preopercular spines are present, more often all of them remain small (Fig. 4.4.7). The preopercular spines are overgrown or resorbed during the post-flexion stage, with no more than a remnant of the spine at the angle persisting until settlement. Small spines or serrations form on the supraocular ridge during the flexion or post-flexion stage in some omobranchins and salariines, but usually do not persist much beyond mid-post-flexion stage. Nasal and ocular cirri form late in the post-flexion stage, and in species having nuchal cirri, these form at settlement.

Larval pigmentation ranges from sparse to heavy but commonly it is light before late in the post-flexion stage (Table 4.4.4); some nemophins are heavily pigmented at hatching or soon thereafter (Fig. 4.4.9). Pigment is present dorsally on the gut, commonly is present dorsally on the head and ventrally on the tail, and often is present on the pectoral fins.

Pigmentation on the head typically is sparse or absent at hatching, but becomes moderate to heavy over at least the midbrain area during the

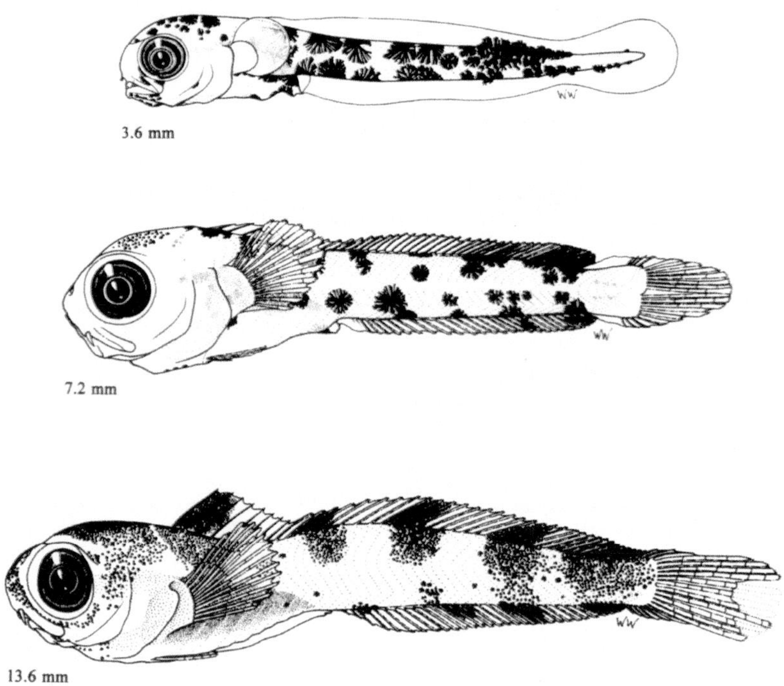

Fig. 4.4.9 Larvae of the nemophin blenniid *Plagiotremus azalea*. Illustrations from Watson (1996f).

pre-flexion or flexion stage in many species. Internal pigment may form at the anterior margin of the forebrain before hatching or during the pre-flexion stage, particularly among the omobranchins and parablenniins. Some *Plagiotremus* species are pigmented on the snout, jaws, and opercular area early in the preflexion stage (Fig. 4.4.9).

The dorsum of the gas bladder and gut apparently always are pigmented and sparse ventral pigment often is present. Pigment forms anteriorly on the gut during the pre-flexion stage in many species, and it spreads ventrolaterally from the dorsum to surround the gut, usually during the post-flexion stage but as early as the pre-flexion stage in some (primarily nemophins).

In most blenniids, melanophores are present on the ventral margin of the tail, commonly about one per myomere, especially in the omobranchins and parablenniins. In some species, these melanophores are initially only on the first few (e.g., *Exallias*, *Ophioblennius*) or last few (e.g., *Enchelyurus*, some *Entomacrodus* and *Istiblennius*) post-anal myomeres, or are absent entirely (e.g., some specimens of *Exallias*). In some blenniids, especially salariins, the ventral melanophores on the tail may disappear by the post-flexion stage. In the others, when the anal fin forms the melanophores associated with the fin-ray bases usually become Y-shaped.

Pigmentation dorsally and laterally on the trunk and tail is uncommon before late in the post-flexion stage, except in nemophins where it is commonly present earlier. Internal melanophores form over or below the notochord/vertebral column late in the flexion stage or during post-flexion stage in some species.

Fin pigmentation is rare before the latter part of the post-flexion stage, except that pectoral-fin pigment is common (absent in all nemophin larvae currently known). Pectoral pigment often is primarily near the distal margin of the fin in larval salariins, but more broadly distributed in blenniins, omobranchins, and parablenniins where it may cover the entire fin, but more commonly is largely limited to its lower 50-75%. In many species, pectoral pigment diminishes late in the post-flexion stage.

During transformation, the larval blenniids typically become heavily pigmented, beginning anteriorly and progressing posteriorly, with the body more or less entirely pigmented at, or soon after, settlement. Saddles, bars, and mottled patterns are common and may extend onto the dorsal and/or anal fins, or other striped or blotched patterns may form on the fins.

SKELETAL DEVELOPMENT

Skeletal development has been described for only about 1% of the blennioid species, representing three families: Tripterygiidae, Clinidae, and Blenniidae. Blennioid skeletal development generally follows a typical perciform pattern (Potthoff *et al.*, 1984, 1987; Watson and Walker, 1992), perhaps with somewhat more similarity between tripterygiids and clinids than between either of those and blenniids. Compared with blenniid larvae, ossification of the neurocranium is delayed in both tripterygiid and clinid larvae.

Ruck (1998) described skeletal development in the tripterygiid *Forsterygion varium*. At hatching, the larval skeleton consists of the chondrocranium, jaws, cartilaginous hyoid and branchial arches, cleithra, and presumably the notochord. The cleithra, and then the jaws are the first to begin ossifying, early in the pre-flexion stage.

The chondrocranium includes the auditory capsules, trabeculae cranii fused anteriorly to form the trabecula communis, the ethmoid plate, synotic tectum, epiphysial bar, lamina orbitonasalis and nasal septum, orbital cartilage, and occipital arch. The first bones of the neurocranium to begin ossification are the basioccipital and the parasphenotic, during the pre-flexion stage. The exoccipitals may begin to ossify during notochord flexion and the parietals and vomer in mid-post-flexion stage, but other bones of the neurocranium apparently do not ossify until the latter part of the post-flexion stage.

The suspensorium, hyoid arch, and jaws include the hyomandibulosymplectic cartilage, the palatopterygoquadrate cartilage, ceratohyal, hypohyal, and basihyal cartilages, and meckel's cartilage. The dentary, angular and articular begin to ossify on the meckel's cartilage early in the pre-flexion stage, simultaneously with ossification of the maxillae. Premaxillae form shortly after, and premaxillary teeth form late in the pre-flexion stage or early in the flexion stage. Dentary teeth apparently form during notochord flexion. The ceratohyal begins to ossify and the first branchiostegal rays form during preflexion stage. The interhyal and epihyal begin to ossify in the flexion stage and the epihyal meets the ceratohyal in an interdigitating joint late in the post-flexion stage. The dorsal and ventral hypohyals ossify and the urohyal forms late in the post-flexion stage. The symplectic begins to ossify during the flexion stage and other bones of the suspensorium begin to ossify during the post-flexion stage; the hyomandibular, and ecto- and endopterygoids are first, by about mid-post-flexion stage. The opercular series bones begin to form during notochord flexion.

Ossification of the gill arches begins during the latter part of the pre-flexion stage with the ceratobranchials and the pharyngeal tooth plates. Ossification of each ceratobranchial begins near its center and spreads toward the ends. Ossification of the basi-, hypo-, and epibranchials is delayed until the latter part of the post-flexion stage.

In the axial skeleton, vertebral centra are first to begin ossifying, after the beginning of notochord flexion. The posterior abdominal centra form first and all centra are present by about mid-post-flexion stage, when cartilaginous neural and haemal arches form and begin to ossify. The anterior arches ossify first and arches are added sequentially in the posterior direction; ossification of each arch spreads distad from its base. Separate cartilaginous dorsal and ventral hypural plates supporting the principal caudal-fin rays are present by the flexion stage and ossifying by the post-flexion stage. Two cartilaginous epurals form during the post-flexion stage. Cartilaginous pterygiophores of the dorsal and anal fins form during the post-flexion stage; posterior (but not the posterior-most) soft-ray pterygiophores form first, and addition is both anteriorly and posteriorly. Fin rays form before the pterygiophores begin to ossify.

The cleithra are ossified early in the pre-flexion stage. Supracleithra and post-temporals form relatively early in the post-flexion stage, and the dorsal and ventral post-cleithra form late in the stage. Pectoral-fin rays are well formed before the four cartilaginous pectoral radials begin to ossify in the latter part of the post-flexion stage. The coracoid and scapula ossify on the coracoscapular cartilage late in the post-flexion stage. The cartilaginous pelvic basipterygia form between the lower ends of the cleithra during the flexion stage and ossify late in the post-flexion stage.

Some limited information is available on skeletal development in the clinid *Heterostichus rostratus* (Watson, 1987). The first bones to begin ossifying, early in pre-flexion stage, are the premaxillae, dentaries, parasphenotic, basioccipital, and cleithra. Ossification in the head region through the flexion stage is largely limited to these, the ceratohyals, ceratobranchials, and opercular series, with most of the neurocranium not ossifying until the post-flexion stage.

In the axial skeleton, the posterior abdominal (and perhaps anterior caudal) vertebral centra begin to ossify first, during notochord flexion, nearly simultaneously with formation of the first cartilaginous neural and haemal arches. Each centrum originates as a ventral, saddle-shaped ossification on the notochord, which spreads upward to surround the notochord. Centra are added anteriorly and posteriorly, except the urostyle and preural centrum 2 ossify before preural centrum 3. In contrast to the sequence in *F. varium*, cartilaginous anterior and middle neural arches form about simultaneously during notochord flexion in *H. rostratus*. Apparently, addition of arches is posteriorly from the anterior arches and anteriorly and posteriorly from the middle arches, and ossification occurs during the post-flexion stage. Cartilaginous haemal arches also form during notochord flexion, with middle arches perhaps forming first, and they ossify in the post-flexion stage. Separate parhypural and ventral hypural plate cartilages form late in the pre-flexion stage, followed by the dorsal hypural plate cartilage. The parhypural and ventral hypural cartilages fuse, and the ventral and dorsal hypural cartilages fuse proximally, during notochord flexion; these ossify and the fifth hypural forms after flexion. Two cartilaginous epurals form during notochord flexion and ossify in the post-flexion stage. An elongate dorsal cartilage supported by the last two neural spines and anterior epural, and a ventral cartilage supported by the last two haemal spines, form during the post-flexion stage and support the procurrent caudal rays.

Cartilaginous dorsal- and anal-fin soft-ray pterygiophores form just before their fin rays. The dorsal pterygiophores may all form about simultaneously, but in the anal fin, the middle pterygiophores appear to form first. Pterygiophores of the dorsal-fin spines are added anteriorly from the posterior-most pterygiophore.

The coracoscapular cartilage, with a long coracoid process, and the cartilaginous pectoral blade are present in the smallest larvae. Four pectoral radials are present by mid-flexion and ossifying in the post-flexion stage. The coracoid process is resorbed during the latter part of the post-flexion stage. The supracleithra, then the post-cleithra, form during notochord flexion. The pelvic basipterygium forms in the post-flexion stage.

Among blennioids skeletal development is best known in the blenniids, although published accounts are available for only a few species (Peters, 1981; Stevens and Moser, 1982; Watson, 1987; Kawaguchi *et al.*, 1999; Ditty *et al.*, 2005). The following synopsis is based primarily on single species of *Chasmodes*, *Hypsoblennius*, *Enchelyurus*, and *Omobranchus*.

At hatching, the chondrocranium consists largely of the trabecula communis, parachordal cartilage, and at least partially formed orbital cartilage, orbital tectum, lamina orbitonasalis, ethmoid plate, occipital arch, and auditory capsule. Immediately at hatching or soon after, the maxilla begins ossifying, as do the dentary and articular on Meckel's cartilage, the hyomandibular and symplectic on their cartilage, the quadrate on the quadrate-pterygoid cartilage, the base of the neurocranium (parasphenotic, basioccipital), posterior branchiostegal rays, and cleithra. Most bones of the neurocranium, jaws and suspensorium are ossifying in the pre-flexion or flexion stage except the basisphenotic, median ethmoid, nasal and circumorbital bones, which ossify in the post-flexion stage. Ossification of some other bones (e.g., vomer, intercalar) also may be delayed in parablenniins. From late in the pre-flexion stage through mid- to late post-flexion stage, some omobranchins have small spinules or ridges on any or all of the frontal, parietal, pterotic, maxillary, dentary, articular, and preopercular bones.

Dentary ossification begins before the formation of dentary teeth; the first two teeth form at the anterolateral corners of the jaw in the pre-flexion stage and teeth are added both mesially and laterally. In contrast, the first two premaxillary teeth form before premaxillary ossification begins. Addition of teeth is mesially and laterally, after the premaxillae begin to ossify. Adult comb-like teeth gradually replace the conical larval teeth in both jaws during the latter part of the post-flexion stage (Peters, 1981; Watson, 1987), or transitional teeth form during this time and are replaced by the comb-like adult teeth shortly before settlement (Ditty *et al.*, 2005). Ophioblennius-stage larvae develop a pair of large, curved, laterally projecting, anterolateral canine teeth in the lower jaw during the flexion stage, and a pair or two of smaller, anteriorly directed hooked teeth may form near the centers of both jaws. These teeth persist until near the end of the ophioblennius stage. Late in the pre-flexion stage or during the flexion stage, nemophin larvae develop a pair of large, posterior dentary teeth that apparently turn into the adult dentary canines.

The hyoid arch initially includes a single ceratohyal-epihyal cartilage with an ossified posterior branchiostegal ray or two. Separate interhyal and hypohyal cartilages also are present initially or soon after hatching. The cerato- and epihyal begin to ossify near the upper and anterior ends of their cartilage, respectively, late in the pre-flexion stage or during the flexion stage, and meet in an interdigitating joint by early in the post-flexion stage. Branchiostegal rays are added anteriorly, with all present by the end of the pre-flexion stage. The ventral hypohyal ossifies on the lower margin of the hypohyal cartilage beginning in the flexion stage, followed by the interhyal and dorsal hypohyal late in the flexion stage or during the post-flexion stage. The urohyal forms during notochord flexion and the basihyal begins to ossify late in the post-flexion stage or at settlement.

Gill arches initially consist of cartilaginous ceratobranchials 1-4 and an infrapharygobranchial cartilage, or they may also include cartilaginous epibranchials 1-3, hypobranchials 1-2, and a single basibranchial cartilage that will become basibranchials 1-3. All of these elements, the cartilaginous fifth ceratobranchials, fourth epibranchials, and fourth basibranchial are present in the pre-flexion stage and all except basibranchial 4 are ossifying by early in the post-flexion stage. The first upper pharyngeal tooth forms at, or soon after, hatching and a few more are added through the flexion stage. Lower pharyngeal teeth form on the fifth ceratobranchials beginning between mid-pre-flexion and early post-flexion stage.

The axial skeleton initially consists only of the notochord. Cartilaginous haemal, or neural and haemal, arches begin forming next in parablenniins, while in omobranchins the hypural cartilage forms about simultaneously with the first arches in the latter part of the pre-flexion stage. Most haemal arches form about simultaneously, and addition is posteriorly. Three sequences of neural arch formation are currently known: (1) the anterior arches form first with addition in the posterior direction; (2) first anterior, then middle arches form, with addition posteriorly from the anterior arches and in both directions from the middle; and (3) anterior and middle arches form simultaneously with addition of arches as in (2) above. Ossification of arches begins during the flexion or post-flexion stage and is anterior to posterior, except in at least one *Chasmodes* species, where the posterior few arches ossify before those located more anteriorly (Peters, 1981). Ossification of each arch begins near its base, spreads distad onto its spine, and is completed at, or after, settlement.

The vertebral centra begin to ossify around the notochord at, or soon after, the beginning of neural and haemal arch ossification. Ossification of centra is anterior to posterior, except the urostyle ossifies before most other caudal vertebrae. In at least one *Chasmodes* species the posterior few centra ossify from posterior to anterior (Peters, 1981). Parapophyses form on the abdominal vertebrae from late flexion to about mid-post-flexion stage. Neural pre- and post-zygapophyses form anteriorly during the post-flexion stage, are added serially from anterior to posterior, and are completed after settlement. Haemal pre- and post-zygapophyses form after settlement. Epineural intermuscular bones form in the myosepta adjacent to the first neural arches, beginning late in the flexion stage to mid-post-flexion stage. Concurrently, pleural ribs originate as cartilage adjacent to parapophyses 3-6 and begin ossifying between mid-post-flexion stage and settlement. Addition is posteriorly in both series and is completed by, or soon after settlement. Cartilaginous hypural plates form late in the pre-flexion stage. In parablenniins, the ventral plate (parhypural + hypurals 1-2) forms just before the dorsal plate (hypurals 3-4), and the two fuse proximally during the flexion or post-flexion stage. In omobranchins, a single cartilage (deeply divided at the outset in *Enchelyurus*, becoming deeply divided in *Omobranchus*) represents both plates. A cartilaginous process, probably representing the fifth hypural, forms at the upper end of the dorsal plate during notochord flexion in some parablenniins but not in omobranchins. Epural cartilage(s) form during notochord flexion. Hypural ossification begins late in the flexion or early post-flexion stage and is completed after settlement. Epurals ossify and uroneurals form after settlement. An elongate dorsal cartilage and another ventrally that support the procurrent caudal-fin rays form during the latter part of the post-flexion stage.

Cartilaginous pterygiophores of the first several dorsal- and anal-fin soft rays form about simultaneously during notochord flexion (before flexion in *Laiphognathus*: Kubo and Sasaki, 2000a), shortly before their fin rays. Soft-ray pterygiophores are added posteriorly and fin-spine pterygiophores are added anteriorly. Distal radials of the soft-ray pterygiophores separate from the proximal radials, and radials begin to ossify during the post-flexion stage; ossification is completed after settlement. Some omobranchins (*Laiphognathus*, some *Omobranchus*) develop a prominent, anterodorsal spine- or blade-like projection on the proximal radial of each dorsal- and anal-fin soft-ray pterygiophore (Watson and Miskiewicz, 1998; Kubo and Sasaki, 2000a). These projections are present until the latter part of the post-flexion stage, then diminish and are absent by settlement (Fig. 4.4.8). They may have a defensive function, or serve as supports for the finfold, improving the larval swimming ability, which should facilitate maintenance of a near-shore larval distribution, in the vicinity of habitat suitable for settlement (Kubo and Sasaki, 2000a).

Initially, the pectoral girdle consists of slender, ossifying cleithra, cartilaginous pectoral blades, and coracoscapular cartilages with a posteroventral coracoid process on each. The supracleithra form early to mid-way through the flexion stage and the post-temporals form late in the flexion to early post-flexion stage. Ventral, and then dorsal, post-cleithra form during the post-flexion stage. The coracoid ossifies ventrally on the coracoscapular cartilage and fuses with the cleithrum from the flexion stage through mid- to late post-flexion stage, concurrently with resorption of the coracoid process. The scapula begins to ossify on the upper part of the coracoscapular cartilage during the latter half of the post-flexion stage and completes ossification after settlement. The pectoral blade originates as a single cartilage that may be deeply incised between future proximal radials 2 and 3. All four radial cartilages, fused at their proximal and distal ends, are visible by mid-way to late in the flexion stage, begin to ossify late in the flexion to early post-flexion stage, and complete ossification after settlement. Small distal radial cartilages form by late in the post-flexion stage. Cartilaginous pelvic basipterygia form between the lower ends of the cleithra late in the pre-flexion to mid-flexion stage, and ossify during the post-flexion stage.

References

Able, K.W. and M.P. Fahay. 1998. *The First Year in the Life of Estuarine Fishes in the Middle Atlantic Bight*. Rutgers University Press, New Brunswick.

Balbontín, F. and R. Perez. 1979. Modalidad de postura, huevos y estados larvales de *Hypsoblennius sordidus* (Bennett) en la Bahía de Valparaíso (Blenniidae: Perciformes). *Revista de Biologia Marina. Vina del Mar* 16: 311-318.

Barnett, A.M., A.E. Jahn, P.D. Sertic and W. Watson. 1984. Distribution of ichthyoplankton off San Onofre, California, and methods for sampling very shallow coastal waters. *Fishery Bulletin* 82: 97-111.

Barnhart, P.S. 1932. Notes on the habits, eggs and young of some fishes of southern California. *Bulletin of the Scripps Institution of Oceanography* 3: 87-99.

Beltrán-León, B.S. and R. Ríos Herrera. 2000. *Estadios tempranos de peces del Pacifico Colombiano*. Instituto Nacional de Pesca y Acuicultura, Buenaventura, Colombia.

Böhlke, J. 1957. A review of the blenny genus *Chaenopsis*, and the description of a related new genus from the Bahamas. *Proceedings of the Academy of Natural Sciences of Philadelphia* 109: 81-122.

Böhlke, J.E. 1968. The descriptions of three new sand stargazers (Dactyloscopidae) from the tropical West Atlantic. *Notulae Naturae* 414: 1-16.

Breder, C.M. 1939. On the life history and development of the sponge blenny, *Paraclinus marmoratus* (Steindachner). *Zoologica* 24: 487-496.

Breder, C.M. 1941. On the reproductive behavior of the sponge blenny, *Paraclinus marmoratus* (Steindachner). *Zoologica* 26: 233-236.

Brogan, M.W. 1992. *Ecology of Larval Fishes Around Reefs in the Gulf of California, Mexico.* Ph.D. Thesis, University of Arizona, Tucson.

Brogan, M.W. 1994. Distribution and retention of larval fishes near reefs in the Gulf of California. *Marine Ecology Progress Series* 115: 1-13.

Catalano, E. 1978. Ritrovamento di *Hypleurochilus bananensis* Poll nel Golfo di Palermo. Osservazioni su alcuni stadi larvali (Perciformes, Blenniidae). *Naturalista Siciliano* 2: 73-83.

Cavalluzzi, M.R. 1997. Larvae of *Gillellus jacksoni, G. uranidea* (Dactyloscopidae), *Stathmonotus stahli tekla*, and *S. hemphilli* (Chaenopsidae), with comments on the use of early life history characters for elucidating relationships within the Blennioidei. *Bulletin of Marine Science* 60: 139-151.

Cavalluzzi, M.R. and J.E. Olney. 2006a. Tripterygiidae: Triplefin blennies. In: *Early Stages of Atlantic Fishes: An Identification Guide for the Western Central North Atlantic*, W.J. Richards (ed.). CRC Press, Boca Raton, pp. 1935-1938.

Cavalluzzi, M.R. and J.E. Olney. 2006b. Dactyloscopidae: Sand stargazers. In: *Early Stages of Atlantic Fishes: An Identification Guide for the Western Central North Atlantic*, W.J. Richards (ed.). CRC Press, Boca Raton, pp. 1939-1946.

Cavalluzzi, M.R. and J.E. Olney. 2006c. Labrisomidae: Labrisomids. In: *Early Stages of Atlantic Fishes, An Identification Guide for the Western Central North Atlantic*, W.J. Richards (ed.). CRC Press, Boca Raton, pp. 1947-1955.

Cavalluzzi, M.R. and J.E. Olney. 2006d. Chaenopsidae: Tube-, pike-, flag blennies. In: *Early Stages of Atlantic Fishes: An Identification Guide for the Western Central North Atlantic*, W.J. Richards (ed.). CRC Press, Boca Raton, pp. 1957-1967.

Chesalina, T.L. 1998. On the identification of the larvae of Black Sea fishes of the family Blenniidae. *Journal of Ichthyology* 38: 140-142.

Cipria, G. 1934. Uova, stadi embrionali e post-embrionali di *Blennius palmicornis* Cuv. *R. Comitato Talassografico Italiano Memoria* 218.

Cipria, G. 1936. Uova, stadi embrionali e post-embrionali nei *Blennidi*. I. *Blennius pavo* Risso. II. *Blennius inaequalis* C.V. *R. Comitato Talassografico Italiano Memoria* 231.

Clarke, T.A. 1993. Larvae of nearshore fishes in oceanic waters near Oahu, Hawaii. *United States Department of Commerce, NOAA Technical Report NMFS 101.* 19p.

Cunningham, J.T. 1889. Studies of the reproduction and development of teleostean fishes occurring in the neighbourhood of Plymouth. *Journal of the Marine Biological Association of the United Kingdom* 1: 10-54.

Dawson, C.E. 1975. Studies on eastern Pacific sand stargazers (Pisces: Dactyloscopidae) 2. Genus *Dactyloscopus*, with descriptions of new species and subspecies. *Natural History Museum of Los Angeles County, Science Bulletin* 22: 1-61.

Dawson, C.E. 1976. Studies on eastern Pacific sand stargazers 3. *Dactylagnus* and *Myxodagnus*, with description of a new species and subspecies. *Copeia* 1976: 13-43.

Dawson, C.E. 1977. Studies on eastern Pacific sand stargazers (Pisces: Dactyloscopidae) 4. *Gillellus, Sindoscopus* new genus, and *Heteristius* with description of new species. *Proceedings of the California Academy of Sciences* 51: 125-160.

Dekhnik, T.V. 1973. *Ichthyoplankton of the Black Sea*. Nauk. Dumka, Kiev. (In Russian).

De Leo, G., E. Catalano and N. Parrinello. 1976. Contributo alla conescenza del *Blennius cristatus* L. 1758 (Perciformes, Blenniidae). *Memorie di Biologia Marina e di Oceanografia* 6: 209-228.

Ditty, J.G., R.F. Shaw and L.A. Fuiman. 2005. Larval development of five species of blenny (Teleostei: Blenniidae) from the western central North Atlantic, with a synopsis of blennioid family characters. *Journal of Fish Biology* 66: 1261-1284.

Ditty, J.G., M.R. Cavalluzzi and J.E. Olney. 2006. Blenniidae: Combtooth blennies. In: *Early Stages of Atlantic Fishes: An Identification Guide for the Western Central North Atlantic*, W.J. Richards (ed.). CRC Press, Boca Raton, pp. 1969-1993.

Dotsu, Y. 1982. The early life history of the combtooth blenny *Dasson trossulus* and spawning of the laboratory-reared fish about three months after hatching. *Bulletin of the Faculty of Fisheries Nagasaki University* 52: 19-27. (In Japanese).

Dotsu, Y. and T. Oota. 1973. The life history of the blenniid fish, *Omobranchus loxozonus*. *Bulletin of the Faculty of Fisheries Nagasaki University* 36: 13-22. (In Japanese).

Dotsu, Y. and S. Moriuchi. 1980. The life history of the blenniid fish, *Blennius yatabei* Jordan et Snyder. *Bulletin of the Faculty of Fisheries Nagasaki University* 49: 17-24. (In Japanese).

Dutt, S. and V. Visweswara Rao. 1960. On the breeding habits and early developmental stages of *Petroscirtes pibunctatus* Day. *Journal of the Zoological Society of India* 12: 158-161.

Eggert, B. 1932. Zur Kenntnis der Biologie, der sekundären Geschlechtsmerkmale und des Eies von *Blennius pavo* (Risso). *Zeitschrift für Morphologie und Ökologie der Tiere* 24: 682-703.

Ehrenbaum, E. 1905. Eier und Larven von Fischen. *Nordisches Plankton* 1. 216 pp.

Fahay, M.P. 2007. *Early Stages of Fishes in the Western North Atlantic Ocean (Davis Strait, Southern Greenland and Flemish Cap to Cape Hatteras). Volume two: Scorpaeniformes through Tetraodontiformes*. Northwest Atlantic Fisheries Organization, Dartmouth, pp. 932-1696.

Faria, C., F. Gil and V.C. Almada. 2005. Ontogenetic development of *Lipophrys trigloides* (Pisces: Blenniidae) with some notes on the spawning behaviour. *Journal of the Marine Biological Association of the United Kingdom* 85: 185-188.

Faria, C., R. Borges, F. Gil, V.C. Almada and E.J. Goncalves. 2002. Embryonic and larval development of *Lipophrys trigloides* (Pisces: Blenniidae). *Scientia Marina (Barcelona)* 66: 21-26.

Feder, H.M., C.H. Turner and C. Limbaugh. 1974. Observations on fishes associated with kelp beds in southern California. *California Department of Fish and Game, Fish Bulletin* 160. 144p.

Fishelson, L. 1963. Observations on littoral fishes of Israel. II. Larval development and metamorphosis of *Blennius pavo* Risso (Teleostei, Blenniidae). *Israel Journal of Zoology* 12: 81-91.

Fishelson, L. 1976. Spawning and larval development of the blennid fish, *Meiacanthus nigrolineatus* from the Red Sea. *Copeia* 1976: 798-800.

Fives, J.M. 1970. Blenniidae of the North Atlantic. In: *Fisches d'identification des oeufs et larves de poisons. Fische no. 3*. Conseil International pour l'Exploration de la Mer, Charlottenlund Slot, 4p.

Fives, J.M. 1980. An account of the eggs and developmental stages of Montagu's blenny, *Coryphoblennius galarita* (L.), with notes on the reproductive behaviour of the adults. *Journal of the Marine Biological Association of the United Kingdom* 60: 749-757.

Ford, E. 1922. On the young stages of *Blennius ocellaris* L., *Blennius pholis* L., and *Blennius gattorugine* L. *Journal of the Marine Biological Association of the United Kingdom* 12: 688-692.

Fritzsche, R.A. 1978. Blenniidae—Combtooth blennies. In: *Development of Fishes of the Mid-Atlantic Bight, an Atlas of Egg, Larval and Juvenile Stages. Volume V. Chaetodontidae Through Ophiididae*. U.S. Department of the Interior, Fish and Wildlife Service FWS/OBS-78/12, pp. 297-309.

Garstang, W. 1900. Preliminary experiments on the rearing of sea-fish larvae. *Journal of the Marine Biological Association of the United Kingdom* 6: 70-93.

Gordina, A.D. and L.A. Duka. 1979. Distribution, morphology, and nutrition of *Tripterygion tripteronotus* Risso larvae in the Black Sea. *Biologia Morya* 51: 65-68.

Graham, D.H. 1939. Breeding habits of the fishes of Otago Harbour and adjacent seas. *Transactions and Proceedings of the Royal Society of New Zealand* 69: 361-372.

Gunn, J.S. and R.E. Thresher. 1991. Viviparity and the reproductive ecology of clinid fishes (Clinidae) from temperate Australian waters. *Environmental Biology of Fishes* 31: 323-344.

Hastings, P.A. 1988. Female choice and male reproductive success in the angel blenny, *Coralliozetus angelica* (Teleostei: Chaenopsidae). *Animal Behavior* 36: 115-124.

Hefford, A.E. 1910. Notes on teleostean ova and larvae observed at Plymouth in spring and summer, 1909. *Journal of the Marine Biological Association of the United Kingdom* 9: 2-58.

Herrera, G.A. and R.J. Lavenberg. 1999. Larval Labrisomidae (Pisces: Blennioidei) from the Galapagos Islands. *Contributions in Science, Natural History Museum of Los Angeles County* 478: 1-14.

Hildebrand, S.F. and L.E. Cable. 1938. Further notes on the development and life history of some teleosts at Beaufort, N.C. *U.S. Department of Commerce, Bulletin of the Bureau of Fisheries* 48: 505-642.

Houde, E.D., S. Almatar, J.C. Leak and C.E. Dowd. 1986. Ichthyoplankton abundance and diversity in the western Arabian Gulf. *Kuwait Bulletin of Marine Science* 1986: 107-393.

Howe, J.C. 1991. Egg surface morphology of *Dialommus fuscus* Gilbert (Pisces: Labrisomidae). *Journal of Fish Biology* 38: 148-152.

Kalinina, E.M. 1976. Embryonal and postembryonal development of Black Sea Blenniidae. *Biologia Morya* 38: 3-18.

Kawaguchi, T., H. Kohno, K. Fujita and Y. Taki. 1999. Early morphological development of *Omobranchus fasciolatoceps* and *O. punctatus* (Blenniidae: Omobranchini) reared in an aquarium. *Ichthyological Research* 46: 163-170.

Kojima, J. and M. Shiogaki. 1988. Blenniidae. In: *An Atlas of the Early Stage Fishes in Japan*, M. Okiyama (ed.). Tokai University Press, Tokyo, pp. 735-751. (In Japanese).

Kubo, M. and K. Sasaki. 2000a. Larvae of *Laiphognathus multimaculatus* (Omobranchini, Blenniidae) with pterygiophore blades: Functional implications. *Ichthyological Research* 47: 193-197.

Kubo, M. and K. Sasaki. 2000b. A pelagic larva of *Enchelyurus kraussi* occurring off Tosa Bay, southern Japan (Pisces, Blenniidae, Omobranchini). *Memoirs of the Faculty of Science, Kochi University* 21: 39-42.

Labell, M. and J.R. Nursall. 1985. Some aspects of the early life history of the redlip blenny, *Ophioblennius atlanticus* (Teleostei: Blenniidae). *Copeia* 1985: 39-49.

Lebour, M.V. 1927. The eggs and newly hatched young of the common blennies from the Plymouth neighbourhood. *Journal of the Marine Biological Association of the United Kingdom* 14: 647-650.

Leis, J.M. and J.M. Miller. 1976. Offshore distributional patterns of Hawaiian fish larvae. *Marine Biology* 36: 355-367.

Leis, J.M. and B.M. Carson-Ewart (eds.). 2000. *The Larvae of Indo-Pacific Coastal Fishes*, Fauna Malesiana Handbooks, Brill, Leiden.

Lindquist, D.G. 1981. Reproduction of the onespot fringehead, *Neoclinus uninotatus*, in Monterey Harbor, California. *Bulletin of the Southern California Academy of Sciences* 80: 12-22.

Lippson, A.J. and R.L. Moran. 1974. *Manual for Identification of Early Developmental Stages of Fishes of the Potomac Estuary*. Maryland Department of Natural Resources, Power Plant Siting Program PPSP-MP-13. 282p.

Matarese, A.C., W. Watson and E.G. Stevens. 1984. Blennioidei: Development and relationships. In: *Ontogeny and Systematics of Fishes*, H.G. Moser, W.J. Richards, D.M. Cohen, M.P. Fahay, A.W. Kendall, Jr. and S.L. Richardson (eds.). American Society of Ichthyologists and Herpetologists Special Publication Number 1, pp. 565-573.

Matarese, A.C., A.W. Kendall, Jr., D.M. Blood and B.M. Vinter. 1989. *Laboratory Guide to Early Life History Stages of Northeast Pacific Fishes*. NOAA Technical Report NMFS 80. 652p.

McGowen, G.A. 1993. Coastal ichthyoplankton assemblages, with emphasis on the Southern California Bight. *Bulletin of Marine Science* 53: 692-722.

Miller, J.M., W. Watson and J.M. Leis. 1979. *An Atlas of Common Nearshore Marine Fish Larvae of the Hawaiian Islands*. University of Hawaii Sea Grant Miscellaneous Report UNIHI-SEAGRANT-MR-80-02. 179p.

M'Intosh, W.C. 1905. On the life history of the shanny (*Blennius pholis* L.). *Zeitschrift für Wissenschaftliche Zoologie* 82: 368-378.

Mito, S. 1954. Breeding habits of a blennioid fish, *Salarias enosimae*. *Japanese Journal of Ichthyology* III(3/4/5): 144-152. (In Japanese).

Mito, S. 1966. Fish eggs and larvae. Volume 7. In: *Illustrations of the marine plankton of Japan*, S. Motoda (ed.). Sôyô-Sha, Tokyo. 74p. (In Japanese).

Mongdard, R.P. 1979. Desarrollo postembrionario de *Tripterygion chilensis* Cancino 1955, en la Bahia de Valparaíso (Tripterygiidae: Perciformes). *Revista de Biología Marina, Departamento de Oceanología, Universidad de Chile* 16: 319-329.

Moser, H.G. (ed.). 1996. *The Early Stages of Fishes in the California Current Region*. California Cooperative Oceanic Fisheries Investigations Atlas 33. Allen Press, Lawrence. 1505p.

Munro, I.S.R. 1955. Eggs and larvae of the sabre-toothed oyster blenny, *Dasson steadi* (Whitley) (Blenniidae). *Australian Journal of Marine and Freshwater Research* 6: 30-34.

Neira, F.J. and M.I. Sporcic. 2002. Use of ichthyoplankton ecology to evaluate ecosystem changes: a case study in a large, semi-enclosed Australian bay. *Australian Journal of Marine and Freshwater Research* 53: 339-354.

Ninos, M. 1984. *Settlement and Metamorphosis in Hypsoblennius (Pisces, Blenniidae)*. Ph.D. Dissertation, University of Southern California, Los Angeles.

Olivar, M.-P. 1986. Development and distribution of *Parablennius pilicornis* (Cuvier) larvae (Teleostei, Blenniidae) off Namibia. *South African Journal of Marine Science* 4: 193-201.

Olivar, M.-P. and J.-M. Fortuño. 1991. Guide to ichthyoplankton of the southeast Atlantic (Benguela Current region). *Scientia Marina* 55: 1-383.

Padoa, E. 1956a. Famiglia 5: Blenniidae. In: *Fauna e Flora del Golfo di Napoli Monograph 38. Uova, larve e stadi giovanili di Teleostei*, pp. 720-745.

Padoa, E. 1956b. Famiglia 6: Clinidae. In: *Fauna e Flora del Golfo di Napoli Monograph 38. Uova, larve e stadi giovanili di Teleostei*, pp. 745-752.

Patzner, R.A. 1984. The reproduction of *Blennius pavo* (Teleostei, Blenniidae) II. Surface structures of the ripe egg. *Zoologischer Anzeiger* 213: 44-50.

Peters, K.M. 1981. Reproductive biology and developmental osteology of the Florida blenny, *Chasmodes saburrae* (Perciformes: Blenniidae). *Northeast Gulf Science* 4: 79-98.

Peters, K.M. 1985. Larval development of *Lupinoblennius nicholsi* with comments on larval Blenniini identification in Tampa Bay, Florida. *Bulletin of Marine Science* 36: 445-453.

Pottoff, T., S. Kelley, M. Moe and F. Young. 1984. Description of porkfish larvae (*Anisotremus virginicus*, Haemulidae) and their osteological development. *Bulletin of Marine Science* 34: 21-59.

Pottoff, T., S. Kelley, V. Saksena, M. Moe and F. Young. 1987. Description of larval and juvenile yellowtail damselfish, *Mircospathodon chrysurus*, (Pomacentridae) and their osteological development. *Bulletin of Marine Science* 40: 330-375.

Prochazka, K. and C.L. Griffiths. 1992. Observations on the distribution patterns, behaviour, diets and reproductive cycles of sand-dwelling clinids (Perciformes: Clinidae) from South Africa. *Environmental Biology of Fishes* 35: 371-379.

Qasim, S.Z. 1956. The spawning habits and embryonic development of the shanny (*Blennius pholis* L.). *Proceedings of the Zoological Society of London* 127: 79-93.

Ré, P. and A.J. Almeida. 1981. Observations on the biology of *Ophioblennius atlanticus atlanticus* (Valenciennes in Cuv. & Val., 1836), (Pisces: Blenniidae) from the Azores. *Arquivos do Museu Bocage* I: 41-52.

Riginos, C. and B.C. Victor. 2001. Larval spatial distributions and other early life-history characteristics predict genetic differentiation in eastern Pacific blennioid fishes. *Proceedings of the Royal Society of London* B268: 1931-1936.

Rosenblatt, R.R. and L.B. Taylor, Jr. 1971. The Pacific species of the clinid fish tribe Starksiini. *Pacific Science* 25: 436-462.

Ruck, J.G. 1973. Development of *Tripterygion capito* and *T. robustum* (Pisces: Tripterygiidae). *Zoology Publications from Victoria University of Wellington* 63: 1-10.

Ruck, J.G. 1980. Early development of *Forsterygion varium*, *Gilloblennius decemdigitatus*, and *G. tripennis* (Pisces: Tripterygiidae). *New Zealand Journal of Marine and Freshwater Research* 14: 313-326.

Ruck, J.G. 1998. Larval, prejuvenile and adult osteology of *Forsterygion varium* Schneider. *Wellington Polytechnic Occasional Paper* 98/2. 72p.

Russell, F.S. 1976. *The Eggs and Planktonic Stages of British Marine Fishes*. Academic Press, London.

Sabates, A. 1994. Larval development of three blenniid species *Aidablennius sphynx*, *Coryphoblennius galarita* and *Lipophrys canavai* (Pisces: Blenniidae: Blenniini) in the western Mediterranean. *Journal of Zoology* 234: 89-103.

Seale, A. 1935. The Templeton Crocker expedition to western Polynesian and Melanesian Islands, 1933. No. 27. Fishes. *Proceedings of the California Academy of Sciences* 21: 337-378.

Shiogaki, M. and Y. Dotsu. 1972. The life history of the blenniid fish, *Neoclinus bryope*. *Bulletin of the Faculty of Fisheries, Nagasaki University* 34: 1-8.

Shiogaki, M. and Y. Dotsu. 1973. The egg development and larva rearing of the tripterygiid blenny, *Tripterygion etheostoma*. *Japanese Journal of Ichthyology* 20: 42-46. (In Japanese).

Shiogaki, M. and Y. Dotsu. 1988a. Tripterygiidae. In: *An Atlas of the Early Stage Fishes in Japan*, M. Okiyama (ed.). Tokai University Press, Tokyo, pp. 733-734.

Shiogaki, M. and Y. Dotsu. 1988b. Chaenopsidae. In: *An Atlas of the Early Stage Fishes in Japan*, M. Okiyama (ed.). Tokai University Press, Tokyo, pp. 733-736.

Springer, V.G. and J.T. Williams. 1994. The Indo-West Pacific blenniid fish genus *Istiblennius* reappraised: a revision of *Istiblennius*, *Blenniella*, and *Paralticus*, new genus. *Smithsonian Contributions to Zoology* 565. 193p.

Stephens, J.S., Jr., E.S. Hobson and R.K. Johnson. 1966. Notes on distribution, behavior, and morphological variation in some chaenopsid fishes from the tropical eastern Pacific, with descriptions of two new species, *Acanthemblemaria castroi* and *Coralliozetus springeri*. *Copeia* 1966: 424-438.

Stepien, C.A. 1986. Life history and larval development of the giant kelpfish, *Heterostichus rostratus* Girard, 1854. *Fishery Bulletin* 84: 809-826.

Stevens, E.G. and H.G. Moser. 1982. Observations on the early life history of the mussel blenny, *Hypsoblennius jenkinsi*, and the bay blenny, *Hypsoblennius gentilis*, from specimens reared in the laboratory. *California Cooperative Oceanic Fisheries Investigations Reports* 23: 269-275.

Sufei, J. and Y. Weibing. 1989. Larvae of *Omobranchus elegans* (Steindachner) found in Luo Yuan Bay and Jiulong River Estuary Fujian. *Journal of Xiamen University (Natural Science)* 28: 671-676.

Suzuki, H. 2003. Reproductive behavior, eggs, larvae and rearing of the blenny, *Scartella cristata* in the aquarium. *Science Report of the Marine Science Museum, Tokai University* 5: 1-10. (In Japanese).

Suzuki, H. and H. Nakayama. 2005. Reproductive behavior, egg, larval development and rearing in the blenny, *Meiacanthus atrodorsalis*, in the aquarium. *Bulletin of the Institute of Oceanic Research and Development, Tokai University* 26: 33-43. (In Japanese).

Suzuki, H., N. Tsuzuki, K. Tanabe, T. Miyake and Y. Tanaka. 2006. Reproductive behavior, egg, larval development and rearing in the blenny, *Meiacanthus grammistes* (Valenciennes) in the aquarium. *Bulletin of the Institute of Oceanic Research and Development, Tokai University* 27: 47-60. (In Japanese).

Tellock, J.A. and F. Alig. 1998. Aquaculture potential of the lyretail blenny, *Meiacanthus atrodorsalis*, for the marine ornamental fish trade. In: *Aquaculture '98 Book of Abstracts*. World Aquaculture Society, Louisiana State University, Baton Rouge, p. 538.

Thomson, J.M. and A.E. Bennett. 1953. The oyster blenny, *Omobranchus anolius* (Valenciennes) (Blenniidae). *Australian Journal of Marine and Freshwater Research* 4: 227-233.

Visweswara Rao, V. 1970. Breeding habits and early development of two blenniid fishes *Omobranchus japonicus* (Bleeker) and *Cruantus smithi* Visweswara Rao from Godavari Estuary. *Journal of the Marine Biological Association of India* 12: 175-182.

Wang, C.S. 1981. *Taxonomy of the Early Life Stages of Fishes—Fishes of the Sacramento-San Joaquin Estuary and Moss Landing Harbor-Elkhorn Slough, California*. Ecological Analysts, Inc., Concord, California.

Watson, W. 1983. Blenniidae—Combtooth blennies. In: *The Larvae of Indo-Pacific Coral Reef Fishes*, J.M. Leis and D.S. Rennis (eds.). New South Wales University Press, Sydney, pp. 184-199.

Watson, W. 1987. Larval development of the endemic Hawaiian blenniid, *Echelyurus brunneolus* (Pisces: Blenniidae: Omobranchini). *Bulletin of Marine Science* 41: 856-888.

Watson, W. 1996a. Tripterygiidae: Triplefin blennies. In: *The Early Stages of Fishes in the California Current Region*, H.G. Moser (ed.). California Cooperative Oceanic Fisheries Investigations Atlas 33. Allen Press, Lawrence, pp. 1148-1150.

Watson, W. 1996b. Dactyloscopidae: Sand stargazers. In: *The Early Stages of Fishes in the California Current Region*, H.G. Moser (ed.). California Cooperative Oceanic Fisheries Investigations Atlas 33. Allen Press, Lawrence, pp. 1177-1181.

Watson, W. 1996c. Labrisomidae: Labrisomid kelpfishes. In: *The Early Stages of Fishes in the California Current Region*, H.G. Moser (ed.). California Cooperative Oceanic Fisheries Investigations Atlas 33. Allen Press, Lawrence, pp. 1151-1163.

Watson, W. 1996d. Clinidae: Clinid kelpfishes. In: *The Early Stages of Fishes in the California Current Region*, H.G. Moser (ed.). California Cooperative Oceanic Fisheries Investigations Atlas 33. Allen Press, Lawrence, pp. 1164-1169.

Watson, W. 1996e. Chaenopsidae: Tube blennies. In: *The Early Stages of Fishes in the California Current Region*, H.G. Moser (ed.). California Cooperative Oceanic Fisheries Investigations Atlas 33. Allen Press, Lawrence, pp. 1170-1175.

Watson, W. 1996f. Blenniidae: Combtooth blennies. In: *The Early Stages of Fishes in the California Current Region*, H.G. Moser (ed.). California Cooperative Oceanic Fisheries Investigations Atlas 33. Allen Press, Lawrence, pp. 1182-1199.

Watson, W. 2000. Blenniidae (Combtooth blennies). In: *The Larvae of Indo-Pacific Coastal Fishes*, J.M. Leis and B.M. Carson-Ewart (eds.). Fauna Malesiana Handbooks, Brill, Leiden, pp. 588-605.

Watson, W. 2001. Larvae of *Enchelyurus ater* (Günther, 1877) and *E. kraussi* (Klunzinger, 1871) (Pisces: Blenniidae: Omobranchini). *Records of the Australian Museum* 53: 57-70.

Watson, W. and A.G. Miskiewicz. 1998. Blenniidae: Blennies. In: *Larvae of Temperate Australian Fishes*, F.J. Neira, A.G. Miskiewicz and T. Trnski (eds.). University of Western Australia Press, Nedlands, pp. 368-381.

Watson, W. and H.J. Walker, Jr. 1992. Larval development of sargo (*Anisotremus davidsonii*) and salema (*Xenistius californiensis*) (Pisces: Haemulidae) from the Southern California Bight. *Bulletin of Marine Science* 51: 360-406.

Wickler, W. 1957. Die Larve von *Blennius fluviatilis* Asso 1784. *Biologisches Zentrablatt* 76: 453-466.

Williams, J.T. 1990. Phylogenetic relationships and revision of the blenniid fish genus *Scartichthys*. *Smithsonian Contributions to Zoology* 492. 30p.

SECTION

5

Ecology and Behaviour

CHAPTER

5.1

Communication in Blennies

Eduardo N. Barata[1] and David M. Gonçalves [2]

INTRODUCTION

Animal communication was traditionally viewed in a dyadic perspective (signaller—receiver dyad), where the evolution of signals and of perceptual systems relied on the fitness consequences for the sender and receiver (Dawkins and Krebs, 1978; Krebs and Dawkins, 1984). Under this perspective, 'true communication' occurs when signal detection is beneficial for both senders and receivers, 'manipulation' when signal detection is only beneficial for the sender, and 'exploitation' when only the receiver gains by detecting the signal (Bradbury and Vehrencamp, 1998). However, this has been recognised as a simplistic approach to understand communication systems and their evolution, since wherever signals travel further than the average spacing between two individuals, there is a

Authors' addresses: [1]CCMAR-CIMAR Laboratório Associado, Universidade do Algarve, Faro, Portugal. and Departamento de Biologia, Universidade de Évora, Portugal.
E-mail: ebarata@ualg.pt
[2]Unidade de Investigação em Eco-Etologia, Instituto Superior de Psicologia Aplicada, Rua Jardim do Tabaco 34, 1149-041 Lisboa, Portugal. E-mail: davidg@ispa.pt

potential for a communication network to exist, including the provision of information to unintended receivers (McGregor and Peake, 2000; McGregor, 2005). Therefore, true communication can be defined as the provision of information by a sender who produces a sensory stimulus or signal to one or more receivers that then respond in a way that is beneficial to both sender and receivers. In this case, signallers and receivers form a mutualism in which signals co-evolve with the sensory biology of receivers. Once established, mutually beneficial communication relationships could be susceptible to deceitful manipulations by signallers, which reduce the receiver fitness, and signal interception by unintended receivers (including predators and conspecifics), the response of which to the stimulus may impose fitness costs to the sender (see Searcy and Nowicki, 2005, for insightful discussion).

How do Fish Communicate?

Fish have radiated to all aquatic habitats and consequently evolved an extraordinary diversity of morphologies, relative development of the sensory systems and behaviours. A parallel diversity of communication systems is also expected. However, probably due to the technical difficulties of studying fish in their natural environment, our current knowledge about fish communication systems is limited and probably overlooks its diversity and sophistication. Nevertheless, enough empirical evidence shows that fish in general use the largest number of sensory modalities, including hearing and vision, mechanical (low-frequency vibrations detectable by the lateral-line system), electric and chemical modalities in their interactions with one another (reviews in Ladich *et al.*, 2006a, b; Rosenthal and Lobel, 2006).

COMMUNICATION IN BLENNIES

Combtooth blennies (Blenniidae) form the largest family of true blennies (suborder Blennioidei), and include more than 300 species of, typically, scaleless demersal marine fishes, living in intertidal and subtidal zones all over the world (Zander, 1986; Nelson, 1994). Males are generally larger than females and have a variety of secondary sexual characteristics, especially during the breeding season (Zander, 1975, 1986; Papaconstantinou, 1979). Such secondary characteristics include crests on the head (e.g., *Salaria* spp.), enlarged supra-orbital tentacles (e.g.,

Aidablennius sphynx and *Parablennius* spp.), colour markings on the head and/or body, club-like glands on the tips of the soft rays of the dorsal, caudal and anal fins (e.g., *Microlipophrys* spp.) and glands on the anal fin close to the urogenital opening (e.g., *Salaria* spp. and *Parablennius* spp.). Blennies are poligynandric (i.e., both sexes are variable in their mate numbers in a breeding season, but males are more variable than females) and their general breeding pattern involves males defending a territory where females deposit adhesive eggs in the inner surface of a cavity (crevices, holes, empty shells, or other nesting cavities), and where parental care is given exclusively by the males, guarding the eggs until they hatch (Gibson, 1969; Wirtz, 1978; Almada and Santos, 1995). In some species, alternative male mating tactics (parasitic males) have been described (Santos, 1985b; Ruchon *et al.*, 1995; Gonçalves *et al.*, 1996; Oliveira *et al.*, 2001b; Neat *et al.*, 2003).

Communication has been studied with respect to visual and olfactory sensory modalities in the context of reproductive behaviour. Studies on communication using other sensory modalities are very scarce or absent, but its use by blennies cannot be discarded. An initial study on sound production by Tavolga (1958) describes courtship sounds emitted by males *Chasmodes bosquianus* in aquaria, but this was not followed up by further studies on the same species. Recently, De Jong *et al.* (2007) described the acoustic behaviour of the rock-pool blenny *Parablennius sanguinolentus parvicornis*, in nature. Males produce a grunt-like call when a female approaches and is inside their nest; the call, the variation of which is related with male size, may affect the ultimate decision of a female to spawn and its emission (occurrence and rate of emission) may express some male mate preference (De Jong *et al.*, 2007).

This chapter reviews the available evidence on how and why blennies communicate, focusing on visual and chemical communication in reproductive behaviour. Since communicating is ultimately an integrated process from a behavioural and cognitive perspective, and partitioning communication by modality is only a useful heuristic tool, we will discuss the possible use of multimodal signals between males and females. Finally, the potential for communication networks in chemical and visual communication of blennies is explored with respect to how, in conjunction with other selective pressures, it may have shaped the evolution of their communication mode.

COMMUNICATION USING VISION

Vision in Blennies

A general model describing vision in blennies is difficult to establish for two reasons. First, the properties of visual systems are mainly the result of species-specific adaptations to environmental light and lifestyle and, thus, it is difficult to establish general properties for visual systems based only on phylogenetic relationships. As an example, among four species of Lake Malawi cichlids, there is a strong variation in spectral sensitivity due to differential expression of primarily only three of the seven available cone opsin-coding genes (Parry *et al.*, 2005). Opsins are proteins that are partially responsible for visual spectral sensitivity and the interspecific divergence in their expression pattern—in spectral sensitivity—is probably the result of species differences in both behaviour and environmental light conditions. Likewise, given the fact that the behavioural repertoire of blennies is both rich and diverse and that its members have a widespread distribution, occupying habitats with variable light conditions, it can be predicted that the properties of their visual systems reflect this diversity. Second, although vision has been extensively studied in fish (reviewed in Douglas and Djamgoz, 1990), particularly in some families (e.g., Cichlidae, Salmonidae), it has been poorly studied in blennies. Despite these limitations to establish a general pattern for the visual system, a brief overview of the available data on blennies' vision is presented.

Similar to other fish living in shallow waters, blennies are assumed to make extensive use of vision and, thus, to have a well-developed visual system. This is supported by histological analyses of their retina. Species investigated so far were found to have a fovea or *area* in the retina, i.e., a region with a high concentration of photoreceptors (Kahmann, 1934; Rochon-Duvigneaud, 1943; Collin and Pettigrew, 1988a, b). These regions are associated with increasing visual acuity and, like in most other benthic fishes, they are located in the medial or dorso-temporal retina, allowing for high visual acuity forward and downward (Kahmann, 1936). Blennies living in low-light environments (*Blennius ocellaris* at depths between 10 and 400 m and the cave-dwelling *Microlipophrys nigriceps*) have larger eyes and larger cone outer segments in order to increase the sensitivity of the photonic system, i.e., to decrease the light threshold level necessary for cone-based vision (Zaunreiter, 1990). In contrast, shallow water blennies seem to have smaller eyes and higher cone densities (Zander, 1972; Zaunreiter *et al.*, 1985; Zaunreiter, 1990), presumably having a higher visual acuity. In addition, some intertidal species have evolved adaptations for aerial vision. For instance, the supratidal blenny *Alticus kirkii*, has a *cornea propria*, a structure separated from the anterior part of the cornea and connected to the lens, which defines an additional eye chamber that can be extended by contracting the lens, thus, allowing a further adjustment of the focal point during aerial vision (Zander, 1974).

In 14 species of Mediterranean blennies, the retina contains rods, single cones and paired cones arranged in a regular mosaic (Zaunreiter *et al.*, 1989). This arrangement is typical of species that greatly rely on vision for their lifestyle (Wagner, 1990). The presence of several cone classes usually predicts the occurrence of different photopigments, thus creating the potential for colour vision. Accordingly, multiple cone photopigments have been found in the two blenny species studied so far, the shanny, *Lipophrys pholis* (Loew and Lythgoe, 1978) and the peacock blenny, *Salaria pavo* (White *et al.*, 2004). In the shanny, apart from a rod photopigment, microspectrophotometry has revealed three cone photopigments sensitive to short, mid and long-wavelengths (Loew and Lythgoe, 1978). The spectral sensitivity of the peacock blenny was studied using both microspectrophotometry and behavioural methods. In this species, the lenses exhibit a short-wavelength cut-off (<400 nm) and, unlike the shanny, short-wavelength sensitive photopigments were not detected in the retina (White *et al.*, 2004). The difference in cone classes between these two species may relate to differences in their retinal cone mosaics. The cone mosaics of blennies have paired cones and single central cones and, in some species, accessory single cones (Ali and Anctil, 1976). *S. pavo* lacks the accessory single cones (Zaunreiter *et al.*, 1985) which are, possibly, the short-wavelength sensitive cones of *L. pholis* and other blennies. The lack of short-wavelength sensitive cones and the presence of an ultra-violet cut-off filter in the lens of *S. pavo* is surprising, as these blennies present several bluish spots in the body and an iridescent blue eye-spot in the head that reflect both in the blue and in the ultra-violet part of the spectrum (M. Cummings, D. Gonçalves and R.F. Oliveira, unpublished data). However, measures of spectral sensitivity using the optomotor response showed that *S. pavo* was able to detect wavelengths in the bluish part of the spectrum, possibly using the mid-wavelength sensitive cones which have a wide response spectrum (White *et al.*, 2004).

From the above examples, and keeping in mind the previously mentioned limitations on drawing general properties of visual systems, it seems that blennies have well-developed visual systems of high sensitivity and acuity, capable of colour discrimination and movement detection, as predicted by Kotrschal (1999). Interspecific variation in the visual system may be related with behavioural and lifestyle differences as also with the light properties of their particular environments.

Visual Displays in Blennies

Studies from an ethological perspective show that blennies have complex and stereotyped sexual and agonistic behaviours, and that there is significant interspecific variation in their rich repertoire of visual displays. Therefore, rather than trying to establish general behavioural patterns, examples are given illustrating the diversity of visual displays in blennies.

Most male blennies establish reproductive territories around a crevice or hole in the substrate—which is used as a nest—and undertake aggressive displays against intruding males and courtship displays towards females. In some species, males exhibit conspicuous swimming behaviours in the water column, which may signal their territories to females and other male competitors. These displays include the 'vertical loop swim' of *Istiblennius zebra* (Phillips, 1977), the 'hovering swim' of *Parablennius pilicornis* (Denoix, 1984), or the 'loop swimming' of *P. sanguinolentus parvircornis* (Santos, 1985a). Intertidal blennies living in more turbulent environments do not display in the water columns, instead displaying by head movements like 'nodding' or 'lateral head displays', usually associated with an increase in intensity of head- and body-colour markings, with the male partially inside the nest (Abel, 1964; Zander, 1975; Papaconstantinou, 1979; Almada *et al.*, 1990; Almada and Santos, 1995; Gonçalves and Almada, 1998).

Agonistic interactions may include ritualised threatening displays such as 'lateral displays' (male erecting the unpaired fins and showing the flank to the opponent, and may show lateral turnings of the head), 'undulation swings' (males slowly undulate their bodies from the front to the back), or 'horizontal loops' (male swim around in front of the opponent in a horizontal circle) (Gibson, 1968; Phillips, 1977; Almada *et al.*, 1983; Santos and Barreiros, 1993). In some species, e.g., *S. pavo* (pers. obs.) and *Hypsoblennius brevipinnis* (Losey, 1976), males and sometimes females may turn on, within seconds, an agonistic coloration during aggressive displays. Submission to opponents may also be signalled with visual displays that involve retracting the unpaired fins and lowering the body towards the substrate, and may also include lateral body rotations like the 'lateral presentation' of *P. sanguinolentus parvicornis* males (Santos and Barreiros, 1993). Often, submission postures are accompanied by a general lightening of body colours like in *Hypsoblennius* spp. (Losey, 1976).

In general, males take the initiative in courtship while females adopt a more passive role. Males usually leave the nest and perform conspicuous behaviours towards the female such as 'circling', '8-figure swimming', 'zig-zag swimming', 'lateral body jerkings' or 'quivering' (Fishelson, 1963; Phillips, 1977; Patzner *et al.*, 1986; Santos and Barreiros, 1993). During these displays, males can turn on a nuptial coloration within seconds. As an example, male *S. pavo* increase the conspicuousness of dark bars in the head, crest and chin as the yellow background becomes more intense, with blue spots and lines in the body also increasing in conspicuousness (Patzner *et al.*, 1986).

Females may also exhibit courtship displays. For instance, *S. pavo* females can display to nest holders by presenting their belly while flicking the pectoral fins and rapidly open-and-closing the mouth in synchrony (Patzner *et al.*, 1986). During these displays, females turn on a conspicuous nuptial coloration that consists of vertical dark stripes contrasting with a light background throughout the body (Fishelson, 1963; Patzner *et al.*, 1986). In some populations of *S. pavo*, female courtship is more common than male courtship due to ecological constraints. In the Ria Formosa (southern Portugal), nest sites are scarce and females compete for the access to nesting males, taking the initiative in courtship and playing the major role during sexual interactions (Almada *et al.*, 1995). In contrast, in an Adriatic population, where nest sites are abundant, males display more courtship acts than females (J. Saraiva, D. Gonçalves and R.F. Oliveira, unpublished data). These results suggest that both males and females adjust their visual displays according to ecological settings.

Male blennies may also display to other males in a sexual context. In *P. sanguinolentus parvicornis*, small males achieve parasitic fertilisations of the eggs either by sneaking into the nest of territorial males (i.e., bourgeois males *sensu* Taborsky, 1997) during spawning events or by helping in territory defence and gain a privileged access to the nest during spawning episodes (Santos, 1985b; Oliveira *et al.*, 2001a). In *S. pavo*, small males reproduce using a different strategy. These males have a female-like morphology, and mimic female courtship and nuptial coloration in order to approach the nests during spawning events and parasitically fertilise the eggs (Gonçalves *et al.*, 1996). As discussed below, the occurrence of female-mimics is likely to have a strong impact on male and female courtship displays and, potentially, on the properties of sensorial systems.

In summary, the examples described above clearly demonstrate that blennies make extensive use of visual displays to communicate reinforcing the idea that vision is well developed in this group.

COMMUNICATION USING OLFACTION

Pheromones and Their Detection

Pheromones are molecules (single or in a mixture) released to the environment that cause specific and adaptive physiological and behavioural responses in conspecifics (Stacey and Sorensen, 2002). Vertebrates, in general, release a wide range of different molecules, from relatively small volatile compounds to water-soluble peptides, which are involved in a multitude of different social behaviours such as finding and identifying a mate, regulating the level of aggression and social dominance, mediating the recognition of kin and non-kin, and triggering of alarm reactions (Wyatt, 2003; Kikuyama *et al.*, 2005; Brennan and Zufall, 2006).

In fishes including the Cyclostomata (hagfish and lamprey), there has been growing evidence for water-soluble compounds acting as pheromones and playing a role in all the above-mentioned varieties of social behaviours, as well as in migratory behaviour (Liley, 1982; Wisenden and Stacey, 2005; Rosenthal and Lobel, 2006; Stacey and Sorensen, 2006; Barata *et al.*, 2007). However, with a few exceptions, the chemical structures of the compounds involved remain largely unknown. In many teleosts, reproductive hormones (steroids and prostaglandins) and/or their metabolites released through the urine and/or gills act as hormonal pheromones inducing dramatic changes in the reproductive behaviour and physiology of receivers (Stacey and Sorensen, 2006 and references therein). Other chemical classes of compounds have also been shown to function as sex pheromones in a few fish species. In masu salmon, *Oncorhynchus masou*, L-kynurenine (a major metabolite of the amino acid L-tryptophan in vertebrates) is released through the urine of mature females and acts as a male-attracting pheromone (Yambe *et al.*, 2006). In sea lampreys, *Petromyzon marinus*, males release large amount of bile acids through the gills, which act as sex pheromones attracting females at long distances (Li *et al.*, 2002, 2003; Siefkes *et al.*, 2003). Also, a mixture of sulphated sterols from sea lamprey larvae functions as a migratory pheromone for the adults (Sorensen *et al.*, 2005).

Behavioural and physiological responses to pheromones are mediated by the olfactory system that is, in general, well developed in fishes (for a recent review of the olfactory system, see Zielinski and Hara, 2007). The olfactory organ is a paired structure located in nasal or olfactory cavities located on the dorsal surface of the head to the *rostrum*. The anatomy of the olfactory organ varies across fish taxa, reflecting also differences in lifestyle and behaviour. However, in general, a few or several lamellae, each bearing the olfactory epithelium, form an olfactory rosette located in the bottom of the olfactory cavity; in addition, certain fish have accessory or ventilation cavities which mainly aid in directing the water in and out of the nasal cavity (Zeiske *et al.*, 1992; Kasumyan, 2004).

Receptor cells (sensory neurons) in the olfactory epithelium are of two main types, ciliated and microvillous cells, and a third less abundant type, the crypt cells. These three cell types have clear structural differences and seem to detect different classes of chemical stimuli (Kasumyan, 2004; Zielinski and Hara, 2006; Hamdani and Doving, 2007; Hamdani *et al.*, 2008). Axons of all receptor cells join into the olfactory nerve that conveys olfactory information to paired olfactory bulbs that are the primary olfactory centres for information processing. The olfactory information is further relayed to secondary centres in the telencephalon by the olfactory tract and information causing different behavioural responses (food search, alarm reaction, sexual behaviour) is carried by different nerve fibres (Hamdani and Doving, 2007 and references therein). In the telencephalon, the olfactory information is thought to be integrated with other sensory modalities resulting in appropriate physiological and behavioural responses.

To the best of our knowledge, the olfactory system of blennies has not been studied. Preliminary observations in *S. pavo* females using scanning electron microscopy indicate the presence of two olfactory lamellae bearing ciliated, microvillous and, possibly crypt cells, and a well-developed ventilation sac in each nasal cavity (J. Araújo, O. Lopes and E.N. Barata, unpublished data).

Pheromones in Blennies

The involvement of pheromones in the reproduction of blennies has been hypothesised for a long time. One of the first empirical studies was by Losey (1969), who showed that water conditioned by sexually active conspecifics (males and females) attracted sexually active males in three *Hypsoblennius* species. Glandular complexes on the fins of male blennies were thought to be the source of the putative pheromones (Eggert, 1931; Wickler, 1957; Losey, 1969; Blüm, 1972). Further empirical evidence was obtained by Laumen *et al.* (1974), showing that administration of mammalian luteinising hormone to *S. pavo* males induces size-increase in the glands on the first two anal-fin rays (anal glands), and that fully developed anal glands are a source of putative pheromones effective in attracting females. Later, Lahnsteiner *et al.* (1993) showed that the testicular blind pouches (tubular evaginations of the spermatic ducts) of male blennies, which do not function as sperm reservoirs, secrete steroid glucuronides during the breeding season, and speculated that such compounds could act as pheromones by analogy with the black goby (*Gobius niger*) and the African catfish (*Clarias gariepinus*) (see Patzner and Lahnsteiner, this book). The mesorchial gland of the black goby (Colombo *et al.*, 1980) and the seminal vesicle of the African catfish (Resink, 1987; Resink *et al.*, 1989) produce large amounts of steroid glucuronides which have been suggested to act as sex pheromones that attract females.

Only recently, further progress was achieved on the issue of male pheromones in blennies by making use of behavioural and endocrinological assays combined with the recording of the electro-olfactogram to identify chromatographic fractions of male-conditioned water and anal glands containing putative pheromones. Clearly, the anal gland of male blennies is a source of a sex pheromone that attracts reproductive females from a distance (Barata *et al.*, 2008; Serrano *et al.*, 2008a), thereby enhancing male reproductive success (Barata *et al.*, 2008). In *S. pavo*, each of the two anal glands is a highly folded glandular epithelium containing numerous clusters of secretory cells. These cells differentiate during the breeding season in response to rising plasma-levels of androgens (11-ketotestosterone from the glandular region of the testis—testicular gland) and produce neutral mucins (glycoproteins) (Serrano *et al.*, 2008b, c). Furthermore, the clusters of secretory cells likely produce highly hydrophilic compounds (mostly smaller than 500 Da but may include also compounds larger than 10 kDa) that are odorants for females and may constitute the anal gland pheromone (Serrano *et al.*, 2008a).

The odorants produced by anal glands seem to be different from those derived from hormonal pheromones described in other fishes, and may include amino acids and/or peptides that could be related with mucins produced by the secretory cells. Whether the anal gland pheromones of male blennies are water-soluble peptides or other type of hydrophilic molecules, production of pheromones together with mucous would imply a sustained slow release of sex pheromones for attracting females from a distance to the nesting site throughout the breeding season (Serrano *et al.*, 2008a).

In addition to the anal gland pheromone, sexually mature *S. pavo* males release to the water less hydrophilic compounds which are odorous to the females. These odorants may be responsible for arresting reproductive females in the vicinity of the sending male and, thereby, act as sex pheromones (Serrano *et al.*, 2008a). Although further research is needed, these putative pheromones may originate in testicular accessory organs, i.e., blind pouches and/or testicular glands that are a source of potent male odorants for females (Serrano *et al.*, 2008c) and may include steroid glucuronides. The steroid maturation hormone 17,20β-dihydroxy-4-pregnen-3-one and its sulphate and glucuronide conjugates, and the androgen 11-ketotestosterone are excluded as possible candidates since the olfactory system of females is not sensitive to neither of these sex hormones (R.M. Serrano, P.C. Hubbard and E.N. Barata, unpublished data) which are hormonal pheromones in other teleost species (see Stacey and Sorensen, 2006 and references therein).

A multi-component male pheromone, comprising different compounds of different physiological origin, provides chemical information to the females, the effect of which in their reproductive behaviour may depend on the distance to the sending male (Serrano *et al.*, 2008a, b, c). The anal gland pheromones may disperse through long distances and convey information to females about species identity and location of potential mates. At close distance from the male in its nesting site, putative pheromones from the testicular accessory organs, especially the blind pouches, may arrest the females in the vicinity of a male that is likely to fertilise the eggs and invest in parental care. The mixture of anal-gland and gonad pheromones may affect female's ultimate decision to spawn.

The hypothesis of a multi-component male pheromone in *S. pavo* requires further studies on the chemical identities of male odorants and their effect on female reproductive behaviours. However, given that all combtooth male blennies studied so far have testicular blind pouches (Patzner and Lahnsteiner, 1999; Richtarski and Patzner, 2000; Patzner and Lahnsteiner, this book) and glands on the fins (either anal glands or club-like glands on the tips of rays in the several fins) (Northcott and Bullock, 1991 and references therein), it is reasonable to predict that multi-component pheromones may be widespread in this fish group.

It is unknown whether *S. pavo* males also respond to male pheromones or if females produce pheromones that could affect the behaviour of males or females. In the black goby, which has a mating system similar to that of combtooth blennies, territorial males respond aggressively to putative pheromones from conspecific males but parasitic males are pheromonally inconspicuous (Locatello *et al.*, 2002).

Evolution of Chemical Communication

As noted first by Døving (1976), hormonal products are pre-adapted to serve as pheromones because they are produced and released at relevant times (Stacey and Sorensen, 2006). Indeed, several teleost species use reproductive hormones and/or their metabolites as reproductive pheromones and its use is thought to be widespread among teleosts (Stacey and Sorensen, 2006 and references therein). Although senders can have some control as to the timing, context, and quantity of release of hormonal pheromones, e.g., accumulation and release *via* the urine (Appelt and Sorensen, 2007), remains to be determined in most cases to what extent waterborne hormones are acting as pheromonal signals, rather than as pheromonal cues incidentally released by reproductive individuals.

Evolution of true chemical communication (pheromonal signals) would be expected when individual donors improve their reproductive success due to change(s) in the way they produce and/or release a pheromonal cue (Stacey and Sorensen, 2006). However, sex hormones function endogenously in senders; therefore, their function as a pheromonal signal seems to be constraint by their function as hormones. Consequently, evolution of sex pheromone signals in teleosts would require strong selective pressures (e.g., sexual selection) imposed by receivers upon senders which in turn would find ways to specialise their release of pheromones; e.g., through increase of diversity, amount and water-solubility of sex steroid metabolites released to the water by specialised cells and/or differentiation of external glands under control of key sex hormones, that would produce and secrete specialised chemical compounds.

Sexual selection is a selective pressure that likely has been operating in the evolution of blennies' reproductive behaviour and may have driven the evolution of their putative multi-component pheromone system. The androgen 11-ketotestosterone (presumably from the testicular gland) is a key hormone that organises the production of putative pheromones from the anal gland and testicular blind pouches (Serrano *et al.*, 2008 c) and is positively correlated with the expression of parental/bourgeois male behaviour (Oliveira *et al.*, 2001c, d). Therefore, the hypothesised multi-component male pheromone reflects accurately the physiological condition of a male and can convey reliable information to females about the location of a conspecific mate that is likely to fertilise the eggs and invest in parental care (i.e., fanning and guarding the eggs). Consequently, female reproductive success may have been an important driving force for the evolution of the proposed chemical communication system. In addition, a possible role of male-male competition in shaping this evolutionary process cannot be excluded.

MULTIMODAL COMMUNICATION

Communication is ultimately an integrated process from a behavioural and cognitive perspective. Although communication is partitioned by modality as a useful heuristic tool of study, the outcome of the process integrates several sensory modalities. For example, to track an odour source, Banded kokopu (*Galaxias fasciatus*) use hydrodynamic cues detected by the lateral line system rather than bilateral comparison of olfactory stimuli (Baker *et al.*, 2002). In the Mozambique tilapia (*Oreochromis mossambicus*), males use both visual displays and chemical signals released via the urine to advertise their aggressive motivation and social status (Barata *et al.*, 2007) and in the cichlid *Pseudotropheus emmiltos*, both chemical and visual information are necessary for recognition of conspecific males (Plenderleith *et al.*, 2005).

In blennies, since visual, olfactory and apparently, acoustic signals are sent by males, there is potential for females to use various sensory modalities in their reproductive behaviour. Although male pheromones seem to play no role in triggering female courtship which is exhibited in close visual contact with a nesting male in *S. pavo* (Gonçalves *et al.*, 2002), the relative importance of different sensory modalities can change with the context or distance to the sender. At long distance without the male in sight, the anal gland pheromone may lead the female to a closer distance and then—together with putative pheromones from the testicular blind pouches—the full pheromonal 'bouquet' may arrest the female in the vicinity of the potential mate. Furthermore, at close distance visual and pheromonal signals may reinforce the male's features causing a step further, e.g., female courting the male and entering the nest. Finally, acoustic signals may also be involved in final steps of choosing the mate given their apparent importance in the reproductive behaviour of male blennies, as suggested by De Jong *et al.* (2007).

In the peacock blenny (*S. pavo*), females exert mate choice by choosing larger males with a larger head crest, anal gland and genital papilla, that have more eggs in the nest and court females (Oliveira *et al.*, 1999; Fagundes *et al.*, 2007). Pheromones from the anal gland seem to be crucial in female mate-choice, at least when eggs are not present in the nest (Barata *et al.*, 2008). Also, it has been suggested that females use a 'one-step decision' in searching for mates, i.e., females mate with males that satisfy an adjustable threshold criterion, balancing the quality of the mates expected to find in the next step of the search and the effort of finding them (Fagundes *et al.*, 2007). Therefore, the use of several sensory (e.g., vision and olfaction) modalities to assess mate quality (i.e., likelihood of egg fertilisation and efficient paternal investment) may increase the amount of information gained and decrease the time or energy spent on male assessment by females. Moreover, multi-component signals may also aid in discriminating males and in recognition through associative learning. For reviews on the use of multi-component signals in animal communication, see Rowe (1999), Candolin (2003), Partan (2005) and Harper (2006).

In summary, the reproduction of blennies seems to involve communication using multiple sensory modalities and multi-component signals within a sensory modality. However, more empirical evidence is necessary to make generalisations and understand the consequences that the use of multiple signals may have in the process of sexual selection and speciation in blennies.

COMMUNICATION NETWORKS

Signals may travel distances large enough to be detected by third parties and thus are usually available for several receivers simultaneously (McGregor, 1993; McGregor and Dabelsteen, 1996; Peake, 2005). The concept that animals typically communicate in networks (involving at least one signaller and more than one receiver) and the ecological and evolutionary consequences of such networks have been explored thoroughly for visual, acoustic and chemical signals by McGregor (2005). Wisenden and Stacey (2005) have discussed the chemical communication networks in fishes. Here, some general predictions on the evolution of fish signals and sensory systems will be presented under the conceptual framework of communication networks and hypothesis will be drawn with respect to communication networks in blennies.

Eavesdropping refers to 'the use of information in signals by individuals other than the primary target' (Peake, 2005). Eavesdropping may have costs for both senders and receivers. For example, the interception of courtship signals by parasitic males can have negative fitness consequences for both the bourgeois male and the female. For the bourgeois male because egg fertilisations can be lost for the parasitic male, thus decreasing its own reproductive success; for the female either because bourgeois males will decrease parental investment when the probability of paternity decreases and/or because larger bourgeois males can offer higher genetic benefits to females than small parasitic males (see Taborsky, 1999). At this point, it is useful to make a distinction between species where the parasitic males are morphologically distinct from females and reproduce by sneaking or darting into the nest and species where parasitic males are similar to females.

In species with morphologically distinct parasitic males, the females and bourgeois males are expected to be able to detect them and should evolve inconspicuous signals, i.e., 'conspiratorial whispers' (Dawkins and Krebs, 1978; Maynard Smith, 1991; Johnstone, 2000), or adjust their signalling behaviour when the parasitic males are in the vicinity in order to decrease the probability of their signals being intercepted. For instance, in the Mediterranean wrasse *Symphodus ocellatus*, bourgeois males decrease courtship displays towards females when the number of parasitic males is experimentally increased and the reverse occurs when the number of parasitic males is decreased (van den Berghe and Warner, 1989; Alonzo and Warner, 1999; 2000). Likewise, females of this species prefer to spawn when parasitic males are absent, increasing their spawning rate several fold when parasitic male were experimentally removed (van den Berghe and Warner, 1989; Alonzo and Warner, 2000). Thus, when the parasitic males are morphologically distinct from females and impose costs on both the bourgeois male and the female, courtship signals are expected to be inconspicuous or not to be displayed in the presence of eavesdroppers, and parasitic males are expected to counteract by evolving strategies and perceptual systems that favour signal interception.

In the case of female-mimicry, different predictions can be drawn. In these species, parasitic males reproduce by mimicking females in order to approach the nests of bourgeois males and achieve parasitic fertilisations of the eggs. Thus, female-mimics rely on deception to reproduce and it can be predicted that bourgeois males will be under a strong selective pressure to discriminate females from female-mimics. Females, on the other hand, are expected to gain by signalling to the males that they are females and not female-mimics in order to be accepted into the nest. This could lead into an increase in the complexity of female displays in order to difficult its mimicry by parasitic males. Because parasitic males rely on female-mimicry to reproduce, an evolutionary arms race between the females and the female-like male displays is predicted (Dawkins and Krebs, 1979).

In blennies, female-mimicry has only been described for *S. pavo* (Gonçalves *et al.*, 1996). Small males (sneakers), with higher gonadosomatic index than bourgeois males, reproduce by mimicking female morphology and behaviour, and sneaking fertilisation of eggs during spawning events (Gonçalves *et al.*, 1996; Oliveira *et al.*, 2001e). Sneaker males have lower plasma concentration of androgens than the nesting males (Oliveira *et al.*, 2001b) and do not differentiate the anal gland. Interestingly, this species presents one of the most elaborated female courtship displays described in blennies. It is unknown whether this is a consequence of the predicted evolutionary arms race between females and female-like males, or of selective pressure imposed by the perceptual system of bourgeois males, or both. Small sneaker-males size-matching females are able to deceive bourgeois males as these direct an equivalent proportion of courtship and aggressive acts towards females and sneakers (Gonçalves *et al.*, 2005). However, with increasing body size of the sneakers, the efficacy of female-mimicry seems to decrease as bourgeois males direct more aggressive acts and less courtship displays towards these males when compared with females of the same size (Gonçalves *et al.*, 2005). Presumably, and considering that bourgeois males should be under a strong selective pressure to discriminate females from female-like males, larger parasitic males are easier to discriminate from females but the cues used by bourgeois males to achieve this have not been identified.

Also, sneaker males are more sensitive to longer wavelengths than both bourgeois males and females (White *et al.*, 2004). Whether this relates to the different behavioural tasks that bourgeois males, females and sneaker males need to accomplish remains to be tested.

In addition, although not tested yet, what can be predicted is the eavesdropping of sneaker males on the chemical communication between bourgeois males and females in order to find the bourgeois males more attractive to females. This may also impose a selective pressure shaping the elaboration of the chemical signal from bourgeois males, the perceptual system of females to male chemical signal and the male's perceptual system to possible odours from conspecifics.

Eavesdropping can also confer fitness advantages to both senders and receivers, although this is less intuitive. In some species, the females copy the mate choice of other females by observing which males are chosen by other females (Dugatkin, 1992). Thus, a high quality male (i.e., a male that is usually chosen by the females) should gain by exhibiting conspicuous signals that favour eavesdropping by other females. The primary female, i.e., the female being directly courted by the male may also gain by having other females eavesdropping on the interaction. For example, in some fish species, the males provide more or better parental care to the eggs when the number of eggs increases (Coleman *et al.*, 1985). The primary female can gain direct benefits in terms of increasing male parental care over her eggs by having its courtship interaction eavesdropped by other females. In such a scenario, female and male courtship signals should be conspicuous and promote eavesdropping. Presently, no empirical evidence is available to support this hypothesis in blennies.

Finally, eavesdropping can have different fitness consequence for senders and receivers and thus generate conflicts between them. For instance, in the bitterling *Rhodeus sericeus*, females increase the fertilisation success of their eggs if sneakers participate in spawning (Smith and Reichard, 2005). Thus, females perform conspicuous behaviours in order to increase eavesdropping probability and participation in spawning by sneakers (Smith and Reichard, 2005). Bourgeois males, however, decrease their reproductive success when sneakers participate in spawning e.g., *Lepomis punctatus* (Dewoody *et al.*, 2000) and, thus, are expected to avoid eavesdropping. This creates a conflict between females and males. In this case, both male and female sexual signals result from a compromise between costs and benefits for both. Males may reject or punish females that solicit spawning with signals above a certain threshold, as this would attract eavesdroppers, and females may not respond to males that display below a certain threshold, as this would not attract eavesdroppers. A similar argument can be predicted for situations where eavesdropping favours males but has costs for females. In the above example on female copying, males may gain by having the interaction intercepted by other females but the primary female may suffer costs from attracting other females (for example, egg fertilisation rate may be lower if other females participate in spawning). Again, male and female signals are expected to result from an evolutionary compromise between the sexes.

In summary, communicating in a network has evolutionary consequences for signal and perceptual systems design. Fish are not an exception and the properties of their signals and sensory systems are best understood in the evolutionary context of communication networks. However, there is still a long way to go in order to fully understand the proximate and ultimate causes of communication in fish and especially in blennies.

SUMMARY AND CONCLUSIONS

The scarcity of studies limits the validity of generalisations on how and why blennies communicate. However, it is clear that blennies use signal and perceptual systems involving hearing, vision and olfaction for communication in the context of their reproduction.

Very little is known about sound production and reception in blennies, since there is only one study showing that male rock-pool blennies produce grunt-like calls during courtship which may play a role in female spawning decision.

Vision is clearly important in blennies as they have well-developed visual systems with interspecific variation that may relate to behavioural and lifestyle differences. Moreover, males have visual features clearly distinct from females and there is great intra- and interspecific variation in the rich repertoire of courtship and agonistic visual displays of blennies, reinforcing the idea that visual communication is important in their reproduction.

Female blennies use their sense of olfaction to detect male odours that originate in external epidermal glands and testicular accessory organs, the production of which depends on androgen plasma levels in correlation with the seasonal development of males' reproductive organ. These odours are putative sex pheromones, likely comprised of a mixture of different chemical compounds, which attract females and may affect their mating decision. However, only the peacock blenny has been studied. Therefore, establishing the general use of pheromones in blennies and how and why pheromones are involved in their reproduction requires additional studies with more species and chemical identification of the active compounds.

In conclusion, communication using multiple sensory modalities is involved in the reproduction of blennies. Females may use vision and olfaction to find males in their nesting sites and integration of visual, olfactory and acoustic signals from males may affect their ultimate decision to spawn; the relative importance of different sensory modalities affecting female reproductive behaviour may change with the context and/or distance to male signallers. Communication within sexes and eavesdropping on signalling systems has not been studied. Clearly, proximate and ultimate factors underlying communication deserve further attention in order to understand how and why blennies, as well as fish in general, communicate to achieve their reproduction.

Acknowledgements

Ministry of Science and Higher Education and European Social Funds through the Portuguese National Science Foundation (FCT) has funded research coordinated by E.N. Barata on sex pheromones in male blennies (projects Praxis/P/BIA/11157/98 and POCTI/BSE/45843/2002) and D. Gonçalves through a post-doctorate fellowship (SFRH/BPD/7188/2001).

References

Abel, E.F. 1964. Freiwasserstudien zur Fortpflanzungsethologie zweier Mittelmeerfische, *Blennius canevae* Vinc. und *Blennius inaequalis* C. V. *Zeitschrift für Tierphysiologie* 21: 205-222.

Ali, M. and M. Anctil. 1976. *Retinas of Fishes: An Atlas*, Springer-Verlag, Berlin.

Almada, V.C. and R.S. Santos. 1995. Parental care in the rocky intertidal: A case study of adaptation and exaptation in Mediterranean and Atlantic blennies. *Reviews in Fish Biology and Fisheries* 5: 23-37.

Almada, V.C., E.J. Gonçalves, R.F. Oliveira and A.J. Santos. 1995. Courting females—Ecological constraints affect sex-roles in a natural population of the blenniid fish *Salaria pavo*. *Animal Behaviour* 49: 1125-1127.

Almada, V., J. Dores, A. Pinheiro, M. Pinheiro and R.S. Santos. 1983. Contribuição para o estudo do comportamento de *Coryphoblennius galerita* (L.) (Pisces: Blenniidae). *Memórias do Museu do Mar, Série Zoológica* 2: 1-165.

Almada, V., R.F. Oliveira, E.N. Barata, E.J. Gonçalves and A.P. Rito. 1990. Field observations on the behaviour of the breeding males of *Lipophrys pholis* (Pisces: Blenniidae). *Portugaliae Zoologica* 1: 27-36.

Alonzo, S.H. and R.R. Warner. 1999. A trade-off generated by sexual conflict: Mediterranean wrasse males refuse present mates to increase future success. *Behavioral Ecology* 10: 105-111.

Alonzo, S.H. and R.R. Warner. 2000. Dynamic games and field experiments examining intra- and intersexual conflict: Explaining counterintuitive mating behavior in a Mediterranean wrasse, *Symphodus ocellatus*. *Behavioral Ecology* 11: 56-70.

Appelt, C.W. and P.W. Sorensen. 2007. Female goldfish signal spawning readiness by altering when and where they release a urinary pheromone. *Animal Behaviour* 74: 1329-1338.

Baker, C.F., J.C. Montgomery and T.E. Dennis. 2002. The sensory basis of olfactory search behavior in banded kokopu (*Galaxias fasciatus*). *Journal of Comparative Physiology* A188: 553-560.

Barata, E.N., P.C. Hubbard, O.G. Almeida, A. Miranda and A.V.M. Canário. 2007. Male urine signals social rank in the Mozambique tilapia (*Oreochromis mossambicus*, Peters 1852). *BMC Biology* 5:54 doi:10.1186/1741-7007-5-54.

Barata, E.N., R.M. Serrano, A. Miranda, R. Nogueira, P.C. Hubbard and A.V.M. Canário. 2008. Putative pheromones from the anal glands of male blennies attract females and enhance male reproductive success. *Animal Behaviour* 75: 379-389.

Blüm, V. 1972. Influence of ovine follicle stimulating hormone (FSH) and luteinizing-hormone (LH) on male reproductive system and skin of Mediterranean blenniid fish *Blennius sphinx* (Valenciennes). *Journal of Experimental Zoology* 181: 203-216.

Bradbury, J.W. and S.L. Vehrencamp. 1998. *Principles of Animal Communication*. Sinauer Associates, Inc., Sunderland.

Brennan, P.A. and F. Zufall. 2006. Pheromonal communication in vertebrates. *Nature (London)* 444: 308-315.

Candolin, U. 2003. The use of multiple cues in mate choice. *Biological Reviews* 78: 575-595.

Coleman, R.M., M.R. Gross and R.C. Sargent. 1985. Parental investment decision rules: a test in bluegill sunfish. *Behavioural Ecology and Sociobiology* 18: 295-299.

Collin, S.P. and J.D. Pettigrew. 1988a. Retinal topography in reef teleosts. I. Some species with well-developed areae but poorly developed streaks. *Brain, Behavior and Evolution* 31: 269-282.

Collin, S.P. and J.D. Pettigrew. 1988b. Retinal topography in reef teleosts. II. Some species with prominent horizontal streaks and high-density areae. *Brain, Behavior and Evolution* 31: 283-295.

Colombo, L., A. Marconato, P.C. Belvedere and C. Frisco. 1980. Endocrinology of teleost reproduction. A testicular steroid pheromone in the black goby, *Gobius jozo* L. *Bollettino Zoologico* 47: 355-364.

Dawkins, M.S. 1993. Are there general principles of signal design? *Philosophical Transactions–of the Royal Society of London* B340: 251-255.

Dawkins, R. and J.R. Krebs. 1978. Animal signals: information or manipulation? In: *Behavioural Ecology*, J.R. Krebs and N.B. Davies (eds.), Blackwell Scientific Publications, Oxford, pp. 282-309.

Dawkins, R. and J.R. Krebs. 1979. Arms races between and within species. *Proceedings of the Royal Society of London* B205: 489-511.

De Jong, K., N. Bouton and H. Slabbekoorn. 2007. Azorean rock-pool blennies produce size-dependent calls in a courtship context. *Animal Behaviour* 74: 1285-1292.

Denoix, M. 1984. *Zur Biologie des Schleimfisches Parablennius pilicornis Cuvier 1829 (Blenniidae, Perciformes) unter besonderer Berücksichtigung der sekundären Geschlechtsmerkmale des Männchens nach Hormonbehandlung.* Ph.D. Thesis, University of Tübingen.

Dewoody, J.A., D.E. Fletcher, M. MacKiewicz, S.D. Wilkins and J.C. Avise. 2000. The genetic mating system of spotted sunfish (*Lepomis punctatus*): mate numbers and the influence of male reproductive parasites. *Molecular Ecology* 9: 2119-2128.

Douglas, R.H. and M.B.A. Djamgoz. 1990. *The Visual System of Fish*. Chapman and Hall, London.

Døving, K. 1976. Evolutionary trends in olfaction. In: *The Structure–Activity Relationships in Olfaction*, G. Benz (ed.), IRL Press, London, pp. 149-159.

Dugatkin, L.A. 1992. Sexual selection and imitation: Females copy the mate choice of others. *American Naturalist* 139: 1384-1389.

Eggert, B. 1931. Die Geschlechtsorgane der Gobiiformes und Blenniiformes. *Zeitschrift für Wissenschaftliche Zoologie* 139: 249-559.

Fagundes, T., D.M. Goncalves and R.F. Oliveira. 2007. Female mate choice and mate search tactics in a sex role reversed population of the peacock blenny *Salaria pavo* (Risso, 1810). *Journal of Fish Biology* 71: 77-89.

Fishelson, L. 1963. Observations on littoral fishes of Israel I. Behaviour of *Blennius pavo* Risso (Teleostei, Blenniidae). *Israel Journal of Zoology* 12: 67-91.

Gibson, R.N. 1968. The agonistic behaviour of juvenile *Blennius pholis* L. (Teleostei). *Behaviour* 30: 192-217.

Gibson, R.N. 1969. The biology and behavior of littoral fish. *Oceanography and Marine Biology: An Annual Review* 7: 367-410.

Gonçalves, D., R. Matos, T. Fagundes and R. Oliveira. 2005. Bourgeois males of the peacock blenny, *Salaria pavo*, discriminate female mimics from females? *Ethology* 111: 559-572.

Gonçalves, D.M., E.N. Barata, R.F. Oliveira and A.V.M. Canário. 2002. The role of male visual and chemical cues on the activation of female courtship behaviour in the sex-role reversed peacock blenny. *Journal of Fish Biology* 61: 96-105.

Gonçalves, E.J. and V.C. Almada. 1998. A comparative study of territoriality in intertidal and subtidal blennioids (Teleostei: Blennioidei). *Environmental Biology of Fishes* 51: 257-264.

Gonçalves, E.J., V.C. Almada, R.F. Oliveira and A.J. Santos. 1996. Female mimicry as a mating tactic in males of the blenniid fish *Salaria pavo*. *Journal of the Marine Biological Association of the United Kingdom* 76: 529-538.

Hamdani, E.H. and K.B. Doving. 2007. The functional organization of the fish olfactory system. *Progress in Neurobiology* 82: 80-86.

Hamdani, E.H., S. Lastein, F. Gregersen and K.B. Doving. 2008c. Seasonal variations in olfactory sensory neurons—Fish sensitivity to sex pheromones explained? *Chemical Senses* 33: 119-123.

Harper, D.G.C. 2006. Maynard Smith: Amplifying the reasons for signal reliability. *Journal of Theoretical Biology* 239: 203-209.

Johnstone, R.A. 2000. Conflicts of interests in signal evolution. In: *Animal Signals*, Y. Espmark, T. Amundsen and G. Rosenqvist (eds.), Tapir Academic Press, Trondheim, pp. 465-485.

Kahmann, H. 1934. Über die Vorkommen einer fovea centralis im Knochenfischauge. *Zoologischer Anzeiger* 106: 49-55.

Kahmann, H. 1936. Über das foveale Sehen der Wirbeltiere. I. Über die fovea centralis und die fovea lateralis bei einigen Wirbeltieren. *Albrecht von Graefes Archiv für Ophthalmologie* 135: 265-276.

Kasumyan, A.O. 2004. The olfactory system in fish: Structure, function, and role in behavior. *Journal of Ichthyology* 44: S180-S223.

Kikuyama, S., T. Nakada, F. Toyoda, T. Iwata, K. Yamamoto and J.M. Conlon. 2005. Amphibian pheromones and endocrine control of their secretion. *Annals of the New York Academy of Sciences* 1040: 123-130.

Kotrschal, K. 1999. Sensory systems. In: *Intertidal Fishes: Life in Two Worlds*, M.H. Horn, K.L.M. Martin and M.A. Chotkowsky (eds.), Academic Press, San Diego, pp. 126-142.

Krebs, J.R. and R. Dawkins. 1984. Animal signals: Mind reading and manipulation. In: *Behavioural Ecology: An Evolutionary Approach*, J.R. Krebs and N.B. Davies (eds.), Blackwell Scientific Publications, Oxford, pp. 380-402.

Ladich, F., S.P. Collin, P. Moller and B.G. Kapoor. 2006a. *Communication in Fishes. I. Acoustic and Chemical Communication*. Science Publishers, Enfield (NH), USA.

Ladich, F., S.P. Collin, P. Moller and B.G. Kapoor. 2006b. *Communication in Fishes. II. Visual Communication and Electrical Communication*. Science Publishers, Enfield (NH), USA.

Lahnsteiner, F., B. Nussbaumer and R.A. Patzner. 1993. Unusual testicular accessory organs, the testicular blind pouches of blennies (Teleostei, Blenniidae). Fine structure, (enzyme-) histochemistry and possible functions. *Journal of Fish Biology* 42: 227-241.

Laumen, J., U. Pern and V. Blüm. 1974. Investigations on the function and hormonal regulation of the anal appendices in *Blennius pavo* (Risso). *Journal of Experimental Zoology* 190: 47-56.

Li, W., A.P. Scott, M.J. Siefkes, H. Yan, Q. Liu, S.-S. Yun and D.A. Gage. 2002. Bile acid secreted by male sea lamprey that acts as a sex pheromone. *Science* 296: 138-141.

Li, W., A.P. Scott, M.J. Siefkes, S.-S. Yun and B. Zielinski. 2003. A male pheromone in the sea lamprey (*Petromyzon marinus*): an overview. *Fish Physiology and Biochemistry* 28: 259-262.

Liley, N.R. 1982. Chemical communication in fish. *Canadian Journal of Fisheries and Aquatic Sciences* 39: 22-35.

Locatello, L., C. Mazzoldi and M.B. Rasotto. 2002. Ejaculate of sneaker males is pheromonally inconspicuous in the black goby, *Gobius niger* (Teleostei, Gobiidae). *Journal of Experimental Zoology* 293: 601-605.

Loew, E.R. and J.N. Lythgoe. 1978. The ecology of cone pigments in teleost fishes. *Vision Research* 18: 715-722.

Losey, G.S., Jr. 1969. Sexual pheromone in some fishes of the genus *Hypsoblennius* Gill. *Science* 163: 181-183.

Losey, G.S., Jr. 1976. The significance of coloration in fishes of the genus *Hypsoblennius* Gill. *Bulletin of the Southern California Academy of Sciences* 75: 183-198.

Maynard Smith, J. 1991. Honest signalling: The Philip Sydney game. *Animal Behaviour* 42: 1034-1035.

McGregor, P. 1993. Signalling in territorial systems: A context for individual identification, ranging and eavesdropping. *Philosophical Transactions of the Royal Society of London* B340: 237-244.

McGregor, P.K. 2005. *Animal Communication Networks*. Cambridge University Press, Cambridge.

McGregor, P.K. and T. Dabelsteen. 1996. Communication networks. In: *Ecology and Evolution of Acoustic Communication among Birds*, D.E. Kroodsma and E.H. Miller (eds.), Cornell University Press, Ithaca, New York, pp. 409-425.

McGregor, P.K. and T.M. Peake. 2000. Communication networks: Social environments for receiving and signalling behaviour. *Acta Ethologica* 2: 71-81.

Neat, F.C., L. Locatello and M.B. Rasotto. 2003. Reproductive morphology in relation to alternative male reproductive tactics in *Scartella cristata*. *Journal of Fish Biology* 62: 1381-1391.

Nelson, J.S. 1994. *Fishes of the World*. Wiley, New York.

Northcott, S.J. and A.M. Bullock. 1991. The morphology of the club glands on the dorsal fin of mature male shannies, *Lipophrys pholis* (L). (Blenniidae, Teleostei). *Journal of Fish Biology* 39: 795-806.

Oliveira, R.F., V.C. Almada, E. Forsgren and E.J. Gonçalves. 1999. Temporal variation in male traits, nesting aggregations and mating success in the peacock blenny. *Journal of Fish Biology* 54: 499-512.

Oliveira, R.F., A.V. Canário, M.S. Grober and R.S. Santos. 2001a. Endocrine correlates of male polymorphism and alternative reproductive tactics in the Azorean rock-pool blenny, *Parablennius sanguinolentus parvicornis*. *General and Comparative Endocrinology* 121: 278-288.

Oliveira, R.F., A.V.M. Canario and M.S. Grober. 2001b. Male sexual polymorphism, alternative reproductive tactics, and androgens in combtooth blennies (Pisces: Blenniidae). *Hormones and Behavior* 40: 266-275.

Oliveira, R.F., L.A. Carneiro and A.V.M. Canário. 2001c. Effects of androgens on social behaviour and morphology of alternative reproductive males of Azorean rock-pool blenny. *Hormones and Behavior* 39: 157-166.

Oliveira, R.F., L.A. Carneiro, D.M. Gonçalves, A.V.M. Canário and M.S. Grober. 2001d. 11-Ketotestosterone inhibits the alternative mating tactic in sneaker males of the peacock blenny, *Salaria pavo*. *Brain, Behavior and Evolution* 58: 28-37.

Oliveira, R.F., E.J. Gonçalves and R.S. Santos. 2001e. Gonadal investment of young males in two blenniid fishes with alternative mating tactics. *Journal of Fish Biology* 59: 459-462.

Papaconstantinou, C.A. 1979. Secondary sex characters of Blennioid fishes (Pisces: Blenniidae). *Thalassographica* 1: 57-75.

Parry, J.W.L., K.L. Carleton, T. Spady, A. Carboo, D.M. Hunt and J.K. Bowmaker. 2005. Mix and match color vision: Tuning spectral sensitivity by differential opsin gene expression in Lake Malawi Cichlids. *Current Biology* 15: 1734-1739.

Partan, S.R. and P. Marler. 2005. Issues in the classification of multimodal communication signals. *American Naturalist* 166: 231-245.

Patzner, R.A. and F. Lahnsteiner. 1999. The accessory organs of the male reproductive system in Mediterranean blennies (Blenniidae) in comparison with those of other blenniid fishes (tropical Blenniidae, Tripterygiidae, Labrisomidae, Clinidae, Chaenopsidae, Dactyloscopidae). In: *Behaviour and Conservation of Littoral Fishes*, V.C. Almada, R.F. Oliveira and E.J. Gonçalves (eds.), ISPA, Lisboa, pp. 179-228.

Patzner, R.A., M. Seiwald, M. Adlgasser and G. Kaurin. 1986. The reproduction of *Blennius pavo* (Teleostei, Blenniidae) V. Reproductive behavior in natural environment. *Zoologischer Anzeiger* 216: 338-350.

Peake, T.M. 2005. Eavesdropping in communication networks. In: *Animal Communication Networks*, P.K. McGregor (ed.), Cambridge University Press, Cambridge, pp. 13-37.

Phillips, R.R. 1977. Behavioral field study of the Hawaiian Rockskipper, *Istiblennius zebra* (Teleostei, Blenniidae). *Zeitschrift für Tierpsychologie* 43: 1-22.

Plenderleith, M., C. van Oosterhout, R.L. Robinson and G.F. Turner. 2005. Female preference for conspecific males based on olfactory cues in a Lake Malawi cichlid fish. *Biology Letters* 1: 411-414.

Resink, J.W., P.K. Voorthuis, R. Van den Hurk, R.C. Peters and P.G.W.J. Vanoordt. 1989. Steroid glucuronides of the seminal vesicle as olfactory stimuli in African catfish, *Clarias gariepinus*. *Aquaculture* 83: 153-166.

Resink, J.W., R. Van den Hurk, R.F.O.G. Van Zoelen, E.A. Huisman. 1987. The seminal vesicle as source of sex attracting substances in the African catfish, *Clarias gariepinus*. *Aquaculture* 63: 115-127.

Richtarski, U. and R.A. Patzner. 2000. Comparative morphology of male reproductive systems in Mediterranean blennies (Blenniidae). *Journal of Fish Biology* 56: 22-36.

Rochon-Duvigneaud, A. 1943. *Les Yeux et la Vision des Vertébrés*. Masson, Paris.

Rosenthal, G.G. and P.S. Lobel. 2006. Communication. In: *Behaviour and Physiology of Fish*, K.A. Sloman, R.W. Wilson and S. Balshine (eds.), Academic Press, London, pp. 39-78.

Rowe, C. 1999. Receiver psychology and the evolution of multicomponent signals. *Animal Behaviour* 58: 921-931.

Ruchon, F., T. Laugier and J.P. Quignard. 1995. Alternative male reproductive strategies in the peacock blenny. *Journal of Fish Biology* 47: 826-840.

Santos, R.S. 1985a. Estrutura e função dos territórios em machos parentais de *Blennius sanguinolentus* Pallas (Pisces: Blenniidae). *Memórias do Museu do Mar* 3: 1-46.

Santos, R.S. 1985b. Parentais e satélites: Tácticas alternativas de acasalamento nos machos de *Blennius sanguinolentus* Pallas (Pisces: Blenniidae). *Arquipélago* 6: 119-146.

Santos, R.S. and J.P. Barreiros. 1993. The ethogram of *Parablennius sanguinolentus parvicornis* (Valenciennes in Cuvier & Valenciennes, 1836) (Pisces: Blenniidae) from the Azores. *Arquipélago* A 11: 73-90.

Searcy, A.A. and S. Nowicki. 2005. *The Evolution of Animal Communication. Reliability and Deception in Signalling Systems*. Princeton University Press, Princeton.

Serrano, R.M., E.N. Barata, M.A. Birkett, P.C. Hubbard, P.S. Guerreiro and A.V.M. Canário. 2008a. Behavioral and olfactory responses of female *Salaria pavo* (Pisces: Blenniidae) to a putative multi-component male pheromone from the anal gland of males. *Journal of Chemical Ecology* 34: 647-658.

Serrano, R.M., O. Lopes, P.C. Hubbard, J. Araújo, A.V.M. Canário and E.N. Barata. 2008b. Seasonal cell differentiation and olfactory potency of secretions by the anal glands of male peacock blenny *Salaria pavo* (Risso, 1810). *Journal of Fish Biology* 73: 1790-1798.

Serrano, R.M., P.C. Hubbard, O. Lopes, J. Araújo, A.V.M. Canário and E.N. Barata. 2008c. 11-ketotestosterone stimulates putative sex pheromone production in the male peacock, *Salaria pavo* (Risso 1810). *Biology of Reproduction* 79: 861-868.

Siefkes, M.J., A.P. Scott, B. Zielinski, S.-S. Yun and W. Li. 2003. Male sea lampreys, *Petromyzon marinus* L., excrete a sex pheromone from gill epithelia. *Biology of Reproduction* 69: 125-132.

Smith, C. and M. Reichard. 2005. Females solicit sneakers to improve fertilization success in the bitterling fish (*Rhodeus sericeus*). *Proceedings of the Royal Society* B272: 1683-1688.

Sorensen, P.W., J.M. Fine, V. Dvornikovs, C.S. Jeffrey, F. Shao, J. Wang, L.A. Vrieze, K. Anderson and T.R. Hoye. 2005. Mixture of new sulphated steroids functions as a migratory pheromone in the sea lamprey. *Nature Chemical Biology* 1: 324-328.

Stacey, N. and P. Sorensen. 2002. Hormonal pheromones in fish. In: *Hormones, Brain and Behavior*, D.W. Pfaff, A.P. Arnold, A.M. Etgen, S.E. Fahrbach and R.T. Rubin (eds.), Academic Press, New York, pp. 375-434.

Stacey, N. and P. Sorensen. 2005. Reproductive pheromones. In: *Behaviour and Physiology of Fish*, K.A. Sloman, R.W. Wilson and S. Balshine (eds.), Academic Press, London, pp. 359-412.

Taborsky, M. 1997. Bourgeois and parasitic tactics: Do we need collective, functional terms for alternative reproductive behaviours? *Behavioral Ecology and Sociobiology* 41: 361-362.

Taborsky, M. 1999. Conflict or cooperation: what determines optimal solutions to competition in fish reproduction? In: *Behaviour and Conservation of Littoral Fishes*, V.C. Almada, R.F. Oliveira and E.J. Gonçalves (eds.), ISPA, Lisboa, pp. 301-349.

Tavolga, W.N. 1958. Underwater sounds produced by males of the blennid fish, *Chasmodes bosquianus*. *Ecology* 39: 759-760.

van den Berghe, E.P. and R.R. Warner. 1989. The effects of mating system on male mate choice in a coral reef fish. *Behavioral Ecology and Sociobiology* 24: 409-415.

Wagner, H.-J. 1990. Retinal structure of fishes. In: *The Visual System of Fishes*, R.H. Douglas and M.B.A. Djamgoz (eds.), Chapman and Hall, London, pp. 109-157.

White, E.M., D.M. Goncalves, J.C. Partridge and R.F. Oliveira. 2004. Vision and visual variation in the peacock blenny. *Journal of Fish Biology* 65: 227-250.

Wickler, W. 1957. Vergleichende Verhaltensstudien an Grundfischen. I. Beiträge zur Biologie, besonders zur Ethologie von *Blennius fluviatilis* Asso, im Vergleich zu einigen anderen Bodenfischen. *Zeitschrift für Tierpsychologie* 14: 393-428.

Wirtz, P. 1978. The behaviour of the Mediterranean *Tripterygion* species (Pisces, Blennioidei). *Zeitschrift für Tierphysiologie* 48: 142-174.

Wisenden, B.D. and N.E. Stacey. 2005. Fish semiochemicals and the evolution of communication networks. In: *Animal Communication Networks*, P.K. McGregor (ed.), Cambridge University Press, Cambridge, pp. 540-567.

Wyatt, T.D. 2003. *Pheromones and Animal Behaviour. Communication by Smell and Taste*. Cambridge University Press, Cambridge.

Yambe, H., S. Kitamura, M. Kamio, M. Yamada, S. Matsunaga, N. Fusetani and F. Yamazaki. 2006. L-Kynurenine, an amino acid identified as a sex pheromone in the urine of ovulated female masu salmon. *Proceedings of the National Academy of Sciences of the United States of America* 103: 15370-15374.

Zander, C.D. 1972. Beiträge zur Ökologie und Biologie von Blennidae (Pisces) des Mittelmeeres. *Helgoländer wissenschaftliche Meersuntersuchungen* 23: 193-231.

Zander, C.D. 1974. Beziehungen zwischen Körperbau und Lebensweise bei Blenniidae (Pisces) aus dem Roten Meer. III. Morphologie des Auges. *Marine Biology* 28: 61-71.

Zander, C.D. 1975. Secondary sex characteristics of blennioid fishes (Perciformes). *Pubblicazioni della Stazione Zoologica di Napoli* 39: 717-727.

Zander, C.D. 1986. Blenniidae. In: *Fishes of the North-eastern Atlantic and the Mediterranean*, P.J.P. Whitehead, M.L. Bauchot, J.C. Hureau, J. Nielsen and E. Tortonese (eds.), UNESCO, Paris, pp. 1096-1112.

Zaunreiter, M. 1990. Zwischenartliche Differenzierung des Optischen Systems bei mediterranen Blenniiden (Blenniidae, Perciformes, Teleostei), Ph.D. Thesis, University of Salzburg, Salzburg.

Zaunreiter, M., K. Kotrschal, A. Goldschmid and H. Adam. 1985. Ecomorphology of the optic system in 5 species of blennies (Teleostei). *Fortschritte der Zoologie* 30: 731-734.

Zaunreiter, M., A. Goldschmid and K. Kotrschal. 1989. Comparative retinal morphology of 14 species of Mediterranean blennies (Teleostei: Perciformes: Blenniidae). *Fortschritte der Zoologie* 35: 367-369.

Zeiske, E., B. Theisen and H. Breucker. 1992. Structure, development and evolutionary aspects of the peripheral olfactory system. In: *Fish Chemoreception*, T.J. Hara (ed.), Chapman and Hall, London, pp. 13-39.

Zielinski, B.S. and T.J. Hara. 2007. Molecular and neural mechanisms in olfaction. In: *Fish Physiology: Sensory Systems Neuroscience*, J.H. Toshiaki and B.S. Zielinski (eds.), Elsevier, Amsterdam, pp. 1-43.

CHAPTER

5.2

Interspecific Relationships in Blennies

Karen L. Cheney

INTRODUCTION

In 1999, John Thompson wrote that, 'the history of evolution and biodiversity is fundamentally a history of the evolution of species interaction. Species living in pure isolation simply don't make sense' (Thompson, 1999). Indeed, most living organisms depend on one or more species if they are to survive and reproduce. Interspecific interactions occur through processes such as predation, competition, mimicry and symbioses. Moreover, it has long been recognised that interspecific interactions provide a framework for functional ecosystems, determines the abundance of populations, and increases biodiversity (Gause and Witt, 1935; Thompson, 1996). For the family Blenniidae, interspecific interactions have a marked effect on the behaviour, physiology, distribution and abundance of individuals, and since blennies reside in

Author's address: School of Integrative Biology, The University of Queensland, St Lucia QLD 4072, Australia. E-mail: k.cheney@uq.edu.au

almost every underwater environment imaginable–from depths of over 100 m, freshwater rivers, coral reefs, and intertidal environments–blennies interact with a wide variety of species for many different purposes. This chapter provides an overview of some of these interactions.

MIMICRY IN BLENNIIDAE

Mimicry is often used as a simple and intuitive example of adaptive evolution by natural selection (Turner, 1987; Brower, 1988). Mimetic species evolve to resemble another species to escape predation, to capture prey or to gain other benefits such as pollination or transportation (Wickler, 1968). Mimicry systems have been classified in many different ways (e.g. Pasteur, 1982), but here I consider five main types: Batesian, aggressive, social, sexual and Müllerian depending on the nature of the fitness advantage in terms of who is being deceived and why (Wickler, 1965). The first four mimicry types typically involve three participants: a model, a mimic and a dupe (Pasteur, 1982). However, in Müllerian mimicry, systems also exist where multiple mimic and model species are present, which are often described as 'mimicry rings'.

In fish, the most published examples of mimicry are found in tropical marine environments (Moland *et al.*, 2005; Randall, 2005). On coral reefs, the family Blenniidae contribute a relatively large proportion of mimetic fish species: 28% (7/25) of reported Batesian mimics; 25% (8/32) of aggressive mimics; and 38% (3/8) of social mimics (Moland *et al.*, 2005; Table 5.2.1). Müllerian mimicry is uncommon in fish however, Müllerian-type mimicry systems have been suggested for some complexes involving blennies (Moland *et al.*, 2005). Blenniidae also act as models in numerous model-mimic complexes: 52% (11/21) in Batesian and 10% (5/48) aggressive mimics. There are currently no reported blenny species that act as models in social mimicry systems (0/10); however, blennies rarely exhibit schooling behaviour.

Batesian Mimicry

In classical Batesian mimicry, a palatable mimetic species copies the warning signal of a defended, aposematic model species (Bates, 1862). The mimic benefits from the resemblance as predators learn to avoid the signal emitted by the model and hence also avoid the mimic. Whereas the predator incurs costs related to lost foraging opportunities. Batesian mimicry systems are considered to be frequency-dependent: if predators

Table 5.2.1 Blenniidae involved in mimicry complexes (adapted from Moland *et al.*, 2005). Bold names indicate species from the family Blenniidae. Other family names in parentheses after species name.

Mimic species	*Model species*	*Type of mimicry*	*Reference*
Amblygobius linki (Gobiidae)	***Meiacanthus anema***	Batesian	Springer and Smith-Vaniz (1972)
Aspidontus taeniatus	*Labroides dimidiatus* (Labridae)	Aggressive	Randall and Randall (1960); Kuwamara (1983)
Aspidontus tractus	*Labroides dimidiatus* (Labridae)	Aggressive	Springer and Smith-Vaniz (1972)
Cheilodipterus nigrotaeniatus (Apogonidae)	***Meiacanthus grammistes***	Batesian	Smith-Vaniz (1987)
Cheilodipterus parazonatus (Apogonidae)	***Meiacanthus vittatus***	Batesian	Smith-Vaniz *et al.* (2001)
Cheilodipterus zonatus (Apogonidae)	***Meiacanthus geminatus***	Batesian	Gon (1993)
	Meiacanthus vittatus	Batesian	Lieske and Myers (2001)
Ecsenius bicolor	***Meiacanthus atrodorsalis***	Batesian	Losey (1972); Springer and Smith-Vaniz (1972)
Ecsenius gravieri	***Meiacanthus nigrolineatus***	Batesian	Springer and Smith-Vaniz (1972)
Ecsenius midas	*Luzonichthys whitleyi* (Serranidae)	Social	Randall and McCosker (1993)
	Pseudanthias bartlettorumi (Serranidae)	Social	Randall and McCosker (1993)
	Pseudanthias dispar (Serranidae)	Social	Randall and McCosker (1993)
	Pseudanthias squamipinnis	Social	Strack (1969)
Hemiemblemaria similes	*Thalasomma bifasciatum* (Labridae)	Aggressive	Longley and Hildebrand (1940)
Meiacanthus nigrolineatus	*Cheilodipterus* spp. (Apogonidae)	Social	Dafni and Diamant (1984)
Meiacanthus urostigma	*Cheilodipterus quinquelineatus* (Apogonidae)	Social	Smith-Vaniz *et al.* (2001)
Pentapodus trivittatus (Nemipteridae)	***Meiacanthus crinitus***	Batesian	Smith-Vaniz *et al.* (2001)
Petroscirtes breviceps	***Meiacanthus grammistes***	Batesian	Smith-Vaniz (1987)
	Meiacanthus kamoharai	Batesian	Smith-Vaniz *et al.* (2001)
	Meiacanthus vittatus	Batesian	Smith-Vaniz *et al.* (2001)
Petroscirtes fallax	***Meiacanthus lineatus***	Batesian	Springer and Smith-Vaniz (1972); Russell *et al.* (1976); Smith-Vaniz *et al.* (2001)

Table 5.2.1 Contd.

Table 5.2.1 Contd.

Plagiotremus azaleas	*Thalassoma lucasanum* (Labridae)	Aggressive and Social	Hobson (1968); Hobson (1969)
Plagiotremus laudandus flavus	***Meiacanthus oualanensis***	Batesian and aggressive	Losey (1972); Springer and Smith-Vaniz (1972);Smith-Vaniz *et al.* (2001)
Plagiotremus laudandus laudandus	***Meiacanthus atrodorsalis***	Batesian and aggressive	Springer and Smith-Vaniz (1972); Smith-Vaniz *et al.* (2001)
	Ecsenius bicolor	Aggressive	
Plagiotremus phenax	***Meiacanthus smithi***	Batesian	Springer and Smith-Vaniz (1972); Smith-Vaniz *et al.* (2001)
Plagiotremus rhinorhynchos	*Labroides dimidiatus* (Labridae)	Aggressive	Wickler (1961)
	Pseudanthias squamipinnis (Serranidae)		Russell *et al.* (1976)
	Pseudanthias huchtii (Serranidae)		Kuwamara (1981); Kuwamara (1983); Randall *et al.* (1997); Côté and Cheney (2004); Moland and Jones (2004);
Plagiotremus tapeinosoma	*Thalassoma amblycephalum* (Labridae)	Aggressive	Russell *et al.* (1976)
	Trachinops taeniatus (Plesiopidae)	Aggressive	
	Forsterygion sp. (Tripterygiidae)	Aggressive	
Plagiotremus townsendi	***Meiacanthus nigrolineatus***	Aggressive	Springer and Smith-Vaniz (1972); Smith-Vaniz *et al.* (2001)
Scolopsis bilineatus (Nemipteridae)	***Meiacanthus lineatus***	Batesian	Springer and Smith-Vaniz (1972)
	Meiacanthus oualanensis	Batesian	Russell *et al.* (1976); Randall (2005)
	Meiacanthus smithi	Batesian	Smith-Vaniz *et al.* (2001); Randall (2005)
	Meiacanthus grammistes	Batesian	Randall (2005)
Scolopsis margaritifer (Nemipteridae)	***Meiacanthus vittatus***	Batesian	Russell *et al.* (1976)
	Ecsenius gravieri		
Valencienna helsdingenii (Gobiidae)	***Meiacanthus anema***	Batesian	Springer and Smith-Vaniz (1972)

encounter palatable mimics too frequently, then avoidance learning is disrupted and the protection of both model and mimic can be lost (Bates, 1862; Lindström *et al.*, 1997). In blennies, examples of Batesian mimics are found in the genera *Ecsenius*, *Petroscirtes* and *Plagiotremus*, all of which are mimics of the *Meiacanthus* genera (Table 5.2.1). The enlarged canine teeth of *Meiacanthus* possess a venom gland at the base of each tooth, which secretes a toxic milky fluid (Randall *et al.*, 1997). The model can deliver toxic bites when threatened, and during experiments in which living *Meiacanthus* spp. were fed to predators, the majority of individuals were taken into the predator's mouth and then promptly rejected (Losey, 1972; Springer and Smith-Vaniz, 1972). If a palatable mimic was then subsequently offered to the predator, it would also be rejected (Losey, 1972; Springer and Smith-Vaniz, 1972).

Cardinal fish (family Apogonidae) and juvenile breams (family Nemipteridae) are also reported to be Batesian mimics of the *Meiacanthus* genera (Russell, 1977; Table 5.2.1). Breams have very similar colour patterns and swimming behaviour to *Meiacanthus* blennies, however they have a much deeper body (Russell, 1977). Juvenile two-lined monocle bream (*Scolopsis bilineatus*) exhibit parallel geographic variation of colour patterns depending on the *Meiacanthus* species or colour morph present. Throughout the majority of the Indo-West Pacific, juvenile *S. bilineatus* are white on the lower body, with alternating black and yellow stripes on the top half of the body, which is thought to mimic the lined fangblenny (*Meiacanthus lineatus*). In Fiji, where the entirely yellow *M. oualanensis* exists, juvenile *S. bilineatus* are also entirely yellow. In the Indian Ocean *M. smithi* has a white body with a broad black lateral stripe along their back and is mimicked by a similar colour morph of *S. bilineatus* (Randall, 2005).

Aggressive Mimicry

Aggressive mimics are defined as the resemblance of a predatory or parasitic species to another harmless or even attractive species in order to approach and attack prey, or to gain other benefits such as pollination or transportation (Wickler, 1968). In aggressive mimicry, the victim pays costs when deceived, either in terms of injury, death or a loss of other resources. The mimic benefits from its resemblance to the model by being more attractive or less noticeable to potential victims. The model also incurs costs such as decreased foraging or sexual encounters. Mimicry of the blue streak cleaner wrasse (*Labroides dimidiatus*) by the bluestriped fangblenny (*Plagiotremus rhinorhynchos*) and false cleanerfish (*Aspidontus taeniatus*) are perhaps the best known and cited examples of mimicry in reef fish. The cleaner wrasses (*Labroides* sp.) remove ectoparasites and mucus from larger fish (for review on cleaning symbioses see Côté, 2000). *L. dimidiatus* is the most widespread and abundant cleaner fish in the Indo-Pacific, and are visited by up to 190 fish clients per hour (Grutter, 1995). Adult *L. dimidiatus* can ingest up to 1200 gnathid ectoparasites per day from the surface of their fish clients (Grutter, 1996), and have been shown to significantly reduce ectoparasites loads of coral reef fish (Grutter, 1999). *P. rhinorhynchos* copies the black body and neon blue lateral body stripe of juvenile *L. dimidiatus* (Figure 5.2.1), and are usually found within close proximity of their model (Côté and Cheney, 2004). However, instead of cleaning reef fish, *P. rhinorhynchos* (in a similar manner to other members of the *Plagiotremus* genera) attack passing reef fish removing scales and dermal tissue (Randall and Randall, 1960; Kuwamara, 1983). Close spatial association of mimic to model reduced foraging rates by juvenile cleaner wrasse and increased access to victims for mimics (Côté and Cheney, 2004). Like Batesian mimicry, aggressive mimicry systems are also frequency-dependent; Cheney and Côté (2005) found that an increase in the number of fangblenny mimics in relation to the number of cleaner wrasse models altered the relationship between victim and model, and decreased the success of the mimic. As numbers of mimics increase, potential victims may become more cautious in approaching a model or mimic, and may learn to take evasive action to avoid attacks by aggressive mimics (Cheney and Côté, 2005).

Interestingly, *P. rhinorhynchos* does not always adopt the mimic coloration of juvenile cleaner wrasse, but has the ability to switch between mimetic and non-mimetic colouration at will (Côté and Cheney, 2005). When near a juvenile cleaner wrasse, *P. rhinorhynchos* will often adopt a black and neon blue colouration (Fig. 5.2.2a; Cheney *et al.*, 2008). However, when the juvenile cleaner is removed, or the fangblenny is transplanted to a different habitat, the fangblenny adopts a non-mimetic form: brown, olive or orange body colour with two pale blue lateral stripes (Moland and Jones, 2004; Côté and Cheney, 2005). Non-mimic colour forms are often found associated with schooling fish, such as anthias (*Pseudanthias huchti*, *P. tuka*, and *P. squamipinnis*) and blue-green chromis (*Chromis viridis*) (Randall *et al.*, 1997; Côté and Cheney, 2005). It appears that the colour of the fangblenny blend in with colours of their shoal mates

Fig. 5.2.1 Mimicry in Blenniidae: **a)** juvenile bluestreak cleaner wrasse, *Labroides dimidiatus* mimicked by **b)** bluestriped fangblenny, *Plagiotremus rhinorhynchos* and **c)** juvenile false cleanerfish, *Aspidontus taeniatus*; **d)** adult *L. dimidiatus* mimicked by **e)** adult *A. taeniatus*; **f)** yellowtail fangblenny, *Meiacanthus atrodorsalis* mimicked by **g)** bicolor fangblenny *P. laudandus*. All photographs © Karen Cheney.

(Cheney *et al.*, 2008). By living within a shoal of similar coloured fish, they may increase their attack success as prey fish are less likely to notice the predatory blenny, or they may decrease their risk of predation. *Plagiotremus azaleus* and *P. tapeinosoma*, which share a similar feeding mode, also regularly occur amongst aggregations of similarly coloured fish which they superficially resemble, including *Thalassoma lucasanum* (Hobson, 1968, 1969) and *T. amblecephalum* (Russell *et al.*, 1976), respectively.

Mimicry of *Labroides dimidiatus* by *Aspidontus taeniatus* was first described in detail by Randall and Randall (1960). *A. taeniatus* is an almost exact replica of both juvenile and adult *L. dimidiatus* to all but a specialised

Fig. 5.2.2 Colour variations found in *Plagiotremus rhinorhynchos:* **a**) olive, **b**) orange, **c**) black and blue 'mimic', and **d**) brown. All photographs © Karen Cheney.

underwater observer. *A. taeniatus* can be distinguished from *L. dimidiatus* by the position of the mouth, which is terminal in the wrasse but sub-terminal in the blenny. *A. taeniatus* also has a longer dorsal fin base compared to the wrasse, which can be erect when swimming, and will enter abandoned tube holes or other suitable holes in the reef. Mimicry of cleaner wrasse by *A. taeniatus* is perhaps considered to be a combination of mimicry types, which may benefit the blenny in one of the two following ways. First, the blenny mimic occasionally nips scales and fins from unsuspecting fish; therefore, resembling a cleaner wrasse may help them approach and attack their victims. However, recent observations of *A. taeniatus* have indicated that they attack fish infrequently. Kuwamara (1983) reported that only 2.3% of feeding incidents were by attacking reef fish. Furthermore, during 30 hours of observations on the Great Barrier Reef, twelve adult *A. taeniatus* were observed to attack fish on average 0.08 attacks hour^{-1} (unpublished data). Indeed, studies which have investigated the diet of *A. taeniatus* suggest that scales and fins are perhaps a minor portion of the blennies diet. Although Randall and Randall (1960) found that 4 from 8 of the fish examined (50%) had fish parts in their diets, other more recent studies report that *A. taeniatus* diet (n = 11) consisted mainly of demersal fish eggs and polychaete tentacles, without any fish parts in their stomachs (Kuwamara, 1983). Ontogenetic and temporal shifts may occur in the mode of feeding by *A. taeniatus*. During three hours of observing one juvenile *A. taeniatus* exhibiting the black and blue colouration of the juvenile cleaner wrasse, the individual was observed to attack reef fish a mean 7.2 times hour^{-1} (unpublished data). However, further observations on juvenile *A. taeniatus* are needed. Second, the resemblance of *A. taeniatus* to the cleaner wrasse may reduce the risk of predation for the mimic. Cleaners are thought to be relatively immune from predation, with very few reports of predation reported in the literature (Feder, 1966; Côté, 2000). In the few cases where it has been reported, it was in a non-cleaning context (Hobson, 1971). However, this protection from predation for the mimic is yet to be shown.

The bicolor fangblenny (*Plagiotremus laudandus*) is also thought to benefit from a combination of Batesian and aggressive mimicry in its resemblance to *M. atrodorsalis* (Losey, 1972; Moland *et al.*, 2005). In addition to gaining protection from predation, *P. laudandus* also attacks passing coral reef fish to nip on scales, fins and tissue. To most reef fish *M. atrodorsalis* is harmless, as they will only attack fish if under threat. By closely spatially associating with *M. atrodorsalis* individuals, *P. laudandus* may increase the proportion of successful foraging attacks, as victims will not exhibit an avoidance response to individuals.

Fewer examples of mimicry have been reported beyond the Indo-Pacific. In the Western Atlantic, the wrasse blenny (*Hemiemblemaria simulus*) has been described as an aggressive mimic of the bluehead wrasse (*Thalassoma bifasciatum*). *H. simulus* resembles the bluehead wrasse almost perfectly, and can be observed to leave the bottom and swim freely within schools of *T. bifasciatum* in order to feed on small fish and crustaceans. Mimicry is thought to help *H. simulus* approach small fish and shrimps on which it preys. In the Mediterranean sea, possible mimic relationships exist between *Parablennius rouxi* and *Gobius vittatus* (Heymer and Zander, 1978; Zander and Nieder, 1997); however, this requires further investigation.

Social Mimicry

Social mimicry was originally proposed for birds by Moynihan (1968), but has since been shown to be relatively common in fish (Dafni and Diamant, 1984; Moland *et al.*, 2005), and refers to two or more similarly coloured species that associate with each other to escape the attention of predators (Moynihan, 1968; Randall and McCosker, 1993). In comparison to aggressive mimics, social mimics forage on the same food or trophic level as their putative model. The foraging advantage of living within an interspecific group has been demonstrated for fish (Foster, 1985), and they may also benefit from a reduced predation risk. *Ecsenius midas* is considered a social mimic as it schools with similarly coloured anthias and both species are plankton feeders (Randall and McCosker, 1993; Table 5.2.1). *Meiacanthus urostigma* is also considered a social mimic of the cardinalfish *Cheilodipterus quinquelineatus*—juvenile *M. urostigma* have been observed associated with juvenile *C. quinquelineatus* (Smith-Vaniz *et al.*, 2001). Both species have a similar body colouration comprising of black stripes, an orange-yellow caudal peduncle and a black spot at the base of the caudal peduncle.

Müllerian Mimicry

Müllerian mimicry is usually viewed as a system of mutual warning colouration arising from the convergence of two or more warningly coloured species (Müller, 1879). Whilst there appears to be no single Müllerian-type mimicry in coral reef fish, many of the mimicry systems involving *Meiacanthus* may involve a mixed-type of mimicry, which may include elements of Müllerian mimicry (Springer and Smith-Vaniz, 1972; Smith-Vaniz *et al.*, 2001). *Meiacanthus* blennies possess grooved canines associated with toxic buccal glands that are used to inject toxic fluid into their predators, which may cause some predators to avoid them (Losey, 1972; Springer and Smith-Vaniz, 1972). Members of the genus *Plagiotremus*, which mimic members of *Meiacanthus*, also possess fangs that are not associated with toxic glands, but maybe sufficient to cause some element of predator protection. Furthermore, predators have been shown to reject both live and dead specimens of *Plagiotremus townsendi* (Smith-Vaniz *et al.*, 2001), which may indicate a level of unpalatability. However, *P. rhinorhynchos* has been observed to be readily predated on by slingjaw wrasse (*Epibulus insidiator*) and lizardfish (*Synodus* sp.) in the field (per. obs.). Further investigations would be required to elucidate the palatability of *Plagiotremus* genera to determine whether this was Müllerian-type mimicry.

Sexual Mimicry

Intraspecific sexual mimicry has been reported in *Salaria pavo* (Gonçalves *et al.*, 1996). Smaller, younger males mimicked females in order to approach spawning fish and achieve sneak fertilisations. Sneaker males resembled females in their body colouration, lack of secondary sexual characteristics and courtship behaviour. However, sneaker males could still be distinguished from females by a less vivid colouration and reduced courtship behaviour. Many teleost fish have adopted alternative reproductive strategies in which the smaller males adopt female characteristics in order to gain access to spawning females; however, female mimicry is relatively uncommon (Taborsky, 1994).

PREDATOR–PREY INTERACTIONS

Predation is a predominant form of mortality in most coral reef fish. In addition to affecting prey population size and community patterns, predators may influence morphology, physiology, behaviour and life histories of their prey (Edmunds, 1974). Blennies are certainly an integral part of many marine food web systems in both predator and prey roles, and are important in regulating marine community food web structure and ecosystem function (Depczynski and Bellwood, 2003; Pinnegar and Polunin, 2004; Bruno and O'Connor, 2005).

Blennies as Predators

Many blennies are herbivorous; however, the family also includes planktivores and carnivores that feed on a variety of food items including crustaceans, gastropods, brittle stars, chironomid larvae, sea urchins, and polychaete worms (Randall, 1983). Planktivorous blennies include members of the genus *Acanthemblemaria*, which tend to live in small holes in hard substrata. They visually examine the surrounding water and substrate for small invertebrates that serve as their food (Greenfield and Greenfield, 1982). When a suitable prey item is within striking distance, the blenny will dart from its refuge, capture the prey item and re-enter the shelter tail first. Differences in predatory behaviour between *Acanthemblemaria* blenny species have been demonstrated. For example, *A. spinosa* attacks at a greater distance and is more inclined to strike at prey that are behind it, whereas *A. aspera* approaches prey more slowly and modulates strike speed with changes in water movement (Clarke *et al.*, 2005). Slight variations in predatory behaviour may explain differences in prey preference and help in niche partitioning between species (Greenfield and Greenfield, 1982). Capture success may depend on water turbidity and movement and whether the prey exhibits an avoidance response (Clarke *et al.*, 2005). Calanoid copepods have some of the most vigorous escape response found in the aquatic environment, with initial accelerations of over 200 m s^{-2} and reaching speeds of over 500 mm s^{-1} (Buskey *et al.*, 2002). Other escape responses include passive sinking by cladocerans (Kerfoot, 1980).

Micro carnivorous blennies, feeding on amphipods and isopods in algal mats, have been shown to be an integral part of trophic cascades that occur in marine environments due to human interference or other forms of disturbance (Pinnegar and Polunin, 2004; Bruno and O'Connor, 2005). If top fish predators such as Serranidae are removed from an ecosystem due to increased fishing pressure, the biomass of micro carnivorous fish, including blennies, will increase. This in turn will lead to a reduction in herbivorous invertebrates and consequently, an increase in the biomass of plants (Pinnegar and Polunin, 2004).

Macro carnivorous blennies, such as the arrow blenny (*Lucayablennius zingaro*), feed on small fish or crustaceans. *L. zingaro* live either solitary or in unorganised groups, usually in association with food masked goby *Coryphopterus personatus* or *Coryphopterus hyalinus* on which it feeds (Greenfield, 1972). *L. zingaro* swims, for the most part, in a J or U-shape with tail bent and propels itself with pectoral fins with the rest of the body motionless. They will drift in this position until they shoot forward to prey on small fish and shrimp (Greenfield, 1972). Unlike other blennies, they rarely rest on their pectoral fins. However, they do rest in small holes or empty worm tubes. The sabre-tooth fangblennies (discussed in the mimicry section of this chapter) feed on scales, mucus, and tissue of larger fish. *P. rhinorhynchos* have been observed to strike various reef fish up to 20 attacks per 15 min (Moland and Jones, 2004; Cheney and Côté, 2005). Fangblennies usually attack from behind and will often nip the ventral surface of the fish, often beneath the eyes, and around the head area. After an attack, victims will often retaliate and chase the predatory blenny. However, damage to the blenny is rarely observed.

Blennies as Prey

Blennies are preyed upon by a number of generalist fish predators (Table 5.2.2), and mammals including otters (*Lutra lutra*) in Portugal (Clavero *et al.*, 2004), and the introduced American Mink (*Mustela vison*) in SW Europe (Delibes *et al.*, 2004). In intertidal areas, birds will also prey on blennies such as the shanny (*Lipophrys pholis*) or zebrablenny (*Istiblennius zebra*) (Hobson, 1968; Lehikoinen, 2005). The presence of potential predators may markedly affect the behaviour of blennies. Gulf signal

Table 5.2.2 Fish predators of Blenniidae that have been reported in the literature. Note: this list does not claim to be complete.

Prey species	*Predator family*	*Predator species*	*References*
Blenniella gibbifrons	Mullidae	*Parupeneus cyclostomus*	Hobson (1974)
Blenniella periophthalmus	Blenniidae	*Blenniella periophthalmus*	Hiatt and Strasburg (1960)
Chasmodes bosquianus	Moronidae	*Morone saxatilis*	Hartman and Brandt (1995)
	Pomatomidae	*Pomatomus saltator*	
	Sciaenidae	*Cynoscion regalis*	
Cirripectes variolosus	Serranidae	*Cephalopholis argus*	Randall and Brock (1960)
	Serranidae	*Plectropomus leopardus*	Goeden (1978)
	Muraenidae	*Gymnothorax flavimarginatus*	Hiatt and Strasburg (1960)
Coryphoblennius galerita	Serranidae	*Serranus atricauda*	Morato *et al.* (2000)
Ecsenius yaeyamaensis	Fistulariidae	*Fistularia commersonii*	Takeuchi *et al.* (2002)
Entomacrodus nigricans	Carangidae	*Caranx ruber*	Randall (1967)
Hypsoblennius hentz	Lobotidae	*Lobotes surinamensis*	Franks *et al.* (2003)
	Moronidae	*Morone saxatilis*	Walter *et al.* (2003)
Hypsoblennius sordidus	Pinguipedidae	*Pinguipes chilensis*	González and Oyarzún (2003)
Lipophrys pholis	Serranidae	*Serranus atricauda*	Morato *et al.* (2000)
Ophioblennius atlanticus	Aulostomidae	*Aulostomus maculatus*	Randall (1967)
	Carangidae	*Caranx ruber*	
	Serranidae	*Mycteroperca tigris*	
	Serranidae	*Serranus atricauda*	
Parablennius incognitus	Serranidae	*Serranus atricauda*	Morato *et al.* (2000)
Parablennius laticlavius	Zeidae	*Zeus faber*	Russell (1983)
Parablennius ruber	Serranidae	*Serranus atricauda*	Morato *et al.* (2000)
Paralipophrys trigloides	Serranidae	*Serranus atricauda*	Morato *et al.* (2000)
Plagiotremus goslinei	Synodontidae	*Synodus variegatus*	Hobson (1974)
	Lutjanidae	*Aphareus furca*	
	Mullidae	*Parupeneus cyclostomus*	
Salaria fluviatilis	Esocidae	*Esox lucius*	Elvira *et al.* (1996)
	Cyprinidae	*Rutilus rubilio*	Daoulas and Economidis (1984)
Salarias fasciatus	Serranidae	*Plectropomus leopardus*	Kingsford (1992)

blennies (*Emblemaria hypacanthus*) were shown to retreat into their shelter when potential predators such as lutjanids (snappers) and serranids (sea basses) swam past. In contrast, pomacentrids (damselfish) and gerreids (mojarras) were ignored. Both male and female *E. hypacanthus* spent more time in their shelter, were less active and used their shelters more frequently when predators were more common (Hastings, 1991). Males were also darker and courted the females less vigorously at low risk sites (Hastings, 1991). Courtship colouration is known to increase exposure of males to predation risk (Haas, 1976).

Avoiding Predation

Blennies have evolved many characteristics in order to protect themselves against predators, either by reducing the chance of detection, or reducing the chances that the prey will be eaten once detected (Edmunds, 1974; Wootton, 1990). Defense mechanisms include behavioural or morphological modifications including: mimicry, cryptic or warning colouration, secondary defense structures, and the use of refuges.

Colouration and body patterns. Organisms that exhibit cryptic colouration evolve to closely resemble their background (Armbruster and Page, 1996), physiologically adapt their appearance to changes in their current background (Chiao and Hanlon, 2001), or behaviourally match their background depending on their colour (Endler, 1984). In each case, crypsis is achieved by animals exhibiting a random sample of the background (Endler, 1988). Most blennies use blotches, vertical bars, and spots in order to closely blend in with their habitat. For example, the scaly-headed triplefin (*Karalepis stewarti*) is a mottled pattern of red, white, and brown, with irregular red-brown vertical bars on the body. This camouflage appears to be extremely effective against encrusting growth. Some species of blennies can also change their body colour to become cryptic with their background or mimic other fish species (Abel, 1993; Cheney and Côté, 2005; Cheney *et al.*, 2008).

Conspicuous colour patterns often act as warning signals to predators that a species is chemically or structurally defended. Members of the *Meiacanthus* genera are brightly coloured or have conspicuous stripes, which may warn predators that individuals have secondary defense mechanisms. In the case of *Meiacanthus* spp., these are enlarged canine teeth with associated venom glands that secrete a toxic fluid (Randall *et al.*, 1997). The toxicity of the bite has been tested in mice and in humans, and is reported to cause bleeding and irritation in mice, and reddening and swelling in human subjects (Losey, 1972; Springer and Smith-Vaniz, 1972). Hairtail blenny (*Xiphasia setifer*) may also use their long curved pointed teeth at the front of the lower jaw for defensive purposes. However, these are not associated with a venom gland (Randall *et al.*, 1997) and their function is unknown.

Other colour patterns which may help in the protection against predators include ocellated spots, which are found in a few species of blennies including the butterfly blenny (*Blennius ocellaris*), blackflap blenny (*Cirripectes auritus*), seram blenny (*Salarias ceramensis*), spotcheek blenny (*Labrisomus nigricinctus*), diamond blenny (*Malacoctenus boehlkei*), and quillfin blenny (*Labrisomus filamentosus*). In fish and lower animals, 'eyespots' are thought to misdirect predator attacks towards less vital body areas, intimidate predators by appearing as the eyes of larger individuals, and confuse predators into misjudging the distance of the eyespot bearer (Neudecker, 1989). However, eyespots may also signal to conspecifics factors such as territory location or direction of resources.

Behavioural adaptations. By using shelters or holes, blennies reduce their chances of being detected or captured by predators. Many blennies, particularly those from the family Chaenopsidae, use abandoned worm tubes or clam shells as shelters. Other blennies find shelter in rock crevices, under coral and in oyster beds. The striped blenny (*Chasmodes bosquianus*) relies on oyster reefs for nesting sites, feeding grounds, and protection from predators (Harding and Mann, 2000). The possession of refuge holes for some blennies appear critical for survival and blennies in the Mediterranean select shelters based on their entrance aperture in relation to their head and body size (Koppel, 1988). Spinyhead blennies (*Acantheblemaria spinosa)* without holes were predated upon by the redspotted hawkfish (*Amblycirrhitus pinos*) and lizardfish (*Synodus* sp.) (Buchheim and Hixon, 1992). Furthermore, resident blennies fiercely defended holes from intruders suggesting that significant benefits were obtained when residing in a hole (Buchheim and Hixon, 1992).

A number of blenny species also facultatively associate with sea anemones, including the clinids *Malacoctenus macropus, M. boehlkei, M. gilli, M. triangulates, Hemiemblemaria simulus* and *Starksia hassi,* which have all been found associated with the pink-tipped sea anemone *Condylactis gigantean* (Hanlon and Kaufmann, 1976; Albrecht, 1977; Hanlon *et al.*, 1983). In laboratory experiments with *Malacoctenus* spp., individuals were initially stung when they came in contact with anemone's tentacles, but within a few minutes to a couple of hours became acclimatised to the anemone after which they could move freely within its tentacles (Hanlon and Kaufmann, 1976). Whether the anemone incurs costs or benefits from the blenny's presence is unclear (Lieske and Myers, 2001).

Reducing activity levels, including the amount of time that blennies are active and speed of movement, may also help to reduce the chance that predators detect the blennies. However, there is a clear tradeoff between high rates of activity whilst foraging and increased predation risk.

Animals are predicted to have lower activity levels when the risk of predation is high (Werner and Anholt, 1993; Hastings, 1991).

In shallow water environments, some blennies have evolved to leave the water for extended periods of time and are able to move across the land using a variety of locomotory modes. Amphibious blennies may leave the water to escape aquatic predators, but also can escape environmental degradation such as low oxygen levels, or to feed on terrestrial algal mats (Sayer and Davenport, 1991).

SYMBIOTIC RELATIONSHIPS

Symbiosis is defined as the interaction between two different organisms living together in close association, and occurs in all forms of life from the intracellular organisation of eukaryotic cells, to parasites on the largest of mammals. Symbiotic interactions can be sub-divided into three main categories defined by the benefit or cost that is gained by each participant: parasitism, when one species benefits whilst the other incurs costs; mutualism, when both species benefit; and commensalism, when one species benefits, whilst there is neither a cost nor benefit to the other.

Parasitism

The parasite fauna of blennies include digeneans, monogeneans, marine leeches and protists (Lester and Sewell, 1989; Diaz and George-Nascimento, 2002). Digeneans are the largest and most diverse group in the Class Trematoda and are primarily parasites of the gut. In fish they can also occur under the scales, in the gills and in other parts of the body. In blennies, digeneans have been reported in a number of species including *Plagiotremus* spp., *Meiacanthus grammistes*, *Salaria fasciatus* and *Istiblennius* spp. (Dyer *et al.*, 1988; Lester and Sewell, 1989; Cribb *et al.*, 1994). On the Great Barrier Reef, the marine leech *Richardsonobdella lineatae* (Hirudinea) was reported on yellowtail fangblenny (*Meiacanthus lineatus*) (Burreson and Dybdahl, 1989). Interestingly, the majority of leeches were collected from male hosts, and prevalence was high (63% of males; 32/51), whilst on females infection was lower. The difference in parasite numbers between sexes may be due to the breeding behaviour of *Meiacanthus*. Males guard eggs in nest holes until they hatch; therefore, leeches may deposit cocoons in the host-nesting hole, and when larvae hatch, infect fish in the nest. Female *M. lineatus* were also observed at cleaning stations, which means that leeches on females may have been removed by cleaner organisms. In contrast, males may be reluctant to leave nesting holes unguarded to visit cleaning stations (Burreson and Dybdahl, 1989). Interestingly, the leech was never found on other fish species sampled, including *Petroscirtes fallax*, the Batesian mimic of *M. lineatus* (Springer and Smith-Vaniz, 1972), which was present at the study site.

Haemogregarine blood parasites are apicomplexan protozoa found in both erythrocytes and leukocytes of vertebrates, and have been reported in over 90 species of host fishes (Davies, 1995; Davies *et al.*, 2004). *Haemogregarine bigemina* was first described in the intertidal blenniid species *Lipophrys pholis* and *Coryphoblennius galerita* in northern France (Laveran and Mesnil, 1901). It has since been reported in *Ecsenius bicolor, Parablennius cornutus, P. gattorugine* and *Blenniella periophthalmus* (Davies *et al.*, 2004). Vectors that may transmit haemogregarines include leeches (in the United Kingdom, potential vectors for *H. bigemina* have been identified on *L. pholis* as the leech *Oceanobdella blenni* (Davies *et al.*, 2004)) and gnathid isopod larvae (Davies *et al.*, 2004; Smit *et al.*, 2006).

Gnathids are one of the most common ectoparasites of marine fish (Grutter and Poulin, 1998a). During their 3 larval stages, gnathids feed from the body surface, buccal cavity and the gills of fishes. Once engorged on fish blood and other fish fluids, they will drop off and return to the substrate to digest their blood meal, molt into the next juvenile stage and then return to the water column to infect other fish (Smit *et al.*, 2003a). Gnathids are thought to damage fish tissues and in high numbers can kill captive fish (Paperna and Por, 1977; Mugridge and Stallybrass, 1983). *Gnathia africana* and *G. maxillaries* have been reported to infect the intertidal horned blenny (*Parablennius cornutus*) (Smit *et al.*, 2003b). However, the prevalence and abundance of gnathids on other species of blennies are unknown as many studies have failed to investigate blennies in their assessment of reef fish ectoparasites (e.g., Grutter and Poulin, 1998a).

Mutualism

In the marine environment, gnathid isopod larvae are the main food of cleaner organisms. Cleaning interactions are defined as the removal of ectoparasites and other material from the body surface of larger fish by small fish or shrimp (for review, see Côté, 2000). Both participants gain benefits from this remarkable cooperative relationship: cleaners gain a valuable food source, whilst visitors gain a reduction in ectoparasites loads (Grutter, 1999; Cheney and Côté, 2001). Members of the Blenniidae family appear to be rare visitors to cleaning stations. During behavioural observations on coral reefs in the Caribbean, the hairy blenny (*Labrisomus nuchipinnis*) and redlip blenny (*Ophioblennius atlanticus*) were not observed visiting cleaning stations run by cleaning gobies (*Elacatinus* sp.) during 25 hours of observations (Arnal *et al.*, 2000), despite being relatively dense in the study site (Rakitin and Kramer, 1996). In the Indopacific, *Ecsenius* sp. was observed to be cleaned by the cleaner wrasse (*Labroides dimidiatus*), but inspection times were short (Grutter and Poulin, 1998b). Ectoparasites are thought to be the main stimuli that motivate fish to seek cleaners (Grutter, 2001). Therefore, if blennies generally harbour few ectoparasites, they may be reluctant to leave territories to seek cleaners, which has been shown in territorial damselfish (Cheney and Côté, 2001).

Commensalism

Commensalistic relationships involving blennies are uncommon in the literature; however, one such example involves the saddle blenny (*Malacoctenus triangulates*) and beaugregory damselfish (*Stegastes leucostictus*), both of which defend territories from potential egg predators. Animals are expected to be territorial when the benefits of maintaining a territory, such as increased access to food sources and protection of nests, are outweighed by costs of defending it, such as energetic costs and increased risk of predation. Many blennies defend their territories from egg predators and algal eaters, which is a costly activity in terms of energy expended chasing intruders and time spent guarding nests. However, sharing territory defense with another non-competing species may reduce the costs to both participants, and may be considered a form of commensalism. In Jamaica, the saddle blenny defends small territories and is often found within territories of the beaugregory damselfish. Both species are susceptible to egg predation from the bluehead wrasse (*Thalassoma bifasciatum*) and defend the same area against this species (Draud *et al.*, 1990). However, little aggression occurs between the two species. In the presence of a beaugregory damselfish, saddle blennies significantly reduced their aggression towards egg predators; therefore, there appears to be advantages (in terms of reduced energy expenditure and time spent chasing egg predators), at least to the saddle blenny, of sharing territory with another aggressive territorial species (Draud *et al.*, 1990).

INTERSPECIFIC COMPETITION

Interspecific competition has an important role in structuring ecological communities: species that compete for one or more essential resource, such as space, food or shelters, should competitively exclude one another. Interspecific aggression may help dominant species to compete for suitable shelters or resources (Faria *et al.*, 1998). Competition for shelter and nest sites occurs between *Acanthemblemaria* species, which can reach high densities averaging up to 60 individuals per m^2 (Lindquist, 1985). Small-scale habitat partitioning was shown between *Acanthemblemaria spinosa* and *A. greenfieldi* (Greenfield and Greenfield, 1982), with *A. greenfieldi* occupying holes in the horizontal surfaces of dead coral, whilst, *A. spinosa* occupying holes of vertical surfaces. *A. spinosa* appeared to be competitively excluded from the horizontal habitat by the larger *A. greenfieldi*, possibly because of a size advantage. In St. Croix, *A. spinosa* occupied higher locations on dead coral heads of *Acropora palmata* than *A. aspersa* (Clarke 1989). After a large mortality event, which increased the number of shelter holes available, population numbers of *A. spinosa* substantially increased, suggesting that suitable holes were a limiting factor for these species (Buchheim and Hixon, 1992).

In rocky intertidal fish communities, the blennies *Lipophrys pholis* and *Coryphoblennius galerita* are major components of the rocky intertidal fish fauna on western European shores, and are frequently found within the same tide pools. Competition between species can occur when predators are present if availability of suitable shelters is limited. Faria and Almada (2001) found that blennies, especially *C. galerita,* tended to win agonistic encounters with gobiids of similar size. Agonistic interactions between blennies can quickly lead to death; therefore, the agility of competitors to escape aggressive attacks, may determine how much the blenny is affected by such competition (Faria and Almada, 2001).

Herbivorous blennies may compete with other algal grazers, such as damselfish. However, Roberts (1987) demonstrated that whitetail damselfish (*Pomacentrus flavicauda*) did not exclude the jewelled blenny (*Salarias fasciatus*) from their territories, which is a competitor for algal resources. The benefit gained from reducing the amount of algal removed from their territory was considered far less than the energetic expensive of excluding blennies from their territories.

CONCLUSION

This chapter has attempted to draw together the fragmented literature on interspecific relationships in blennies. It has demonstrated that the use of blennies as model systems to test theoretical predictions in a wide range of topics has been successful, especially in studies conducted on mimicry. Most of the work reviewed here has been conducted on coral reef blennies. Examples of interspecific relationships in other environments were less obvious. Although there have been many interesting studies on the biology of blennies, much needs to be done to gain a general understanding of this family, especially in the less common species.

References

Abel, E.F. 1993. Coloration phenomena of Mediterranean blennies (Pisces, Blenniidae). *P.S.Z.N. Marine Ecology* 14: 291-312.

Albrecht, H. 1977. Einige Beobachtungen an Anemonenfischen in der Karibischen See. *Bijdragen tot de Dierkunde* 47: 109-119.

Armbruster, J.W. and L.M. Page. 1996. Convergence of a cryptic saddle pattern in benthic freshwater fishes. *Environmental Biology of Fishes* 45: 249-257.

Arnal, C., I.M. Côté, P. Sasal and S. Morand. 2000. Cleaner–client interactions on a Caribbean reef: Influence of correlates of parasitism. *Behavioral Ecology and Sociobiology* 47: 353-358.

Bates, H.W. 1862. Contributions to an insect fauna of the Amazon Valley. Lepidoptera: Heliconidae. *Transactions of the Linnean Society of London* 23: 495-566.

Brower, L.P. 1988. *Mimicry and the Evolutionary Process.* University of Chicago Press, Chicago.

Bruno, J.F. and M.I. O'Connor. 2005. Cascading effects of predator diversity and omnivory in a marine food web. *Ecology Letters* 8: 1048-1056.

Buchheim, J.R. and M.A. Hixon. 1992. Competition for shelter holes in the coral reef fish *Acanthemblemaria spinosa* Metzelaar. *Journal of Experimental Marine Biology and Ecology* 164: 45-54.

Burreson, E.M. and R.E. Dybdahl. 1989. *Richardsonobdella lineatae,* gen. et sp. nov. (Hirudinea), a parasite of *Meiacanthus lineatus* (Pisces: Blenniidae) from Heron Island, Great Barrier Reef. *Australian Journal of Zoology* 37: 89-93.

Buskey, E.J., P.H. Lenz and D.K. Hartline. 2002. Escape behaviour of planktonic copepods in response to hydrodynamic disturbances: High speed video analysis. *Marine Ecology Progress Series* 235: 135-146.

Cheney, K.L. and I.M. Côté. 2001. Are Caribbean cleaning symbioses mutualistic? Costs and benefits of visiting cleaning stations to longfin damselfish. *Animal Behaviour* 62: 927-933.

Cheney, K.L. and I.M. Côté. 2005. Frequency-dependent success of aggressive mimics in a cleaning symbiosis. *Proceedings of the Royal Society of London* B272: 2635-2639.

Cheney, K.L., A.S. Grutter and N.J. Marshall. 2008. Facultative mimicry: cues for colour change and colour accuracy in a coral reef fish. *Proceedings of the Royal Society of London* B275: 117-122.

Chiao, C.C. and R.T. Hanlon. 2001. Cuttlefish camouflage: Visual perception of size, contrast and number of white squares on artificial checkerboard substrata initiates disruptive coloration. *Journal of Experimental Biology* 204: 2119-2125.

Clarke, R.D. 1989. Population fluctuation, competition and microhabitat distribution of 2 species of tube blennies, *Acanthemblemaria* (Teleostei, Chaenopsidae). *Bulletin of Marine Science* 44: 1174-1185.

Clarke, R.D., E.J. Buskey and K.C. Marsden. 2005. Effects of water motion and prey behaviour on zooplankton capture by two coral reef fishes. *Marine Biology* 146: 1145-1155.

Clavero, M., J. Prenda and M. Delibes. 2004. Influence of spatial heterogeneity on coastal otter (*Lutra lutra*) prey consumption. *Annales Zoologici Fennici* 41: 551-561.

Côté, I.M. 2000. Evolution and ecology of cleaning symbioses in the sea. *Oceanography and Marine Biology Annual Review* 38: 311-355.

Côté, I.M. and K.L. Cheney. 2004. Distance-dependent costs and benefits of aggressive mimicry in a cleaning symbiosis. *Proceedings of the Royal Society of London* B271: 2627-2630.

Côté, I.M. and K.L. Cheney. 2005. Choosing when to be a cleaner-fish mimic. *Nature (London)* 433: 211-212.

Cribb, T.H., R.A. Bray and S.C. Barker. 1994. Bivesiculidae and Haplosplanchnidae (Digenea) from Fishes of the Southern Great-Barrier-Reef, Australia. *Systematic Parasitology* 28: 81-97.

Dafni, J. and A. Diamant. 1984. School-oriented mimicry, a new type of mimicry in fishes. *Marine Ecology Progress Series* 20: 45-50.

Daoulas, C.C. and P.S. Economidis. 1984. The feeding of *Rutilus rubilio* (Bonaparte) (Pisces, Cyprinidae) in Lake Trichonis, Greece. *Cybium* 8: 29-38.

Davies, A.J. 1995. The biology of fish haemogregarines. *Advances in Parasitology* 36: 188-203.

Davies, A.J., N.J. Smit, P.M. Hayes, A.M. Seddon and D. Wertheim. 2004. *Haemogregarina bigemina* (Protozoa: Apicomplexa: Adeleorina)—past, present and future. *Folia Parasitologica* 51: 99-108.

Delibes, M., M. Clavero, J. Prenda, M.D. Blazquez and P. Ferreras. 2004. Potential impact of an exotic mammal on rocky intertidal communities of northwestern Spain. *Biological Invasions* 6: 213-219.

Depczynski, M. and D.R. Bellwood. 2003. The role cryptobenthic reef fishes in coral reef trophodynamics. *Marine Ecology Progress Series* 256: 183-191.

Diaz, F. and M. George-Nascimento. 2002. Temporal stability in parasite infracommunities of the blenny *Scartichthys viridis* (Valenciennes, 1836) (Pisces: Blenniidae) on the central coast of Chile. *Revista Chilena De Historia Natural* 75: 641-649.

Draud, M., D.E. Itzkowitz and M. Itzkowitz. 1990. Co-defense of territory space by 2 species of coral reef fishes. *Bulletin of Marine Science* 47: 721-724.

Dyer, W.G., E.H. Williams and L.B. Williams. 1988. Digenetic trematodes of marine fishes of Okinawa, Japan. *Journal of Parasitology* 74: 638-645.

Edmunds, M. 1974. *Defence in Animals*. Longman, New York.

Elvira, B., G.G. Nicola and A. Almodovar. 1996. Pike and red swamp crayfish: a new case on predator–prey relationship between aliens in central Spain. *Journal of Fish Biology* 48: 437-446.

Endler, J.A. 1984. Progressive background matching in moths, and a quantitative measure of crypsis. *Biological Journal of the Linnean Society* 22: 187-231.

Endler, J.A. 1988. Frequency-dependent predation, crypsis and aposematic coloration. *Philosophical Transactions of the Royal Society of London* B319: 505-523.

Faria, C. and V. Almada. 2001. Microhabitat segregation in three rocky intertidal fish species in Portugal: does it reflect interspecific competition? *Journal of Fish Biology* 58: 145-159.

Faria, C., V. Almada and M.D. Nunes. 1998. Patterns of agonistic behaviour, shelter occupation and habitat preference in juvenile *Lipophrys pholis, Coryphoblennius galerita* and *Gobius cobitis*. *Journal of Fish Biology* 53: 1263-1273.

Feder, H.M. 1966. Cleaning symbiosis in the marine environment. In: *Symbiosis,* S. Mark Henry (ed.), Academic Press, New York, pp. 327-380.

Foster, S.A. 1985. Group foraging by a coral reef fish: A mechanism for gaining access to defended resources. *Animal Behaviour* 33: 782-792.

Franks, J.S., K.E. VanderKooy and N.M. Garber. 2003. Diet of tripletail, *Lobotes surinamensis*, from Mississippi coastal waters. *Gulf and Caribbean Research* 15: 27-32.

Gause, G.F. and A.A. Witt. 1935. Behavior of mixed populations and the problem of natural selection. *American Naturalist* 69: 596-609.

Goeden, G.B. 1978. A monograph of the coral trout, *Plectropomus leopardus* (Lacepède). *Queensland Fisheries Service Research Bulletin* 1: 1-42.

Gon, O. 1993. Revision of the cardinalfish genus *Cheilodipterus* (Perciformes: Apogonidae), with description of five new species. *Indo-Pacific Fishes* 22: 1-59.

González, P. and C. Oyarzún. 2003. Diet of the Chilean sandperch, *Pinguipes chilensis* (Perciformes, Pinguipedidae) in the southern Chile. *Journal of Applied Ichthyology* 19: 371-375.

Greenfield, D.W. 1972. Notes on biology of Arrow Blenny, *Lucayablennius zingaro* (Bohlke) from British-Honduras. *Copeia* 1972: 590-592.

Greenfield, D.W. and T.A. Greenfield. 1982. Habitat and resource partitioning between two species of *Acanthemblemaria* (Pisces: Chaenopsidae), with comments on the chaos hypothesis. *Smithsonian Contributions to Marine Science* 12: 499-507.

Grutter, A.S. 1995. Relationship between cleaning rates and ectoparasite loads in coral-reef fishes. *Marine Ecology Progress Series* 118: 51-58.

Grutter, A.S. 1996. Parasite removal rates by the cleaner wrasse *Labroides dimidiatus*. *Marine Ecology Progress Series* 130: 61-70.

Grutter, A.S. 1999. Cleaner fish really do clean. *Nature (London)* 398: 672-673.

Grutter, A.S. 2001. Parasite infection rather than tactile stimulation is the proximate cause of cleaning behaviour in reef fish. *Proceedings of the Royal Society of London* B268: 1361-1365.

Grutter, A.S. and R. Poulin. 1998a. Intraspecific and interspecific relationships between host size and the abundance of parasitic larval gnathiid isopods on coral reef fishes. *Marine Ecology Progress Series* 164: 263-271.

Grutter, A.S. and R. Poulin. 1998b. Cleaning of coral reef fishes by the wrasse *Labroides dimidiatus*: Influence of client body size and phylogeny. *Copeia* 1998: 120-127.

Haas, R. 1976. Sexual selection in *Nothobranchius guentheri* (Pisces: Cyprinodontidae). *Evolution* 30: 614-622.

Hanlon, R.T. and L. Kaufmann. 1976. Associations of seven West Indian reef fishes with the sea anemone *Condylactis gigantean* in the Dry Tortugas, Florida. *Bulletin of Marine Science* 26: 225-232.

Hanlon, R.T., R.F. Hixon and D.G. Smith. 1983. Behavioural associations of seven West Indian reef fishes with sea anemones at Bonaire, Netherlands Antilles. *Bulletin of Marine Science* 33: 928-934.

Harding, J.M. and R. Mann. 2000. Estimates of naked goby (*Gobiosoma bosc*), striped blenny (*Chasmodes bosquianus*) and eastern oyster (*Crassostrea virginica*) larval production around a restored Chesapeake Bay oyster reef. *Bulletin of Marine Science* 66: 29-45.

Hartman, K.J. and S.B. Brandt. 1995. Trophic resource partitioning, diets, and growth of sympatric estuarine predators. *Transactions of the American Fisheries Society* 124: 520-537.

Hastings, P.A. 1991. Flexible responses to predators in a marine fish. *Ethology, Ecology and Evolution* 3: 177-184.

Heymer, A. and C.D. Zander. 1978. Morphology and ecology of *Gobius vittatus* and its possible mimicry relationship to *Blennius rouxi* in the Mediterranean. Zeitschrift fur Zoologische Systematik und Evolutionsforschung 16: 132-143.

Hiatt, R.W. and D.W. Strasburg. 1960. Ecological relationships of the fish fauna on coral reefs of the Marshall Islands. *Ecological Monographs* 30: 65-127.

Hobson, E.S. 1968. Predatory behaviour of some shore fishes in the Gulf of California. *United States Bureau of Sports Fisheries and Wildlife, Research Report* 73: 1-92.

Hobson, E.S. 1969. Possible advantages to the blenny *Runula azalea* in aggregating with the wrasse *Thalassoma lucasanam* in the tropical eastern Pacific. *Copeia* 1969: 191-193.

Hobson, E.S. 1971. Cleaning symbiosis among California inshore fishes. *Fishery Bulletin* 69: 491-523.

Hobson, E.S. 1974. Feeding relationships of teleostean fishes on coral reefs in Kona, Hawaii. *Fisheries Bulletin* 72: 915-1031.

Kerfoot, W.C. 1980. Visual observation of live zooplankters: evasion, escape and chemical defenses. *American Society of Limnology and Oceanography—Special Symposium* 3: 10-27.

Kingsford, M. 1992. Spatial and temporal variation in predation on reef fishes by coral trout (*Plectropomus leopardus*, Serranidae). *Coral Reefs* 11: 193-198.

Koppel, V.H. 1988. Habitat selection and space partitioning among 2 Mediterranean blenniid species. *P.S.Z.N. Marine Ecology* 9: 329-346.

Kuhlmann, M.L. 1994. Indirect effects of a predatory gastropod in a seagrass community. *Journal of Experimental Marine Biology and Ecology* 183: 163-178.

Kuwamara, T. 1981. Mimicry of the cleaner wrasse *Labroides dimidiatus* by the blennies *Aspidontus taeniatus* and *Plagiotremus rhynorhynchos*. *Nanki Seibutu* 23: 61-70.

Kuwamara, T. 1983. Reexamination on the aggressive mimicry of the cleaner wrasse *Labroides dimidiatus* by the blenny *Aspidontus taeniatus*. *Journal of Ethology* 1: 22-33.

Laveran, A. and F. Mesnil. 1901. Deux hémogrégarines nouvelles des poissons. C. R. *Académie des Sciences Paris* 133: 572-577.

Lehikoinen, A. 2005. Prey-switching and diet of the great cormorant during the breeding season in the Gulf of Finland. *Waterbirds* 28: 511-515.

Lester, R.J.G. and K.B. Sewell. 1989. Checklist of parasites from Heron Island, Great Barrier Reef. *Australian Journal of Zoology* 37: 101-128.

Lieske, E. and R. Myers. 2001. *Coral Reef Fishes*. Princeton University Press, Princeton.

Lindquist, D.G. 1985. Depth zonation, microhabitat, and morphology of three species of *Acanthemblemaria* (Pisces: Blennioidea) in the Gulf of California, Mexico. *Marine Ecology* 6: 329-344.

Lindström, L., R.V. Alatalo and J. Mappes. 1997. Imperfect Batesian mimicry—The effects of the frequency and the distastefulness of the model. *Proceedings of the Royal Society of London* B264: 149-153.

Longley, W.H. and S.F. Hildebrand. 1940. New genera and species of fishes from Tortugas, Florida. *Papers from Tortugas Laboratory* (Carnegie Inst. Wash.) 32: 225-285.

Losey, G.S. 1972. Predation protection in the poison-fang blenny *Meiacanthus atrodorsalis*, and its mimics, *Ecsenius bicolor* and *Runula laudandus* (Blenniidae). *Pacific Science* 26: 129-139.

Moland, E. and G.P. Jones. 2004. Experimental confirmation of aggressive mimicry by a coral reef fish. *Oecologia* 140: 676-683.

Moland, E., J.A. Eagle and G.P. Jones. 2005. Ecology and evolution of mimicry in coral reef fishes. *Oceanography and Marine Biology Annual Review* 43: 455-482.

Morato, T., R.S. Santos and J.P. Andrade. 2000. Feeding habits, seasonal and ontogenetic diet shift of blacktail comber, *Serranus atricauda* (Pisces: Serranidae), from the Azores, north-eastern Atlantic. *Fisheries Research* 49: 51-59.

Moynihan, M. 1968. Social mimicry; character convergence versus character displacement. *Evolution* 22: 315-331.

Mugridge, R.E.R. and H.G. Stallybrass. 1983. A mortality of eels, *Anguilla anguilla* L. attributed to Gnathiidae. *Journal of Fish Diseases* 6: 81-82.

Müller, F. 1879. Ituna and Thyridia: a remarkable case of mimicry in butterflies. (R. Meldola, trans.). *Proceedings of the Entomological Society of London* 1879: 20-29.

Neudecker, S. 1989. Eye camouflage and false eyeposts: chaetodontoid responses to predators. *Environmental Biology of Fishes* 25: 143-157.

Paperna, I. and F.D. Por. 1977. Preliminary data on the Gnathiidae (Isopoda) of the northern Red Sea, the Bitter Lakes and the eastern Mediterranean and the biology of *Gnathia piscivora* n.sp. *Rapports du Comité International de la Mer Méditerrannée* 24: 195-197.

Pasteur, G. 1982. A classificatory review of mimicry systems. *Annual Review of Ecological Systems* 13: 169-199.

Pinnegar, J.K. and N.V.C. Polunin. 2004. Predicting indirect effects of fishing in Mediterranean rocky littoral communities using a dynamic simulation model. *Ecological Modelling* 172: 249-267.

Rakitin, A. and D.L. Kramer. 1996. Effect of a marine reserve on the distribution of coral reef fishes in Barbados. *Marine Ecology Progress Series* 131: 97-113.

Randall, J.E. 1967. Food habits of reef fishes of the West Indies. *Studies in Tropical Oceanography, Miami* 5: 665-847.

Randall, J.E. 1983. *Caribbean Reef Fishes*. T.F.H. Publications Inc., Neptune City, New Jersey.

Randall, J.E. 2005. A review of mimicry in marine fishes. *Zoological Studies* 44: 299-328.

Randall, J.E. and H.A. Randall. 1960. Examples of mimicry and protective resemblance in tropical marine fishes. *Bulletin of Marine Science of the Gulf and Caribbean* 10: 444-459.

Randall, J.E. and V.E. Brock. 1960. Observations on the ecology of Epinephelinae and lutjanid fishes of the Society Islands, with emphasis on food habits. *Transactions of the American Fisheries Society*. 89: 9-16.

Randall, J.E. and J.E. McCosker. 1993. Social mimicry in fishes. *Revue Francaise d'Aquariologie* 20: 5-8.

Randall, J.E., G.R. Allen and R. Steene. 1997. *Fishes of the Great Barrier Reef and Coral Sea*. Crawford House, Bathurst.

Roberts, C.M. 1987. Experimental analysis of resource sharing between herbivorous damselfish and blennies on the Great Barrier Reef. *Journal of Experimental Marine Biology and Ecology* 111: 61-75.

Russell, B.C. 1977. Sabre-teeth and poison-fanged blennies. *Oceans* 1: 66-69.

Russell, B.C. 1983. The food and feeding habits of rocky reef fish of north-eastern New Zealand. *New Zealand Journal of Marine and Freshwater Research* 17: 121-145.

Russell, B.C., G.R. Allen and H.R. Lubbock. 1976. New cases of mimicry in marine fishes. *Journal of Zoology (London)* 180: 407-423.

Sayer, M.D.J. and J. Davenport. 1991. Amphibious fish: why do they leave the water? *Reviews in Fish Biology and Fisheries* 1: 159-181.

Smit, N.J., L. Basson and J.G. Van As. 2003a. Life cycle of the temporary fish parasite, *Gnathia africana* (Crustacea : Isopoda : Gnathiidae). *Folia Parasitologica* 50: 135-142.

Smit, N.J., J.G. Van As and A.J. Davies. 2003b. Taxonomic re-evaluation of the south African fish haemogregarine, *Desseria fragilis*. *Journal of Parasitology* 89: 151-153.

Smit, N.J., A.S. Grutter, R.D. Adlard and A.J. Davies. 2006. Hematozoa of teleosts from Lizard Island, Australia, with some comments on their possible mode of transmission and the description of a new haemogregarine species. *Journal of Parasitology* 92: 778-788.

Smith-Vaniz, W.F. 1987. The saber-toothed blennies, tribe Nemophini (Pisces: Blenniidae): an update. *Proceedings of the Academy of Natural Sciences of Philadelphia* 139: 1-52.

Smith-Vaniz, W.F., U. Satapoomin and G.R. Allen. 2001. *Meiacanthus urostigma*, a new fangblenny from the northeastern Indian Ocean, with a discussion and examples of mimicry in species of *Meiacanthus* (Teleostei: Blenniidae: Nemophini). *Aqua* 5: 25-43.

Springer, V.G. and W.F. Smith-Vaniz. 1972. Mimetic relationships involving fishes of the family Blenniidae. *Smithsonian Contributions to Zoology* 112: 1-36.

Strack, W.A. 1969. *Ecsenius (Anthiiblennius) midas* a new subgenus and species of mimic blenny from the western Indian Ocean. *Notulae Naturae* 419: 1-9.

Taborsky, M. 1994. Sneakers, satellites and helpers: parasitic and cooperative behaviour in fish reproduction. *Advances in the Study of Behaviour* 23: 1-100.

Takeuchi, N., H. Hashimoto and K. Gushima. 2002. Short-term foraging patterns of individual cornetfish, *Fistularia commersonii*, based on stomach content analysis. *Ichthyological Research* 49: 76-80.

Thompson, J.N. 1996. Evolutionary ecology and the conservation of biodiversity. *Trends in Ecology and Evolution* 11: 300-303.

Thompson, J.N. 1999. The evolution of species interactions. *Science* 284: 2116-2118.

Topolski, M.F. and S.T. Szedlmayer. 2004. Vertical distribution, size structure, and habitat associations of four Blenniidae species on gas platforms in the north-central Gulf of Mexico. *Environmental Biology of Fishes* 70: 193-201.

Turner, J.R.G. 1987. The evolutionary dynamics of Batesian and Müllerian mimicry: similarities and differences. *Ecological Entomology* 12: 81-95.

Walter, J.F., III. and H.M. Austin., 2003. Diet composition of large striped bass (*Morone saxatilis*) in Chesapeake Bay. *Fisheries Bulletin* 101: 414-423.

Werner, E.E. and B.R. Anholt. 1993. Ecological consequences of the trade-off between growth and mortality-rates mediated by foraging activity. *American Naturalist* 142: 242-272.

Wickler, W. 1961. Über das Verhalten der Blenniiden *Runula* und *Aspidontus* (Pisces; Blenniidae). *Zeitschrift für Tierpsychologie* 20: 657-679.

Wickler, W. 1965. Mimicry and evolution of animal communication. *Nature (London)* 208: 519-521.

Wickler, W. 1966. Mimicry in tropical fishes. *Philosophical Transactions of the Royal Society of London* B251: 473-474.

Wickler, W. 1968. *Mimicry in Plants and Animals*. McGraw-Hill, New York.

Wootton, R.J. 1990. *Ecology of Teleost Fishes*. Chapman and Hall, London.

Zander C.D. and J. Nieder. 1997. Interspecific associations in Mediterranean fishes: feeding communities, cleaning symbioses and cleaner mimics. *Vie et Milieu* 47: 203-212.

CHAPTER

5.3

Patterns of Microhabitat Utilisation in Blennies

Emanuel J. Gonçalves[1] and Cláudia Faria[2]

INTRODUCTION

Most blennioids are small bottom-dwelling site-attached fishes, many of them territorial, showing strong associations with the microhabitats they inhabit and spending a lot of time keeping close to the substrate, often in cryptic microhabitats (e.g., holes, crevices, under overhangs or small boulders). They are generally highly sedentary, swimming only for short distances and spending very little time in open water (Gibson, 1969). This is especially true for the intertidal species, for which a tight association with the substratum can prevent the risk of displacement caused by water turbulence (Gibson, 1969; Gonçalves and Almada, 1998). Territoriality is quite common in many species, not only during the breeding season (e.g., Almada *et al.*, 1987, 1992, 1996; Almada and Santos, 1995; Gonçalves and Almada, 1998), but also throughout the year, probably to ensure an

Authors' address: Eco-Ethology Research Unit, Instituto Superior de Psicologia Aplicada, Rua Jardim do Tabaco 34, 1149-041 Lisboa, Portugal.
E-mails: [1]emanuel@ispa.pt (*Corresponding author*); [2]cbmfaria@clix.pt

efficient use of food and shelter resources (Gibson, 1969, 1982; Lindquist, 1985; Koppel, 1988; Mayr and Berger, 1992).

There is little literature dealing with the morphological adaptations of these benthic fishes to their habitat (but see Gibson, 1969, 1982 for reviews). Zander (1972a, b, 1974, 1980) has made an assessment of the external morphology, anatomy of fins, and eye morphology in relation to the habitats and behaviour of some Mediterranean and Red Sea blennies. Stephens *et al.* (1970) suggested that body size in some species could be the result of an evolutionary matching with the sizes of available shelters. These authors found that tube size controlled fish size in *Hypsoblennius jenkinsi*, a small tubiculous blenny living in two separate habitats. One population of *H. jenkinsi* inhabiting vermetid gastropod refuges was significantly smaller than an adjacent population inhabiting larger pholadid mollusk holes. The authors further suggested a habitat-matching mechanism-based on the assumption that as the blennies outgrow their tubes they are forced into less protective areas where they become subjected to higher predation levels.

Microhabitat partitioning between ecological related reef fish species has been well documented (Stephens *et al.*, 1966, 1970; Greenfield and Greenfield, 1982; Lindquist, 1985; Mayr and Berger, 1992; Ormond *et al.*, 1996; Davies, 2000; Munday, 2004; Feary and Clements, 2006). Moreover, many fishes either actively select specific microhabitats at settlement (Carr, 1991; Levin, 1991) or emigrate to them during juvenile or adult stages (Lewis, 1997; Faria and Almada, 2001). These preferences for distinct microhabitats may be advantageous by mediating the effects of competition and predation (Steele, 1999; Forrester and Steele, 2000; Hyndes *et al.*, 2003; Willis and Anderson, 2003), providing greater food resources (Clarke, 1992; Wilson, 2001; Hyndes *et al.*, 2003) or nesting habitats (Koppel, 1988; Feary and Clements, 2006).

Owing to their small size and relatively small home ranges, the habitat choice in blennioid fishes should be mainly related to those environmental features that operate on a comparable spatial scale (La Mesa *et al.*, 2004, 2006; La Mesa and Vacchi, 2005). One would expect that smaller benthic species, that tend to be cryptic and closely associated with the substrate, such as most blennioids, show greater habitat specificity. Greenfield and Johnson (1990) demonstrated that different habitats, which can be adjacent to each other, presented distinctive blennioid assemblages. The chaenopsids provide many examples of high habitat specificity within the blennioidei (e.g., Greenfield and Greenfield, 1982; Lindquist, 1985; Hastings 1986, 1988, 1992; Clarke, 1992; Tyler and Tyler, 1997, 1999; Smith *et al.*, 1998; Clarke and Tyler, 2003). An example of this very high level of microhabitat specificity is the segregation of the species *Emblemariopsis pricei* and *Emblemaria pandionis*. At Glovers Reef, Belize, whereas the former has only been found in holes of living corals, the later inhabits only holes in coral rubble on sandy bottoms, never being found in living coral (Greenfield, 1975; Tyler and Tyler 1999). Non-breeding males and females of *E. pricei* have been found exclusively at the surface of sea fans (Tyler and Tyler, 1999). Another example of this high specificity in chaenopsids is the association between *Emblemariopsis diaphanus* and the stony coral *Meandrina meandrites* (Butter *et al.*, 1980).

According to many authors, the patterns of habitat distribution in reef fish populations may be dependent on the spatial scale of observations (Tolimieri, 1995; Chittaro, 2000; La Mesa *et al.*, 2004, 2006; La Mesa and Vacchi, 2005; Feary and Clements, 2006). Large-scale habitats and zones can support specific assemblages of fish species (Choat and Ayling, 1987; Green, 1996; Ormond *et al.*, 1996; Gibson and Yoshiyama, 1999; Townsend and Tibbetts, 2000; Arbuto-Oropeza and Balart, 2001; Curley *et al.*, 2002). At the microhabitat scale, a small coral colony, an algal holdfast or some other specific substrata, for example, will attract a distinct fish assemblage (Macpherson, 1994; Munday *et al.*, 1997; Angel and Ojeda, 2001). Similarly, unique fish assemblages may aggregate in habitat patches, such as monospecific stands of coral beds or kelp forests (Williams and Sale, 1981; Jones, 1988; Anderson, 1994; Levin and Hay, 1996; Willis and Anderson, 2003).

Here we explore the available information concerning patterns of microhabitat utilisation in blennioids and discuss some of the general trends derived from the data. Most information was retrieved from FishBase (Froese and Pauly, 2007) and a word of caution is needed since some information could not be confirmed using the primary sources. Nevertheless, we do believe that the general patterns extracted from the data realistically describe the general trends for the different groups and habitats.

GENERAL PATTERNS OF HABITAT USE

The Blennioidei consists of mainly small marine reef-associated shallow water species of temperate and tropical waters, which explore a wide range of microhabitats. There are however some exceptions to this description. For instance, some species may occur in freshwater environments. This seems to be the case of the blennies *Meiacanthus anema* (Séret 1997; Myers, 1999), *Omobranchus ferox* (Springer, 1986), *Omox biporos* (Masuda *et al.*, 1984), *Salaria fluviatilis* (Neat *et al.*, 2003), the clinid *Clinus spatulatus* (Smith, 1986; Whitfield 1996), and the tripterygiid *Grahamina nigripenne* (Fricke, 1994). There is a single described species that occurs in the polar environments: *Helcogrammoides antarcticus* known only from the type locality, Paradise Bay, Antarctic Peninsula (Tomo, 1981; Gon, 1990; Williams and Springer, 2001).. All other known blennioids are distributed in temperate and tropical seas.

Looking at the general geographical distribution of the different families, one can found that, regarding the percentage of species in each family with a tropical affinity, all but the Clinidae present a great majority of species in this region (Blenniidae 85%, Chaenopsidae 86%, Clinidae 6%, Dactyloscopidae 98%, Labrisomidae 90%, Tripterygiidae 66%). The approximate percentage of species with a subtropical distribution in each family is: Blenniidae 12%, Chaenopsidae 14%, Clinidae 74%, Dactyloscopidae 2%, Labrisomidae 10%, Tripterygiidae 15%. The proportion of species in the temperate regions is: Blenniidae 3%, Clinidae 20%, Tripterygiidae 19%. The families Chaenopsidae, Dactyloscopidae and Labrisomidae have only a few species represented in the temperate regions (see Hastings, Chapter 2.1. this book).

The general description of the Blennioidei state that most are small cryptic fishes with medium sizes smaller than 15 cm. Based on the available information on maximum sizes provided in FishBase (further complemented with literature data for some missing values), information on standard length (SL) for 313 species and total length (TL) for 432 species exist, with 140 species without any of these measures. In order to make the SL and TL data comparable, we have extrapolated SL values to TL for those 313 species, assuming an average 15 % difference between these measures. This value was obtained from the few published information on SL/TL relationships in blennioids as well as for several temperate northeast Atlantic blennioid species for which we have data. This extrapolation should however be taken with caution since this relationship is quite variable depending on the general anatomy of different species. Nevertheless, we found that it would be interesting to explore the general patterns of length distributions in these fishes in trying to understand their patterns of microhabitat use.

With the assumptions and words of caution outlined above, there are significant differences in size distribution within the six blennioid families (Kruskal-Wallis ANOVA: H (5, N = 745) = 177.54; $p < 0.001$) (Fig. 5.3.1). Clinids are significantly larger than all other blennioids. Mean (± S.D.) length values obtained are: blenniids = 9.24 ± 0.32 cm, N = 328; chaenopsids = 6.38 ± 0.69 cm, N = 69; clinids = 14.48 ± 0.65 cm, N = 78; dactyloscopids = 6.33 ± 0.94 cm, N = 37; labrisomids = 7.48 ± 0.64 cm, N = 80; and tripterygiids = 5.37 ± 0.46 cm, N = 153. Blenniids are smaller than clinids but larger than all other families, while chaenopsids, dactyloscopids and labrisomids are not significantly different from each other. Tripterygiids are the smallest, being significantly smaller than clinids, blenniids and labrisomids.

Some outliers are present, notably within the Clinidae with several species reaching over 30 cm TL, and *Heterostichus rostratus* reaching 61 cm. In the Blenniidae, the two species of the genus *Xiphasia* reach

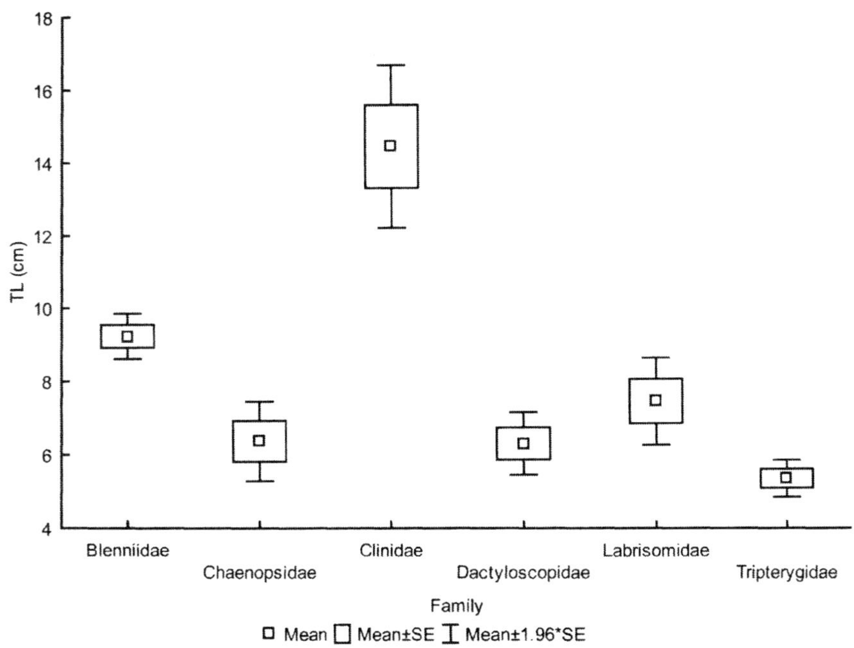

Fig. 5.3.1 Size distribution (total length) for the six blennioid families (N = 745 species). Maximum total length (TL) was obtained from the available information in FishBase (Froese and Pauly, 2007) (N = 432 species) and extrapolated for species for which only standard length (SL) information was available (N = 313 species) (see text for details).

30 cm SL (*Xiphasia matsubarai*) and 53 cm SL (*Xiphasia setifer*). In the Chaenopsidae, species of the genus *Neoclinus* are the largest, with *Neoclinus uninotatus* and *Neoclinus blanchardi* attaining 25 cm and 30 cm TL, respectively. Dactyloscopids are smaller than 10 cm, with the exception of *Dactylagnus mundus* that reaches a size of 15 cm SL. Labrisomids present several examples of species larger than 10 cm TL, with *Labrisomus philippii* reaching 35 cm. Finally, all tripterygiids are smaller than 15 cm, with the exception of *Blennodon dorsale* and *Notoclinus fenestratus* with 15.1 cm and 18.2 cm SL, respectively. Variation within each family of mean TL for those genera which present more than five species with available information is presented in Table 5.3.1. In spite of the above mentioned outliers, no pattern of larger size can be ascribed to any specific genus within any of the families. This suggests that the largest representatives in each family are probably exceptions in the general size trend of each genus.

Depth distribution patterns of blennioids agree with the general description which states that these fishes occur almost exclusively in shallow water. The available data on depth distributions reveals that no significant differences exist among the six families of blennioids (Kruskal-Wallis ANOVA: H (5, N= 435) = 9.24, $p = 0.10$) (Fig. 5.3.2). Mean depth values for all families lie within the first 40 m. Mean (± S.D.) maximum depth values obtained are: blenniids = 21.63 ± 43.28 m, N = 163; chaenopsids = 24.37 ± 35.51 m, N = 49; clinids = 21.58 ± 48.75 m, N = 26; dactyloscopids = 28.93 ± 45.38 m, N = 30; labrisomids = 32.32 ± 37.47 m, N = 44; and tripterygiids = 25.94 ± 22.41 m, N = 123.

Again, some exceptions to this general pattern also occur. Within the blenniids, a few species have been collected in deep waters: e.g., *Blennius normani* (to 200 m), *Blennius ocellaris* (to 400 m) and *Petroscirtes marginatus* (to 300 m). The two species of genus *Xiphasia*, *Xiphasia setifer* and *Xiphasia matsubarai*, are reported to occur from shallow water to, respectively, 1190 m and 4960 m (Froese and Pauly, 2007). These records need however confirmation since this represents not only a wide range of depths for a single species, but also because it departs from the general pattern described for the Blenniidae. Given that we could not confirm these data using mainly primary sources and due to the great departure from the described pattern for the family, we have not included this two species in the depth analysis. The only chaenopsids exceeding 100 m are

Table 5.3.1 Mean maximum total length (TL) obtained from the available information in FishBase (Froese and Pauly, 2007) and extrapolated for species for which only standard length (SL) information was available (see text for details). Only genera with more than five species with available data are represented.

Family	*Genus*	*Mean maximum TL (cm)*	*N*
Blenniidae	*Ecsenius*	5.56	53
	Praealticus	6.63	6
	Antennablennius	6.71	6
	Omobranchus	7.00	14
	Meiacanthus	7.63	23
	Alticus	8.49	6
	Entomacrodus	8.78	23
	Petroscirtes	9.46	11
	Scartella	9.53	6
	Hypsoblennius	9.54	13
	Salarias	10.04	9
	Plagiotremus	11.12	9
	Parablennius	11.52	22
	Cirripectes	11.60	22
	Blenniella	12.04	9
	Istiblennius	13.88	14
Chaenopsidae	*Acanthemblemaria*	4.62	13
	Emblemariopsis	4.76	8
	Emblemaria	5.26	13
	Chaenopsis	9.92	7
	Neoclinus	11.67	9
Clinidae	*Ophiclinus*	10.08	6
	Pavoclinus	13.81	8
	Heteroclinus	14.46	15
	Clinus	16.40	16
Dactyloscopidae	*Gillellus*	5.27	9
	Platygillellus	5.32	6
	Dactyloscopus	6.98	12
Labrisomidae	*Starksia*	3.50	22
	Paraclinus	5.15	17
	Malacoctenus	7.38	17
	Labrisomus	13.04	16
Tripterygiidae	*Enneapterygius*	3.46	54
	Axoclinus	3.63	6
	Enneanectes	3.79	8
	Helcogramma	4.30	31

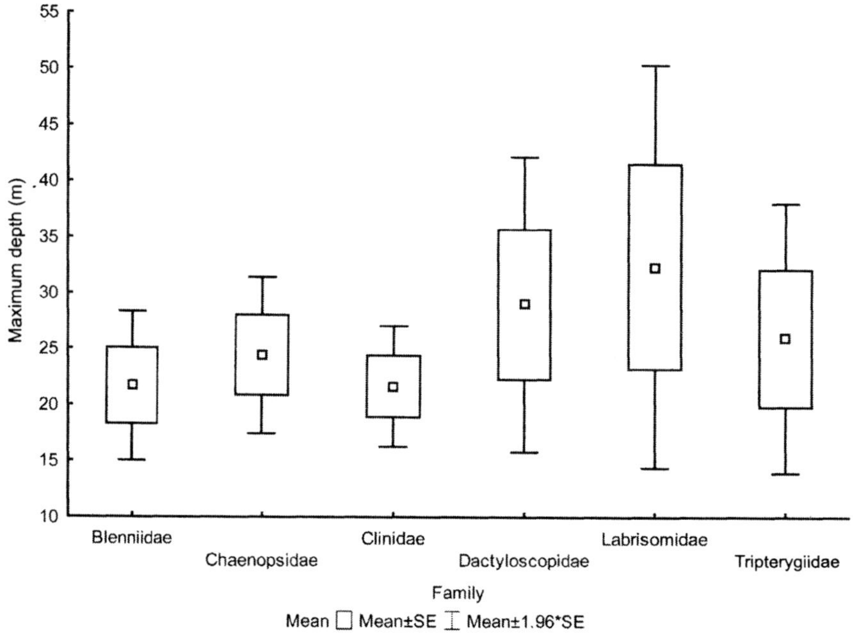

Fig. 5.3.2 Maximum depth (m) registered for species within each family. Total number of species with available information are: blenniids, N = 163; chaenopsids, N = 49; clinids, N = 26; dactyloscopids, N = 30; labrisomids, N = 44; and tripterygiids, N = 123.

Lucayablennius zingaro (106 m) and *Emblemaria atlantica* (to 110 m), while the deepest record for the clinids is 60 m (*Cristiceps argyropleura*). In dactyloscopids also only two species exceed 100 m, *Gillellus arenicola* (to 137 m) and *Gillellus semicinctus* (to 140 m), the same being true for the labrisomids, *Starksia hassi* (to 175 m) and *Haptoclinus apectolophus* (to 366 m). Four tripterygiids have been recorded in waters deeper than 100 m: *Forsterygion flavonigrum* (to 110 m), *Apopterygion oculus* (to 186 m), *Matanui profundum* (to 500 m) and *Matanui bathytaton* (to 550 m) (Fricke, 1994).

Variation within each family of mean maximum depth for those genera, which present more than five species with available information, is presented in Table 5.3.2. Again, no pattern for occurring at deeper waters can be ascribed to any specific genus with only the dactyloscopid genus *Gillellus* being found at average depths of over 40 m (average = 60.7 m, N = 7).

Table 5.3.2 Information on mean maximum depth (m) for different genera within each family of the Blennioidei. Only genera with more than five species with available data are represented.

Family	*Genus*	*Mean maximum depth (m)*	*N*
Blenniidae	*Entomacrodus*	4.4	7
	Salarias	6.5	6
	Hypsoblennius	13.2	6
	Parablennius	14.0	14
	Cirripectes	15.6	22
	Ecsenius	18.0	32
	Meiacanthus	20.0	14
	Plagiotremus	36.7	7
Chaenopsidae	*Acanthemblemaria*	13.9	11
	Chaenopsis	29.3	8
	Emblemaria	34.5	11
Clinidae	*Heteroclinus*	21.0	6
Dactyloscopidae	*Dactyloscopus*	14.3	10
	Platygillellus	19.3	6
	Gillellus	60.6	7
Labrisomidae	*Labrisomus*	13.4	8
	Paraclinus	15.1	10
	Malacoctenus	18.6	9
	Starksia	31.8	13
Tripterygiidae	*Helcogramma*	10.2	28
	Enneapterygius	13.1	45

It should be however noted that the depth range in between the normal use of SCUBA diving and the use of small submersibles, notably between 40 m and 200 m, is one of the least studied zones with regard to coastal reef communities (both in the tropics and in temperate regions) which means that the patterns presented above may change with the recent advent of sophisticated breeding apparatus, as for example rebreathers. They allow divers to explore deeper reefs for much longer periods and several new species and new depth ranges have been recently described and characterised using this technique (Pyle, 1996a, b, 1999, 2000; Randall and Pyle, 2001a, b). The recent advances in other underwater equipment such as the greater availability of remote operated vehicles (ROVs) and autonomous underwater vehicles (AUVs) promises to bring new discoveries on the deeper coastal fish assemblages.

Looking at the main habitats occupied by blennioids at the system level (Table 5.3.3), one can find that most species occur on rocky reefs

Table 5.3.3 Percentage of occurrence in the main habitats occupied by species in each blennioid family (N = 752 species).

Main habitats	*Blenniidae*	*Chaenopsidae*	*Clinidae*	*Dactyloscopidae*	*Labrisomidae*	*Tripterygiidae*
Coral reefs	46	44	3	0	31	40
Estuaries	2	0	4	0	0	2
Freshwater	0	0	0	0	0	0
Mangroves	2	0	0	0	0	0
Rocky reefs	45	46	77	3	51	56
Sand/mud	1	8	7	97	5	0
Seagrasses	2	1	8	0	13	2

(N = 384, corresponding to 51% of blennioid species) and coral reefs (N = 266, corresponding to 35%). The other habitats where blennioids occur are estuaries, freshwater systems, mangroves, sand/mud and seagrasses (all with less than 10% of occurrences). At the family level, coral reefs and rocky reefs are the main habitats where species occur for all families, with the exception of clinids (with only 3% of species on coral reefs) and dactyloscopids (which occur almost exclusively on sand/mud habitats—97% of the species).

SPECIFIC PATTERNS OF MICROHABITAT UTILISATION

A variety of studies have been carried out on the effects of environmental variables on the structure of fish communities, trying to solve the important question of what factors might explain the differential habitat and microhabitat use in fish assemblages. Several investigations have demonstrated the relevant role of habitat complexity (Luckhurst and Luckhurst, 1978; Roberts and Ormond, 1987; Connell and Jones, 1991; Friedlander and Parrish, 1998; Garcia Charton and Pérez Ruzafa, 1998; Ferreira *et al.*, 2001; Gratwicke and Speigth, 2005), depth (Bell, 1983; Roberts and Ormond, 1987; McGehee, 1994, Patzner, 1999; Feary and Clements, 2006; La Mesa *et al.*, 2006) and algal canopy (Carr, 1989, 1991; Chabanet *et al.*, 1997; Prochazka, 1998; Hyndes *et al.*, 2003; Gratwicke and Speigth, 2005) in structuring of coastal fish communities. On the other hand, as pointed out by many authors, microhabitat features of the substratum, such as complexity, heterogeneity and types of cover, have a strong influence on the patterns of microhabitat use for many small cryptobenthic fishes (Costello, 1992; Wilkins and Myers, 1992; Macpherson, 1994; Syms, 1995; Macpherson and Zika, 1999; Feary and Clements, 2006; La Mesa *et al.*, 2006).

In both coral and rocky reefs, species richness, diversity and density of benthic fishes are often positively correlated with habitat complexity (Luckhurst and Luckhurst, 1978; Roberts and Ormond, 1987; Connell and Jones, 1991; Garcia Charton and Pérez Ruzafa, 1998, 2001; Willis and Anderson, 2003) and habitat heterogeneity measures (Roberts and Ormond, 1987; Garcia Charton and Pérez Ruzafa, 2001). Such an effect has been suggested to rely upon different mechanisms, for instance by enhancing shelter opportunities and protection against predators (Hixon and Beets, 1993; Macpherson, 1994; Caley and St. John, 1996; La Mesa *et al.*, 2004). Concerning the algal canopy, the potential benefit for fish species richness and increased abundance of a dense algal canopy with respect to unvegetated habitats remains controversial (La Mesa *et al.*, 2006). For instance, Syms and Jones (1999) observed only a weak and very transient effect of kelp removal on a tripterygiid guild. However, the three dimensional structure of these habitats are certainly of great importance for some species of clinids which live associated to kelp beds (Coyer, 1982; Stepien 1987; Stepien *et al.*, 1988).

Information on the available data on microhabitat utilisation in blennioids is presented in Table 5.3.4. Microhabitats were ascribed to 23 groups. Overall, the three most used microhabitats are intertidal pools (N = 107 species, corresponding to 17.6%), rock covered with seaweeds (N = 87 species, corresponding to 14.3%), and small shelters (N = 79 species, corresponding to 13%). However, these numbers vary greatly within families.

Some present a high specificity on microhabitat use, as is the case of the dactyloscopids (with 83% of species occurring on sand and gravel and 10% on intertidal pools) and chaenopsids (with 70% of species occurring on small shelters and 8% on coral rubble). Clinids seem to occur preferentially on intertidal pools (30%), rock covered with seaweeds (32%), kelp beds (12%) and seagrasses (12%), while labrisomids occur in rock covered with seaweeds (32%), seagrasses (15%) and intertidal pools (15%). Blenniids and tripterygiids occur in almost all microhabitats identified. There seems however to be a preference for blenniids to occur in small shelters (17%), intertidal pools (14%) and in the surge zone of coral reefs (11%), while tripterygiids are mainly present in intertidal pools (25%), overhangs (16%) and boulders or rock (11%).

Interestingly, some microhabitats are mostly used by one or a few families. This is the case for instance of bivalve beds, used by blenniids and

Table 5.3.4 Percentage of occurrence in the main microhabitats occupied by species in each blennioid family (N = 608 species).

Microhabitats	*Family*					
	Blenniidae	*Chaenopsidae*	*Clinidae*	*Dactyloscopidae*	*Labrisomidae*	*Tripterygiidae*
N	*210*	*60*	*84*	*29*	*73*	*152*
Biogenic	2	2	2	0	10	3
Bivalve beds	5	3	0	0	0	0
Boulders or rock	7	0	4	0	5	11
Caves	1	0	0	3	0	0
Coral branches	6	3	0	0	0	2
Coral heads	4	0	0	0	1	1
Coral rubble	7	8	0	0	8	7
Coralline algae	1	0	0	3	0	1
Crevices	3	2	0	0	5	5
Floating weeds	1	2	0	0	0	0
Free-swimming	3	2	0	0	0	0
Intertidal pools	14	0	30	10	15	25
Kelp beds	0	0	12	0	0	3
Open rock surface	0	0	0	0	0	9
Overhangs	0	0	0	0	0	16
Rock covered with seaweeds	10	5	32	0	32	9
Sand and gravel	1	3	8	83	5	1
Seagrasses	4	0	12	0	15	2
Small shelters	17	70	0	0	1	1
Spray zone	1	0	0	0	0	0
Surge zone	11	0	0	0	0	4
Urchins and urchin barrens	0	0	0	0	1	2

chaenopsids; caves, used by blenniids and dactyloscopids; kelp beds, used by clinids and tripterygiids; open rock surfaces and overhangs, used mainly by tripterygiids; small shelters like narrow holes, empty tubes, holes and tunnels of worms and clams, and empty barnacle shells, used by blenniids and chaenopsids; the spray zone used almost exclusively by blenniids (labrisomids of the genus *Dialommus* also occur in this habitat; Martin and Bridges, 1999; Nieder, 2001); and the surge zone used by blenniids and tripterygiids. Some examples on microhabitats used by different blennioids are presented in Figure 5.3.3.

Many blennioids utilise specific microhabitats as nests (Table 5.3.5) and, for the 156 species for which we could compile data, roughly 70%

Fig. 5.3.3 Some examples of microhabitat use by different blennioids. **a**. *Lipophrys pholis* male in a small rocky intertidal shelter (Arrábida Marine Park, Portugal); **b**. *Parablennius gattorugine* among algal tufts (Arrábida Marine Park, Portugal); **c**. *Tripterygion delaisi* male and female on a nest (clean algal surface) (Arrábida Marine Park, Portugal); **d**. *Malacoctenus boehlkei* on coral (photo by Robert A. Patzner, Cuba); **e**. *Acanthemblemaria macrospilus* occupying a hole in live coral (photo by Alicia Hermosillo, Bahia Bandaras, Mexico); **f**. *Clinus superciliosus* on algae (photo by Dennis King, False Bay ,Cape town on sand/rubble bottom with algae).

(110 species) use nests in highly protected places, like crevices, shells or invertebrate holes (categorised as 'small shelters'). The chaenopsids revealed again a high level of microhabitat specialisation with all 20 species for which data were available nesting in small shelters. Blenniids also nest predominantly in small shelters (79 out of 85 species).

Table 5.3.5 Number of species breeding in each microhabitat type for each blennioid family (N = 156 species).

Nest habitats	*Blenniidae*	*Chaenopsidae*	*Clinidae*	*Labrisomidae*	*Tripterygiidae*
N	*85*	*20*	*10*	*6*	*35*
Algal tufts and turfs	1		4		9
Coral heads	1				
Crevices	10				9
Kelp			1		
Open rock surfaces				1	7
Seagrasses	1				
Shells	14				1
Small shelters	55	20		1	
Sponges				2	
Under stones and large crevices	3			2	9
Viviparous			5		

Tripterygiids (35 species) use as nests different types of microhabitats such as crevices, open rock surfaces, under stones and in overhangs and seaweeds. They do not seem to breed however in small shelters. For clinids and labrisomids, there are no sufficient data to get any pattern. Some clinids are viviparous (Gunn and Thresher, 1991), some others breed on kelp beds and algal tufts. For dactyloscopids information is only available for a few species (e.g., *Myxodagnus opercularis*) and it was observed that males carry a clump of eggs behind the pectoral fins (Böhlke and Chaplin, 1968; Petti, 1969).

There is an overall higher number of species in small shelters and related microhabitats, rocks covered with seaweeds and tide pools than on more exposed habitats, like boulders and rock platforms. Small shelters and boulders seem to represent the two extremes of cryptobenthic microhabitat utilisation in relation to species diversity. Small shelters have the highest number of species, including a number of habitat specialists. In contrast, boulders and rock platforms are depauperate microhabitats in what regards the occurrence of blennioids. A number of distinct physical differences exist between these two types of microhabitats, potentially impacting on the numbers of individuals and species each can support.

CORAL REEFS VERSUS ROCKY REEFS

Coral reefs support the co-existence of many species of fishes, and both coral diversity and habitat complexity are considered to be important factors governing the species richness and population abundance of fishes (Clarke, 1977; Williams 1991; Green, 1996; Chabanet *et al.*, 1997). The distribution of fish species within coral reef fish assemblages have been extensively studied (Greenfield and Johnson, 1990; Wilson, 2001; Depczynski and Bellwood, 2004), and the role of species-specific use of particular coral habitats in the coexistence of large, highly diverse reef fish assemblages has been well documented (Alevizon *et al.*, 1985; Greenfield and Johnson, 1990; Bean *et al.*, 2002). For instance, Ault and Johnson (1998) have shown that up to 75% of the diversity of coral reef assemblages can be explained by microhabitat structure. Several studies have also examined the relationship between various substratum variables and species richness and abundance of resident fish communities on coral reefs (Risk, 1972; Gladfelter and Gladfelter, 1978; Luckhurst and Luckhurst, 1978; Carpenter *et al.*, 1981; McManus *et al.*, 1981; Bell and Galzin, 1984; Sano *et al.*, 1984; Greenfield and Johnson, 1990; Friedlander and Parrish, 1998; Townsend and Tibbetts, 2000; Depczynski and Bellwood, 2004). Several of these studies have found positive significant correlations between substratum structural complexity and fish species richness or diversity (Risk, 1972; Gladfelter and Gladfelter, 1978; Luckhurst and Luckhurst, 1978; Carpenter *et al.*, 1981; Friedlander and Parrish, 1998; Depczynski and Bellwood, 2004). However, correlations between structural complexity and fish abundance have been less consistent. For example, Carpenter *et al.* (1981) found a strong positive correlation of fish biomass but not abundance with substratum complexity, and Risk (1972) found that fish abundance and substratum complexity were not significantly correlated.

The arrangement of coral reefs may be either continuous or patchy, with different coral colonies providing diversified microhabitats to which different fish species may associate. This affects the spatial distribution of fishes, recruitment patterns of juveniles and also predation intensity. It is known that the main mechanisms underlying organisation of fish communities at continuous habitat sites differ from those at isolated habitat sites (Nanami and Nishihira, 2002). In continuous habitats, diverse coral communities provide a greater area and diversity of living spaces for fishes, with relatively lower predation pressures than at isolated habitat sites. This is explained by the fact that in continuous reefs, the abundance of foliose and branching corals provides more refuges against predators.

Many studies showed that suitable living space appears to be the resource most likely to be in short supply for many coral-reef fishes (Sale, 1972, 1975; Smith and Tyler, 1972; Buchheim and Hixon, 1992). For example, Buchheim and Hixon (1992) suggested that holes are a limiting resource determining the number of spiny head blennies that can live on a coral head. By increasing the number of holes on a coral head, the number of resident blennies also increased, whereas without vacant holes the number of resident fish remained relatively constant, regardless of the number of fish added to the coral heads. These patterns suggest that the same coral head would support a larger number of blennies if more suitable shelter holes were available.

Among the coastal environments in temperate regions, rocky reefs host the most diverse and abundant benthic fish populations, owing to their high spatial heterogeneity and complexity (Gibson, 1969), offering many niche types and shelter opportunities. The distribution of rocky habitats can generally be regarded as patchy and multiscalar (Horn *et al.*, 1999). Physical barriers, such as sandy beaches, often isolate different stretches of rocky shores, while entire regions can be separated by vast expanses of ocean. At large spatial scales, differences in fish assemblages have been related to the presence of different habitats at each location that are utilised by different species (Marsh *et al.*, 1978; Yoshiyama, 1981; Bennett and Griffiths, 1984; Prochazka and Griffiths, 1992; Gibson and Yoshiyama, 1999). Within rocky reefs, the abundance of fish species has often been associated with particular physical and biological characteristics of the reef substratum (Choat and Ayling, 1987; Arbuto-Oropeza and Balart, 2001).

The main environmental factors known to influence distribution patterns of blennioids in temperate regions are depth, bottom slope, substratum complexity and heterogeneity, and type of cover (Abel, 1962; Heymer and Zander, 1975; Kotrschal, 1983, 1988; Patzner, 1985, 1999; Koppel, 1988; Illich and Kotrschal, 1990; Nieto and Alberto, 1993-1994; Macpherson, 1994; Syms, 1995; Macpherson and Zika, 1999; Nieder *et al.*, 2000; La Mesa *et al.*, 2004, 2006; La Mesa and Vacchi, 2005; Feary and Clements, 2006). Among-location differences in the relative abundance of species may be attributed to the different microhabitats available and to species-specific preferences for particular microhabitats (Gibson, 1972; Gibson and Yoshyiama, 1999; Griffiths, 2003). At smaller spatial scales, for example, differences in rockpool fish assemblages may be influenced by environmental variables, namely tidal height of rockpools, rockpool volume, algal cover, substrate heterogeneity and exposure to wave action (Silberschneider and Booth, 2001; Griffiths, 2003). Increased algal cover in South African rockpools (Marsh *et al.*, 1978; Bennett and Griffiths, 1984) and substrate rugosity in southern California rockpools (Davies, 2000) have been suggested as factors responsible for increased numbers of fishes, possibly due to the increased cover and number of microhabitats available for fishes to exploit.

For blenniids, chaenopsids, labrisomids and tripterygiids, which are abundant both on rocky and coral reefs, the available information reveals different patterns of habitat use in these environments both within and among families. Taking into consideration that the same species may occupy more than one microhabitat and that information is scarce for many genera within these families, one can however look at the general patterns of microhabitat use by different genera derived from the available data.

On coral reefs, blennies occupy preferentially the surge zone (*Andamia*, *Cirripectes*, *Entomacrodus*, *Nannosalarias*, *Rhabdoblennius* and *Stanulus*), coral rubble (*Ecsenius*, *Enchelyurus*, *Hypleurochilus*, *Meiacanthus*, *Omobranchus* and *Salarias*), the branches of corals (*Aspidontus*, *Atrosalarias*, *Cirripectes*, *Ecsenius*, *Exallias* and *Salarias*) and small shelters (*Aspidontus*, *Blenniella*, *Istiblennius*, *Plagiotremus*, *Salarias*) (with, respectively, 15.8%, 14.7%, 12.6% and 10.5% of the total number of occurrences). On rocky reefs, on the contrary, the preferred microhabitats for blennies are intertidal pools (*Blenniella*, *Coryphoblennius*, *Entomacrodus*, *Hypsoblennius*, *Istiblennius*, *Lipophrys*, *Omobranchus*, *Ophioblennius*, *Parablennius*, *Scartella* and *Scartichthys*) small shelters (*Aidablennius*, *Coryphoblennius*, *Hypleurochilus*, *Hypsoblennius*, *Lipophrys*, *Microlipophrys*, *Ophioblennius*, *Parablennius*, *Plagiotremus*, *Salaria* and *Scartella*), boulders or rock (*Entomacrodus*, *Hypsoblennius*, *Ophioblennius*, *Parablennius* and *Parenchelyurus*) and rock covered with seaweeds (*Aidablennius*, *Cirripectes*, *Entomacrodus*, *Hypsoblennius*, *Laiphognathus*, *Parablennius* and *Scartella*) (with, respectively, 22.5%, 22.5%, 12.2% and 10.2% of the total number of occurrences).

For chaenopsids, the preferred microhabitats are restricted to small shelters (*Acanthemblemaria*, *Chaenopsis*, *Emblemaria*, *Emblemariopsis*, *Hemiemblemaria* and *Lucayablennius*) and coral rubble (*Chaenopsis*, *Emblemaria* and *Stathmonotus*) on coral reefs (with, respectively, 54.2%

and 20.8% of the total number of occurrences), and to small shelters (*Acanthemblemaria*, *Cirriemblemaria*, *Coralliozetus*, *Ekemblemaria*, *Emblemaria*, *Mccoskerichthys*, *Neoclinus* and *Protemblemaria*) on rocky reefs (with 76.7% of occurrences).

Labrisomids, on coral reefs, occupy mainly coral rubble (*Malacoctenus*, *Paraclinus* and *Starksia*), rock covered with seaweeds (*Malacoctenus*, *Paraclinus* and *Starksia*) and biogenic microhabitats, such as tube sponges and sea-anemones (*Malacoctenus* and *Starksia*) (with, respectively, 26.1%, 21.7%, and 17.4% of the total number of occurrences), while on rocky reefs the main microhabitats used by this group are rock covered with seaweeds (*Exerpes*, *Labrisomus*, *Malacoctenus*, *Paraclinus* and *Xenomedea*) and intertidal pools (*Cottoclinus*, *Dialommus*, *Malacoctenus*, *Paraclinus* and *Xenomedea*) (with, respectively, 50.0% and 21.9% of the total number of occurrences).

Finally, tripterygiids occur mostly on intertidal pools (*Enneapterygius*, *Helcogramma*, *Norfolkia* and *Notoclinops*), coral rubble (*Enneapterygius* and *Helcogramma*) and overhangs (*Enneapterygius*, *Helcogramma* and *Norfolkia*) on coral reefs (with, respectively, 27.0%, 15.9%, and 14.3% of the total number of occurrences), and on intertidal pools (*Bellapiscis*, *Blennodon*, *Brachynectes*, *Cryptichthys*, *Enneanectes*, *Enneapterygius*, *Forsterygion*, *Grahamina*, *Helcogramma*, *Lepidoblennius*, *Notoclinops*, *Notoclinus*, *Springerichthys*, *Trinorfolkia* and *Tripterygion*), overhangs (*Bellapiscis*, *Crocodilichthys*, *Cryptichthys*, *Forsterygion*, *Helcogramma*, *Karalepis*, *Lepidonectes*, *Norfolkia*, *Notoclinops* and *Tripterygion*), boulders or rock (*Axoclinus*, *Cryptichthys*, *Enneapterygius*, *Forsterygion*, *Helcogramma*, *Lepidonectes*, *Notoclinops*, *Ruanoho* and *Tripterygion*) and rock covered with seaweeds (*Axoclinus*, *Enneanectes*, *Enneapterygius*, *Forsterygion*, *Gilloblennius*, *Helcogramma*, *Lepidoblennius* and *Tripterygion*), on rocky reefs (with, respectively, 24.4%, 18.6%, 12.8% and 11.6% of the total number of occurrences).

The genera which occupy the greatest number of microhabitats in each environment (coral reefs and rocky reefs) are the blenniids *Cirripectes* and *Ecsenius* on coral reefs and *Parablennius*, *Entomacrodus* and *Hypsoblennius* on rocky reefs, being found in more different types of microhabitats. The chaenopsids are clearly specialists, as shown above, with only one or two types of microhabitats where each genus can be found (most occurring in small shelters). For clinids, the same pattern applies and they are mainly found on rocky reef environments and associated habitats. In labrisomids, the genera *Malacoctenus* and *Starksia* on coral reefs and *Malacoctenus* on rocky reefs occur in the greatest number of microhabitats. Finally, in tripterygiids, the genera *Enneapterygius* and *Helcogramma* on coral reefs and *Enneapterygius*, *Forsterygion*, *Helcogramma* and *Tripterygion* on rocky reefs are also generalists.

ONTOGENETIC SHIFTS OF MICROHABITAT USE

Through the different life history stages after recruitment, fishes may associate with particular characteristics of the substratum and to diverse microhabitats. Many species select and settle into distinct habitats as early-recruits (Booth and Beretta, 1994; Tolimieri, 1995), while post-settlement movements in juvenile and adult fish allow habitat choices to be further refined throughout adult life (McCormick, 1995; Lewis, 1997; Gonçalves *et al.* 2002). The cues used by fish to choose different microhabitats as well as the reasons why, at certain stages, intraspecific shifts in microhabitat use occur, are largely unknown (Gibson, 1982).

Habitat use often reflects behavioural decisions associated with the demands of feeding, avoiding predators or reproducing (Sutherland, 1996). Because these demands change during ontogeny (e.g., due to an increase in body size or shifts in diet), animals often move between different microhabitats in ways that meet their changing needs (Dahlgren and Eggleston, 2000). There are however few studies that address the microhabitat use by juveniles of blennioid species and/or the microhabitat shifts which occur during ontogeny in these fishes (Qasim, 1957; Connell and Jones, 1991; Macpherson, 1994; Faria and Almada, 2001; Hastings, 2002; Dempster and Kingsford, 2004). Nevertheless, Faria and Almada (2001), in a study on two intertidal blenniid species in Portugal, showed that whereas *Coryphoblennius galerita* remains in intertidal pools as juveniles and adults, only males leave the pools to nest in small shelters during the breeding season, returning after the end of this period (Almada *et al.*, 1996). *Lipophrys pholis* use the same intertidal pools as juveniles and move to large crevices (Faria and Almada, 2001) or larger pools when adults (Qasim, 1957), both situated at lower shore levels. The blenny *Petroscirtes mitratus* inhabits shallow protected lagoons and reef flats. Juveniles however are often found associated with floating *Sargassum* weeds (Kuiter and Tonozuka, 2001). The tripterygiid *Forsterygion malcolmi* occurs in a variety of microhabitats. While juveniles are found in shallow waters, adults are more abundant in deeper waters (Paulin and Roberts,

1992). The same applies to *Matanui bathytaton* (Fricke, 1994). Hastings (2002) describes ontogenetic microhabitat changes in several species of chaenopsids (genus *Coralliozetus*), with females retaining the juvenile behavior of habitat use, while males occupy small shelters.

Connell and Jones (1991) showed that for the tripterygiid *Forsterygion varium*, although recruitment occurs roughly evenly across all microhabitats studied by these authors, complex habitats support a greater proportion of older fish, which is probably related to a higher survival in these habitats. The processes through which topographic complexity influences the survivorship, movement and growth of this species are unknown. However, adequate shelter sites may be an important factor contributing to juvenile survival (Sale, 1976; Sale *et al.*, 1980; Doherty, 1983). In fact, early juvenile survivorship has been experimentally demonstrated to be strongly influenced by the number of available refuges (Shulman, 1984).

CONTROL MECHANISMS

Habitat structure plays an important role in structuring ecological communities, affecting both adult and juvenile fishes through its effects on processes such as guaranteeing refuge against predation (Hixon and Beets, 1993; Feary and Clements, 2006) or adverse environmental conditions (hydrodynamics, light or extreme temperatures) (Kotrschal, 1983; Feary and Clements, 2006), and finding food or nesting sites (Koppel, 1988; Kotrschal, 1988). Microhabitat types can play a crucial role in determining the abundance, dynamics and distribution of cryptobenthic fishes (Syms, 1995; Gonçalves and Almada, 1998; Gonçalves *et al.*, 2002; Willis and Andersen, 2003; La Mesa *et al.*, 2004; Beldade *et al.*, 2006).

Several hypotheses may explain the correlation between fish assemblage variables, such as abundance, richness, diversity or composition, and particular factors describing microhabitat structure. One of those hypothesis states that a particular habitat may be in short supply ('saturated') and that competition for space when recruits are in excess could be an important structuring force for those assemblages (Smith and Tyler, 1972; Gladfelter and Gladfelter, 1978; Sale, 1978).

According to Depczynski and Bellwood (2004), the high densities at which cryptobenthic fishes coexist, their small home ranges, and their sedentary behaviour suggest that competition may be a significant factor structuring microhabitat distribution of these fishes. Many studies have shown that intense inter- and intra-specific competition for specific microhabitats plays a leading role in determining assemblage structure, densities and habitat usage patterns of small cryptic reef fishes (Critchlow, 1972; Greenfield and Greenfield, 1982; Freeman and Alevizon, 1983; Clarke, 1989; Munday *et al.*, 2001).

Critchlow (1972) reported that *Coralliozetus angelicus* probably limits the distribution and abundance of the young of *Hypsoblennius brevipinnis* at Guaymas, Sonora. Stephens (1963) reported that the chaenopsid *Acanthemblemaria balanorum* appears to have a high affinity for barnacles and successfully compete for this resource when it becomes limiting. Buchheim and Hixon (1992), in an experimental study using the chaenopsid *Acanthemblemaria spinosa* on coral-reefs located at the Virgin Islands, showed that resident fish redistribute themselves among holes on a coral head after neighbours were removed, suggesting that the holes are of different quality. Overall, their experimental results suggest that, over the limited spatial and temporal scale of the study, spiny head blennies compete intraspecifically for shelter holes.

However, other studies have also demonstrated that shelter is not a limiting resource (Robertson and Sheldon, 1979; Robertson *et al.*, 1981; Kotrschal, 1988). According to Kotrschal (1988), the hypothesis that the number and size of blennies in a certain area are controlled by the numbers and sizes of available shelter may not apply for example to the Adriatic Sea populations, where there seems to be many more holes available than those utilised by the studied species.

Macpherson (1994), in a study also in the Mediterranean, demonstrated that the preference for a specific substratum used by different species was unrelated to substratum availability in the environment. It was suggested that, in this case, interspecific space competition may not be a major determining factor affecting the assemblage structure, but rather that most species occupy the substratum without any adverse influence on the others.

Competition for space between blennioid species that live in the same biotope can be avoided in different ways (La Mesa and Vacchi, 2005). Differences in depth distribution or microhabitat use between species may substantially reduce their spatial overlap (Abel, 1962; Illich and Kotrschal, 1990; Mayr and Berger, 1992; Macpherson, 1994; Syms, 1995). On the other hand, where species overlap in their vertical distribution there is considerable evidence that the actual microhabitats occupied by each species differ (Critchlow, 1972; Gibson, 1972; Lindquist, 1985). Critchlow (1972) found that eight species of rocky shore fishes differ considerably in at least one of the three niche dimensions studied (depth distribution, diet and feeding periodicity).

Faria and Almada (2001) showed that, in a Portuguese rock intertidal fish assemblage there was a marked microhabitat segregation between species, both for types of shelters and breeding sites, and therefore interspecific competition does not seem to be a major structuring force in these assemblages. The authors suggested that differences between species could have arisen by slight differences in anatomy and behaviour of the ancestors that allowed more efficient capture of larger prey by the larger species, leading to benefits from faster growth and shifts into niches not used by the smaller members of the assemblage, even without competition.

The relative importance of habitat complexity and its interaction with predation levels in predicting fish species abundances is variable (Behrents, 1987; Hixon and Beets, 1993; Levin and Hay, 1996; Beukers and Jones, 1997; Tupper and Boutilier, 1997; Steele, 1999; Syms and Jones, 1999). Complex habitats may provide more shelter, reducing the impact of predation, physical disturbance, or both, on fish populations (Talbot and Goldman, 1972; Luckhurst and Luckhurst, 1978; Behrents, 1987; Hixon and Beets, 1989; Willis and Anderson, 2003; Feary and Clements, 2006).

During the 1970's, competition for space was seen as being the predominant process structuring populations and communities of reef fishes (Smith and Tyler, 1972). This paradigm was challenged in the 1980's, by the advocates of recruitment limitation, where larval settlement was seen as being so sparse that reef populations seldom reached densities where competition was important (see e.g., Doherty and Williams, 1988). In recent years, a pluralistic view has emerged, where multiple processes, including competition, mortality and recruitment patterns, are acknowledged as being important factors regulating fish communities and populations (see Doherty, 1991, 2002; Hixon, 1991; Jones, 1991; Sale, 1991; Williams, 1991; Hixon and Webster, 2002).

It has often been suggested that the mechanism causing higher densities of fish in more physically complex habitats is a reduction of predation pressure, related to the greater availability of refuges (Connell and Jones, 1991; Hixon and Beets, 1993; Caley and St. John, 1996; Feary and Clements, 2006). Microhabitat features may influence juvenile demography, but whether such effects are then translated to adult densities at larger spatial scales can depend on species-specific biological characteristics of adults (Tolimieri, 1998). According to Willis and Anderson (2003) many small size benthic cryptic species (such as most blennioids), probably remain vulnerable to predation by larger piscivores throughout their entire life cycle.

The potential effects of cover bymacrophytic algae on assemblage structure are still not clear (Willis and Anderson, 2003). Hindell *et al.* (2000) found that effects of predation on small fishes in seagrass beds were reduced compared to unvegetated habitats, but Levin *et al.* (1997) detected no difference in the rates of predation between sand and seagrass habitats. According to Levin and Hay (1996) the algal cover seems to have positive effects on the density of post-settlement reef fishes. However, Syms and Jones (1999) showed that removing algal cover had only weak and short-lived effects on the composition of a guild of tripterygiid fishes.

According to Wilson (2001) and Depczynski and Bellwood (2004) fish assemblages could also be affected by differences in the distribution of small prey species or food detritus, which in turn depends on the structural features of available microhabitats. Depczynski and Bellwood (2004) suggested that, for instance, for the cave-dwelling species, carnivores are dominant since the outer-edge ceilings and walls of cave openings provided an elevated and concealed position in the water column from which they can prey on passing zooplankton.

Wilson (2001) also suggested that the exposed reef crests seem to be the favoured habitat for a range of detritivorous fishes. The presence of sediments with a high percentage of detritus on exposed reef tops at Lizard Island, Australia, coincides with high estimates of blenny abundance and biomass in this environment. The association of high blenny abundance with exposed reef tops is consistent over a large geographical range of coral reefs (Odum and Odum, 1955; Townsend and Tibbetts, 2000). According to Wilson (2001), it is possible that the association of blennies with non-living corals was related to the suitability of these microhabitats for accumulating food, as coral skeletons are generally overgrown with filamentous algae that incorporate a matrix of material, including aggregates of detritus.

Greenfield and Johnson (1990) suggested that different selective pressures could be operating on the small site-attached species present in the reef assemblage predicting that species that obtain their food from the reef itself might be more site-specific than those that obtain their food from the plankton.

However, according to Wilson (2001), looking at broad-scale fish distributions, other factors than resource availability and predation are likely to influence the fine-scale distribution of blennies. Mechanisms relating to predator avoidance or reproductive behaviour might also be important factors in determining the site fidelity of these small fishes.

Theoretically, increasingly complex environments increase habitat heterogeneity, dimensionality and living area, restrict predators and allow compartmentalisation of space, thereby reducing competitive interactions and opening up new possibilities for habitat exploitation (Miller, 1996). Numerous studies have shown a direct positive correlation between the availability of shelter and fish abundance (Caley and St. John, 1996; Friedlander and Parrish, 1998), lowered mortality rates in fish prey (Beukers and Jones, 1997; Forrester and Steele, 2000), and a reduction in the foraging efficiency of predators (e.g., Schofield, 2003). More complex habitats are expected to increase speciation of niche specialists adapted to living in these environments, allowing these habitats to be shared by a larger number of species (Luckhurst and Luckhurst, 1978; Roberts and Ormond, 1987). Blennioids are a fruitful ground to further explore and test these ideas.

FUTURE RESEARCH

The positive relationship between habitat structure and species composition can emerge from several processes acting on the fish assemblages, perhaps simultaneously in a 'multiple causality' model (Jones, 1991). Future research should focus on such relationships at the population (or even individual) level, in search of mechanistic explanations that may be extrapolated to the community level (Schoener, 1986; May, 1989; García Charton and Pérez Ruzafa, 1998). The important role of microhabitat characteristics on assemblage's structure and species distributions would appear to exert an effect right from the time of settlement. The role of habitat in limiting recruitment has been reported for different assemblages (Eckert, 1985; Bodkin, 1988; Carr, 1989; Connell and Jones, 1991; Jones, 1991; Buchheim and Hixon, 1992). Nevertheless, further work needs to be carried out to analyse the influence of habitat characteristics regulating settlement and their effects on the structure of the adult population, as well as the degree to which these early habitat associations are modified throughout the life cycle resulting in important shifts in habitat use. The relationship between size and microhabitat use also deserves further analysis.

Although it is widely recognised that populations respond to both physical and biological changes to habitats (Shulman, 1984, Choat and Ayling, 1987; Jones, 1988; Hixon and Beets, 1989), the magnitude of the response at different life history stages is poorly understood (Connell and Jones, 1991). Finally, while spatial and temporal variation in juvenile survivorship has been extensively documented (DeMartini and Anderson, 1980; Kotrschal, 1983; Jones, 1984, 1986; Aldenhoven, 1986; Eckert, 1987; Sale and Ferrell, 1988; Connell and Jones, 1991), the degree to which habitat characteristics are causing these variations is not often known, and there have been few attempts to experimentally test these habitat effects (Shulman, 1984; Jones, 1988; Connell and Jones, 1991). Quantifying the types of response to habitat complexity is crucial for applied aspects such as the effects of artificial reefs and the effectiveness of marine reserves, especially when these management tools aim to maximise species abundances and richness (García Charton and Pérez Ruzafa, 1998). The potential of using artificial reefs or replicate structures simulating different habitats as tools for field experiments (Seaman and Sprague, 1991) is greatly improving research on these issues. Experimental artificial reefs could be used to search for correlations between a wide series of microhabitat descriptors and fish assemblages parameters (García Charton and Pérez Ruzafa, 1998).

Finally, blennioids are very interesting subjects to explore in greater detail the relationships outlined in this chapter since the widespread distribution of these fishes throughout the world allows for an integrated evolutionary and ecological approach of these issues.

References

Abel, E.F. 1962. Freiwasserbeobachtungen an Fischen im Golf von Neapel als Beitrag zur Kenntnis ihrer Ökologie und ihres Verhaltens. *Internationale Revue der gesamten Hydrobiologie und Hydrographie* 47: 219-290.

Aldenhoven, J.M. 1986. Local variation in mortality rates and life expectancy estimates of the coral-reef fish *Centropyge bicolour* (Pisces: Pomacentridae). *Marine Biology* 92: 237-244.

Alevizon, W., R. Richardson, P. Pitts and G. Serviss. 1985. Coral zonation and patterns of community structure in Bahamian reef fishes. *Bulletin of Marine Science* 36: 304-318.

Almada, V.C. and R.S. Santos. 1995. Parental care in the rocky intertidal: a case study of adaptation and exaptation in Mediterranean and Atlantic blennies. *Reviews in Fish Biology and Fisheries* 5: 23-37.

Almada, V.C., G. Garcia and R.S. Santos. 1987. Padrões de actividade e estrutura dos territórios dos machos parentais de *Parablennius pilicornis* Cuvier (Pisces: Blenniidae) da costa portuguesa. *Análise Psicológica* 2: 261-280.

Almada, V.C., E.J. Gonçalves, R.F. Oliveira and E.N Barata. 1992. Some features of the territories in the breeding males of the intertidal blenny *Lipophrys pholis* (Pisces: Blenniidae). *Journal of the Marine Biological Association of the United Kingdom* 72: 187-197.

Almada, V.C., H. Carreiro, C. Faria and E.J. Gonçalves. 1996. The breeding season of *Coryphoblennius galerita* in Portuguese waters. *Journal of Fish Biology* 48: 295-297.

Anderson, T.W. 1994. Role of macroalgal structure in the distribution and abundance of a temperate reef fish. *Marine Ecology Progress Series* 113: 279-290.

Angel, A. and F.P. Ojeda. 2001. Structure and trophic organization of subtidal fish assemblages on the northern Chilean coast: The effect of habitat complexity. *Marine Ecology Progress Series* 217: 81-91.

Arburto-Oropeza, O. and E.F. Balart. 2001. Community structure of reef fish in several habitats of a rocky reef in the Gulf of California. *Marine Ecology* 22: 283-305.

Ault, T.R. and C.R. Johnson. 1998. Spatially and temporally predictable fish communities on coral reefs. *Ecological Monographs* 68: 25-50.

Bean, K., G.P. Jones and M.J. Caley. 2002. Relationships among distribution, abundance and microhabitat specialisation in a guild of coral reef triggerfish (Balistidae). *Marine Ecology Progress Series* 233: 263-272.

Behrents, K.C. 1987. The influence of shelter availability on recruitment and early survivorship of *Lythrypnus dalli* Gilbert (Pisces: Gobiidae). *Journal of Experimental Marine Biology and Ecology* 107: 45-59.

Beldade, R., K. Erzini and E.J. Gonçalves. 2006. Temporal dynamics of a temperate rocky cryptobenthic fish assemblage. *Journal of the Marine Biological Association of the United Kingdom* 86: 1221-1228.

Bell, J.D. 1983. Effects of depth and marine reserve fishing restrictions on the structure of a rocky reef fish assemblage in the northwestern Mediterranean Sea. *Journal of Applied Ecology* 20: 357-369.

Bell, J.D. and R. Galzin. 1984. Influence of live coral cover on coral-reef fish communities. *Marine Ecology Progress Series* 15: 265-274.

Bennett, B.A. and C.L. Griffiths. 1984. Factors affecting the distribution, abundance and diversity of rock-pool fishes on the Cape Peninsula, South Africa. *South African Journal of Zoology* 19: 97-104.

Beukers, J.S. and G.P. Jones. 1997. Habitat complexity modifies the impact of piscivores on a coral reef fish population. *Oecologia* 114: 50-59.

Bodkin, J.L. 1988. Effects of kelp forest removal on associated fish assemblages in central California. *Journal of Experimental Marine Biology and Ecology* 117: 227-238.

Böhlke, J.E. and C.C.G. Chaplin. 1968. *Fishes of the Bahamas and Adjacent Tropical Waters*. Livingston Publ. Co., Wynnewood, PA.

Booth, D.J. and G.A. Beretta. 1994. Seasonal recruitment, habitat associations and survival of pomacentrid reef fish in the US Virgin Islands. *Coral Reefs* 13: 81-89.

Buchheim, J.R. and M.A. Hixon. 1992. Competition for shelter holes in the coral-reef fish *Acanthemblemaria spinosa* Metzelaar. *Journal of Experimental Marine Biology and Ecology* 164: 45-54.

Butter, M.E., M. Wapstra and E. van Dijk. 1980. *Meandrina meandrites* and *Emblemariopsis diaphana*, first record of an association between a stony coral and a fish, similar to anemone/fish relationships. *Bijdragen tot de Dierkunde* 50: 87-95.

Caley, J.M. and J. St. John. 1996. Refuge availability structures assemblages of tropical reef fishes. *Journal of Animal Ecology* 65: 414-428.

Carpenter, K.E., R.I. Miclat, V.D. Albaladejo and V.T. Corpuz. 1981. The influence of substrate structure on the local abundance and diversity of Philippine reef fishes. *Proceedings of the Fourth International Coral Reef Symposium, Manila* 2: 497-502.

Carr, M.H. 1989. Effects of macroalgal assemblages on the recruitment of temperate zone reef fishes. *Journal of Experimental Marine Biology and Ecology* 126: 59-76.

Carr, M.H. 1991. Habitat selection and recruitment of an assemblage of temperate reef fishes. *Journal of Experimental Marine Biology and Ecology* 146: 113-137.

Chabanet, P., H. Ralambondrainy, M. Amanieu, F. Faure and R. Galzin. 1997. Relationships between coral reef substrata and fish. *Coral Reefs* 16: 93-102.

Chittaro, P.M. 2000. Fish-habitat associations across multiple spatial scales. *Coral Reefs* 23: 235-244.

Choat, J.H. and A.M. Ayling. 1987. The relationship between habitat structure and fish faunas on New Zealand reefs. *Journal of Experimental Marine Biology and Ecology* 110: 257-284.

Clarke, R.D. 1977. Habitat distribution and species diversity of chaetodontid and pomacentrid fishes near Bimini, Bahamas. *Marine Biology* 40: 277-289.

Clarke, R.D. 1989. Population fluctuation, competition and microhabitat distribution of two species of tube blennies, *Acanthemblemaria* (Teleostei: Chaenopsidae). *Bulletin of Marine Science* 44: 1174-1185.

Clarke, R.D. 1992. Effects of microhabitat and metabolic rate on food intake, growth and fecundity of two competing coral reef fish. *Coral Reefs* 11: 199-205.

Clarke, R.D. and J.C. Tyler. 2003. Differential space utilization by male and female spiny head Blennies, *Acanthemblemaria spinosa* (Teleostei: Chaenopsidae). *Copeia* 2003: 241-247.

Connell, S.D. and G.P. Jones. 1991. The influence of habitat complexity on postrecruitment processes in a temperate reef population. *Journal of Experimental Marine Biology and Ecology* 151: 271-294.

Costello, M.J. 1992. Abundance and spatial overlap of gobies (Gobiidae) in Lough Hyne. *Environmental Biology of Fishes* 33: 239-248.

Coyer, J.A. 1982. Observations on the reproductive behaviour of the Giant Kelpfish, *Heterostichus rostratus* (Pisces: Clinidae). *Copeia* 1982: 344-350.

Critchlow, K.R. 1972. *Resource Utilization in Some Rocky Shore Fishes in the Gulf of California*. Ph.D. Thesis. University of California, Los Angeles.

Curley, B.G., M.J. Kingsford and B.M. Gillanders. 2002. Spatial and habitat-related patterns of temperate reef fish assemblages: implications for the design of marine protected areas. *Marine and Freshwater Research* 53: 1197-1210.

Dahlgren, C.P. and D.B. Eggleston. 2000. Ecological processes underlying ontogenetic habitat shifts in a coral reef fish. *Ecology* 8: 2227-2240.

Davies, J.L.D. 2000. Spatial and seasonal patterns of habitat partitioning in a guild of southern California tidepool fishes. *Marine Ecology Progress Series* 196: 253-268.

DeMartini, E.E. and M.E. Anderson. 1980. Comparative survivorship and life history of painted greenling. *Environmental Biology of Fishes* 5: 33-47.

Dempster, T. and M.J. Kingsford. 2004. Drifting objects as habitat for pelagic juvenile fish off New South Wales, Australia. *Australian Journal of Marine and Freshwater Research* 55: 675-687.

Depczynski, M. and D.R. Bellwood. 2004. Microhabitat utilisation in cryptobenthic coral reef fish communities. *Marine Biology* 145: 455-463.

Doherty, P.J. 1983. Tropical territorial damselfishes: Is density limited by aggression or recruitment? *Ecology* 64: 176-190.

Doherty, P.J. 1991. Spatial and temporal patterns in recruitment. In: *The Ecology of Fishes on Coral Reefs*, P.F. Sale (ed.). Academic Press, San Diego, pp. 261-293.

Doherty, P.J. 2002. Variable replenishment and the dynamics of reef fish populations. In: *Coral Reef Fishes: Dynamics and Diversity in a Complex Ecosystem*, P.F. Sale (ed.). Academic Press, San Diego, pp. 327-355.

Doherty, P.J. and D.M. Williams. 1988. Are local populations of coral reef fishes equilibrial assemblages? The empirical database. *Proceedings of the 6th International Coral Reef Symposium* 1: 131-139.

Eckert, G.J. 1985. Settlement of coral reef fishes to natural substrata at different depths. *Proceedings of the Fifth International Coral Reef Congress, Tahiti* 5: 385-390.

Eckert, G.J. 1987. Estimates of adult and juvenile mortality for labrid fishes at One Tree Reef, Great Barrier Reef. *Marine Biology* 95: 167-171.

Faria, C. and V.C. Almada. 2001. Microhabitat segregation in three rocky intertidal fish species in Portugal: Does it reflect interspecific competition? *Journal of Fish Biology* 58: 145-159.

Feary, D.A. and K.D. Clements. 2006. Habitat use by triplefin species (Tripterygiidae) on rocky reefs in New Zealand. *Journal of Fish Biology* 69: 1031-1046.

Ferreira, E.L., J.E.A. Gonçalves and R. Coutinho. 2001. Community structure and fish habitat complexity on a tropical rocky shore. *Environmental Biology of Fishes* 61: 353-369.

Forrester, G.E. and M.A. Steele. 2000. Variation in the presence and cause of density-dependent mortality in three species of reef fishes. *Ecology* 81: 2416-2427.

Freeman, S. and W. Alevizon. 1983. Aspects of territorial behaviour and habitat distribution of the fairy basket *Gramma loreto*. *Copeia* 1983: 829-832.

Fricke, R. 1994. *Tripterygiid Fishes of Australia, New Zealand and the Southwest Pacific Ocean (Teleostei)*. Koeltz Scientific Books, Königstein.

Friedlander, A.M. and J.D. Parrish. 1998. Habitat characteristics affecting fish assemblages on a Hawaiian reef. *Journal of Experimental Marine Biology and Ecology* 224: 1-30.

Froese, R. and D. Pauly. (eds.). 2007. *FishBase*. World Wide Web electronic publication. www.fishbase.org, version (10/2007).

García Charton, J.A. and A. Pérez Ruzafa. 1998. Correlation between habitat structure and a rocky reef fish assemblage in the southwest Mediterranean. *Marine Ecology* 19: 111-128.

García Charton, J.A. and A. Pérez Ruzafa. 2001. Spatial patterns and the habitat structure of a Mediterranean rocky reef fish local assemblage. *Marine Biology* 138: 917-934.

Gibson, R.N. 1969. The biology and behaviour of littoral fish. *Oceanography and Marine Biological Annual Review* 7: 367-409.

Gibson, R.N. 1972. The vertical distribution and feeding relationships of intertidal fish on the Atlantic coast of France. *Journal of Animal Ecology* 41: 189-207.

Gibson, R.N. 1982. Recent studies on the biology of intertidal fishes. *Oceanography and Marine Biological Annual Review* 20: 363-414.

Gibson, R.N. and R.M. Yoshiyama. 1999. Intertidal fish communities. In: *Intertidal Fishes. Life in Two Worlds*, M.H. Horn, K.L.M. Martin and M.A. Chotkowski (eds.). Academic Press, San Diego, pp. 264-296.

Gladfelter, W.B. and E.H. Gladfelter. 1978. Fish community structure as a function of habitat structure on West Indian patch reefs. *Review of Tropical Biology* 26: 65-84.

Gon, O. 1990. Tripterygiidae. In: *Fishes of the Southern Ocean*, O. Gon and P.C. Heemstra (eds.). J.L.B. Smith Institute of Ichthyology, Grahamstown, South Africa, pp. 400-401.

Gonçalves, E.J. and V.C. Almada. 1998. A comparative study of territoriality in intertidal and subtidal blennioids (Teleostei, Blennioidei). *Environmental Biology of Fishes* 51: 257-264.

Gonçalves, E.J., M. Barbosa, H.N. Cabral and M. Henriques. 2002. Ontogenetic shifts in patterns of microhabitat utilization in the small-headed clingfish, *Apletodon dentatus* (Gobiesocidae). *Environmental Biology of Fishes* 63: 333-339.

Gratwicke, B. and M.R. Speight. 2005. The relationship between fish species richness, abundance and habitat complexity in a range of shallow tropical marine habitats. *Journal of Fish Biology* 66: 650-667.

Green, L.A. 1996. Spatial, temporal and ontogenetic patterns of habitat use by coral reef fishes (family Labridae). *Marine Ecology Progress Series* 133: 1-11.

Greenfield, D.W. 1975. *Emblemariopsis pricei*, a new species of chaenopsid blenny from Belize. *Copeia* 1975: 713-715.

Greenfield, D.W. and T.A. Greenfield. 1982. Habitat and resource partitioning between two species of *Acanthemblemaria* (Pisces: Chaenopsidae), with comments on the chaos hypothesis. *Smithsonian Contributions to Marine Science* 12: 499-507.

Greenfield, D.W. and R.K. Johnson. 1990. Community structure of Western Caribbean blennioid fishes. *Copeia* 1990: 433-448.

Griffiths, S.P. 2003. Spatial and temporal dynamics of temperate Australian rockpool ichthyofaunas. *Australian Journal of Marine and Freshwater Research* 54: 163-176.

Gunn, J.S. and R.E. Thresher. 1991. Viviparity and the reproductive ecology of clinid fishes (Clinidae) from temperate Australian waters. *Environmental Biology of Fishes* 31: 323-344.

Hastings, P.A. 1986. Habitat selection, sex ratio and sexual selection in *Coralliozetus angelica* (Blennioidei: Chaenopsidae). In: *Indo-Pacific Fish Biology: Proceedings of the Second International Conference on Indo-Pacific Fishes*, T. Uyeno, R. Arai, T. Taniuchi and K. Matsuura (eds.), Ichthyological Society of Japan, Tokyo, pp. 758-793.

Hastings, P.A. 1988. Female choice and male reproductive success in the angel blenny, *Coralliozetus angelica* (Teleostei: Chaenopsidae). *Animal Behaviour* 36: 115-124.

Hastings, P.A. 1992. Nest-site size as a short-term constraint on the reproductive success of paternal fishes. *Environmental Biology of Fishes* 34: 213-218.

Hastings, P.A. 2002. Evolution of morphological and behavioral ontogenies in females of a highly dimorphic clade of blennioid fishes. *Evolution* 56: 1644-1654.

Heymer, A. and C.D. Zander. 1975. Morphologische und ökologische Untersuchungen an *Blennius rouxi*, Cocco 1833. (Pisces; Perciformes; Blenniidae). *Vie Milieu* 25: 311-333.

Hindell, J.S., G.P. Jenkins and M.J. Keough. 2000. Evaluating the impact of predation by fish on the assemblage structure of fishes associated with seagrass (*Heterozostera tasmanica*) (Martens ex Aschen) den Hartog, and unvegetated sand habitats. *Journal of Experimental Marine Biology and Ecology* 255: 153-174.

Hixon, M.A. 1991. Predation as a process structuring coral reef fish communities. In: *The Ecology of Fishes on Coral Reefs*, P.F. Sale (ed.). Academic Press, San Diego, pp. 475-508.

Hixon, M.A. and J.P. Beets. 1989. Shelter characteristics and Caribbean fish assemblages: experiments with artificial reefs. *Bulletin of Marine Science* 44: 666-680.

Hixon, M.A. and J.P. Beets. 1993. Predation, prey refuges and the structure of coral reef fish assemblages. *Ecological Monographs* 63: 77-101.

Hixon, M.A. and M.S. Webster. 2002. Density dependence in reef fish populations. In: *Coral Reef Fishes: Dynamics and Diversity in a Complex Ecosystem*, P.F. Sale (ed.). Academic Press, San Diego, pp. 303-325.

Horn, M.H., K.L.M. Martin and M.A. Chotkowski. 1999. Introduction. In: *Intertidal Fishes. Life in Two Worlds*, M.H. Horn, K.L.M. Martin and M.A. Chotkowski (eds.). Academic Press, San Diego, pp. 1-6.

Hyndes, G.A., A.J. Kendrick, L.D. MacArthur and E. Stewart. 2003. Differences in the species- and size-composition of fish assemblages in three distinct seagrass habitats with differing plant and meadow structure. *Marine Biology* 142: 1195-1206.

Illich, I.P. and K. Kotrschal. 1990. Depth distribution and abundance of Northern Adriatic littoral rocky reef blennioid fishes (Blenniidae and Tripterygion). *Marine Ecology* 11: 277-289.

Jones, G.P. 1984. Population ecology of the temperate reef fish *Pseudolabrus celidotus* Bloch and Schneider (Pisces: Labridae). I. Factors influencing recruitment. *Journal of Experimental Marine Biology and Ecology* 75: 257-276.

Jones, G.P. 1986. Food availability affects growth in a coral reef fish. *Oecologia* 70: 136-139.

Jones, G.P. 1988. Experimental evaluation of the effects of habitat structure and competitive interactions on the juveniles of two coral reef fishes. *Journal of Experimental Marine Biology and Ecology* 123: 115-126.

Jones, G.P. 1991. Postrecruitment processes in the ecology of coral reef fish populations: a multifactorial perspective. In: *The Ecology of Fishes on Coral Reefs*, P.F. Sale (ed.). Academic Press, New York, pp. 294-328.

Koppel, V.H. 1988. Habitat selection and space partitioning among two Mediterranean blenniid species. *Marine Ecology* 9: 329-346.

Kotrschal, K. 1983. Northern Adriatic rocky reef fishes at low winter temperatures. *Marine Ecology* 4: 275-286.

Kotrschal, K. 1988. Blennies and endolithic bivalves: differential utilization of shelter in Adriatic Blenniidae (Pisces: Teleostei). *Marine Ecology* 9: 253-269.

Kuiter, R.H. and T. Tonozuka. 2001. *Pictorial Guide to Indonesian Reef Fishes. Part 2. Fusiliers—Dragonets, Caesionidae—Callionymidae*. Zoönetics, Australia.

La Mesa, G. and M. Vacchi. 2005. Analysis of the blennioid assemblages associated with different rocky shore habitats in the Ligurean Sea. *Journal of Fish Biology* 66: 1300-1327.

La Mesa, G., M. Micalizzi, G. Giaccone and M. Vacchi. 2004. Cryptobenthic fishes of the "Ciclopi Islands" marine reserve (central Mediterranean Sea): assemblage composition, structure and relations with habitat features. *Marine Biology* 145: 233-242.

La Mesa, G., S. Di Muccio and M. Vacchi. 2006. Structure of a Mediterranean cryptobenthic fish community and its relationships with habitat characteristics. *Marine Biology* 149: 149-167.

Levin, P.S. 1991. Effects of microhabitat on recruitment variation in a Gulf of Maine reef fish. *Marine Ecology Progress Series* 75: 183-189.

Levin, P.S. and M.E. Hay. 1996. Responses of temperate reef fishes to alterations in algal structure and species composition. *Marine Ecology Progress Series* 134: 37-47.

Levin, P., R. Petrik and J. Malone. 1997. Interactive effects of habitat selection, food supply and predation on recruitment of an estuarine fish. *Oecologia* 112: 55-63.

Lewis, A.R. 1997. Effects of experimental coral disturbance on the structure of fish communities on large patch reefs. *Marine Ecology Progress Series* 161: 37-50.

Lindquist, D.G. 1985. Depth zonation, microhabitat, and morphology of three species of *Acanthemblemaria* (Pisces: Blennioidea) in the Gulf of California, Mexico. *Marine Ecology* 6: 329-344.

Luckhurst, B.E. and K. Luckhurst. 1978. Analysis of the influence of substrate variables on coral reef fish communities. *Marine Biology* 49: 317-323.

McCosker, J.E., J.S. Stephens and R.H. Rosenblatt. 2003. *Cottoclinus canops*, a new genus and species of blenny (Perciformes: Labrisomidae) from the Galápagos Islands. *Proceedings of the California Academy of Sciences* 54: 155-160.

Macpherson, E. 1994. Substrate utilisation in a Mediterranean littoral fish community. *Marine Ecology Progress Series* 114: 211-218.

Macpherson, E. and U. Zika. 1999. Temporal and spatial variability of settlement success and recruitment level in the three blennioid fishes in the northwestern Mediterranean. *Marine Ecology Progress Series* 182: 269-282.

Marsh, B., T.M. Crowe and W.R. Siegfried. 1978. Species richness and abundance of clinid fish (Teleostei: Clinidae) in intertidal rock pools. *African Zoology* 13: 283-291.

Martin, K.L.M. and C.R. Bridges. 1999. Respiration in water and air. In: *Intertidal Fishes. Life in Two Worlds*, M.H. Horn, K.L.M. Martin and M.A. Chotkowski (eds.). Academic Press, San Diego, pp. 54-78.

Masuda, H., K. Amaoka, C. Araga, T. Uyeno and T. Yoshino. 1984. *The Fishes of the Japanese Archipelago*. Vol. 1. Tokai University Press, Tokyo, Japan.

May, R.M. 1989. Levels of organization in ecology. In: *Ecological Concepts. The Contribution of Ecology to an Understanding of the Natural World*, J.M. Cherrett (ed.). Blackwell Science, Oxford, pp. 339-363.

Mayr, M. and A. Berger. 1992. Territoriality and microhabitat selection in two intertidal New Zealand fish. *Journal of Fish Biology* 40: 243-256.

McCormick, M.I. 1995. Fish feeding on mobile benthic invertebrates: influence of spatial variability in habitat associations. *Marine Biology* 121: 627-637.

McGehee, M.A. 1994. Correspondence between assemblages of coral reef fishes and gradients of water motion, depth, and substrate size off Puerto Rico. *Marine Ecology Progress Series* 105: 243-255.

McManus, J.W., R.I. Micla and V.P. Palaganas. 1981. Coral and fish community structure of Sombrero island, Batangas, Philippines. *Proceedings of the Fourth International Coral Reef Symposium, Manila* 2: 271-280.

Miller, P.J. 1996. The functional ecology of small fish: some opportunities and consequences. *Symposium of Zoological Society of London* 68: 175-199.

Munday, P.L. 2004. Competitive coexistence of coral-dwelling fishes: the lottery hypothesis revisited. *Ecology* 85: 623-628.

Munday, P.L., G.P. Jones and M.J. Caley. 1997. Habitat specialisation and the distribution and abundance of coral-dwelling gobies. *Marine Ecology Progress Series* 152: 227-239.

Munday, P.L., G.P. Jones and M.J. Caley. 2001. Interspecific competition and coexistence in a guild of coral-dwelling fishes. *Ecology* 82: 2177-2189.

Myers, R.F. 1999. *Micronesian Reef Fishes: A Comprehensive Guide to the Coral Reef Fishes of Micronesia*, Coral Graphics, Barrigada, Guam, 3rd Edition.

Nanami, A. and M. Nishihira. 2002. The structure and dynamics of fish communities in an Okinawan coral reef: effects of coral-based habitat structures at sites with rocky and sandy sea bottoms. *Environmental Biology of Fishes* 63: 353-372.

Neat, F.C., W. Lengkeek, E.P. Westerbeek, B. Laarhoven and J.J. Videler. 2003. Behavioural and morphological differences between lake and river populations of *Salaria fluviatilis*. *Journal of Fish Biology* 63: 374-387.

Nieder, J. 2001. Amphibious behaviour and feeding ecology of the four-eyed blenny *Dialommus fuscus* (Labrisomidae) in the intertidal zone of the island of Santa Cruz (Galapagos, Ecuador). *Journal of Fish Biology* 58: 755-767.

Nieder, J., G. La Mesa and M. Vacchi. 2000. Blenniidae along the Italian coasts of the Ligurian and the Tyrrhenian Sea: community structure and new records of *Scartella cristata* for northern Italy. *Cybium* 24: 359-369.

Nieto, P. and L.J. Alberto. 1993-1994. Gobiidae Y Blenniidae de la zona intermareal de la costa suroccidental de Andalucia (España). *Miscellanea Zoológica* 17: 179-188.

Odum, H.T. and E.P. Odum. 1955. Trophic structure and productivity of a windward coral reef community on Eniwetok Atoll. *Ecological Monographs* 2583: 291-320.

Ormond, R.F.G., J.M. Roberts and R.Q. Jan. 1996. Behavioural differences in microhabitat use by damselfishes (Pomacentridae): implications for reef fish biodiversity. *Journal of Experimental Marine Biology and Ecology* 202: 85-95.

Patzner, R.A. 1985. The blennies (Pisces, Blennioidea) at the marine biological station of Aurisina (Gulf of Trieste, Italy). *Nova Thalassia* 7: 107-119.

Patzner, R.A. 1999. Habitat utilization and depth distribution of small cryptobenthic fishes (Blenniidae, Gobiesocidae, Gobiidae, Tripterygiidae) in Ibiza (Western Mediterranean Sea). *Environmental Biology of Fishes* 55: 207-214.

Paulin, C. and C. Roberts. 1992. The rockpool fishes of New Zealand (Te ika aaria o Aotearoa). *Museum of New Zealand* (Te Papa Tongarewa).

Petti, J.C. 1969. *Behavioral and Morphological Adaptations to Burrowing of Two Species of Dactyloscopid Fishes from the Northern Gulf of California*. Unpubl. Master's thesis. University of Arizona, Tucson.

Prochazka, K. 1998. Spatial and trophic partitioning in cryptic fish communities of shallow subtidal reefs in False Bay, South Africa. *Environmental Biology of Fishes* 51: 201-220.

Prochazka, K. and C.L. Griffiths. 1992. Observations on the distribution patterns, behaviour, diets and reproductive cycles of sand-dwelling clinids (Perciformes: Clinidae) from South Africa. *Environmental Biology of Fishes* 35: 371-379.

Pyle, R.L. 1996a. Exploring deep coral reefs: How much coral reef biodiversity are we missing? *Global Biodiversity* 6: 3-7.

Pyle, R.L. 1996b. The twilight zone. *Natural History* 105: 59-62.

Pyle, R.L. 1999. Mixed-gas, closed-circuit rebreather use for identification of new reef fish species from 200–500 fsw. In: *Assessment and Feasibility of Technical Diving Operations for Scientific Exploration*, R.W. Hamilton, D.F. Pence and D.E. Kesling (eds.). American Academy of Underwater Sciences, Nahant, Massachusetts, pp. 53-65.

Pyle, R.L. 2000. Assessing undiscovered fish biodiversity on deep coral reefs using advanced self-contained diving technology. *Marine Technology Society Journal* 34: 82-91.

Qasim, S.Z. 1957. The biology of *Blennius pholis* L. (Teleostei). *Proceedings of the Zoological Society of London* 128: 161-208.

Randall, J.E. and R.L. Pyle. 2001a. Three new species of labrid fishes of the genus *Cirrhilabrus* from islands of tropical Pacific. *Aqua* 4: 89-98.

Randall, J.E. and R.L. Pyle. 2001b. Four new serranid fishes of the anthiine genus *Pseudanthias* from the South Pacific. *Raffles Bulletin Zoology* 49: 19-34.

Risk, M.J. 1972. Fish diversity on a coral reef in the Virgin Islands. *Atoll Research Bulletin* 153: 1-4.

Roberts, C.M. and R.F.G. Ormond. 1987. Habitat complexity and coral reef fish diversity and abundance on red Sea fringing reefs. *Marine Ecology Progress Series* 41: 1-8.

Robertson, D.R. and J.M. Sheldom. 1979. Competitive interactions and the availability of sleeping sites for a diurnal coral reef fish. *Journal of Experimental Marine Biology and Ecology* 40: 285-298.

Robertson, D.R., S.G. Hoffman and J.M. Sheldon. 1981. Availability of space for the territorial Caribbean damselfish *Eupomacentrus planiforns*. *Ecology* 62: 1162-1169.

Sale, P.F. 1972. Influence of corals in the dispersion of the pomacentrid fish *Dascyllus aruanus*. *Ecology* 53: 741-744.

Sale, P.F. 1975. Patterns of use of space in a guild of territorial reef fish. *Marine Biology* 29: 89-97.

Sale, P.F. 1976. The effect of territorial adult pomacentrid fishes on the recruitment and survival on patches of coral rubble. *Journal of Experimental Marine Biology and Ecology* 24: 297-306.

Sale, P.F. 1978. Coexistence of coral reef fishes: a lottery for living space. *Environmental Biology of Fishes* 3: 85-102.

Sale, P.F. 1991. Introduction. In: *The Ecology of Fishes on Coral Reefs*, P.F. Sale (ed.). Academic Press, San Diego, pp. 3-15.

Sale, P.F. and D.J. Ferrell. 1988. Early survivorship of juvenile coral reef fish. *Coral Reefs* 7: 117-124.

Sale, P.F., P.J. Doherty and W.A. Douglas. 1980. Juvenile recruitment strategies and the coexistence of territorial pomacentrid fishes. *Bulletin of Marine Science* 30: 147-158.

Sano, M., M. Shimizu and Y. Nose. 1984. Changes in the structure of coral reef fish communities by destruction of hermatypic corals: observational and experimental views. *Pacific Science* 38: 51-79.

Schoener, T.W. 1986. Mechanistic approaches to community ecology: a new reductionism? *American Naturalist* 26: 81-106.

Schofield, P.J. 2003. Habitat selection of two gobies (*Microgobius gulosus*, *Gobiosoma robustum*): influence of structural complexity, competitive interactions, and presence of a predator. *Journal of Experimental Marine Biology and Ecology* 288: 125-137.

Seaman, W. and L.M. Sprague (eds.). 1991. *Artificial Habitats for Marine and Freshwater Fisheries*. Academic Press, New York.

Séret, B. 1997. Les poissons d'eau douce de Nouvelle-Calédonie: implications biogéographiques de récentes découvertes. *Memoirs du Museum National d'Histoire Naturelle* 171: 369-378.

Shulman, M.J. 1984. Resource limitation and recruitment patterns in a coral reef fish assemblage. *Journal of Experimental Marine Biology and Ecology* 74: 85-109.

Silberschneider, V. and D.J. Booth. 2001. Resource use by *Enneapterygius rufopileus* and other rockpool fishes. *Environmental Biology of Fishes* 61: 195-204.

Smith, C.L. and J.C. Tyler. 1972. Space resource sharing in a coral reef fish community. Los Angeles County. *Natural History Museum Science Bulletin* 14: 125-170.

Smith, C.L., J.C. Tyler, H. Andreyko and D.M. Tyler. 1998. Behavioral ecology of the sailfin blenny, *Emblemaria pandionis* (Pisces: Chaenopsidae), in the Caribbean off Belize. *American Museum Novitates* 3232: 1-40.

Smith, M.M. 1986. Clinidae. In: *Smiths' Sea Fishes*, M.M. Smith and P.C. Heemstra (eds.). Springer-Verlag, Berlin, pp. 758-769.

Springer, V.G. 1986. Blenniidae. In: *Smiths' Sea Fishes*, M.M. Smith and P.C. Heemstra (eds.). Springer-Verlag, Berlin, pp. 742-755.

Steele, M.A. 1999. Effects of shelter and predators on reef fishes. *Journal of Experimental Marine Biology and Ecology* 233: 65-79.

Stephens, J.S. 1963. A revised classification of the blennioid fishes of the American family Chaenopsidae. *University of California Publications in Zoology* 68: 1-135.

Stephens, J.S., E.S. Hobson and R.K. Johnson. 1966. Notes on distribution, behaviour, and morphological variation in some Chaenopsid fishes from the Tropical Eastern Pacific, with descriptions of two new species, *Acanthemblemaria castroi* and *Coralliozetus springeri*. *Copeia* 1966: 424-438.

Stephens, J.S., R.K. Johnson, G.S. Key and J.E. McCosker. 1970. The comparative ecology of three sympatric species of California blennies of the genus *Hypsoblennius* Gill (Teleostomi, Blenniidae). *Ecological Monographs* 40: 213-233.

Stepien, C.A. 1987. Color pattern and habitat differences between male, female and juvenile giant kelpfish (Blennioidei: Clinidae). *Bulletin of Marine Science* 41: 45-58.

Stepien, C.A., M. Glattke and K.M. Fink. 1988. Regulation and significance of color patterns of the spotted kelpfish, *Gibbonsia elegans* Cooper, 1864 (Blennioidei: Clinidae). *Copeia* 1988: 7-15.

Sutherland, W.J. 1996. *From Individual Behaviour to Population Ecology*. Oxford University Press, Oxford.

Syms, C. 1995. Multi-scale analysis of habitat association in a guild of blennioid fishes. *Marine Ecology Progress Series* 125: 31-43.

Syms, C. and G.P. Jones. 1999. Scale of disturbance and the structure of a temperate fish guild. *Ecology* 80: 921-940.

Talbot, F.H. and B. Goldman. 1972. A preliminary report on the diversity and feeding relationships of the reef fishes of One Tree Island, Great Barrier Reef System. *Proceedings Symposium of Coral Reefs* 1: 425-442.

Tomilieri, N. 1995. Effects of microhabitat characteristics on the settlement and recruitment of a coral reef fish at two spatial scales. *Oecologia* 102: 52-63.

Tolimieri, N. 1998. Contrasting effects of microhabitat use on large-scale adult abundance in families of Caribbean reef fishes. *Marine Ecology Progress Series* 167: 227-239.

Tomo, A.P. 1981 Contribución al conocimiento de la fauna ictiological del sector Antárctico Argentino. *Publicacion Dirección Nacional Instituto Antárctico Argentina* 14: 1-242.

Townsend, K.A. and I.R. Tibbetts. 2000. Biomass and distribution of herbivorous blennies in the southern Great Barrier Reef. *Journal of Fish Biology* 56: 774-791.

Tupper, M. and R.G. Boutilier. 1997. Effects of habitat on settlement, growth, predation risk and survival of a temperate reef fish. *Marine Ecology Progress Series* 151: 225-236.

Tyler, D.M. and J.C. Tyler. 1997. A new species of chaenopsid fish, *Emblemariopsis ruetzleri*, from the western Caribbean off Belize (Blennioidei), with notes on its life history. *Proceedings of the Biological Society of Washington* 110: 24-38.

Tyler, J.C. and D.M. Tyler. 1999. Natural history of the sea fan blenny, *Emblemariopsis pricei* (Teleostei: Chaenopsidae), in the western Caribbean. *Smithsonian Contributions to Zoology* 601: 1-24.

Whitfield, A.K. 1996. Threatened fishes of the world: *Clinus spatulatus* Bennett, 1983 (Clinidae). *Environmental Biology of Fishes* 46: 26.

Wilkins, H.K.A. and A.A. Myers. 1992. Microhabitat utilisation by an assemblage of temperate Gobiidae (Pisces: Teleostei). *Marine Ecology Progress Series* 90: 103-112.

Williams, D.M. 1991. Patterns and processes in the distribution of coral reef fishes. In: *The Ecology of Fishes on Coral Reefs*, P.F. Sale (ed.). Academic Press, San Diego, pp. 437-474.

Williams, D.M. and P.F. Sale. 1981. Spatial and temporal patterns of recruitment of juvenile coral reef fishes to coral habitats within "One Tree Lagoon", Great Barrier Reef. *Marine Biology* 65: 245-253.

Willis, T.J. and M.J. Anderson. 2003. Structure of cryptic reef fish assemblages: relationships with habitat characteristics and predator density. *Marine Ecology Progress Series* 257: 209-221.

Wilson, S. 2001. Multiscale habitat associations of detrivorous blennies (Blenniidae: Salariini). *Coral Reefs* 20: 245-251.

Yoshiyama, R.M. 1981. Distribution and abundance of rock intertidal fishes in Central California. *Environmental Biology of Fishes* 6: 315-332.

Zander, C.D. 1972a. Beziehungen zwischen Körperbau und Lebensweise bei Blenniidae (Pisces) aus dem Roten Meer. I. Äußere Morphologie. *Marine Biology* 13: 238-246.

Zander, C.D. 1972b. Beziehungen zwischen Körperbau und Lebensweise bei Blenniidae (Pisces) aus dem Roten Meer. II. Bau der Flossen und ihrer Muskulatur. *Zeitschrift für Tierepsychologie* 71: 299-327.

Zander, C.D. 1974. Beziehungen zwischen Körperbau und Lebensweise bei Blenniidae (Pisces) aus dem Roten Meer. III. Morphologie des Auges. *Marine Biology* 28: 61-72.

Zander, C.D. 1980. Morphological and ecological investigations on sympatric *Lipophrys* species (Blenniidae, Pisces). *Helgoländer Meeresuntersuchungen* 34: 91-110.

SECTION 6

Species of Blennies

CHAPTER

6

List of Valid Species of Blennies

Robert A. Patzner[1], Philip A. Hastings[2], Victor G. Springer[3], Peter Wirtz[4] and Emanuel J. Gonçalves[5]

FAMILY TRIPTERYGIIDAE

Acanthanectes hystrix Holleman and Buxton, 1993
Acanthanectes rufus Holleman and Buxton, 1993

Apopterygion alta Kuiter, 1986
Apopterygion oculus Fricke and Roberts, 1994

Authors' addresses: [1]Dept. Organismic Biology, University of Salzburg, Austria.
E-mail: robert.patzner@sbg.ac.at.
[2]Scripps Institution of Oceanography, University of California, San Diego, USA.
E-mail: phastings@ucsd.edu.
[3]National Museum of Natural History, Smithsonian Institution, Washington, D.C., USA.
E-mail: springer@si.edu.
[4]Centro de Ciencias do Mar, Univ. Algarve, Faro, Portugal.
E-mail: peterwirtz2004@yahoo.com.
[5]Eco-Ethology Research Unit, Institute Superior Psicologia Aplicada, Rua Jardim do Tabaco 34, 1149–041 Lisboa, Portugal.
E-mail: emanuel@ispa.pt

Axoclinus cocoensis Bussing, 1991
Axoclinus lucillae Fowler, 1944
Axoclinus multicinctus Allen and Robertson, 1992
Axoclinus nigricaudus Allen and Robertson, 1991
Axoclinus rubinoffi Allen and Robertson, 1992
Axoclinus storeya (Brock, 1940)

Bellapiscis lesleyae Hardy, 1987
Bellapiscis medius (Günther, 1861)

Blennodon dorsale (Clarke, 1879)

Brachynectes fasciatus Scott, 1957

Ceratobregma acanthops (Whitley, 1964)
Ceratobregma helenae Holleman, 1987

Cremnochorites capensis (Gilchrist and Thompson, 1908)

Crocodilichthys gracilis Allen and Robertson, 1991

Cryptichthys jojettae Hardy, 1987

Enneanectes altivelis Rosenblatt, 1960
Enneanectes atrorus Rosenblatt, 1960
Enneanectes boehlkei Rosenblatt, 1960
Enneanectes carminalis (Jordan and Gilbert, 1882)
Enneanectes jordani (Evermann and Marsh, 1899)
Enneanectes pectoralis (Fowler, 1941)
Enneanectes reticulatus Allen and Robertson, 1991
Enneanectes smithi Lubbock and Edwards, 1981

Enneapterygius abeli (Klausewitz, 1960)
Enneapterygius atriceps (Jenkins, 1903)
Enneapterygius atrogulare (Günther, 1873)
Enneapterygius bahasa Fricke, 1997
Enneapterygius cheni Wang, Shao and Shen, 1996

Enneapterygius clarkae Holleman, 1982
Enneapterygius clea Fricke, 1997
Enneapterygius destai Clark, 1980
Enneapterygius elaine Holleman, 2005
Enneapterygius elegans (Peters, 1876)
Enneapterygius etheostoma (Jordan and Snyder, 1902)
Enneapterygius fasciatus (Weber, 1909)
Enneapterygius flavoccipitis Shen, 1994
Enneapterygius fuscoventer Fricke, 1997
Enneapterygius genamaculatus Holleman, 2005
Enneapterygius gracilis Fricke, 1994
Enneapterygius gruschkai Holleman, 2005
Enneapterygius hemimelas (Kner and Steindachner, 1867)
Enneapterygius hollemani Randall, 1995
Enneapterygius howensis Fricke, 1997
Enneapterygius kermadecensis Fricke, 1994
Enneapterygius kosiensis Holleman, 2005
Enneapterygius larsonae Fricke, 1994
Enneapterygius melanospilus Randall, 1995
Enneapterygius mirabilis Fricke, 1994
Enneapterygius miyakensis Fricke, 1987
Enneapterygius namarrgon Fricke, 1997
Enneapterygius nanus (Schultz, 1960)
Enneapterygius niger Fricke, 1994
Enneapterygius nigricauda Fricke, 1997
Enneapterygius obscurus Clark, 1980
Enneapterygius ornatus Fricke, 1997
Enneapterygius pallidoserialis Fricke, 1997
Enneapterygius pallidus Clarke, 1980
Enneapterygius paucifasciatus Fricke, 1994
Enneapterygius philippinus (Peters, 1868)
Enneapterygius pusillus Rüppell, 1835
Enneapterygius pyramis Fricke, 1994
Enneapterygius randalli Fricke, 1997

Enneapterygius rhabdotus Fricke, 1994
Enneapterygius rhothion Fricke, 1997
Enneapterygius rubicauda Shen, 1994
Enneapterygius rufopileus (Waite, 1904)
Enneapterygius senoui Motomira, Haruzai and Hardy, 2005
Enneapterygius signicauda Fricke, 1997
Enneapterygius similis Fricke, 1997
Enneapterygius triserialis Fricke, 1994
Enneapterygius trisignatus Fricke, 2001
Enneapterygius tutuilae Jordan and Seale, 1906
Enneapterygius unimaculatus Fricke, 1994
Enneapterygius ventermaculus Holleman, 1982
Enneapterygius vexillarius Fowler, 1946
Enneapterygius williamsi Fricke, 1997
Enneapterygius ziegleri Fricke, 1994

Forsterygion flavonigrum Fricke and Roberts, 1994
Forsterygion lapillum Hardy, 1989
Forsterygion malcolmi Hardy, 1987
Forsterygion varium (Forster, 1801)

Gilloblennius abditus Hardy, 1986
Gilloblennius tripennis (Forster, 1801)

**Grahamina capito* (Jenyns, 1842)
Grahamina gymnota (Scott, 1977)
Grahamina nigripenne (Valenciennes, 1836)

Helcogramma albimacula Williams and Howe, 2003
Helcogramma alkamr Holleman, 2007
Helcogramma aquilum Williams and McCormick, 1990
Helcogramma billi Hansen, 1986
Helcogramma capidata Rosenblatt, 1960
Helcogramma cerasina Williams and Howe, 2003

---

*After Jawad (2008. Journal of Natural History 42, 2943-2989). This genus is now included in *Forsterygion*.

Helcogramma chica Rosenblatt, 1960
Helcogramma decurrens McCulloch and Waite, 1918
Helcogramma desa Williams and Howe, 2003
Helcogramma ellioti (Herre, 1944)
Helcogramma ememes Holleman, 2007
Helcogramma fuscipectoris (Fowler, 1946)
Helcogramma fuscopinna Holleman, 1982
Helcogramma gymnauchen (Weber, 1909)
Helcogramma hudsoni (Jordan and Seale, 1906)
Helcogramma inclinata (Fowler, 1946)
Helcogramma kranos Fricke, 1997
Helcogramma lacuna Williams and Howe, 2003
Helcogramma larvata Fricke and Randall, 1992
Helcogramma maldivensis Fricke and Randall, 1992
Helcogramma microstigma Holleman, 2006
Helcogramma nesion Williams and Howe, 2003
Helcogramma nigra Williams and Howe, 2003
Helcogramma novaecaledoniae Fricke, 1994
Helcogramma obtusirostre (Klunzinger, 1871)
Helcogramma randalli Williams and Howe, 2003
Helcogramma rharhabe Holleman, 2007
Helcogramma rhinoceros Hansen, 1986
Helcogramma rosea Holleman, 2006
Helcogramma serendip Holleman, 2007
Helcogramma shinglensis Lal Mohan, 1971
Helcogramma solorensis Fricke, 1997
Helcogramma springeri Hansen, 1986
Helcogramma steinitzi Clark, 1980
Helcogramma striata Hansen, 1986
Helcogramma trigloides (Bleeker, 1858)
Helcogramma vulcana Randall and Clark, 1993

Helcogrammoides antarcticus (Tomo, 1982)
Helcogrammoides chilensis (Cancino, 1960)
Helcogrammoides cunninghami (Smitt, 1898)

Karalepis stewarti Hardy, 1984

Lepidoblennius haplodactylus Steindachner, 1867
Lepidoblennius marmoratus (Macleay, 1878)

Lepidonectes bimaculata Allen and Robertson, 1992
Lepidonectes clarkhubbsi Bussing, 1991
Lepidonectes corallicola (Kendall and Radcliffe, 1912)

Matanui bathytaton (Hardy, 1989)
Matanui profundum (Fricke and Roberts, 1994)

Norfolkia brachylepis (Schultz, 1960)
Norfolkia leeuwin Fricke, 1994
Norfolkia squamiceps (McCulloch and Waite, 1916)
Norfolkia thomasi Whitley, 1964

Notoclinops caerulepunctus Hardy, 1989
Notoclinops segmentatus (McCulloch and Phillipps, 1923)
Notoclinops yaldwyni Hardy, 1987

Notoclinus compressus (Hutton, 1872)
Notoclinus fenestratus (Forster, 1801)

**Obliquichthys maryannae* Hardy, 1987

Ruanoho decemdigitatus (Clarke, 1879)
Ruanoho whero Hardy, 1986

Springerichthys bapturus (Jordan and Snyder, 1902)
Springerichthys kulbickii (Fricke and Randall, 1994)

Trianectes bucephalus McCulloch and Waite, 1918

Trinorfolkia clarkei (Morton, 1888)
Trinorfolkia cristata (Kuiter, 1986)
Trinorfolkia incisa (Kuiter, 1986)

*After Jawad (2008. Journal of Natural History 42, 2943-2989). This genus is now included in *Forsterygion*.

Tripterygion delaisi Cadenat and Blache, 1970
Tripterygion melanurus Guichenot, 1850
Tripterygion minor Kolombatovic, 1892
Tripterygion tartessicum Carreras-Carbonell, Pascual, and Macpherson 2007
Tripterygion tripteronotus (Risso, 1810)
Tripterygion xanthosoma Zander & Heymer, 1970

Ucla xenogrammus Holleman, 1993

FAMILY BLENNIIDAE

Aidablennius sphynx (Valenciennes, 1836)

Alloblennius anuchalis (Springer and Spreitzer, 1978)
Alloblennius jugularis (Klunzinger, 1871)
Alloblennius parvus Springer and Spreitzer, 1978
Alloblennius pictus (Lotan, 1969)

Alticus aldabrensis (Regan, 1912)
Alticus anjouanae (Fourmanoir, 1955)
Alticus arnoldorum (Curtiss, 1938)
Alticus kirkii (Günther, 1868)
Alticus monochrus Bleeker, 1869
Alticus montanoi (Sauvage, 1880)
Alticus orientalis Tomiyama, 1955
Alticus saliens (Lacepède, 1800)
Alticus sertatus (Garman, 1903)
Alticus simplicirrus Smith-Vaniz and Springer, 1971

Andamia amphibius (Walbaum, 1792)
Andamia cyclocheilus Weber, 1909
Andamia expansa Blyth, 1858
Andamia heteroptera (Bleeker, 1857)
Andamia pacifica Tomiyama, 1955
Andamia reyi (Sauvage, 1880)
Andamia tetradactylus (Bleeker, 1858)

Antennablennius adenensis Fraser-Brunner, 1951
Antennablennius australis Fraser-Brunner, 1951
Antennablennius bifilum (Günther, 1861)
Antennablennius ceylonensis Bath, 1983
Antennablennius hypenetes (Klunzinger, 1871)
Antennablennius sexfasciatus von Bonde, 1923
Antennablennius simonyi (Steindachner, 1902)
Antennablennius variopunctatus (Jatzow and Lenz, 1898)
Antennablennius veliger (Smith, 1959)

Aspidontus dussumieri (Valenciennes, 1836)
Aspidontus taeniatus Quoy and Gaimard, 1834
Aspidontus tractus Fowler, 1903

Atrosalarias fuscus (Rüppell, 1838)
Atrosalarias holomelas (Günther, 1872)
Atrosalarias hosokawai Suzuki and Senou, 1999

Bathyblennius antholops (Springer and Smith-Vaniz, 1970)

Blenniella bilitonensis (Bleeker, 1858)
Blenniella caudolineata (Günther, 1877)
Blenniella chrysospilos (Bleeker, 1857)
Blenniella cyanostigma (Bleeker, 1849)
Blenniella gibbifrons (Quoy and Gaimard, 1824)
Blenniella interrupta (Bleeker, 1857)
Blenniella leopardus (Fowler, 1904)
Blenniella paula (Bryan and Herre, 1903)
Blenniella periophthalmus (Valenciennes, 1836)

Blennius normani Poll, 1949
Blennius ocellaris Linnaeus, 1758

Chalaroderma capito (Valenciennes, 1836)
Chalaroderma ocellata (Gilchrist and Thompson, 1908)

Chasmodes bosquianus (Lacepède, 1800)
Chasmodes longimaxilla Williams, 1983
Chasmodes saburrae Jordan and Gilbert, 1882

Cirripectes alboapicalis (Ogilby, 1899)
Cirripectes alleni Williams, 1993
Cirripectes auritus Carlson, 1981
Cirripectes castaneus (Valenciennes, 1836)
Cirripectes chelomatus Williams and Maugé, 1984
Cirripectes filamentosus (Alleyne and Macleay, 1877)
Cirripectes fuscoguttatus Strasburg and Schultz, 1953
Cirripectes gilberti Williams, 1988
Cirripectes hutchinsi Williams, 1988
Cirripectes imitator Williams, 1985
Cirripectes jenningsi Schultz, 1943
Cirripectes kuwamurai Fukao, 1984
Cirripectes obscurus (Borodin, 1927)
Cirripectes perustus Smith, 1959
Cirripectes polyzona (Bleeker, 1868)
Cirripectes quagga (Fowler and Ball, 1924)
Cirripectes randalli Williams, 1988
Cirripectes springeri Williams, 1988
Cirripectes stigmaticus Strasburg and Schultz, 1953
Cirripectes vanderbilti (Fowler, 1938)
Cirripectes variolosus (Valenciennes, 1836)
Cirripectes viriosus Williams, 1988

Cirrisalarias bunares Springer, 1976

Coryphoblennius galerita (Linnaeus, 1758)

Crossosalarias macrospilus Smith-Vaniz and Springer, 1971

Dodekablennos fraseri Springer and Spreitzer, 1978

Ecsenius aequalis Springer, 1988
Ecsenius alleni Springer, 1988
Ecsenius aroni Springer, 1971
Ecsenius australianus Springer, 1988
Ecsenius axelrodi Springer, 1988
Ecsenius bandanus Springer, 1971
Ecsenius bathi Springer, 1988
Ecsenius bicolor (Day, 1888)
Ecsenius bimaculatus Springer, 1971
Ecsenius caeruliventris Springer and Allen, 2004
Ecsenius collettei Springer, 1972
Ecsenius dentex Springer, 1988
Ecsenius dilemma Springer, 1988
Ecsenius fijiensis Springer, 1988
Ecsenius fourmanoiri Springer, 1972
Ecsenius frontalis (Valenciennes, 1836)
Ecsenius gravieri (Pellegrin, 1906)
Ecsenius isos McKinney and Springer, 1976
Ecsenius kurti Springer, 1988
Ecsenius lineatus Klausewitz, 1962
Ecsenius lividanalis Chapman and Schultz, 1952
Ecsenius lubbocki Springer, 1988
Ecsenius mandibularis McCulloch, 1923
Ecsenius melarchus McKinney and Springer, 1976
Ecsenius midas Starck, 1969
Ecsenius minutus Klausewitz, 1963
Ecsenius monoculus Springer, 1988
Ecsenius nalolo Smith, 1959
Ecsenius namiyei (Jordan and Evermann, 1902)
Ecsenius niue Springer, 2002
Ecsenius oculatus Springer, 1988
Ecsenius oculus Springer, 1971

Ecsenius ops Springer and Allen, 2001
Ecsenius opsifrontalis Chapman and Schultz, 1952
Ecsenius pardus Springer, 1988
Ecsenius paroculus Springer, 1988
Ecsenius pictus McKinney and Springer, 1976
Ecsenius polystictus Springer and Randall, 1999
Ecsenius portenoyi Springer, 1988
Ecsenius prooculis Chapman and Schultz, 1952
Ecsenius pulcher (Murray, 1887)
Ecsenius randalli Springer, 1991
Ecsenius schroederi McKinney and Springer, 1976
Ecsenius sellifer Springer, 1988
Ecsenius shirleyae Springer and Allen, 2004
Ecsenius stictus Springer, 1988
Ecsenius stigmatura Fowler, 1952
Ecsenius taeniatus Springer, 1988
Ecsenius tessera Springer, 1988
Ecsenius tigris Springer, 1988
Ecsenius tricolor Springer and Allen, 2001
Ecsenius trilineatus Springer, 1972
Ecsenius yaeyamaensis (Aoyagi, 1954)

Enchelyurus ater (Günther, 1877)
Enchelyurus brunneolus (Jenkins, 1903)
Enchelyurus flavipes Peters, 1868
Enchelyurus kraussii (Klunzinger, 1871)
Enchelyurus petersi (Kossmann and Räuber, 1877)

Entomacrodus cadenati Springer, 1967
Entomacrodus caudofasciatus (Regan, 1909)
Entomacrodus chapmani Springer, 1967
Entomacrodus chiostictus (Jordan and Gilbert, 1882)
Entomacrodus corneliae (Fowler, 1932)
Entomacrodus cymatobiotus Schultz and Chapman, 1960

Entomacrodus decussatus (Bleeker, 1858)
Entomacrodus epalzeocheilos (Bleeker, 1859)
Entomacrodus lemuria Springer and Fricke, 2000
Entomacrodus lighti (Herre, 1938)
Entomacrodus longicirrus Springer, 1967
Entomacrodus macrospilus Springer, 1967
Entomacrodus nigricans Gill, 1859
Entomacrodus niuafoouensis (Fowler, 1932)
Entomacrodus randalli Springer, 1967
Entomacrodus rofeni Springer, 1967
Entomacrodus sealei Bryan and Herre, 1903
Entomacrodus stellifer (Jordan and Snyder, 1902)
Entomacrodus strasburgi Springer, 1967
Entomacrodus striatus (Valenciennes, 1836)
Entomacrodus textilis (Valenciennes, 1836)
Entomacrodus thalassinus (Jordan and Seale, 1906)
Entomacrodus vermiculatus (Valenciennes, 1836)
Entomacrodus vomerinus (Valenciennes, 1836)
Entomacrodus williamsi Springer and Fricke, 2000

Exallias brevis (Kner, 1868)

Glyptoparus delicatulus Smith, 1959

Haptogenys bipunctata (Day, 1876)

Hirculops cornifer (Rüppell, 1830)

Hypleurochilus aequipinnis (Günther, 1861)
Hypleurochilus bananensis (Poll, 1959)
Hypleurochilus bermudensis Beebe and Tee-Van, 1933
Hypleurochilus caudovittatus Bath, 1994
Hypleurochilus fissicornis (Quoy and Gaimard, 1824)
Hypleurochilus geminatus (Wood, 1825)
Hypleurochilus langi (Fowler, 1923)

Hypleurochilus multifilis (Girard, 1858)
Hypleurochilus pseudoaequipinnis Bath, 1994
Hypleurochilus springeri Randall, 1966

Hypsoblennius brevipinnis (Günther, 1861)
Hypsoblennius caulopus (Gilbert, 1898)
Hypsoblennius exstochilus Böhlke, 1959
Hypsoblennius gentilis (Girard, 1854)
Hypsoblennius gilberti (Jordan, 1882)
Hypsoblennius hentz (Lesueur, 1825)
Hypsoblennius invemar Smith-Vaniz and Acero P., 1980
Hypsoblennius ionthas (Jordan and Gilbert, 1882)
Hypsoblennius jenkinsi (Jordan and Evermann, 1896)
Hypsoblennius maculipinna (Regan, 1903)
Hypsoblennius paytensis (Steindachner, 1876)
Hypsoblennius proteus (Krejsa, 1960)
Hypsoblennius sordidus (Bennett, 1828)
Hypsoblennius striatus (Steindachner, 1876)

Istiblennius bellus (Günther, 1861)
Istiblennius colei (Herre, 1934)
Istiblennius dussumieri (Valenciennes, 1836)
Istiblennius edentulus (Forster and Schneider, 1801)
Istiblennius flaviumbrinus (Rüppell, 1830)
Istiblennius lineatus (Valenciennes, 1836)
Istiblennius meleagris (Valenciennes, 1836)
Istiblennius muelleri (Klunzinger, 1879)
Istiblennius pox Springer and Williams, 1994
Istiblennius rivulatus (Rüppell, 1830)
Istiblennius spilotus Springer and Williams, 1994
Istiblennius steindachneri (Pfeffer, 1893)
Istiblennius unicolor (Rüppell, 1838)
Istiblennius zebra (Vaillant and Sauvage, 1875)

Laiphognathus longispinus Murase, 2007
Laiphognathus multimaculatus Smith, 1955

Lipophrys pholis (Linnaeus, 1758)
Lipophrys trigloides (Valenciennes, 1836)

Litobranchus fowleri (Herre, 1936)

Lupinoblennius nicholsi (Tavolga, 1954)
Lupinoblennius paivai (Pinto, 1958)
Lupinoblennius vinctus (Poey, 1867)

Medusablennius chani Springer, 1966

Meiacanthus abditus Smith-Vaniz, 1987
Meiacanthus anema (Bleeker, 1852)
Meiacanthus atrodorsalis (Günther, 1877)
Meiacanthus bundoon Smith-Vaniz, 1976
Meiacanthus crinitus Smith-Vaniz, 1987
Meiacanthus ditrema Smith-Vaniz, 1976
Meiacanthus fraseri Smith-Vaniz, 1976
Meiacanthus geminatus Smith-Vaniz, 1976
Meiacanthus grammistes (Valenciennes, 1836)
Meiacanthus kamoharai Tomiyama, 1956
Meiacanthus limbatus Smith-Vaniz, 1987
Meiacanthus lineatus (De Vis, 1884)
Meiacanthus luteus Smith-Vaniz, 1987
Meiacanthus mossambicus Smith, 1959
Meiacanthus naevius Smith-Vaniz, 1987
Meiacanthus nigrolineatus Smith-Vaniz, 1969
Meiacanthus oualanensis (Günther, 1880)
Meiacanthus phaeus Smith-Vaniz, 1976
Meiacanthus procne Smith-Vaniz, 1976
Meiacanthus reticulatus Smith-Vaniz, 1976
Meiacanthus smithi Klausewitz, 1962

Meiacanthus tongaensis Smith-Vaniz, 1987
Meiacanthus urostigma Smith-Vaniz, Satapoomin and Allen, 2001
Meiacanthus vicinus Smith-Vaniz, 1987
Meiacanthus vittatus Smith-Vaniz, 1976

Microlipophrys adriaticus (Steindachner and Kolombatovic, 1883)
Microlipophrys bauchotae Wirtz and Bath, 1982
Microlipophrys caboverdensis (Wirtz and Bath, 1989)
Microlipophrys canevae (Vinciguerra, 1880)
Microlipophrys dalmatinus (Steindachner and Kolombatovic, 1883)
Microlipophrys nigriceps (Vinciguerra, 1883)
Microlipophrys velifer (Norman, 1935)

Mimoblennius atrocinctus (Regan, 1909)
Mimoblennius cas Springer and Spreitzer, 1978
Mimoblennius cirrosus Smith-Vaniz and Springer, 1971
Mimoblennius lineathorax Fricke, 1999
Mimoblennius rusi Springer and Spreitzer, 1978

Nannosalarias nativitatis (Regan, 1909)

Oman ypsilon Springer, 1985

Omobranchus anolius (Valenciennes, 1836)
Omobranchus aurosplendidus (Richardson, 1846)
Omobranchus banditus Smith, 1959
Omobranchus elegans (Steindachner, 1876)
Omobranchus elongatus (Peters, 1855)
Omobranchus fasciolatoceps (Richardson, 1846)
Omobranchus fasciolatus (Valenciennes, 1836)
Omobranchus ferox (Herre, 1927)
Omobranchus germaini (Sauvage, 1883)
Omobranchus hikkaduwensis Bath, 1983
Omobranchus loxozonus (Jordan and Starks, 1906)
Omobranchus mekranensis (Regan, 1905)
Omobranchus obliquus (Garman, 1903)

Omobranchus punctatus (Valenciennes, 1836)
Omobranchus robertsi Springer, 1981
Omobranchus rotundiceps (Macleay, 1881)
Omobranchus smithi (Rao, 1974)
Omobranchus steinitzi Springer and Gomon, 1975
Omobranchus verticalis Springer and Gomon, 1975
Omobranchus woodi (Gilchrist and Thompson, 1908)
Omobranchus zebra (Bleeker, 1868)

Omox biporos Springer, 1972
Omox lupus Springer, 1981

Ophioblennius atlanticus (Valenciennes, 1836)
Ophioblennius clippertonensis Springer, 1962
Ophioblennius macclurei (Silvester, 1915)
Ophioblennius steindachneri Jordan and Evermann, 1898
Ophioblennius trinitatis Miranda Ribeiro, 1919

Parablennius cornutus (Linnaeus, 1758)
Parablennius cyclops (Rüppell, 1830)
Parablennius dialloi Bath, 1990
Parablennius gattorugine (Linnaeus, 1758)
Parablennius goreensis (Valenciennes, 1836)
Parablennius incognitus (Bath, 1968)
Parablennius intermedius (Ogilby, 1915)
Parablennius laticlavius (Griffin, 1926)
Parablennius lodosus (Smith, 1959)
Parablennius marmoratus (Bennett, 1828)
Parablennius marmoreus (Poey, 1876)
Parablennius maoricus (Kner, 1864)
Parablennius opercularis (Murray, 1887)
Parablennius parvicornis (Valenciennes, 1836)
Parablennius pilicornis (Cuvier, 1829)
Parablennius rouxi (Cocco, 1833)
Parablennius ruber (Valenciennes, 1836)

Parablennius salensis Bath, 1990
Parablennius sanguinolentus (Pallas, 1814)
Parablennius serratolineatus Bath and Hutchins, 1986
Parablennius sierraensis Bath, 1990
Parablennius tasmanianus (Richardson, 1842)
Parablennius tentacularis (Brünnich, 1768)
Parablennius thysanius (Jordan and Seale, 1907)
Parablennius verryckeni (Poll, 1959)
Parablennius yatabei (Jordan and Snyder, 1900)
Parablennius zvonimiri (Kolombatovic, 1892)

Parahypsos piersoni (Gilbert and Starks, 1904)

Paralticus amboinensis (Bleeker, 1857)

Parenchelyurus hepburni (Snyder, 1908)
Parenchelyurus hyena (Whitley, 1953)

Pereulixia kosiensis (Regan, 1908)

Petroscirtes ancylodon Rüppell, 1835
Petroscirtes breviceps (Valenciennes, 1836)
Petroscirtes fallax Smith-Vaniz, 1976
Petroscirtes lupus (De Vis, 1885)
Petroscirtes marginatus Smith-Vaniz, 1976
Petroscirtes mitratus Rüppell, 1830
Petroscirtes pylei Smith-Vaniz, 2005
Petroscirtes springeri Smith-Vaniz, 1976
Petroscirtes thepassii (Bleeker, 1853)
Petroscirtes variabilis Cantor, 1849
Petroscirtes xestus Jordan and Seale, 1906

Phenablennius heyligeri (Bleeker, 1858-59)

Plagiotremus azaleus (Jordan and Bollman, 1890)
Plagiotremus ewaensis (Brock, 1948)

Plagiotremus flavus Smith-Vaniz, 1976
Plagiotremus goslinei (Strasburg, 1956)
Plagiotremus iosodon Smith-Vaniz, 1976
Plagiotremus laudandus (Whitley, 1961)
Plagiotremus phenax Smith-Vaniz, 1976
Plagiotremus rhinorhynchos (Bleeker, 1852)
Plagiotremus spilistius Gill, 1865
Plagiotremus tapeinosoma (Bleeker, 1857)
Plagiotremus townsendi (Regan, 1905)

Praealticus bilineatus (Peters, 1868)
Praealticus caesius (Seale, 1906)
Praealticus dayi (Whitley, 1929)
Praealticus labrovittatus Bath, 1992
Praealticus margaritatus (Kendall and Radcliffe, 1912)
Praealticus multistriatus Bath, 1992
Praealticus natalis (Regan, 1909)
Praealticus oortii (Bleeker, 1851)
Praealticus poptae (Fowler, 1925)
Praealticus semicrenatus (Chapman, 1951)
Praealticus striatus Bath, 1992
Praealticus tanegasimae (Jordan and Starks, 1906)
Praealticus triangulus (Chapman, 1951)

Rhabdoblennius nitidus Günther, 1861
Rhabdoblennius nigropunctatus Bath, 2004
Rhabdoblennius papuensis Bath, 2004
Rhabdoblennius rhabdotrachelus (Fowler and Ball, 1924)
Rhabdoblennius snowi (Fowler, 1928)

Salaria basilisca (Valenciennes, 1836)
Salaria economidisi Kottelat, 2004
Salaria fluviatilis (Asso, 1801)
Salaria pavo (Risso, 1810)

Salarias alboguttatus Kner, 1867
Salarias ceramensis Bleeker, 1852
Salarias fasciatus (Bloch, 1786)
Salarias guttatus Valenciennes, 1836
Salarias luctuosus Whitley, 1929
Salarias nigrocinctus Bath, 1996
Salarias obscurus Bath, 1992
Salarias patzneri Bath, 1992
Salarias ramosus Bath, 1992
Salarias segmentatus Bath and Randall, 1991
Salarias sexfilum Günther, 1861
Salarias sibogai Bath, 1992
Salarias sinuosus Snyder, 1908

Scartella caboverdiana Bath, 1990
Scartella cristata (Linnaeus, 1758)
Scartella emarginata (Günther, 1861)
Scartella nuchifilis (Valenciennes, 1836)
Scartella poiti Rangel, Gasparini and Guimarães, 2004
Scartella springeri (Bauchot, 1967)

Scartichthys crapulatus Williams, 1990
Scartichthys gigas (Steindachner, 1876)
Scartichthys variolatus (Valenciennes, 1836)
Scartichthys viridis (Valenciennes, 1836)

Spaniblennius clandestinus Bath and Wirtz, 1989
Spaniblennius riodourensis (Metzelaar, 1919)

Stanulus seychellensis Smith, 1959
Stanulus talboti Springer, 1968

Xiphasia matsubarai Okada and Suzuki, 1952
Xiphasia setifer Swainson, 1839

FAMILY LABRISOMIDAE

Alloclinus holderi (Lauderbach, 1907)

Auchenionchus crinitus (Jenyns, 1841)
Auchenionchus microcirrhis (Valenciennes, 1836)
Auchenionchus variolosus (Valenciennes, 1836)

Calliclinus geniguttatus (Valenciennes, 1836)
Calliclinus nudiventris Cervigón and Pequeño, 1979

Cottoclinus canops McCosker, Stephens and Rosenblatt, 2003

Cryptotrema corallinum Gilbert, 1890
Cryptotrema seftoni Hubbs, 1954

Dialommus fuscus Gilbert, 1891
Dialommus macrocephalus (Günther, 1861)

Exerpes asper (Jenkins and Evermann, 1889)

Haptoclinus apectolophus Böhlke and Robins, 1974

Labrisomus albigenys Beebe and Tee-Van, 1928
Labrisomus bucciferus (Poey, 1868)
Labrisomus cricota Sazima, Gasparini and Moura, 2002
Labrisomus dendriticus (Reid, 1935)
Labrisomus fernandezianus (Guichenot, 1848)
Labrisomus filamentosus Springer, 1960
Labrisomus gobio (Valenciennes, 1836)
Labrisomus guppyi (Norman, 1922)
Labrisomus haitiensis Beebe and Tee-Van, 1928
Labrisomus jenkinsi (Heller and Snodgrass, 1903)
Labrisomus kalisherae (Jordan, 1904)
Labrisomus multiporosus Hubbs, 1953
Labrisomus nigricinctus Howell Rivero, 1936
Labrisomus nuchipinnis (Quoy and Gaimard, 1824)

Labrisomus philippii (Steindachner, 1866)
Labrisomus pomaspilus Springer and Rosenblatt, 1965
Labrisomus socorroensis Hubbs, 1953
Labrisomus striatus Hubbs, 1953
Labrisomus wigginsi Hubbs, 1953
Labrisomus xanti Gill, 1860

Malacoctenus africanus Cadenat, 1951
Malacoctenus aurolineatus Smith, 1957
Malacoctenus boehlkei Springer, 1959
Malacoctenus costaricanus Springer, 1959
Malacoctenus delalandii (Valenciennes, 1836)
Malacoctenus ebisui Springer, 1959
Malacoctenus erdmani Smith, 1957
Malacoctenus gigas Springer, 1959
Malacoctenus gilli (Steindachner, 1867)
Malacoctenus hubbsi Springer, 1959
Malacoctenus macropus (Poey, 1868)
Malacoctenus margaritae (Fowler, 1944)
Malacoctenus mexicanus Springer, 1959
Malacoctenus polyporosus Springer, 1959
Malacoctenus sudensis Springer, 1959
Malacoctenus tetranemus (Cope, 1877)
Malacoctenus triangulatus Springer, 1959
Malacoctenus versicolor (Poey, 1876)
Malacoctenus zacae Springer, 1959
Malacoctenus zonifer (Jordan and Gilbert, 1882)
Malacoctenus zonogaster Heller and Snodgrass, 1903

Nemaclinus atelestos Böhlke and Springer, 1975

Paraclinus altivelis (Lockington, 1881)
Paraclinus arcanus Guimarães and Bacellar, 2002
Paraclinus barbatus Springer, 1955
Paraclinus beebei Hubbs, 1952

Paraclinus cingulatus (Evermann and Marsh, 1899)
Paraclinus ditrichus Rosenblatt and Parr, 1969
Paraclinus fasciatus (Steindachner, 1876)
Paraclinus fehlmanni Springer and Trist, 1969
Paraclinus grandicomis (Rosén, 1911)
Paraclinus infrons Böhlke, 1960
Paraclinus integripinnis (Smith, 1880)
Paraclinus magdalenae Rosenblatt and Parr, 1969
Paraclinus marmoratus (Steindachner, 1876)
Paraclinus mexicanus (Gilbert, 1904)
Paraclinus monophthalmus (Günther, 1861)
Paraclinus naeorhegmis Böhlke, 1960
Paraclinus nigripinnis (Steindachner, 1867)
Paraclinus rubicundus (Starks, 1913)
Paraclinus sini Hubbs, 1952
Paraclinus spectator Guimarães and Bacellar, 2002
Paraclinus stephensi Rosenblatt and Parr, 1969
Paraclinus tanygnathus Rosenblatt and Parr, 1969
Paraclinus walkeri Hubbs, 1952

Starksia atlantica Longley, 1934
Starksia brasiliensis (Gilbert, 1900)
Starksia cremnobates (Gilbert, 1890)
Starksia culebrae (Evermann and Marsh, 1899)
Starksia elongata Gilbert, 1971
Starksia fasciata (Longley, 1934)
Starksia fulva Rosenblatt and Taylor, 1971
Starksia galapagensis Rosenblatt and Taylor, 1971
Starksia grammilaga Rosenblatt and Taylor, 1971
Starksia guadalupae Rosenblatt and Taylor, 1971
Starksia guttata (Fowler, 1931)
Starksia hassi Klausewitz, 1958
Starksia hoesei Rosenblatt and Taylor, 1971
Starksia lepicoelia Böhlke and Springer, 1961

Starksia lepidogaster Rosenblatt and Taylor, 1971
Starksia leucovitta Williams and Mounts, 2003
Starksia melasma Williams and Mounts, 2003
Starksia multilepis Williams and Mounts, 2003
Starksia nanodes Böhlke and Springer, 1961
Starksia occidentalis Greenfield, 1979
Starksia ocellata (Steindachner, 1876)
Starksia posthon Rosenblatt and Taylor, 1971
Starksia rava Williams and Mounts, 2003
Starksia sella Williams and Mounts, 2003
Starksia sluiteri (Metzelaar, 1919)
Starksia smithvanizi Williams and Mounts, 2003
Starksia spinipenis (Al-Uthman, 1960)
Starksia starcki Gilbert, 1971
Starksia variabilis Greenfield, 1979
Starksia y-lineata Gilbert, 1965

Xenomedea rhodopyga Rosenblatt and Taylor, 1971

FAMILY CLINIDAE

Blennioclinus brachycephalus (Valenciennes, 1836)
Blennioclinus stella Smith, 1946

Blennophis anguillaris (Valenciennes, 1836)
Blennophis striatus (Gilchrist and Thompson, 1908)

Cancelloxus burrelli Smith, 1961
Cancelloxus elongatus Heemstra and Wright, 1986
Cancelloxus longior Prochazka and Griffiths, 1991

Cirrhibarbis capensis Valenciennes, 1836

Climacoporus navalis Barnard, 1935

Clinitrachus argentatus (Risso, 1810)

Clinoporus biporosus (Gilchrist and Thompson, 1908)

Clinus acuminatus (Bloch and Schneider, 1801)
Clinus agilis Smith, 1931
Clinus arborescens Gilchrist and Thompson, 1908
Clinus berrisfordi Penrith, 1967
Clinus brevicristatus Gilchrist and Thompson, 1908
Clinus cottoides Valenciennes, 1836
Clinus helenae (Smith, 1946)
Clinus heterodon Valenciennes, 1836
Clinus latipennis Valenciennes, 1836
Clinus nematopterus Günther, 1861
Clinus robustus Gilchrist and Thompson, 1908
Clinus rotundifrons Barnard, 1937
Clinus spatulatus Bennett, 1983
Clinus superciliosus (Linnaeus, 1758)
Clinus taurus Gilchrist and Thompson, 1908
Clinus venustris Gilchrist and Thompson, 1908
Clinus woodi (Smith, 1946)

Cologrammus flavescens (Hutton, 1872)

Cristiceps argyropleura Kner, 1865
Cristiceps aurantiacus Castelnau, 1879
Cristiceps australis Valenciennes, 1836

Ericentrus rubrus (Hutton, 1872)

Fucomimus mus (Gilchrist and Thompson, 1908)

Gibbonsia elegans (Cooper, 1864)
Gibbonsia metzi Hubbs, 1927
Gibbonsia montereyensis Hubbs, 1927

Heteroclinus adelaidae Castelnau, 1873
Heteroclinus antinectes (Günther, 1861)
Heteroclinus eckloniae (McKay, 1970)

Heteroclinus equiradiatus (Milward, 1960)
Heteroclinus heptaeolus (Ogilby, 1885)
Heteroclinus johnstoni (Saville-Kent, 1886)
Heteroclinus kuiteri Hoese and Rennis, 2006
Heteroclinus macrophthalmus Hoese, 1976
Heteroclinus nasutus (Günther, 1861)
Heteroclinus perspicillatus (Valenciennes, 1836)
Heteroclinus puellarum (Scott, 1955)
Heteroclinus roseus (Günther, 1861)
Heteroclinus tristis (Klunzinger, 1872)
Heteroclinus whiteleggii (Ogilby, 1894)
Heteroclinus wilsoni (Lucas, 1891)

Heterostichus rostratus Girard, 1854

Muraenoclinus dorsalis (Bleeker, 1860)

Myxodes cristatus Valenciennes, 1836
Myxodes ornatus Stephens and Springer, 1974
Myxodes viridis Valenciennes, 1836

Ophiclinops hutchinsi George and Springer, 1980
Ophiclinops pardalis (McCulloch and Waite, 1918)
Ophiclinops varius (McCulloch and Waite, 1918)

Ophiclinus antarcticus Castelnau, 1872
Ophiclinus brevipinnis George and Springer, 1980
Ophiclinus gabrieli (Waite, 1906)
Ophiclinus gracilis (Waite, 1906)
Ophiclinus ningulus George and Springer, 1980
Ophiclinus pectoralis George and Springer, 1980

Pavoclinus caeruleopunctatus Zsilavecz, 2001
Pavoclinus graminis (Gilchrist and Thompson, 1908)
Pavoclinus laurentii (Gilchrist and Thompson, 1908)
Pavoclinus litorafontis Penrith, 1965

Pavoclinus mentalis (Gilchrist and Thompson, 1908)
Pavoclinus myae Christensen, 1978
Pavoclinus pavo (Gilchrist and Thompson, 1908)
Pavoclinus profundus Smith, 1961
Pavoclinus smalei Heemstra and Wright, 1986

Peronedys anguillaris Steindachner, 1883

Ribeiroclinus eigenmanni (Jordan, 1888)

Smithichthys fucorum (Gilchrist and Thompson, 1908)

Springeratus caledonicus (Sauvage, 1874)
Springeratus polyporatus Fraser, 1972
Springeratus xanthosoma (Bleeker, 1857)

Sticharium clarkae George and Springer, 1980
Sticharium dorsale Günther, 1867

Xenopoclinus kochi Smith, 1948
Xenopoclinus leprosus Smith, 1961

FAMILY CHAENOPSIDAE

Acanthemblemaria aspera (Longley, 1927)
Acanthemblemaria atrata Hastings and Robertson, 1999
Acanthemblemaria balanorum Brock, 1940
Acanthemblemaria betinensis Smith-Vaniz and Palacio, 1974
Acanthemblemaria castroi Stephens and Hobson, 1966
Acanthemblemaria chaplini Böhlke, 1957
Acanthemblemaria crockeri Beebe and Tee-Van, 1938
Acanthemblemaria exilispinus Stephens, 1963
Acanthemblemaria greenfieldi Smith-Vaniz and Palacio, 1974
Acanthemblemaria hancocki Myers and Reid, 1936
Acanthemblemaria harpeza Williams, 2002
Acanthemblemaria johnsoni Almany and Baldwin, 1996
Acanthemblemaria macrospilus Brock, 1940

Acanthemblemaria mangognatha Hastings and Robertson, 1999
Acanthemblemaria maria Böhlke, 1961
Acanthemblemaria medusa Smith-Vaniz and Palacio, 1974
Acanthemblemaria paula Johnson and Brothers, 1989
Acanthemblemaria rivasi Stephens, 1970
Acanthemblemaria spinosa Metzelaar, 1919
Acanthemblemaria stephensi Rosenblatt and McCosker, 1988

Chaenopsis alepidota (Gilbert, 1890)
Chaenopsis coheni Böhlke, 1957
Chaenopsis deltarrhis Böhlke, 1957
Chaenopsis limbaughi Robins and Randall, 1965
Chaenopsis megalops Smith-Vaniz, 2000
Chaenopsis ocellata Gill, 1865
Chaenopsis resh Robins and Randall, 1965
Chaenopsis roseola Hastings and Shipp, 1981
Chaenopsis schmitti Böhlke, 1957
Chaenopsis stephensi Robins and Randall, 1965

Cirriemblemaria lucasana (Stephens, 1963)

Coralliozetus angelicus (Böhlke and Mead, 1957)
Coralliozetus boehlkei Stephens, 1963
Coralliozetus cardonae Evermann and Marsh, 1899
Coralliozetus micropes (Beebe and Tee-Van, 1938)
Coralliozetus rosenblatti Stephens, 1963
Coralliozetus springeri Stephens and Johnson, 1966

Ekemblemaria lira Hastings, 1992
Ekemblemaria myersi Stephens, 1963
Ekemblemaria nigra (Meek and Hildebrand, 1928)

Emblemaria atlantica Jordan and Evermann, 1898
Emblemaria australis Ramos, Rocha and Rocha, 2003
Emblemaria biocellata Stephens, 1970

Emblemaria caldwelli Stephens, 1970
Emblemaria caycedoi Acero P., 1984
Emblemaria culmenis Stephens, 1970
Emblemaria diphyodontis Stephens and Cervigón, 1970
Emblemaria hudsoni Evermann and Radcliffe, 1917
Emblemaria hyltoni Johnson and Greenfield, 1976
Emblemaria hypacanthus (Jenkins and Evermann, 1889)
Emblemaria nivipes Jordan and Gilbert, 1883
Emblemaria pandionis Evermann and Marsh, 1900
Emblemaria piratica Ginsburg, 1942
Emblemaria piratula Ginsburg and Reid, 1942
Emblemaria vitta Williams, 2002
Emblemaria walkeri Stephens, 1963

Emblemariopsis bahamensis Stephens, 1961
Emblemariopsis bottomei Stephens, 1961
Emblemariopsis dianae Tyler and Hastings, 2004
Emblemariopsis diaphanus Longley, 1927
Emblemariopsis leptocirris Stephens, 1970
Emblemariopsis occidentalis Stephens, 1970
Emblemariopsis pricei Greenfield, 1975
Emblemariopsis ramirezi (Cervigón, 1999)
Emblemariopsis randalli Cervigón, 1965
Emblemariopsis ruetzleri Tyler and Tyler, 1997
Emblemariopsis signifera (Ginsburg, 1942)
Emblemariopsis tayrona (Acero P., 1987)

Hemiemblemaria simulus Longley and Hildebrand, 1940

Lucayablennius zingaro (Böhlke, 1957)

Mccoskerichthys sandae Rosenblatt and Stephens, 1978

Neoclinus blanchardi Girard, 1858
Neoclinus bryope (Jordan and Snyder, 1902)
Neoclinus chihiroe Fukao, 1987

Neoclinus lacunicola Fukao, 1980
Neoclinus nudus Stephens and Springer, 1971
Neoclinus okazakii Fukao, 1987
Neoclinus stephensae Hubbs, 1953
Neoclinus toshimaensis Fukao, 1980
Neoclinus uninotatus Hubbs, 1953

Protemblemaria bicirrus (Hildebrand, 1946)
Protemblemaria perla Hastings, 2001
Protemblemaria punctata Cervigón, 1966

Stathmonotus culebrai Seale, 1940
Stathmonotus gymnodermis Springer, 1955
Stathmonotus hemphillii Bean, 1885
Stathmonotus lugubris Böhlke, 1953
Stathmonotus sinuscalifornici (Chabanaud, 1942)
Stathmonotus stahli (Evermann and Marsh, 1899)
Stathmonotus tekla Nichols, 1910

Tanyemblemaria alleni Hastings, 1992

FAMILY DACTYLOSCOPIDAE

Dactylagnus mundus Gill, 1862
Dactylagnus parvus Dawson, 1976
Dactylagnus peratikos Böhlke and Caldwell, 1961

Dactyloscopus amnis Miller and Briggs, 1962
Dactyloscopus boehlkei Dawson, 1982
Dactyloscopus byersi Dawson, 1969
Dactyloscopus comptus Dawson, 1982
Dactyloscopus crossotus Starks, 1913
Dactyloscopus elongatus Myers and Wade, 1946
Dactyloscopus fallax Dawson, 1975
Dactyloscopus fimbriatus (Reid, 1935)
Dactyloscopus foraminosus Dawson, 1982

Dactyloscopus heraldi Dawson, 1975
Dactyloscopus insulatus Dawson, 1975
Dactyloscopus lacteus (Myers and Wade, 1946)
Dactyloscopus lunaticus Gilbert, 1890
Dactyloscopus metoecus Dawson, 1975
Dactyloscopus minutus Dawson, 1975
Dactyloscopus moorei (Fowler, 1906)
Dactyloscopus pectoralis Gill, 1861
Dactyloscopus poeyi Gill, 1861
Dactyloscopus tridigitatus Gill, 1859
Dactyloscopus zelotes Jordan and Gilbert, 1896

Gillellus arenicola Gilbert, 1890
Gillellus chathamensis Dawson, 1977
Gillellus greyae Kanazawa, 1952
Gillellus healae Dawson, 1982
Gillellus inescatus Williams, 2002
Gillellus jacksoni Dawson, 1982
Gillellus ornatus Gilbert, 1896
Gillellus searcheri Dawson, 1977
Gillellus semicinctus Gilbert, 1890
Gillellus uranidea Böhlke, 1968

Heteristius cinctus (Osburn and Nichols, 1916)

Leurochilus acon Böhlke, 1968

Myxodagnus belone Böhlke, 1968
Myxodagnus macrognathus Hildebrand, 1946
Myxodagnus opercularis Gill, 1861
Myxodagnus sagitta Myers and Wade, 1946
Myxodagnus walkeri Dawson, 1976

Platygillellus altivelis Dawson, 1974
Platygillellus brasiliensis Feitoza, 2002

Platygillellus bussingi Dawson, 1974
Platygillellus rubellulus (Kendall and Radcliffe, 1912)
Platygillellus rubrocinctus (Longley, 1934)
Platygillellus smithi Dawson, 1982

Sindoscopus australis (Fowler and Bean, 1923)

Storrsia olsoni Dawson, 1982

Index

For Product Safety Concerns and Information please contact our EU representative GPSR@taylorandfrancis.com Taylor & Francis Verlag GmbH, Kaufingerstraße 24, 80331 München, Germany